P9-AEZ-398

Defects in Semiconductors II

MATERIALS RESEARCH SOCIETY SYMPOSIA PROCEEDINGS   VOLUME 14

ISSN 0272-9172

MATERIALS RESEARCH SOCIETY SYMPOSIA PROCEEDINGS   VOLUME 14

Defects in Semiconductors II

Symposium held November 1982 in Boston, Massachusetts, U.S.A.

EDITORS:

Subhash Mahajan

Bell Telephone Laboratories, Murray Hill, New Jersey, U.S.A.

James W. Corbett

Department of Physics, State University of New York, Albany,
New York, U.S.A.

NORTH-HOLLAND
NEW YORK • AMSTERDAM • OXFORD

PHYSICS

7127-1442

This work relates to Department of the Navy Grant N00014-82-G-0112 issued by the Office
of Naval Research. The United States Government has a royalty-free license throughout
the world in all copyrightable material contained herein.

© 1983 by Elsevier Science Publishing Co., Inc.
All rights reserved.

Published by:

Elsevier Science Publishing Company, Inc.
52 Vanderbilt Avenue, New York, New York 10017

Sole distributors outside the USA and Canada:

Elsevier Science Publishers B.V.
P.O. Box 211, 1000 AE Amsterdam, The Netherlands

Library of Congress Cataloging in Publication Data

Main entry under title:

Defects in semiconductors II.
    (Materials Research Society symposia proceedings, ISSN 0272-9172; v.14)
    Includes indexes.
    1. Semiconductors—Defects—Congresses. I. Mahajan, Subhash. II. Corbett,
    James W.   III. Series.
QC611.6.D4D425   1983   537.6'22   83-13266
ISBN 0-444-00812-8
ISSN 0272-9172

Manufactured in the United States of America

$QC\ 611$

v

CONTENTS $.6$

$D4\ D425$

1983

$PHYS$

* Invited Papers

vi

II.   DEFECTS IN SILICON: OXYGEN RELATED

*   Invited Papers

* Invited Papers

---

* Invited Papers

---

* Invited Papers

* Invited Papers

---

* Invited Papers

PREFACE

The study of defects in semiconductors began in earnest with the development of the transistor. Many problems have been solved, but many of the problems encountered many years ago are still unresolved, and new problems arise as a result of the ever decreasing device size and the ever increasing scale of integration. This symposium attempted to present an over-view of the field. Where an area of study was mature, a review was invited to survey that area. Where controversy persists, papers were juxtaposed to high-light this status.

Broadly speaking the symposium emphasis was threefold. First, to assess both from scientific as well as technological viewpoints, the effects of oxygen clustering and precipitation, the behavior and identification of native defects present at high temperatures and the physics of transition metal impurities in silicon. Second, to evaluate the situation regarding fundamental defect studies, impurity levels and dislocation glide and climb in compound semiconductors. Third, the interrelationship between growth- and processing-induced defects and device performance was also emphasized. Invited speakers were responsible for highlighting the three main themes of the symposium, and contributed papers were also solicited. Invited speakers also covered exciting, related areas.

The symposium was scheduled for four days and contained technically very stimulating invited and contributed papers most of which are in this proceedings volume. The symposium was very well attended and was truly international in its character with speakers coming from England, Germany, Hungary, Japan Sweden, and U.S.A.

ACKNOWLEDGEMENTS

The editors gratefully acknowledge the financial support from the Office of Naval Research (Dr. L. Cooper) and the Defense Advanced Research Projects Agency (Dr. S. Roosild) which made the symposium possible. They also wish to acknowledge the help of Mrs. Muriel Hausler in handling the manuscripts and text, and of Ms. Mary Fields in the planning stage of the symposium, of Satya N. Sahu for his help with the indices and the organization of the meeting and A. Jaworowski, R. Kleinhenz, Ge Pei-wen, J. H. Prah, Shi Tian-sheng and You Zhi-pu for their help in running the meeting.

The editors wish to thank all those who participated in the meeting including those whose papers for a variety of reasons do not appear in these Proceedings. The editors also appreciate the great help of the following Session Chairmen in running the meeting: S. M. Hu, L. C. Kimerling, C. G. Kirkpatrick, D. V. Lang, J. W. Mayer, J. R. Patel, H. Queisser and G. Rozgonyi. And the help of the many anonymous reviewers was indispensible.

Defects in Semiconductors II

I
DEFECTS IN SILICON:
GENERAL

MICRODEFECTS AND IMPURITIES IN DISLOCATION-FREE SILICON CRYSTALS

Takao Abe, Hirofumi Harada
Shin-Etsu Handotai Co., 2-13-1, Isobe, Annaka-shi, Gunma-ken, 379-01, Japan

Jun-ichi Chikawa
NHK Broadcasting Science Research Laboratories, Kinuta, Setagaya-ku, Tokyo, 157

ABSTRACT

Microdefects in striated (swirl defects) and non-striated distribution (D-defects) have been observed in float-zoned crystals doped with various impurities by x-ray topography following copper decoration. A new type of defects were found to be present in swirl-free and D-defect-free regions and to become invisible by doping gallium. This gallium effect led to the conclusion that they are microprecipitates produced from residual oxygen impurity in FZ crystals. Effects of various impurities on defect formation indicate that D-defects are of vacancy agglomerates. It was observed that swirl defects are formed when the temperature gradient near the interface is high, and that their formation is suppressed by doping nitrogen. Formation processes of microprecipitates, swirls, and D-defects are discussed on the basis of observation of their mutual interaction and the impurity effects.

1. INTRODUCTION

Swirl defects were first observed by copper-decoration technique in 1965 [1] and by preferential chromic acid etching in 1966 [2]. The name of "swirls" came from observation that they appear in swirl patterns in round slices of float-zoned (FZ) and Czochralski-grown (CZ) crystals. (In longitudinal slices, they are in striated distribution.) De Kock [3, 4] observed the defects by x-ray topography following copper decoration and distinguished swirl defects into two types of larger and smaller ones by their different behavior in lithium decoration, which were named "A-defects (A-swirls) and B-defects (B-swirls), respectively. By electron microscopy, the A-defect was identified to be perfect dislocation-loops or loop clusters of interstitial type, whereas the B-defect was invisible unless it is decorated by copper [5, 6].

Rocksnoer and Van den Boom (1981) [7] found a new type of microdefects by x-ray topography following copper decoration which were present uniformly in FZ crystals with a diameter of 23 mm grown faster than 6 mm/min. They were named "D-defects" and were considered to be of agglomerates of frozen-in vacancies by comparing their nature with that of swirl defects. By increasing the crystal diameter, the present authors [8] found that D-defects are formed at the slower growth rates (slower cooling rates) and formation of D-defects is affected delicately by trace impurities. The observation of D-defects was possible only for the case of the low carbon concentration which was recently achieved.

Point defects in silicon crystals have been investigated extensively with introducing them by radiation effect [9]. Such investigations, however, have not always led to well understanding of defect behavior at high temperatures where the electronic nature of defects fades out. There is an ongoing controversy as to which of vacancies and interstitials is the dominant defect at high temperatures. For instance, diffusion mechanism has been explained by assuming self-interstitials [10] or vacancies [11, 12] as the intrinsic defects. Recently, a coexistence model of vacancies and interstitials was proposed to explain oxidation—enhanced and retarded diffusion [13]. The problem on point defects has also been one of the central issues in formation of swirl defects: although the majority of models [14-17] attributed swirl formation to condensation of self-interstitials, sources of interstitials are different from model to model. A clue to the solution of these problems may be expected from investigation on the microdefects formed at high temperatures.

The purpose of the present paper is to report on a new type of microdefects observed in D—defect-free and swirl-free crystals as well as impurity effect on formation of swirls and D—defects.

2. THERMAL CONDITIONS FOR SWIRL FORMATION

Thermal history from solidification at the growth interface to room temperature is involved as the most important process in formation of microdefects. Temperature measurement of growing crystals by the FZ method was carried out with a radiation thermometer, and the following results were obtained [8]:

(1) The cooling rate is increased with increasing the growth rate V (zone-travelling rate) and decreased by increasing the crystal diameter.

(2) The temperature gradient at the interface is decreased with increasing the growth rate or the crystal diameter.

De Kock et al.[18] found by growing crystals with a diameter of about 20 mm that no swirls are formed for growth rate larger than 5 mm/min. The growth rate required for suppression of swirl

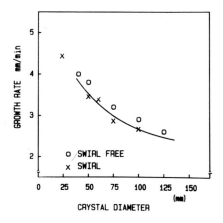

Fig. 1. Crystal diameter vs minimum growth rate for suppression of swirl formation. The datum for 25—mm diameter is by de Kock et al. [18].

formation decreases with increasing the crystal diameter, as shown in Fig. 1. Crystals with a diameter of 5 inches become swirl-free even with a growth rate of 2.5 mm/min. This observation shows that low temperature gradients at the interface or slow cooling after solidification suppresses swirl formation.

To find which of these two factors is responsible for swirl formation, another experiment was carried out, as shown in Fig. 2: During the growth at a growth rate of V=4 mm/min, the zone travelling was stopped for 1, 2, and 4 min in Figs. 2(a), (b), and (c), respectively, and then the crystals were grown again at the same growth rate. These photographs are x-ray topographs taken after Cu decoration for the longitudinal specimens. The position of the pause is indicated by the arrows. Swirls are seen in the region labelled "S" on the both upper and lower sides of the position. (Regions having swirl defects are referred to as "S-regions" hereafter.) The S-region becomes wider with increasing the pausing time from 1 to 2 min. Further increasing does not widen the region on the lower side of the interface position. The width indicates that swirls are grown to visible ones in a temperature range of 1100°C to the melting point (1410°C) from nucleation centers in striated distribution. On the lower side of the pause position, the temperature gradient at the interface increases toward that at V=0 and the cooling was slower by pausing. On the upper side of the position, both the cooling rate and temperature gradient are larger than those of steady-state growth at V=4 mm/min. The swirl generation on both the sides shows that the higher temperature gradient near the interface is responsible for swirl formation.

3. DETECTION OF DEFECTS

The copper decoration followed by x-ray topography has been used for observation of swirl defects [4]. It was found that section topographs of Cu-decorated crystals are effective to detect very small defects. Also, measurement of Cu concentrations in decorated crystals by secondary-ion mass spectroscopy (SIMS) provides information on crystal perfection.

3.1 D-Defects

The areas labelled "V" are seen in Fig. 2, where small Cu precipitates are present uniformly [See Fig. 3]. The defects responsible for the precipitates have been named "D-defects" as a new type of microdefects in Si crystals by Roksnoer and Van Den Boom [7]. They have observed the defects by Cu- and Li-decoration followed by x-ray topographic observation. Since such decoration methods require thermal annealing of crystals, an experiment was made to confirm whether D-defects are present in the as-grown state of crystals.

During growth of a crystal, its diameter was changed. Figure 3(a) is an x-ray topograph of the longitudinal specimen cut from the crystal which was taken after copper decoration. The growth rate was fixed at 4 mm/min in the entire of the growth. Swirl defects are formed in the regions "S" by thermal fluctuation due to both increasing and decreasing the diameter. D-defects appear in the regions "V" with diameters of 40 and 45 mm. (Regions having D-defects are referred to as "V-regions".) In the thinner parts of the crystal, D-defects are not formed, except the small round V-region at the center of the crystal, which happened to form by thermal fluctuation due to the diameter change. Figure 3(b) is a section topograph for the Cu decorated specimen which was taken at the position indicated by the vertical line in Fig. 3(a). In the V-region, many black spots (D-defects) are seen in uniform distribution. Figure 3(c) shows a

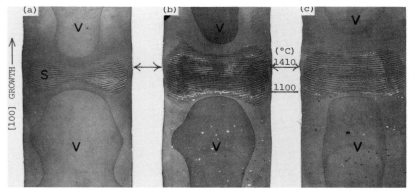

Fig. 2. Effect of in situ annealing during crystal growth at a growth rate of V=4 mm/min. Crystal diameter=42 mm. At the interface position indicated by the arrows, zone travelling was paused for 1 min in (a), 2 min in (b), and 4 min in (c). The observation was made for the longitudinal specimens by x-ray topography following Cu-decoration, using the 400 reflection. A and B-swirl defects are seen white and dark in the regions labelled "S", respectively. The regions "V" contain D-defects.

4

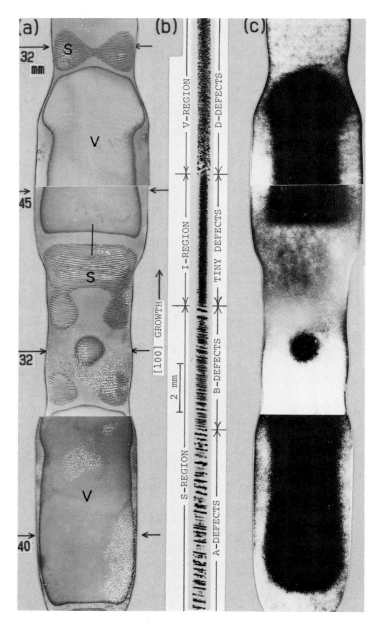

Fig. 3. Dependence of microdefect formation on crystal diameters. During FZ growth, the diameter was changed by adjusting the RF input power. (a) Topograph of Cu-decorated longitudinal specimen. S: Regions with swirl defects. V:Regions with D-defects. (b) Section topograph of the Cu-decorated specimen taken at the position of the vertical line at the center of (a). (All the section topographs in this paper are shown in negative, i.e, white regions have lower diffracted x-ray intensities, but scanning topographs in positive.) (c) Etch pattern of the longitudinal specimen in the as-grown state. The black regions are due to etch depression.

photograph of the longitudinal specimen etched in the as-grown state; the black
regions are due to etching depression [19] and correspond to the V-regions.
This observation shows that D-defects exist in an as-grown state and are responsible for the etch depression.

Figure 3 also shows that formation of D-defects strongly depends on the
crystal diameter.

3.2 Perfection of swirl-free and D-defect-free regions

The section topograph in Fig. 3(b) shows much smaller defects in very high
densities in the swirl-free and D-defect-free region between the S- and V-
regions. Such regions are referred to as "I-regions". The tiny defects are
present in the deep region from the specimen surfaces, as seen from the section
topograph. It was found by SIMS measurement that Cu concentration in this I-
region is the same level as the S-regions and an order of magnitude higher than
that in the V-region. This means that tiny defects still exist in I-regions.
(White regions along both the edges of the section topograph are seen, which had
a Cu concentration higher than that in the central region.

4. DOPING EFFECT ON FORMATION OF MICRODEFECTS

Various impurities were doped during growth of non-doped crystals with
diameters of 42 mm. The transient regions between growth rates V=4 and V=3
mm/min were made in both the non-doped and doped parts of each crystal. Microdefects in the longitudinal specimens were observed by x-ray topography following Cu decoration. A typical result for non-doped parts are shown in Fig. 4.
Swirl defects in striated distribution are seen in the periphery regions label-

led "S" which are pro-
truding from the parts
grown at V=3 mm/min. The
V-region having D-defects
extends over the almost
entire area of the crys-
tal in the steady state
growth at V=4 mm/min and
shrinks in the part grown
at V=3 mm/min. In the
following part (upper
part) grown at 4 mm/min,
the V-region is narrower.
It was confirmed by the
SIMS measurement that the
swirl-free and D-defect-
free region (I-region)
has a copper concentra-
tion in the same level as
that in the S-region
which is an order of
magnitude higher than
that for the V region, as
has been mentioned in
3.2. Figure 4(b) is a
section topograph taken
at the position marked by
the vertical line in Fig.
4(a). Cu precipitates in
the I-region are very
small compared those in
the V-region. Some of

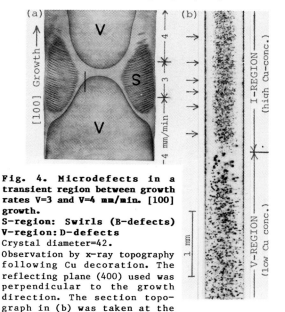

Fig. 4. Microdefects in a
transient region between growth
rates V=3 and V=4 mm/min. [100]
growth.
S-region: Swirls (B-defects)
V-region: D-defects
Crystal diameter=42.
Observation by x-ray topography
following Cu decoration. The
reflecting plane (400) used was
perpendicular to the growth
direction. The section topograph in (b) was taken at the
position indicated by the vertical line in (a),
using (044) reflecting plane parallel to the
growth direction.

6

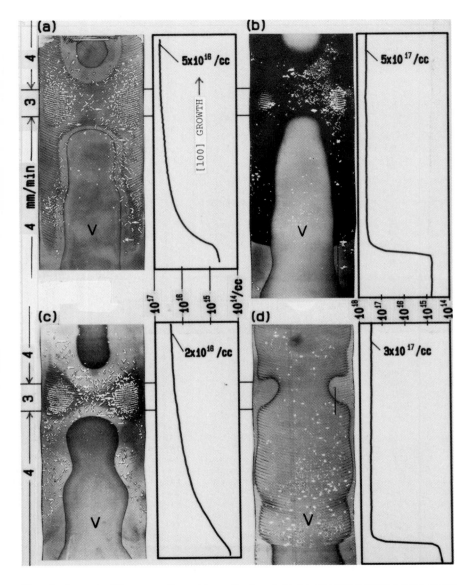

Fig. 5. Dopant effects on formation of microdefects observed with x-ray topographs of Cu-decorated longitudinal specimens. 400 reflection.
(a) B-doped. (b) Ga-doped. (c) P-doped. (d) Sb-doped.

them are a little larger and are in striated distribution, as indicated by the arrows.

4.1 III-group and V-group dopants

Figures 5(a), (b), (c), and (d) are the topographs and concentration profiles for crystals doped with B, Ga, P, and Sb, respectively. The V-regions become narrower by doping impurities except Sb (Compare with Fig. 4). In the B-doped crystal, swirls (B-defects) are seen in the periphery regions of the parts grown even with V=4 mm/min. In the Ga- and P-doped crystals, swirls are formed within and protruding from the regions grown at 3 mm/min, respectively. In the Sb-doped crystal, D-defects are in striated distribution in the V region which extends over the almost entire of the parts grown at V=4 mm/min, and swirl formation is suppressed even in the part grown at V=3 mm/min. The wider V region indicates that vacacies are stabilized by doping Sb, as will be discussed later. Comparison between these photographs suggests that impurities having large covalent radii suppress swirl formation.

In Ga-doped crystals [Fig. 5(b)], the I-regions outside of the V-region are dark, and the boundaries between non-doped and Ga-doped regions are seen clearly. Figure 6(a) is a section topograph which was taken at the position of the vertical line near the right edge of the Cu-decorated crystal in Fig. 5(b). In the V-region, black spots are seen in uniform distribution, and in the I-region of the non-doped part, many tiny dots appear in the central region of the section topograph. In the Ga-doped I-region, Pendel-losung fringes are seen vertically. This observation shows that the Ga-doped I-region [dark region in Fig. 5(b)] is very perfect. The SIMS measurement was also made for the Cu-decorated crystal. It was found that the Cu-concentration is very low in Ga-doped part as shown by figures in Fig. 6. These observations showed that dopant Ga has a special character which is quite different from the other dopants.

Comparison between the non-doped and Ga-doped parts in Fig. 6 shows that the tiny Cu precipitates observed in non-doped I-regions

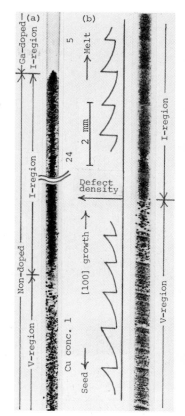

Fig. 6. X-ray section topo-graphs of the Cu-decorated crystals doped with Ga and Sb. (a) and (b) were taken at the positions shown by the verti-cal lines in Figs. 5(b) and (d), respectively. 440 reflec-tion. The figures in (a) indicate Cu concentrations measured by SIMS (arbitrary unit).

are not nucleated statistically by the supersaturated copper atoms, but are due to some kind of defects.

For the Sb-doped crystal, a section topograph was also taken at the position of the vertical line in Fig. 5(d). Both D-defects in the V-region and tiny defects in the I-region are seen in striated distribution with opposite phases: Concentration of D-defects is higher on the seed-side of each growth striation (remelt boundary), similarly to the variation of Sb concentration [20], i.e., defect concentration increases with increase of Sb concentration. Whereas, that

8

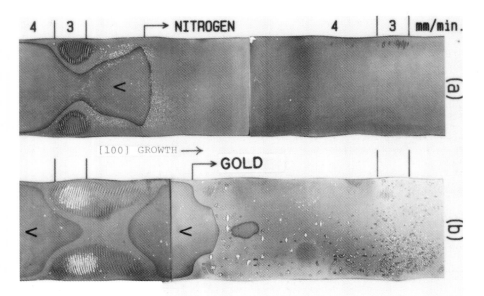

Fig. 7. Impurity effects on formation of microdefects observed with x-ray topographs of Cu-decorated longitudinal specimens. 400 reflection. (a) N-doped. (b) Au-doped. (c) C-doped (d) O-doped.

in the I-region is higher on the melt-side.

4.2 Other impurities: nitrogen, gold, carbon, and oxygen

Nitrogen was doped by adding nitrogen gas into the argon atmosphere during the growth. As seen from Fig. 7(a), effect of nitrogen is remarkable on both swirls and D-defects: By doping nitrogen, the V-region abruptly disappears, and swirls are not formed in the part grown even at a growth rate of 3 mm/min.

Gold impurity shows the similar effect on defect formation as seen from Fig. 7(b): The V-region disappears, and swirl formation is suppressed considerably by doping gold.

Carbon has unique effect on both swirl and D-defect formation [Fig. 7(c)]. The V-region disappears by doping carbon even at a concentration as low as 1.0 ppma. This is the reason why D-defects had not been observed in the earlier work which was carried out with crystls containing carbon at relatively high concentrations. Swirls (B-defects) appear in the periphery regions grown even at a growth rate of V=4 mm/min, i.e., carbon enhances swirl formation, similarly to the case of doping boron [See Fig. 5(a)].

Oxygen was doped by inserting a small quartz bar into the melt during the FZ growth. By doping oxygen, D-defects are in striated distribution similarly to the case of antimony, and the V-region becomes narrower, as shown in Fig. 7(d). Such abrupt shrink of V-regions occurs when oxygen is doped at high concentrations.

In-situ annealing of oxygen-doped crystals was made by pausing the zone travelling similarly to the case of Fig. 2. The zone travelling was stopped for 1, 2, and 4 min for Figs. 8(a), (b), and (c), respectively, and then the crystals were grown again at the same growth rate V=4 mm/min. Comparison with Fig. 2 shows clearly oxygen effect: It is found from the shape of the V-region in Fig. 8(a) that vacancy diffusion is similar to that in non-doped crystals at higher temperatures ($\gtrsim 900°C$), but is slower in lower temperature ranges. It should be noted that the V-region annealed at the higher temperatures shows relatively

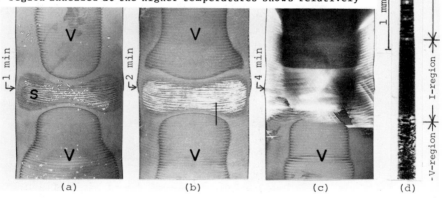

Fig. 8. Oxygen effect on formation of microdefects. In-situ annealing was made during crystal growth at a growth rate of V=4 mm/min. At the intaerface position indicated by the arrows, zone travelling was paused for 1 min in (a), 2 min in (b), and 4 min in (c). Crystal diameter=42 mm. The experimental conditions are the same as the case of Fig. 2 for comparison.

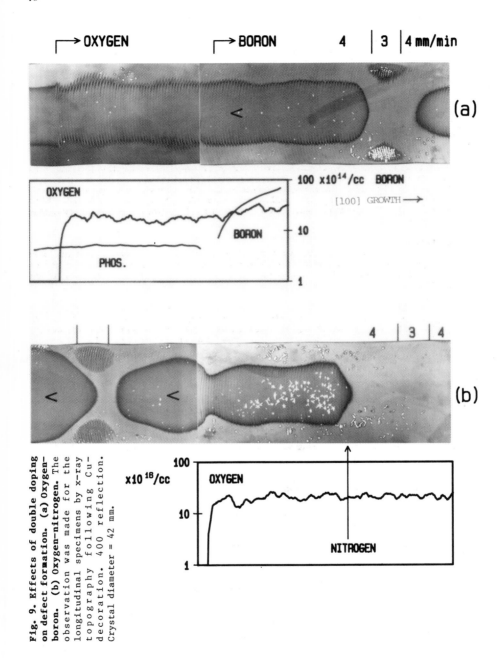

Fig. 9. Effects of double doping on defect formation. (a) Oxygen–boron. (b) Oxygen–nitrogen. The observation was made for the longitudinal specimens by x-ray topography following Cu-decoration. 400 reflection. Crystal diameter = 42 mm.

uniform distribution of D-defects. On the upper side of the pausing position, the cooling rate is larger than that for the steady-state growth, and the V-region is wider in Fig. 8(b). This observation confirms the slower vacancy diffusion at the lower temperatures.

The region having swirls is narrower than that for non-doped crystals. However, sizes of individual swirl defects appear to be larger, and dislocation are generated for a pausing time of 4 min.

The section topograph in Fig. 8(d) was taken at the position indicated by the vertical line in Fig. 8(b). In comparison with section topographs for crystals non-doped and doped other impurities, the defects in the I-regions are smaller with higher densities and appear over the entire of the section. This observation suggests that they are related to oxygen impurity.

4.3 Double doping

For integrated circuits, CZ grown crystals have been used which contain oxygen. Therefore, the effects of various impurities observed in the previous section may differ from that in CZ crystals by the presence of oxygen impurity. In order to know whether our results can be applicable to CZ crystals, both oxygen and other impurities were doped during the FZ growth. Only two examples will be shown here.

Oxygen-Boron

During the growth, oxygen was first doped, and then gas doping of boron was made as shown in Fig. 9(a). The enhancement of swirl formation by impurity boron as observed in Fig. 5(a) is not seen, i.e., the oxygen effect overcomes the boron effect.

Oxygen-Nitrogen

A similar experiment was carried out for oxygen and nitrogen, as shown in Fig. 9(b). The effect of nitogen is so strong that the oxygen effect is not seen: The result is very similar to the case of doping nitrogen only.

As seen from the two examples described above, we may not extend the result for FZ crystals to the case of CZ crystals. Further experiments such as cases of oxygen-carbon impurity are going on.

5. VACANCIES AND D-DEFECTS

By comparing with swirls which have been identified to be interstitial-type dislocation-loops, Roksnoer and Van Den Boom [7] concluded that D-defects are vacancy agglomerates or sponges. We can add some characters of D-defects different from swirls.

(1) D-defects are formed in relatively low temperatures, while swirls are generated near the melting point (Fig. 2)

(2) Carbon impurity suppresses formation of D-defects and enhances generation of B-swirls [Fig. 7(c)].

(3) Antimony impurity enhances formation of D-defects and suppresses swirl formation.

(4) The D-defect formation is similar for B-, Ga-, and P-doped crystals, whereas swirl formation is greatly different between them.

(5) In Cu-decorated specimens, S-regions having swirl defects contain Cu at an order of manitude higher concentration compared with V-regions having D-defects.

In addition, if we assume that the defects in non-striated distribution are due to self-interstitials, swirl formation must be easier for thicker crystals. This conflicts with the result in Fig. 1. From the above argument we can reach the same conclusion that frozen-in vacancies are responsible for D-defect formation, i.e., the V-regions have supersaturated vacancies during the cooling process.

Vacancies near the melting point diffuse much faster than the growth velocity and therefore exist at the equilibrium concentration at the growth interface. When the temperature gradient at the interface is high (low growth rates), vacancies supersaturated in the cooling process escape to the melt by uphill diffusion. For higher growth rates, the temperature gradient is low and cooling rate is fast. In this case, the supersaturated vacancies cannot diffuse out from the interface which is too far from the vacancy-supersaturated region. Thus, crystals grown at high growth rates contain supersaturated vacancies at so high concentrations that they form vacancy agglomerates in relatively low temperatures.

Generally, impurity-doped crystals have V-regions only in their central part, whereas V-regions appear almost over the entire area in non-doped crystals grown at V=4 mm/min in the steady state. One can deduce impurity effect on behavior of vacancies from shapes of the V-regions in the transient region between non-doped and doped part. In equilibrium between both the parts, the vacancy flow from non-doped to doped part must be equal to the reverse flow, i.e.,

$$[V]_n D_n = [V]_d D_d,$$

where D and [V] are the diffusion coefficients and concentrations of vacancies, and the suffixes, n and d, stand for non-doped and doped side of the transient region, respectively. In the part doped with B, P, and Ga, the V-regions are narrower than that in non-doped part, but the shrink occurs gradually on the non-doped side. This means that $D_d = D_n$. Therefore, vacancy concentration is lowered by impurity effect such as formation of vacancy-impuirty complexes. While, for Sb, O, N, and Au-doped crystals, abrupt changes in width of the V-regions take place, indicating that these impurities make vacancy diffusion much slower. Especially, by N and Au impurities, vacancy are immobilized so that vacancy agglomeration is prevented, although vacancy concentration is very high. (If D_d had appreciable values, the V-region on the non-doped side should have shrink more gradually.)

For Sb- and O-doped crystals, out-diffusion of vacancies from the crystal surface is also seen in the parts grown at low growth rates (V=3 mm/min). These impurities expands the lattice spacing, and vacancies are attracted by the stress fields, resulting in the striated distribution of D-defects [See Fig. 6(b)]. Such elastic interaction acts to stabilize vacancies in crystals; it prevent vacancy diffusion perpendicular to the growth striations, whereas the out-diffusion occurs along the growth striations.

As seen from comparison of Figs. 2, 8, and 5(d), vacancy diffusion in O-doped crystals is slower than that in non-doped crystals and faster than that in Sb-doped crystals. Since V-regions in non-doped and Sb-doped crystals grown at V=4 mm/min cover their almost entire area, it is expected that V-regions be extended from rim to rim of O-doped crystals. However, narrow V-regions were often observed. This observation indicates reduction in vacancy concentration by forming oxygen-vacancy complexes and/or some their interaction.

In the case of carbon [Fig. 7(c)], V-region in the transient region on the non-doped side shows a round bondary. This indicates that vacancy diffusion is slower, and vacancy concentration is very low.

The above argument is summarized in Fig. 10. The radial profiles of frozen-in vacacies in crystals by steady-state growth at V=4 mm/min are shown schematically. The vacancy concentration has the maximum at the center of crystals and is low in the periphery owing to out-diffusion from the side surface of the crystal. D-defects are formed in the region where the vacancy concentration exceeds a certain level required for the defect formation. This level is indicated by the horizontal solid line. For B, P, Ga, O, and C impurities, the frozen-in vacancy concentration is decreased as shown by the dotted line, and we have a narrow V-region at the center of the crystals. For impurities which make vacancy diffusion very slow, the minimum level becomes much higher, and V-regions are not formed, as seen for N and Au-doped crystals.

It is concluded from the observations that vacancies are stabilized in compressive stress fields around impurity atoms having covalent radii larger than that of silicon, whereas impurity atoms with smaller covalent radii may assist out-diffusion of vacancies.

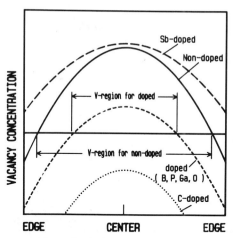

Fig. 10. Radial distribution of frozen-in vacancies depending on doping impurities (schematic).

6. DEFECTS IN D-DEFECT-FREE AND SWIRL-FREE REGIONS

It is clear from comparison of the observation for crystals doped with gallium [Figs. 5(b), 6(a)] and other impurities that D-defect-free and swirl-free regions (I-regions) contain tiny defects. Copper-concentrations in both I- and S-regions of decorated crystals were always in the same level. This observation suggests that the defects in both I- and S-regions have essentially the same. To investigate whether or not mutual annihilation between vacancies and the defects responsible for I and S-regions takes place as expected between vacancies and interstitials, boundaries between V- and I-regions are examined:

(1)   In the boundary regions, in spite of very low densities of Cu precipitates, Pendellösung fringes are not seen [Figs. 3(b) and 4(b)].

(2)   Near the boundary, concentration of D-defects decreases, whereas the average defect density in the I-region is uniform [Figs. 3(b) and 4(b)].

(3)   I-regions having the defects always appear after V-regions recede by annealing at temperatures which are too low to form the defects, as seen from the in-situ annealing in Figs 2 and 8.

(4)   What is most surprising is the sharpness of the boundary between the Ga-doped I-region and non-doped I-region. The defects in I-regions cannot be attributed to intrinsic defects such as interstitials which can diffuse very fast at high temperatures.

These observations lead to the conclusion that the defects in I-regions are microprecipitates of a certain impurity formed at high temperatures; they are present over the entire of a crystal, but their Cu-decoration is prevented by supersaturated vacancies in V-regions: The stress fields around the microprecipitates are relaxed by attracting vacancies, as deduced from the observation for the Sb-doped crystal [Figs. 5(d) and 6(b)]. The stress relaxation makes the microprecipitates invisible by Cu-decoration.* [In Fig. 6(b), therefore, the density profile of Cu-decorated defects in the I region is in the phase opposite to that of D-defects.]

The impurity which forms the microprecipitates should be responsible for the special effect of gallium on I-regions. From this point of view, oxygen is chosen as the candidate reactive with gallium from common residual impurities in silicon. At high temperatures, chemical activity of gallium for oxygen is considered to be higher than that of silicon.** When two substitutional Ga atoms form α-Ga_2O_3 by taking three oxygen atoms, the resulting volume expansion is very small [Volume per Si atoms in the silicon lattice = 20 Å^3. Volume for $1/2(Ga_2O_3)$ = 24 Å^3. Compare with molecule of cristobalite SiO_2 = 43 Å^3]. Consequently, microprecipitates in Ga-doped crystals cannot be observed by Cu-decoration.

Furthermore, formation of the tiny defects in I-regions is enhanced by dopping oxygen as has been seen in Fig. 8(d). It is concluded from the above argument that defects in I-regions are oxygen microprecipitates.

Concentration of residual oxygen impurity in FZ crystals [$\lesssim 10^{16}$ cm^{-3}] is lower than the solubilities [21] which have been reported by many investigators. There are two possible answers for the question why such small precipitates are formed: (1) Since silicon will apparently tolerate very high oxygen supersaturation, the intrinsic solubilities are expected to be much lower than the measured values. (2) Some of oxygen atoms in the silicon melt are incorporated as clusters or crystallites having SiO_x which result in locally high oxygen concentrations [22]. Both are considered as a possible explanation.

7. SWIRL FORMATION

Swirl defects are formed by two processes of nucleation and growth. Their formation will be explained by assuming that they are oxygen precipitates larger compared with ones in I-regions.

7.1 Defect growth

The temperature range in which defects grow is one of the most important thermal conditions for swirl formation and was deduced from the experimental results: Both A and B swirl defects are not formed in 4-mm/min steady-state growth of crystals with a diameter of 40 mm. However, when the zone-travelling rate is decreased to 3 mm/min during the growth, B-defects are usually formed in the periphery regions of the part grown at 4 mm/min, as shown schematically in Fig. 12. This observation shows that B-defects are formed in the regions which are cooled fast in the high-temperature range and slowly at lower temperatures (900-1100°C), whereas A-defects are formed at high temperatures ($\gtrsim$1100°C).

It is well known that swirl formation depends upon remelting during growth [18,24]. The section topograph in Fig. 3(b) shows that

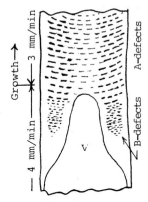

Fig. 11. Schematic illustration of defect formation by changing growth rates.

---

*In Fig. 4(b), the white regions appear along the edges of the section topograph for the I-region. Defects in the surface regions may be invisible by vacancies diffused into the crystal during annealing for Cu decoration.
**Pulsed laser annealing of Si in air does not cause any oxidation for Si, but oxidation of Ga takes place in laser annealing of GaAs [23].))

A-defects are formed along major remelting boundaries, and B-defects along many minor boundaries as well as major ones. Pendellösung fringes are seen between the defects stratified. This observation indicates that microprecipitates are dissolved, and the oxygen atoms are gathered by the nucleation centers along remelt boundaries to contribute to growth of swirl defects.

Next, we consider the stability of microprecipitates in I-regions. Oxygen microprecipitates accompany stress fields around them due to volume expansion which relax by attracting vacancies if vacancy concentration is high enough, i.e., the interfacial free energy of precipitates is lowered with high vacancy concentration. Therefore, when vacancy concentration is decreased by out-diffusion, the microprecipitates become unstable by losing vacancies from their interfaces and dissolve if they are smaller than the critical size. Only larger precipitates as indicated by the arrows in Fig. 4(b) can grow to swirl defects. Therefore, when vacancy out-diffusion takes place at high temperatures where the critical size is much larger, large precipitates along the major remelting boundaries can survive. Their growth accompanies the stress fields to generate interstitial-type dislocation loops around them (A-defects).

In summary, swirl formation depends upon the temperature T at which super-saturation of vacancies vanishes by out-diffusion, i.e., $1100°C \lesssim T \lesssim$ melting point for A-swirls, $900 \lesssim T \lesssim 1100°C$ for B-swirls [See Fig. 2 and 8], and micro-precipitates in I-regions are stable if $T \lesssim 900°C$.

For the vacancy out-diffusion, the temperature gradient at the growth interface associated with the growth rate is critical, especially for large-diameter crystals. High temperature gradients result in low concentrations of vacancies by uphill diffusion toward the melt, as described in 5. Also, high temperature gradients produce strong stress fields near the interface, especial-ly in periphery regions. The temperature gradient reaches 400°C/cm at low growth rates [8]. Such high stress concentrates on the precipitates grown along the major remelting boundaries and generates dislocations having complicated configurations, as reported by Föll and Kolbesen [5].

7.2 Nucleation centers for swirl defects

Formation of the larger precipitates along major and minor remelting boun-daries can be understood by the liquid drop formation just behind the interface [16, 22]. It was observed by in-situ x-ray observation of melting and solidifi-cation of silicon ribbons that, during melting of dislocation-free Si crystals, locally molten regions (drops) are always formed inside the crystals. (Drops are not formed in dislocated crystals). In conventional crystal growth such as the FZ and CZ methods, the temperature fluctuation during growth occurs due to the crystal rotation and melt convection; the fluctuation often reaches 10°C. Therefore, crystals grow by repeating alternatively growth and pausing (or remelting). In the pausing periods, the interface region is superheated highly enough to form drops. In the microscopic regrowth period, the drops solidify. For impurity atoms in the drops, the solidifiction is slow enough to segregate. Most of impurity atoms in Si have segregation coefficients less than unity, and, after solidification of the drops, they segregate at their central region which is solidified finally.

In this process, oxygen impurity has two effects: (1) In order to form drops inside crystals, the free energy barrier due to the solid-liquid inter-facial free energy must be overcome. It was found from the in situ x-ray obser-vation on melting and growth of oxygen-doped crystals [22] that the interfacial free energy for (111) interfaces is greatly decreased by oxygen impurity, and large drops are formed. [From large drops, dislocations are generated easily, as seen from Fig. 8(c).] Conventional float-zoned crystals contain oxygen atoms in order of 10^{16} cm^{-3}. Recently, Abe et al. found that oxygen impurity is in striated distribution [25], similarly to other impurities having segregation coefficients less than unity. Cooperation of the oxygen effect and temperature fluctuation during growth is responsible for nucleation of drops. (2) As expected from the finding by Abe et al.[25] mentioned above, impurity oxygen segregates toward the center of the solidifying drop. Consequently, its concentration at the central region may exceeds the solubility limit even near the melting point to form SiO_2 precipitates larger than the critical sizes. Their growth accompanies volume expansion and result in interstitial type dislocation-loops under the condition decribed in 7.1. The formation of SiO_2 is supported with TEM observation by Ravi and Varker [26] that there is cristobalite SiO_2 inside swirl defects in float-zoned crystals.

7.3 Doping effect

Impurity effect will be discussed for both the processes of the nucleation and growth:

Although oxygen effect has major effect on the nucleation as described above, defect growth is retarded by holding vacancies in the crystal. Particularly, since vacancy diffusion at the lower temperatures in oxygen-doped crystals is slower than that in non-doped crystals [see 4.2], formation of B-defects is retarded.

Stress fields due to impurity atoms having the smaller covalent radii assist vacancy out-diffusion as described in 5. A typical example is carbon impurity, which forms S-and I-regions over the entire of crystals [Fig. 7(c)]. Also, segregation of such impurities due to solidification of drops is favorable for precipitation of SiO_2 which accompanies the lattice expansion. Thus, swirl formation is enhanced by the effects on both the nucleation and growth.

Impurities with the larger covalent radii have the opposite effect: The vacancies attracted by growth striations act to stabilize the microprecipitates in I-regions and to prevent the defect growth. The segregation of such impurities by drop formation suppresses the precipitation of the segregated oxygen impurity which causes further lattice expansion, i.e., swirl formation is prevented as seen for the Sb-doped crystal.

Nitrogen and gold impurity have strong interaction with vacancies to make them immobile. This may stabilize microprecipitates and result in suppression of swirl formation.

8. CONCLUSION

Swirl defects and D-defects were observed in FZ crystals. The former has been identified to be of interstitial-type dislocation loops or loops clusters. It is concluded from the observed difference between D-defects and swirls that D-defects are of vacancy agglomerates and that vacancies are the predominant intrinsic defect near the melting point. A new type of defects were always observed in swirl-free and D-defect-free crystals by copper decoration followed by x-ray topography. They become invisible by doping gallium. This effect leads to the conclusion that they are microprecipitates formed by residual oxygen impurity in FZ crystals. They are stable with supersaturated vacancies. When vacancy out-diffusion takes place, they dissolve, and A- and B-defects are formed in the ranges of high and low temperatures, respectively. Nucleation centers of swirl defects are originated from liquid drops which are formed just behind the interface by temperature fluctuation during crystal growth. Swirl-free crystals can be grown even in the lower growth rates (Fig. 1) by doping nitrogen which has no significant effect on electrical properties of the crystal.

REFERENCES

1. T. S. Plaskett, Trans. Met. Soc. AIME **233**, 809 (1965).

2. T. Abe, T. Samizo, and S. Maruyama, Jpn. J. Appl. Phys. **5**, 458 (1966).

3. A. J. R. de Kock, Appl. Phys. Lett. **16**, 100 (1970).

4. A. J. R. de Kock, Philips Res. Rept. Suppl. No. 1 (1973).

5. H. Föll and B. O. Kolbesen, Appl. Phys. **8**, 319 (1975).

6. P. M. Petroff and A. J. R. de Kock, J. Cryst. Growth **30** 117 (1975).

7. R. J. Roksnoer and M. M. B. Van den Boom, J. Cryst Growth **53**, 563 (1981).

8. T. Abe, H. Harada, and J. Chikawa in: Procedings of 12th International Conference on Defects and Radiation Effects in Semiconductors, to be published.

9. For instance, see J. W. Corbett and P. Karins, Nuclear Instr. Methods **182/183**, 457 (1981).

10. A. Seeger and K. P. Chik, Phys. Stat. Solidi **29**, 455 (1968).

11. M. Yoshida, E. Arai, H. Nakamura, and Y. Terunuma, J. Appl. Phys. **45**, 1498 (1974).

12. S. M. Hu, J. Appl. Phys. **45**, 1567 (1974).

13. S. Mizuo and H. Higuchi, Jpn. J. Appl. Phys. **20**, 739 (1981).

14. S. M. Hu, J. Vac. Sic. Technol. **14** 17 (1977).

15. P. M. Petroff and A. J. R. de Kock, J. Cryst. Growth **35**, 4 (1976).

16. J. Chikawa and S. Shirai, J. Cryst. Growth **39**, 328 (1977).

17. H. Föll, U. Gosele, and B. O. Kolbesen, J. Cryst. Growth **40**, 90 (1977).

18. A. J. R. de Kock, P. J. Roksnoer, and P. G. T. Boonen, J. Cryst. Growth **22**, 311 (1974).

19. W. Keller and A. Mühlbauer, Inst. Phys. Conf. Ser. No. 23, Chap. 8, p. 538 (1975).

20. T. Abe, Y. Abe, and J. Chikawa in: Semiconductor Silicon, H. R. Huff and R. R. Burgess eds. (Electrochemical Soc. Princeton, 1973) pp. 95-106.

21. R. A. Craven in: Semiconductor Silicon, H. R. Huff, R. J. Kriegler, and Y. Takeishi eds. (Electrochemical Soc., Rennington, N. J. 1981) p. 670.

22. J. Chikawa and F. Sato in Defects in Semiconductors, J. Narayan and T. Y. Tan eds. (North-Holland, New York, 1981) pp. 317-332.

23. F. Sato, T. Sunada, and J. Chikawa, Materials Lett. **1**, 111 (1982).

24. J. Chikawa and S. Shirai, Jpn. J. Appl. Phys. **Suppl. 18-1**, 153 (1978).

25. T. Abe, K. Kikuchi, S. Shirai, and S. Muraoka: in Semiconductor Silicon, H. R. Huff, R. J. Kriegler, and Y. Takeishi eds. (Electrochemical Soc., Rennington, N. J. 1981) p. 670.

26. K. V. Ravi and C. J. Varker: in Ref. 20, p. 670.

TRANSITION METAL IMPURITIES IN SILICON

EICKE R. WEBER and NORBERT WIEHL*
II. Physikalisches Institut (* Institut f. Kernchemie)
University of Köln, D5000 Köln 41, West-Germany

ABSTRACT

The properties of transition metals in silicon are reviewed, emphasizing those observations which allow conclusions to be drawn with respect to microscopic defect models. 3d metals diffuse interstitially into silicon and stay predominantly in these sites at high temperatures. 3d elements lighter than Co can be quenched into these interstitial sites, giving rise to well-established energy levels. First theoretic calculations for these ions yield promising results. Co, Ni and Cu vanish out of the interstitial solution during quenching; an appreciable fraction of Cu may form pairs. The understanding of 4d and 5d metals in silicon is much less advanced at present, even for the technologically important elements Au and Pt. Some observations indicate that for Au and Pt pair formation might as well be important.

INTRODUCTION

Transition metals, fast diffusers in silicon, may strongly affect the performance of any silicon-based device. Thus the electrical properties of T-metal doped silicon have been the subject of numerous investigations in the last 25 years, as compiled e.g. in the comprehensive review by Chen and Milnes [1]. Yet in many cases simultaneous presence of several transition metals and uncontrolled complex formation in the samples used, precludes relating these data to well-defined defect configurations.

However, since the work of Ludwig and Woodbury [2] it is known that Electron Paramagnetic Resonance (EPR) is capable of clearly identifying a great variety of transition metal species in silicon, especially from the 3d series. Among these are isolated atoms on interstitial and substitutional sites in various charge states as well as clusters of T-metals and complexes with other dopants. Thus combinations of EPR experiments with additional techniques such as Neutron Activation Analysis (NAA), electrical, or optical investigations, offer the possibility to identify the properties of specific T-metal species.

Those results shall be the subject of this review. Its main part will be devoted to 3d transition metals in silicon for which a number of clear conclusions can be drawn [3]. After this, the state of our knowledge of non 3d-metals, like Au and Pt in Si, will be discussed.

DIFFUSION AND SOLUBILITY OF 3d METALS IN SILICON

A substantial part of the diffusion and solubility data treated in this chapter has been derived from combined NAA/EPR studies of 3d metal diffused and quenched silicon [3-6]. Table 1 lists the EPR species, Table 2 the nuclear reactions used in the course of these investigations. The EPR parameters were found to be in agreement with Ludwig and Woodbury [2].

Mat. Res. Soc. Symp. Proc. Vol. 14 (1983) © Elsevier Science Publishing Co., Inc.

TABLE I
Paramagnetic 3d metal ions used for solubility measurements in silicon [3-6].

metal	EPR species	temperature of measurement	detection limit
chromium	Cr_i^+	35K	10^{11} cm^{-3}
manganese	Mn_i^o	7K	10^{12} cm^{-3}
iron	Fe_i^o	50K	10^{11} cm^{-3}

TABLE II
Nuclear reactions for NAA solubility determinations of 3d metals in silicon

stable isotope	abundance	reaction	capture cross-section	γ-energy (intensity)	half-life	detection limit atoms/cm^3 [a]
^{50}Ti	5.3%	$(n,\gamma)^{51}Ti$	0.18 b	320 keV (95%)	5.76 min	10^{16}
^{51}V	99.8%	$(n,\gamma)^{52}V$	4.96 b	1434 keV (100%)	3.76 min	10^{14}
^{50}Cr	4.4%	$(n,\gamma)^{51}Cr$	16 b	320 keV (9.8%)	27.71 d	$5 \cdot 10^{11}$ [b]
^{55}Mn	100%	$(n,\gamma)^{56}Mn$	13 b	847 keV (99 %)	2.58 h	10^{13}
^{58}Fe	0.3%	$(n,\gamma)^{59}Fe$	1 b	1099 keV (56 %)	44.6 d	10^{12} [b]
^{59}Co	100%	$(n,\gamma)^{60}Co$	37 b	1173/1333 keV (100%/100%)	5.3 a	10^{11}
^{58}Ni	67.9%	$(n,p)^{58}Co$	0.1 b	511/811 keV (30%/99%)	71 d	10^{11} [b]
^{63}Cu	69.1%	$(n,\gamma)^{64}Cu$	4.5 b	511/1346 keV (37%/0.5%)	12 h	10^{13}

[a] for samples of about 200mg, 10^{14} n/cm^2s thermal neutron flux
[b] using an Anticompton-spectrometer [6]

The diffusion time dependence of the total iron concentration in iron plated silicon [4] is shown in Fig. 1. After a short in-diffusion period a maximum occurs which can be explained by insufficient boundary phase ($FeSi_2$) formation [3]. For longer diffusion times an equilibrium concentration is reached. Such behaviour was typical for the 3d metals investigated in that study (Cr to Cu).

From the time necessary to reach ~50% saturation a diffusion coefficient D can be estimated, based, e.g., on the assumption of a stepwise concentration profile moving into the crystal via:

$$a^2 = 4 \cdot D \cdot t \qquad (1)$$

(a: distance from the surface)

Diffusion coefficients thus derived have about a factor of 2 error margin. Yet they are well in accordance with those data available from the literature, see Fig. 2.

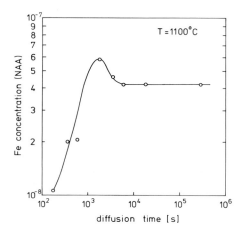

Fig. 1. Fe concentration (NAA) in iron-plated silicon of 4x4x15 mm^3 vs. diffusion time [4].

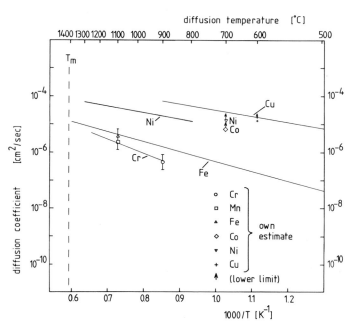

Fig. 2. Diffusion coefficients of 3d metals in silicon: Si:Cu [7], Si:Ni [8], Si:Cr [9]. For Si:Fe various results [10-13] have been combined [3]. Estimated values [3] have been derived from the diffusion time necessary for 50% saturation of NAA samples [4-6].

The line given in Fig. 2 for iron in silicon has been derived [3] from a fit of Struther's high temperature tracer data [10] combined with results obtained from Fe diffusion experiments near room temperature, using various techniques [11-13].

Obviously, the diffusivity of 3d metals increases with atomic number in the 3d row, up to Co, Ni and Cu. Ti and V seem to diffuse markedly slower than e.g. Cr. However, reliable values are not yet available, due mainly to the short half-life of the usable isotopes, see Table 2. For Si:Ti Boltaks [14] determined diffusion coefficients in the 10^{-10} cm^2/s range. Yet there is experimental evidence that these values might be too low by several orders of magnitude [15].

The diffusion coefficients in Fig. 2 are in the 10^{-4} - 10^{-6} cm^2/s range, values typical for the movement of atoms in liquids. Such high diffusion coefficients in silicon can only be understood by diffusion via an interstitial mechanism, independent of native lattice defects [3].

All activation energies of 3d metal diffusion in silicon are in the 0.5eV - 1.0eV range [3], in accordance with simple theoretical models for interstitial diffusion in silicon [16,17].

The fast in-diffusion of 3d metals together with constant metal concentration observed for long diffusion times, Fig. 1, indicates the absence of a noticable change from the interstitial to substitutional sites, as discussed e.g. for Au in Si [18].

That conclusion is supported by the results of EPR solubility measurements: Fig. 3 shows as an example the solubility of Mn in silicon, determined by NAA (Mn_{tot}) and EPR spectroscopy (Mn_i^0) of diffused and quenched silicon [6]. Below the eutectic temperature the saturation concentrations x ("solubilities") can be described by the van't Hoff equation:

$$x = \Delta S^S/k - \Delta H^S/kT \qquad (3)$$

with k being the Boltzmann constant and ΔS^S, ΔH^S the apparent entropy and enthalpy of formation.

Fig. 3 shows the slopes ΔH^S to be the same for both curves, within the scatter of the results. Thus the enthalpy of formation of the total Mn concentration is the same as that of interstitial Mn, supporting the conclusion that practically all Mn has been in interstitial sites before the quench. The absolute difference between the NAA and EPR results can be ascribed primarily to a partial loss of Mn_i during quenching.

Iron and chromium in silicon could as well be found in interstital sites after quenching, in concentrations comparable with the NAA results [3]. For the Cr measurements suitably p-doped silicon was used to ensure complete ionisation to Cr_i^+.

In Si:Co, Si:Ni and Si:Cu no EPR spectra with intensities similar to the total solubilities, see Fig. 4, were found. In p-Si:Ni a weak signal previously ascribed to Ni_i^+ [2] was found, corresponding to 10^{13}-10^{14} defects/cm^3. This spectrum must be ascribed to Ni_i ions stabilized by some additional defect [3]. The high diffusivities observed for these elements (Fig. 2) allows one to deduce that Co_i, Ni_i and Cu_i atoms present during the diffusion treatment vanish out of the interstitial solution even during rapid quenching, presumably by clustering and precipitation, cf. [19] for Si:Co, [20] for Si:Ni.

The driving force for these reactions is the high degree of supersaturation: At room temperature the extrapolated solubility even of Cu in Si is less than one atom per cm^3! Thus it is not surprising that the slower diffusing Fe_i, Mn_i and Cr_i show a typical instability during storage at room temperature: they can be kept in interstitial sites during quenching, but are still mobile near 300K.

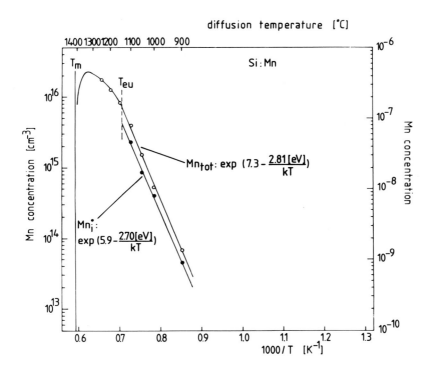

Fig. 3. Solubility of Mn in Si, measured by NAA (Mn_{tot}, open circles)
and EPR (Mn_i^o, full circles) [6].

The total solubilities of 3d metals in silicon, depicted in Fig. 4, have
been calculated [3] using own NAA determinations [4-6], combined in part with
additional results from the literature, esp. [21,22] for Si:Ni and [23,24] for
Si:Cu.

Retrograde solubility, i.e. a solubility maximum occuring at a temperature
higher than the respective eutectic temperature, can be observed for all 3d
elements heavier than Cr; in the latter case the solubility maximum coincides
with the eutectic temperature. The same behaviour can be predicted for Si:Ti
and Si:V [3].

Mn, Fe and Co show almost identical solubilities. Ni and especially Cu show
much higher values than these, whereas Cr and most supposedly Ti and V are
markedly lower. The solubility parameters, Eq. (3), describing solubilities
below the eutectic temperatures, are given in table III.

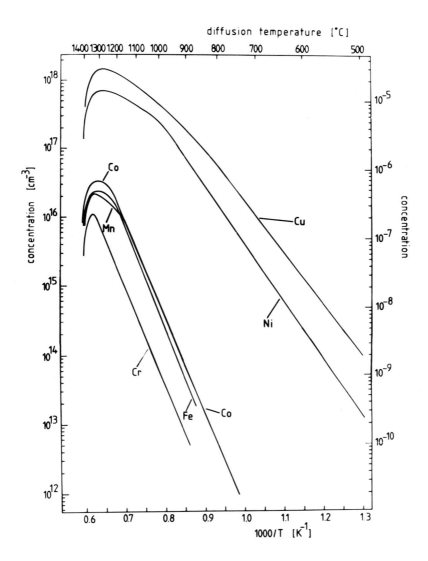

Fig. 4. Solubilities of 3d metals in silicon, determined by NAA [3-6].
The straight lines correspond for Si:Cr to Si:Co to equilibrium with
a silicide of the MSi₂ type, for Si:Ni and Si:Cu to equilibrium with
the metal boundary layer, see [3]. The solubility parameters for this
low-temperature range are listed in table III.

TABLE III
Solubility parameters of 3d metals in silicon (NAA) [3]

metal	$\Delta S^s/k$	ΔH^s	temperature range [a]
Cr	4.7	2.79 eV	900 - 1335 °C
Mn	7.3	2.81 eV	900 - 1142 °C
Fe	8.2	2.94 eV	900 - 1206 °C
Co	7.6	2.83 eV	700 - 1259 °C
Ni	3.2	1.68 eV	500 - 993 °C
Cu	2.4	1.49 eV	500 - 802 °C

[a] upper limits are the respective eutectic temperatures

Solubility values obtained from diffusion temperatures higher than the eutectic temperatures have to be converted into distribution coefficients in order to obtain an Arrhenius-type temperature dependence. From the solubility parameters derived this way, as well as from those given in Table III, effective partial formation enthalpies and entropies of interstitial 3d metals in silicon can be calculated [3], taking into account the respective boundary phase. The results are consistent and show a clear distinction of 3d elements into two groups, according to their effective partial formation enthalpies ΔH^f :

"3d I" elements (Cr, Mn, Fe, Co; Ti, V?) with ΔH^f = 2.1 ± 0.1 eV

"3d II" elements (Ni, Cu)   with ΔH^f = 1.5 ± 0.1 eV

$$(4)$$

These parameters can be envisaged as the energy necessary to transfer a metal atom from the Si surface into an interstitial position in the silicon lattice.

For this distinction two explanations shall be considered:

First, Ni_i^o is supposed to have a closed $3d^{10}$ shell, assuming promotion of the two 4s electrons into the 3d shell as suggested by Ludwig and Woodbury [2]. Such a configuration might experience less interaction with the surrounding Si atoms, compared with an atom with partly filled 3d shell. Thus the effective formation enthalpy might be lower. Cu_i in the positive charge state Cu_i^+ would have the same closed shell electron configuration.

Second, in an approach based on the dielectric two band model, Van Vechten [25] estimated effective formation enthalpies of interstitial metal atoms in silicon. His results were very similar to the values given in (4):

For neutral interstitials: $\Delta H^f \simeq$ 2.4 eV

for singly positively charged interstitials: $\Delta H^f \simeq$ 1.7 eV

$$(5)$$

Thus diffusion as neutral species is suggested for 3d I elements and as positively charged ions for 3d II elements.

The rare drift measurements, i.e. diffusion experiments with an electric field applied [26,27], indicate indeed diffusion as a neutral interstitial for Fe (and Au) in Si and as a singly positively charged interstitial for Cu (and Li) in accordance with the model mentioned last. However, further experiments of this type are necessary for a decision between the two alternatives.

The characterization of 3d elements into these two groups is also important for gettering effects and haze formation [28,29]. In view of the results on diffusivities and solubilities of 3d metals in silicon discussed above, the strong gettering and haze (surface precipitation) formation of Ni and Cu can be explained by their high diffusivities combined with high solubilities [3]. The product of both values determines the flux of metal atoms that can reach gettering sites in a given time, cf. [30]. For example, the solubility, but not the diffusivity of Co in Si is smaller than that of Ni, so that Co gettering or haze formation at least is delayed compared to Ni.

Those 3d elements still present as interstitials after quenching all give rise to a donor level $M_i^{0/+}$ ($M_i^0 \leftrightarrow M_i^+$). Thus they may experience Coulomb attraction from negatively charged shallow acceptors (if present). Indeed, the formation of donor-acceptor pairs is observed at or slightly above room temperature, the kinetics being determined by the appreciable interstitial diffusivities. The EPR spectra of a large number of those pairs have alredy been identified by Ludwig and Woodbury [2].

Metal-acceptor pairs can be dissociated by heating or by illumination [31-33], which may result in instabilities of the electrical properties of such crystals. A careful analysis of the illumination-induced dissociation of FeB pairs led to the proposition of recombination enhanced diffusion of Fe_i in Si [32].

ENERGY LEVELS OF 3d METALS IN SILICON

Electrical levels of interstitial 3d metals in silicon have been collected from the literature [3] using, preferrably, measurements combining different methods like a study of Si:Cr [33]. The DLTS results of Graff and Pieper [34], TSCAP measurements of Lemke [35,36] and Hall effect measurements combined with EPR of Feichtinger et al. [37,38] were especially useful for the establishment of these data.

Fig. 5 shows the result for the 3d elements from Ti to Fe. Obviously, all of these elements introduce single donor levels ($M_i^{0/+}$), Mn, Ti and V additionally introduce an acceptor and a double donor level. Generally the ionization energy, the depth of the (0/+) donor level, measured from the conduction band, increases with increasing nuclear charge, as expected. Yet the $V_i^{0/+}$ donor level shows a clear break. According to the model of Ludwig and Woodbury V_i^0 has a half filled $3d^5$ shell. Thus the pronounced step at $V_i^{0/+}$ may be related with the high ionization energy of that fifth 3d electron.

Literature data for energy levels of Co, Cu or Ni in silicon, see e.g. [1], should be treated with caution as at present they cannot be reliably related to any specific defect configuration. From the considerations described above one can conclude that any data derived for those elements may have to be ascribed to complexes of impurity atoms.

Recently, first cluster calculations yielding energy levels for interstitial 3d metals in silicon have been published [39], see Fig. 5. The results of these calculations support the assumptions of the Ludwig and Woodbury model [2]. The general trend of deeper donor level with increasing atomic number is reproduced well, too.

However, V_i^0 is predicted to be unstable as the donor level comes out to be in the conduction band. Another feature is the prediction of a $Fe_i^{-/0}$ acceptor state below the conduction band that cannot be found experimentally, see [3].

Very recently Lindefelt and Zunger [40] performed pertubation calculations for interstitial 3d metals in silicon, using a quasiband Green's function formalism. The first results show the general trend in the 3d row but they, too, lack an explanation of the deep $V_i^{0/+}$ level.

These calculations do not yet allow one to explain the origin of the 3d I / 3d II distinction. In the near future, however, this might be possible when refined calculations are performed on interstitial 3d metals in silicon. These systems offer a unique possibility to study an experimentally quite well confirmed group of deep level impurities, showing clear chemical trends, but still containing challenging difficulties.

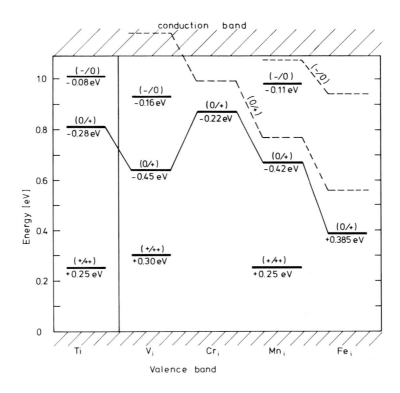

Fig. 5. Energy levels of interstitial 3d metals in silicon (full lines) [3], compared with the results of X_α cluster calculations of DeLeo et al. [39]. The figure contains as well the level positions, given with respect to the adjacent band, respectively.

To conclude this chapter, the various influences of 3d metal contamination on the performance of silicon solar cells, see e.g. [41], shall be discussed. Whereas up to 10^{16}cm^{-3} Cu or Ni might be present in that low-grade material without a serious degradation of cell performance, already 10^{13} cm^{-3} Ti and V seem to be harmful. The above discussion showed that even quenching cannot keep Ni and Cu in electrically active sites so that only a high concentration of clusters or precipitates may alter the electrical characteristics of silicon. The slowly diffusing Ti and V, however, can be expected to stay in a high fraction in electrically active (most supposedly) interstitial sites even during a slow cooling down process. Thus Ti and V which moreover appear in four charge states favourable for recombination processes are indeed expected to be the most harmful 3d elements.

OPTICAL PROPERTIES OF 3d METALS IN SILICON

Optical properties of 3d metals in silicon have been intensely investigated using photo-luminescence (PL) measurements. For isolated interstitial transition metals no luminescence emission could be found. Yet 3d metal-acceptor pairs give rise to strong, characteristic luminescence spectra, as recently reviewed by Sauer and Weber [42].

Uniaxial stress measurements showed the defect symmetries to conform with those known from the corresponding EPR spectra [2].

For CrB pairs PL, EPR, and DLTS investigations were carried out using partly the same specimens [33]. In this way the DLTS level at E_v+0.290 eV and the PL spectrum showing a binding energy of about 300 meV could be unambigously related to the CrB EPR spectrum. For CrB dissociation an activation energy of 0.65 eV was estimated from all three techniques. A simple point charge model of electrostatically attracted, singly charged ions in an nearest-neighbour configuration yields a value of 0.56 eV, as discussed already for FeB pairs [32].

In Cu doped, quenched silicon a strong PL spectrum can be found, which has been ascribed to Cu-Cu pairs with a <111> axis [43]. Its intensity correlates well with the Cu solubility in the 800 - 1200°C diffusion temperature range, suggesting that an appreciable fraction of Cu evolves into this configuration during or after quenching. The result of this analysis is especially interesting in view of the Si:Au and Si:Pt systems to be discussed in the next chapter.

An energy level found in Cu-doped and quenched silicon at E_v+0.10 eV [34] might arise from these Cu-Cu pairs, but this assumption needs further verification.

In conclusion, these photo-luminescence investigations and further optical measurements with transition metals in silicon can now in some cases be related to well-defined defect configurations, thus providing interesting additional information on those defects.

TRANSITION METALS OF THE 4d AND 5d SERIES IN SILICON

Our knowledge of the microscopic state of 4d and 5d transition metals in silicon is inferior to that of 3d metals described in the preceeding chapters. The great number of energy levels ascribed to 4d and 5d atoms [1] indicates the existence of electrically active species the exact nature of which are unknown.

In some cases, especially Si:Au and Si:Pt, the creation of deep levels connected with these impurities is commonly used in device technology although even for those elements reliable defect models could not yet be confirmed.

In general, rapid in-diffusion of an interstitial component is observed for most 4d and 5d transition metals investigated. However, in some cases, a subsequently slower increase of the total metal concentration takes place as has been well investigated for the Si:Au system, e.g. [18,44,45].

This process is explained by a reaction of the interstitial T-metal atom with a native lattice defect - vacancy [44] or self-interstitial [45] - resulting in a change to a substitutional site. Thus a slow increase is observed which is limited by the low equilibrium concentration of native lattice defects in silicon. However, there exist numerous hints that isolated substitutional metal atoms are not producing the observed electrically active defects, e.g. [46,47].

For Au in Si, EPR spectroscopy only detected the presence of Au -Fe pairs after Au diffusion and subsequent quenching [48,49], even when using especially clean diffusion conditions [50,51].

These pairs are thermally unstable above 300°C [48,50]. Using silicon with low shallow doping (10^{12} B/cm^3) an EPR experiment [50] yielded in the 350°C - 400°C temperature range a dissociation time constant of:

$$\tau = 4.7 \cdot 10^{-3} \exp (0.73[eV]/kT) \quad [s] \tag{5}$$

Interstitial iron from a dissociated AuFe pair can be expected to vanish very fast, irreversibly at this temperature. Therefore the activation energy of (5) can be taken as binding energy of AuFe pairs. The value of 0.73 ± 0.1 eV is very similar to that calculated for a nearest-neighbour pair, e.g. $Au_s^- - Fe_i^+$, bound by Coulomb interaction, see above. Thus this pair, too, seems to be bound simply by electrostatic attraction.

Photo-EPR with n-Si proves the existence of an acceptor level of these pairs less than 0.46 eV from the conduction band edge: In n-Si the (AuFe)° EPR is absent in the dark, but can be excited by light of less than 0.46 eV energy; the exact determination was prevented by the quartz-cryostat [50]. A combined investigation showed, in addition, a DLTS level at E_C-0.33 eV to occur in the same concentration as AuFe pairs and to anneal together with them, as measured by EPR, so that this level can be tentatively ascribed to AuFe pairs.

Any relation of AuFe pairs to the "usual" gold donor (E_V+0.34 eV) or acceptor (E_C-0.55 eV) levels, which are thermally more stable, can be excluded. At present a complex of substitutional Au with other defects seems to be the most likely model for these levels [46,47]. It is interesting to note that Ohta [52] describes absorption spectra in Au doped silicon characteristic of Au-Au pairs so that this possibility should as well be further considered.

The latter finding gets more important in view of EPR results from Pt in Si: a spectrum previously ascribed to Pt [2] shows hyperfine satellites from a second Pt nucleus, indicative for a Pt-Pt pair [53,54]. As this occurs in concentrations comparable to the total Pt content of the crystals, the Pt acceptor (E_C-0.23 eV) and donor (E_V+0.32 eV) levels are most likely related with this EPR spectrum.

Thus these first steps towards a microscopic identification of Au and Pt centers in silicon seem to be promising but are still far behind the situation reached for 3d metals in silicon.

CONCLUSION

The understanding of 3d metals in silicon appears to be well advanced [3]: after interstitial in-diffusion with predominant interstitial solution, the 3d metals lighter than Co can be quenched into these sites, whereas isolated Co_i, Ni_i, and Cu_i vanish even during rapid quenching by complexing and precipitation. The solubilities separate the 3d metals into two groups with Ni and Cu in one group and the other 3d metals in the other.

Substitutional 3d metals can only be produced by non-equilibrium methods like irradiation [2,55]. The variety of defects produced this way [55] precludes reliable energy level determinations for substitutional 3d atoms in silicon.

For interstitial 3d metals and for metal-acceptor pairs energy levels have been determined which may help to improve model calculations.

Transition metals of the 4d and 5d series are much less well understood: some of them apparently have a high substitutional solubility besides smaller interstitial solubility. Yet the ideal, isolated substitutional T-metal atom seems not to be predominant. For Si:Au and Si:Pt evidence has been discussed in favour of a pair model for the electrically active defects.

This review should have demonstrated that the properties of well - known transition metal species in silicon can now be analyzed, so as to offer improved understanding of technologically important processes as well as a basis for theoretical progress in this challenging field.

REFERENCES

1. J.W. Chen and A.G. Milnes, Ann. Rev. Mat. Sci. 10, 157 (1980)

2. G.W. Ludwig and H.H. Woodbury, Solid State Phys. 13, 223 (1962)

3. E.R. Weber, Appl. Phys. A, in print

4. E. Weber and H.G. Riotte, J. Appl. Phys. 51, 1484 (1980)

5. N. Wiehl, U. Herpers, and E. Weber, in:
    Nuclear Physics Methods in Materials Research, ed. by K. Bethge,
    H. Baumann, H. Jex, and F. Rauch, (Vieweg, Braunschweig 1980), p. 334

6. N. Wiehl, U. Herpers, and E. Weber, J. Radioanal. Chem. 72, 69 (1982)

7. R.N. Hall and J.H. Racette, J. Appl. Phys. 35, 379 (1964)

8. M.K. Bakhadyrkhanov, S. Zainabidinov, and A. Khamidov,
    Fiz. Tekh. Poluprovodn. 4, 873 (1970) [Sov. Phys. Semicond. 4, 739 (1970)]

9. N.T. Bendik, V.S. Garnyk, and L.S. Milevskii,
    Fiz. Tverd. Tela 12, 190 (1970) [Sov. Phys. Solid State 12, 150 (1970)]

10. J.D. Struthers, J. Appl. Phys. 27, 1560 (1956)

11. L.C. Kimerling, J.L. Benton, and JJ Rubin, in:
    Defects and Radiation Effects in Semiconductors 1980, ed. by R.R. Hasiguti
    (Inst. of Physics, Bristol and London 1981), Conf. Ser. 59, p. 217

12. W.H. Shepherd and J.A. Turner, J. Phys. Chem. Sol. 23, 1697 (1962)

13. G.W. Ludwig and H.H. Woodbury, Proc. Intern. Conf. on Semicond. Physics,
    Prague 1960, p. 596

14. V.P. Boldyrev, I.I. Pokrovskii, S.G. Romanovskaya, A.V. Tkach, and
    I.E. Shimanovich, Fiz. Tekh. Poluprovodn. 11, 1199 (1977)
    [Sov. Phys. Semicond. 11, 709 (1977)]

15. C. Werkhoven, private communication
    K. Graff, private communication

16. R.A. Swalin, J. Phys. Chem. Sol. 23, 153 (1962)

17. M.F. Millea, J. Phys. Chem. Sol. 27, 315 (1966)

18. G.J. Sprokel and J.M. Fairfield, J. Electrochem. Soc. 112, 200 (1965)

19. W. Bergholz, J. Phys. D14, 1099 (1981)

20. M.K. Bakhadyrkhanov and S. Zainoabidinov,
    Fiz. Tekh. Poluprovodn. 12, 683 (1978)
    [Sov. Phys. Semicond. 12, 398 (1978)]

21. J.H. Aalberts and M.L. Verheijke, Appl. Phys. Lett. 1, 19 (1962)

22. M. Yoshida and K. Furusho, Jpn. J. Appl. Phys. 3, 521 (1964)

23. R.C. Dorward and J.S. Kirkaldy, Trans. Metall. Soc. AIME 242, 2055 (1968)

24. C.D. Thurmond and J.D. Struthers, J. Phys. Chem. 57, 831 (1953)

25. J.A. Van Vechten, in: Handbook on Semiconductors, Vol. 3, ed. by
    S.P.Keller (North Holland, New York and Oxford 1980), p. 1

26. C.S. Fuller and J.C. Severins, Phys. Rev. 96, 21 (1954)

27. C.J. Gallagher, J. Phys. Chem. Sol. 3, 82 (1957)

28. W.T. Stacy, D.F. Allison, and T.-C. Wu, in: Semiconductor Silicon 1981,
    ed. by H.R. Huff, R.J. Kriegler, Y. Takeishi
    (The Electrochemical Society, Pennington 1981), p.344

29. T.M. Buck, J.M. Poate, K.A. Pickar, and C.M. Hsieh,
    Surf. Sci. 35, 362 (1973)

30. R.D. Thompson and K.N. Tu, Appl. Phys. Lett. 41, 440 (1982)

31. K. Graff and H. Pieper, J. Electrochem. Soc. 128, 669 (1981)

32. L.C. Kimerling and J.L. Benton, Physica B, in press

33. H. Conzelmann, K. Graff and E.R. Weber, Appl. Phys. A, to be published

34. K. Graff and H. Pieper,in: Semiconductor Silicon 1981,
    ed. by H.R. Huff, R.J. Kriegler, and Y. Takeishi
    (The Electrochemical Society, Pennington 1981), p. 331

35. H. Lemke, Phys. Stat. Sol. A 64, 549 (1981)

36. H. Lemke, Phys. Stat. Solidi A 64, 215 (1981)

37. H. Feichtinger and R. Czaputa, Appl. Phys. Lett. 39, 706 (1978)

38. H. Feichtinger, J. Waltl, and A. Gschwandtner,
    Solid State Commun. 27, 867 (1978)

39. G.G. DeLeo, G.D. Watkins, and B.W. Fowler, Phys. Rev. B25, 4972 (1982)

40. A. Zunger and U. Lindefelt, Phys. Rev. B, in print

41. R.H. Hopkins, R.G. Seidensticker, J.R. Davies, P. Rai-Choudhury, P.D. Blais
    and J.R. McCormick, J. Cryst. Growth 42, 493 (1977)

42. R. Sauer and J. Weber, Physica B, in press

43. J. Weber, H. Bauch, and R. Sauer, Phys. Rev. B24, 7688 (1982)

44. F.A. Huntley and A.F.W. Willoughby, J. Electrochem. Soc. 120, 414 (1973)

45. N.A. Stolwijk, B. Schuster, J. Hölzl, H. Mehrer, and W. Frank,
    Physica B, in press

46. D.V. Lang, H.G. Grimmeis, E. Meijer, and M. Jaros,
    Phys. Rev. B22, 3917 (1980)

47. J.A. Van Vechten and C.D. Thurmond, Phys. Rev. B14, 3539 (1976)

48. R.L. Kleinhenz, Y.H. Lee, J.W. Corbett, E.G. Sieverts, S.H. Muller and C.A.J. Ammerlaan, Phys. Stat. Solidi B 108, 363 (1981)

49. M. Höhne, Phys. Stat. Solidi B 99, 651 (1980)

50. E.R. Weber, N. Wiehl, G. Borchardt and S.D. Brotherton, to be published

51. G. Borchardt, E. Weber, and N. Wiehl, J. Appl. Phys. 52, 1603 (1981)

52. K. Ohta, Science of Light 22, 12 (1973)

53. J.C.M. Henning, Physica B, in print

54. J.C.M. Henning and E.C.J. Egelmeers, to be publ.

55. S.H. Muller, G.M. Tuynman, E.G. Sieverts, and C.A.J. Ammerlaan, Phys. Rev. B25, 25 (1982)

CHALCOGENIDES IN SILICON

HERMANN G. GRIMMEISS[*] AND ERIK JANZÉN
Department of Solid State Physics, University of Lund, Box 725, S-220 07 LUND,
Sweden

1 INTRODUCTION

If an atom of the host lattice in silicon is replaced by an atom belonging to
the fifth group in the periodic table, the potential binding the extra electron
at the impurity atom can in most cases be approximated by a hydrogen-like
potential [1]. This gives rise not only to an energy level in the bandgap for
the impurity ground state, but, in addition, also to a series of excited states.
In silicon, the ground-state energies of these impurities are of the order of 50
meV. Such centers are therefore called "shallow" impurities, and are widely used
in semiconductor technology for modifying the type and degree of electrical
conductivity. The energy levels of the excited states are almost independent of
the ground-state energies and are well described by effective mass theory (EMT)
[2,3,4] which is one of the reasons for our good understanding of shallow
centers. The assignment of the excited states and of the ground state levels has
been considerably facilitated by the fact that optical absorption spectra of
shallow centers generally exhibit detailed structures [5].

If, on the other hand, a host-atom in silicon is replaced by an atom from the
sixth group in the periodic table, two extra electrons are available, which may
give rise to double donors. Double donors have been treated theoretically quite
extensively in the literature [6-9]. The potential binding these two electrons
of the impurity atoms is frequently compared with that of the two electrons in a
helium atom [6]. In addition, the ground states of chalcogenide atoms in silicon, which have been investigated so far, lie at much greater distance from the
conduction band than impurity atoms from the fifth group in the periodic table.
They are thus creating so-called "deep" impurity levels. Apart from some transition metals, published optical spectra of crystal defects with large binding
energies often show a smooth energy dependence without any of the detailed
structure, which is otherwise characteristic of shallow energy levels. This has
often been considered as an indication of the non-existence of excited states of
deep centers. The situation is quite different for the deep centers caused by
chalcogenides in silicon. Fig. 1 shows the absorption spectrum, which has been
obtained in selenium doped silicon for the neutral version of a double donor in
a tetrahedral surrounding. The spectrum consists of a series of sharp lines,
which – as will be shown later – are all caused by excited states.

Owing to the limited scope of this paper, we prefer first to illustrate the
general properties of chalcogenides in silicon with examples obtained in selenium doped silicon, and then to summarize the individual properties of the three
chalcogenide dopants S, Se and Te in tables.

2 THE ENERGY SPECTRUM OF A SUBSTITUTIONAL CHALCOGENIDE IMPURITY IN SILICON

An isolated substitutional chalcogenide impurity in silicon is expected to
form a double donor. The two energy levels in the bandgap originate from the
ionization of the two extra electrons, which do not participate in the bondings
with their nearest neighbours. If one of these electrons is transferred into an
excited but still bound state, it will move around a singly charged core. The

----

[*] Present address: RIFA AB, IC Division, S-163 81 STOCKHOLM, Sweden

Mat. Res. Soc. Symp. Proc. Vol. 14 (1983) © Elsevier Science Publishing Co., Inc.

part of the excitation spectrum, which is caused by excited states, should therefore be very similar to that observed for the shallow group V donors. The energy spectrum of such a level in a tetrahedral surrounding is shown in fig. 2. All states, except the ground state, are shallow states.

Leaving out the 1s and 2s states, the excited states are well described by an effective mass theory (EMT), which considers only one conduction band and one conduction band valley [4]. However, in order to explain the splittings of the s-states, the multivalley nature of the conduction band must be taken into account [2,3,5].

According to EMT, only odd-parity transitions are allowed for example between the ground state 1s(A_1) and the p-states. However, owing to the large binding energy of the ground state of chalcogenides dopants, symmetry-allowed even-parity transitions are also observed. In fig. 2, both EMT and symmetry-allowed transitions between the ground state and shallow excited states are marked with arrows. It is readily seen from fig. 1 that the EMT-forbidden but symmetry-allowed transitions really occur.

Since the measured binding energies of highly excited states are in excellent agreement with values calculated from EMT, the binding energy of the ground state is easily obtained e.g. by measuring the excitation energy for the transition 1s(A_1) → $2p_\pm$ and adding the EMT binding energy of the $2p_\pm$ state [4]. Once the ionization limit is known, the binding energies of all final states seen in absorption measurements can easily be derived. States which are not allowed as final states in absorption measurements must be determined by other methods. One of these methods is described in section 4.

The energy level giving rise to the line spectrum of fig. 1 has a ground state binding energy of 0.307 eV [10,11]. From the energy spacing of the excited states, we know that the center is neutral when occupied. In order to study the ionized version of the center by absorption, the sample must be counterdoped with shallow acceptor levels. A line spectrum which originates from a doubly charged center is then obtained, giving a ground state binding energy of 0.593 eV [10].

Evidence that these two energy levels originate from the neutral and ionized version of a double donor can be obtained from junction space charge techniques. DLTS measurements of selenium-diffused P^+N diodes show (fig. 3) that the two energy levels have similar concentrations [12]. Furthermore, the thermal activation energies obtained from these measurements are in good agreement with the binding energies deduced from absorption experiments [13]. As further proof, we present, in fig. 4, data which have been obtained from Fourier Photo-Admittance Spectroscopy [14], and which show that this center with a binding energy of about 0.31 eV is neutral when occupied. In fig. 5, we present a spectrum of the photoionization cross section σ_n^0, which was measured employing the photocapacitance technique [13] and which implies that the center with a binding energy of about 0.59 eV is doubly charged when empty. Since all these data were obtained using the same diode, there is no doubt that the two energy levels of similar concentration are identical with the two levels observed in absorption measurements, although the structure at the ionization edges is less clearly seen in junction measurements.

The measurement techniques so far described usually do not give information on the nature of the center, i.e. whether the center is selenium related or not. Such information can often be deduced from ESR measurements [15]. Fig. 6 shows the ESR spectrum of a silicon sample which has been doped with selenium [16]. Depending on the fabrication procedure, different defects and thus different ESR spectra may in principle be expected. The pattern of the ESR spectrum shown in fig. 6 is independent of the orientation of the magnetic field. The doublet splitting observed is consistent with the hf interaction of the I = 1/2 nucleus of ^{77}Se. The strong central line arises from the even Se isotopes with zero nuclear spin. The relative intensities of the lines correspond to the natural abundances of the Se isotopes. From the ESR spectrum, we therefore know that the center consists of an isolated Se atom having spin S = 1/2 and occupying a tetrahedral site.

In order to correlate the ESR spectrum of fig. 6 with the energy position of a particular energy level, we illuminated the sample with monochromatic light of different photon energies [16]. Owing to the photon-induced valency change of the ESR active center, a change in the ESR intensity was observed. Studying the time dependence of the intensity changes at different initial conditions, the spectral distribution of the photoionization cross section for both electrons and holes was obtained (fig. 7). Since these results agree with those spectra which have been obtained from photocapacitance measurements, we have to conclude that the ESR signal originates from the double donor with binding energies of 0.307 eV and 0.593 eV. The ESR measurements only reveal the 0.593 eV energy level, since the center must have an unpaired spin in order to be observed in ESR.

Similar measurements were performed on sulfur and tellurium doped samples [16,17]. In both cases isolated double donors on tetrahedral sites are observed (fig. 8). However, there is strong evidence that different chalcogenide-related centers may be formed in sulfur or selenium doped silicon depending on the doping procedure.

Having presented some general properties of chalcogenides in silicon, we would now like to discuss two special features of these dopants. We will start with the spin-valley splitting of the $1s(T_2)$ state and conclude with a discussion of Fano resonances originating from the interaction between phonon replicas of bound excited states and continuum states.

3 SPIN-VALLEY SPLITTING OF THE $1s(T_2)$ STATE

An additional splitting of the $1s(T_2)$ state due to spin-orbit effects may occur [5,10,18,19] if spin is included. The symmetry representations (as shown in fig. 2) are then changed as follows: $A_1 \to \Gamma_6$, $E \to \Gamma_8$ and $T_2 \to \Gamma_7 + \Gamma_8$. The splitting of the T_2 level is due to its "valley induced" p-character, which makes an interaction with the spin possible. This spin "pseudo-orbit" splitting is sometimes referred to as spin-valley splitting [20,21] (see fig. 9). Due to their different degeneracies, the Γ_8 state is shifted less than the Γ_7 state. If spin-valley interactions were absent, the unsplit $1s(T_2)$ state would lie at an energy of $E(T_2) = \frac{1}{3} E(\Gamma_7) + \frac{2}{3} E(\Gamma_8)$ [22]. Although the $1s(E)$ state cannot split, the $1s\Gamma_8(E)$ state may contain admixtures of T_2 states. At a first approximation, however, spin-valley interaction does not mix $1s \Gamma_8(E)$ states with $1s\Gamma_8(T_2)$ states, especially when the energy separation $1s(T_2) - 1s(E)$ is large [19].

It is not clear whether $1s\Gamma_8(T_2)$ or $1s\Gamma_7(T_2)$ should have the larger binding energy. Γ_8 corresponds to a "pseudo" $p_{3/2}$ state and Γ_7 to a "pseudo" $p_{1/2}$. Hund rules tell us that for a single electron the energy is lowest when the spin is antiparallel to the orbital angular momentum. If Hund rules are applicable for this kind of "pseudo" state, Γ_8 should then lie above Γ_7.

It is readily seen from fig. 10 that the absorption peaks corresponding to transitions from the ground state $1s(A_1)$ to the excited state $1s(T_2)$ are split into two peaks for the Te^+, Se^+ and S^+ center in silicon [10]. There is strong reason to believe that the splitting is due to spin-valley interaction. The integrated absorption of the upper energy component in fig. 10 is larger than the lower one, suggesting that $1s \Gamma_8(T_2)$ with degeneracy 2 (excluding spin) has a smaller binding energy than $1s \Gamma_7(T_2)$ with degeneracy 1. If any random stress or electric field is present in the crystal, $1s \Gamma_8(T_2)$ will split (see fig. 9) and the corresponding peak becomes broader. This effect is clearly seen in fig. 10 for Se^+ and Te^+ [10]. The upper component is considerably broader than the lower one, which indicates that $1s \Gamma_8(T_2)$ does lie above $1s \Gamma_7(T_2)$. The broadening, however, is large enough to imply some uncertainty in the magnitude of the splitting, since it cannot be taken for granted that the peak positions are not changed by the presence of random stress or fields [10]. By using uniaxial stress, which splits the upper component into two peaks, but leaves the lower component unsplit (see fig. 9), Krag et al. have shown that the ordering mentioned above is correct for Bi [19] in Si. Although the data for Sb [5] can be interpreted in the same way, the splitting was not discussed in these terms in ref. 5.

All available data on 1s states for group V and VI impurities in silicon are summarized in table I [10]. It is readily seen that, although the ground state $1s(A_1)$ binding energies vary widely, the energy positions of the $1s(T_2)$ states are very similar and close to the EMT-value for all D^0 centers. Only for neutral tellurium $1s(T_2)$ is slightly deeper. The deviations are larger for the ionized (D^+) chalcogenides. This is not surprising since the Bohr radii for D^+ states are only about 7 Å which should be compared with 20 Å for D^0 ones. Thus, the $1s(T_2)$ states for D^+ centers are more affected by the central cell potential than those for D^0 centers [10].

The experimentally-determined spin-valley splittings vary from 0.3 meV (Sb^0) to 5.4 meV (Te^+). The magnitude of this splitting should be approximately equal to the impurity atomic spin-orbit splitting reduced by the fraction of the donor envelope wavefunction in the central core region [23]. By using spin-orbit parameters and atomic radii from ref. 24 together with Bohr radii deduced from experimentally-determined $1s(T_2)$ energy positions, the spin-valley splittings could be calculated [10]. It turned out that the calculated values were 2 to 3 times smaller than those found experimentally. The best fit between the calcula-ted and measured results was obtained by multiplying the calculated values by 2.4 (see table I) [10]. Although the calculations are very crude, they probably predict hitherto unobserved splittings within a factor of two. In this context, it is interesting to note that unpublished calculations by L.M. Roth suggest a splitting of 0.9 meV for Bi [18].

4 FANO RESONANCES

When photoexcitation can result either in a final discrete state or in a continuum state of the same energy, any small interaction between such states will result in a marked resonance structure in the photoexcitation spectrum, commonly known as a Fano resonance [31]. We have investigated the neutral S^0, Se^0 and Te^0 donors in silicon – which are believed to occupy substitutional sites [16] – using absorption, photoconductivity and Fourier Photo-Admittance Spectroscopy [14], and found structures above the ionization limit which we attribute to Fano resonances. The two interacting processes in our case are a no-phonon excitation from the ground state to the continuum and a transition from the ground state to a bound excited state accompanied by emission of a phonon [28].

Experimental results and discussions

In fig. 11 an absorption spectrum of Si:Te^0 is shown. The transitions to the excited states above $2p_\pm$ are smeared out, possibly due to high tellurium concen-tration. Note that the transitions to the $1s(T_2)$ and $2s(T_2)$ states are observed, but not to the $1s(E)$ and $2s(A_1)$ states in agreement with the symmetry selection rules.

Above the ionization limit (C.B) in fig. 11, clear structure is observed. It is shifted about 60 meV to higher energy compared to the excited states of the Rydberg series. We interpret the structure as due to Fano resonances. It is the one-phonon replicas of the transitions to the excited states that interact with the no-phonon transitions to the continuum. The higher order phonon replicas are not discernible.

It has earlier been shown that in silicon, in zeroth order approximation, only intervalley phonons are allowed in intraband electron-phonon interaction [32]. According to ref. 32 the g LO(Δ_2) phonon of 63.3 meV and the f TO(S_1) phonon of 59.0 meV should couple with comparable strength to electrons. The f LA(S_1) phonon of 46.1 meV should interact less than the other two phonons. It turns out that the Fano resonances in fig. 11 can be interpreted in terms of these phonons. The arrows A, B and C mark the excitation energies to the 1s(T_2) state plus the three phonons mentioned above. The weak structure at A will not be discussed further [28].

In fig. 12 the "Fano part" of fig. 11 has been magnified. The Fano interaction modulates the continuum and shifts the resonance structure in energy compared to the discrete state. Since the degree of modulation as well as the shift may vary for different Fano resonances, the determination of binding energies of bound states from their Fano replicas is somewhat uncertain. The arrows B and G in fig. 12 indicate the excitation energies to the 1s(T_2) and $2p_0$ states respectively plus 59.0 meV. As may be seen, the resonance – antiresonance structure is best developed for the 1s(T_2) replica. The arrow C marks the excitation energy to the 1s(T_2) state plus 63.3 meV. So this phonon replica is probably responsible for the structure at C. Therefore, it now turns out that the energy difference between the antiresonances at B and C is the same as that between the antiresonances at D and E. Using the same effective phonons, one may anticipate that D and E are due to phonon replicas of an excitation to a state 31.6 meV below the conduction band (e.g. binding energy = 198.8 meV (1s(A_1)) − [226.4 meV (E_D) −59.2 meV (eff. phonon)] = 31.6 meV). Since this state is unobserved in no-phonon absorption and close to the 1s EMT-value of 31.27 meV, we believe it is the 1s(E) state [28].

For Si:S^0 and Si:Se^0 similar structures are observed. Since the 1s(T_2) states of these dopants, 34.6 and 34.4 meV respectively, are slightly shallower than for tellurium, 39.1 meV, the 63.3 meV phonon replica of the 1s(T_2) states will interfere with the 59.0 meV replica of the 1s(E) states. The binding energies inferred for the 1s(E) states were 30.9 meV and 30.8 meV for sulfur and selenium, respectively. But due to interference effects these values must be considered less accurate than that for tellurium [28].

At the low energy side of the $2p_0$ Fano replica additional structure occurs (F). It might be the 59 meV phonon-replica of a state lying 14.9 meV below the conduction band (binding energy = 198.8 meV (1s(A_1)) − [243.1 meV (E_F) −59.2 meV (eff. phonon)] = 14.9 meV). Such a state is not visible in no-phonon absorption, so we believe that the structure F is a Fano replica of the 2s(A_1) state. The corresponding binding energies for sulfur and selenium are 18.6 and 18.1 meV, respectively [28].

The results are summarized in table II [28]. The wavefunctions of both the 1s(A_1) states and 2s(A_1) states have non-vanishing amplitudes at the impurity atom. Since it is the central cell that makes the binding energies deviate from the EMT-value, one would expect the same chemical trend for these two states. It will be seen that this is the case for the neutral chalcogenide centers in silicon. It should be noted that the corresponding binding energies for all states in Si:S^0 and Si:Se^0 are very similar.

Fano resonances occur not only for isolated centers, but also for complexes. As an example, figure 13 shows Fano replicas of bound excited states for a neutral selenium pair [33]. It is interesting to note that the same phonons as in the case of the isolated neutral selenium impurity are involved in the spectrum.

Owing to the limited scope of this paper, mostly isolated centers are discussed. However, it is worthwhile mentioning that several different chalcogenide-related centers of a greater complexity have been observed and studied in silicon [34]. It turned out that most of these centers are probably double donors exhibiting excited states which are in good agreement with EMT.

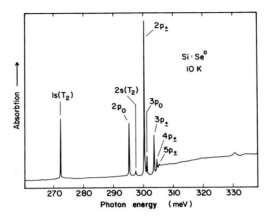

Fig. 1: Absorption spectrum of selenium doped silicon at 10 K showing transitions from the ground state to the excited states of an isolated, neutral impurity occupying a tetrahedral site.

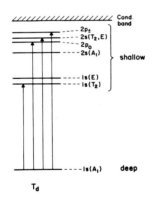

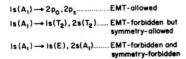

Fig. 2: The lower part of the energy spectrum of a chalcogenide center in silicon. Symmetry-allowed no-phonon transitions are marked with arrows.

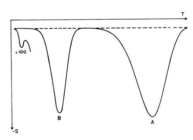

Fig. 3: DLTS spectra of selenium-doped p^+n diodes (solid curve) and of reference diodes not deliberately doped with selenium (dashed curve). The peak labelled B corresponds to thermal ionization of the same neutral impurity which gave rise to the absorption spectrum in figure 1.

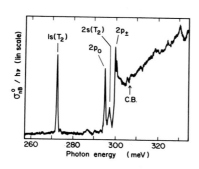

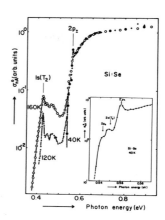

Fig. 4: Spectral distribution of the photoionization cross section σ^0_{nB} of the B-level (cf figs. 1 and 3) in Si:Se measured at 45 K with Fourier Photo-Admittance Spectroscopy.

Fig. 5: The spectral distribution of the photoionization cross section of the A-level in Si:Se measured at different temperatures with photo-capacitance transients. [13]

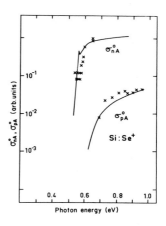

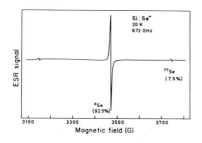

Fig. 6: ESR of the ground-state of the isolated Se$^+$ donor in silicon.[16]

Fig. 7: Energy dependence of the photoionization cross section of Si:Se as determined by ESR. The vertical scales of the experimental data have been adjusted to give optimal coincidence with the photocapacitance data (full lines) reported in reference 13.

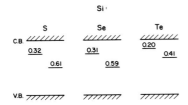

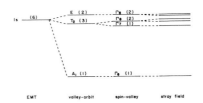

Fig. 8: The binding energies of the isolated double donors in chalcogenide doped silicon.

Fig. 9: Multivalley splitting of 1s donor states in silicon. The numbers in brackets are the degeneracies (excluding spin) of the states (see text).

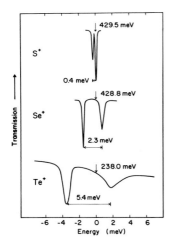

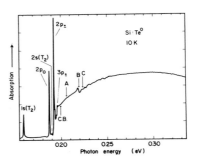

Fig. 10: Excitation spectra due to transitions from the ground states $1s(A_1)$ to the spinvalley split $1s(T_2)$ states for S^+, Se^+ and Te^+ donors in silicon. The component to the right corresponds to $1s\Gamma_8(T_2)$ and that to the left $1s\Gamma_7(T_2)$. The origin of the energy scale is the transition energy to the unsplit $1s(T_2)$ state into which the two components of the doublet would reduce in the absence of the spin-valley interaction.

Fig. 11: Absorption spectrum of $Si:Te^0$. C.B. marks the ionization limit. The arrows A, B and C mark the excitation energy to the $1s(T_2)$ state plus the energies of three different phonons as explained in the text.

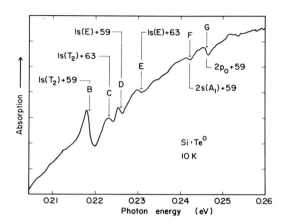

Fig. 12: Enlarged part of figure 11 showing the Fano resonances. B – G represent combinations of different electronic transitions with different phonons.

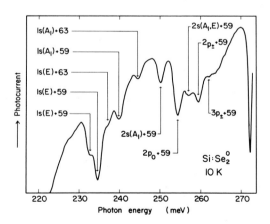

Fig. 13: Fano replicas of bound excited states of a neutral selenium pair in silicon. The sharp dip at 272 meV is due to the transition $1s(A_1)$ $1s(T_2)$ of the isolated neutral selenium impurity.

TABLE I
Available data on the binding energies (in meV) of 1s donor states for Group V and VI impurities in silicon. Data labelled a are from absorption measurements at 10 K in ref. 25, b are from electronic Raman scattering at about 20 K in ref. 26, c and d are from absorption measurements at 30 K and 59 K respectively in ref. 5, f are from absorption measurements at 10 K in ref. 19, and g and k are from photoconductivity measurements in refs. 27 and 28 respectively. Data labelled p, q and r are calculated values from refs. 4, 29 and 30 respectively All unlabelled data are from ref. 10.

	$1s(A_1)$	$1s(T_2)$	$1s\ \Gamma_7(T_2)-1s\ \Gamma_8(T_2)$ exp.	calc.	$1s(E)$	$1s(T_2)-1s(E)$
P^0	45.58[a]	33.90[c]		0.015	32.6[b]$/32.57$[c]	1.33[c]
As^0	53.77[a]	32.65[d]		0.086	31.4[b]$/31.24$[d]	1.41[d]
Sb^0	42.77[a]	32.88[c] 33.17[c]	0.29[c]	0.34	30.6[b]$/30.58$[c]	2.40[c]
Bi^0	71.00[a]	31.92[f] 32.92[f]	1.00[f]	1.03		
S^0	318.4[g]	35.1[g]		0.017	31.6[g]	3.5[g]
Se^0	306.7	34.5		0.13		
Te^0	198.8	39.1		0.55	31.6[k]	7.6
D^0	31.27[p] 47.4 [r]	31.27[p] 31.4 [r]			31.27[p] 30.6 [r]	0.8[r]
S^+	613.5	183.9 184.3	0.4	0.36		
Se^+	593.3	163.7 166.0	2.3	1.6		
Te^+	410.8	171.0 176.4	5.4	6.9		
D^+	125.08[p]	125.08[p] 155[q]		125.08[p]	25 [q] 130 [q]	

Table II
Binding energies in meV of low-lying D^0 states in chalcogenide doped silicon.

	$1s(A_1)$	$1s(T_2)$	$1s(E)$	$2s(A_1)$
S^0	318.3	34.6	30.9	18.6
Se^0	306.6	34.4	30.8	18.1
Te^0	198.8	39.1	31.6	14.9

REFERENCES

1. W. Kohn, Solid State Physics 5, 257 (1957)

2. W. Kohn, Phys. Rev. 98, 915 (1955)

3. S.T. Pantelides, Rev. Mod. Phys. 50, 797 (1978)

4. R.A. Faulkner, Phys. Rev. 184, 713 (1969)

5. R.L. Aggarwal and A.K. Ramdas, Phys. Rev. 140, A1246 (1965)

6. A. Glodeanu, Phys. Status Solidi 19, K 43 (1967)

7. T.H. Ning and C.T. Sah, Phys. Rev. B 4, 3468 and 3482 (1971)

8. S.T. Pantelides and C.T. Sah, Phys. Rev. B 10, 621 and 638 (1974)

9. A.A. Grinberg and E.D. Belorusets, Sov. Phys. Semicond. 12, 1171 (1978)

10. H.G. Grimmeiss, E. Janzén and K. Larsson, Phys. Rev. B 25, 2627 (1982)

11. J.C. Swartz, D.H. Lemmon and R.N. Thomas, Solid State Commun. 36, 331 (1980)

12. H.G. Grimmeiss, E. Janzén and B. Skarstam, J. Appl. Phys. 51, 3740 (1980)

13. H.G. Grimmeiss and B. Skarstam, Phys. Rev. B 23, 1947 (1981)

14. E. Janzén, K. Larsson, R Stedman and H.G. Grimmeiss, J. Appl. Phys. (in press)

15. G.D. Watkins in "Point defects in solids" 2, 333 (1975), J.H. Crawford, L.M. Slifkin eds, Plenum Press

16. H.G. Grimmeiss, E. Janzén, H. Ennen, O. Schirmer, J. Schneider, R. Wörner, C. Holm, E. Sirtl and P. Wagner, Phys. Rev. B (in press)

17. G.W. Ludwig, Phys. Rev. 137, A1520 (1965)

18. L.M. Roth, MIT Lincoln Laboratory, Solid State Research Report, No. 3, p23 (1962) (unpublished)

19. W.E. Krag, W.H. Kleiner and H.J. Zeiger, Proc. 10th Int. Conf. on the Physics of Semicond., Cambridge, Massachusetts, 271 (1970)

20. P.J. Dean, R.A. Faulkner and S. Kimura, Phys. Rev. B 2, 4062 (1970)

21. P.J. Dean, W. Schairer, M.. Lorenz and T.N. Morgan, J. Lumin. 9, 343 (1974)

22 T.G. Castner, Phys. Rev 155, 816 (1967)

23. P.J. Dean, R.A. Faulkner and S. Kimura, Phys. Rev. B 2, 4062 (1970)

24. S. Fraga and J. Karwowski, K.M.S. Saxena, "Handbook of Atomic Data", Elsevier, Amsterdam, 1976

44

25. B. Pajot, J. Kauppinen and R. Anttila, Solid State Commun. 31, 759 (1979)

26. K.L. Jain, S. Lai and M.V. Klein, Phys. Rev. B 13, 5448 (1976)

27. R.G. Humphreys and P. Migliorato (unpublished)

28. E. Janzén, R. Stedman and H.G. Grimmeiss, Proc. 16th Int. Conf. on the Physics of Semicond., Montpellier (1982)

29. M. Altarelli, Proc. 16th Int. Conf. on the Physics of Semicond., Montpellier (1982)

30. M. Altarelli, W.Y. Hsu and R.A. Sabatini, J. Phys. C 10, L 605 (1977)

31. U. Fano, Phys. Rev. 124, 1866 (1961)

32. M. Asche and O.G. Sarbei, Phys. Stat. Sol. (b) 103, 11 (1981)

33. E. Janzén, private communication

34. H.G. Grimmeiss, J. El. Chem. Soc. (in print)

THE NATURE OF POINT DEFECTS AND THEIR INFLUENCE ON DIFFUSION PROCESSES IN SILICON AT HIGH TEMPERATURES

U. GÖSELE
Max-Planck-Institut für Metallforschung, Stuttgart, Fed. Rep. Germany

T.Y. TAN
IBM Thomas J. Watson Research Center, Yorktown Heights, N.Y. 10598, U.S.A.

ABSTRACT

The paper highlights recent progress in understanding the role of vacancies and self-interstitials in self- and impurity diffusion in silicon above about 700°C. How surface oxidation of silicon leads to a perturbation of the point-defect population is described. An analysis of the resulting oxidation-enhanced or -retarded diffusion of group III and group V dopants shows that under thermal equilibrium as well as under oxidation conditions both *vacancies and self-interstitials are present*. For sufficiently long times vacancies and self-interstitials attain dynamical equilibrium which involves their recombination and spontaneous thermal creation in the bulk of silicon crystals. The existence and the nature of a recombination barrier slowing down the recombination process are discussed in this context. Recent experimental and theoretical results on the diffusion of gold in silicon enable us to determine the self-interstitial component of silicon self-diffusion and to obtain an estimate of the respective vacancy contribution. The two components turn out to be of the same order of magnitude from 700°C up to the melting point.

INTRODUCTION

Intrinsic point defects govern material transport (self-diffusion) processes and the diffusion of substitutional impurities. In the case of metals vacancies have been identified as the point defects present under thermal equilibrium conditions. The situation concerning the nature of thermal equilibrium point defects in silicon became highly controversial when Seeger and Chik [1,2] suggested that self-diffusion at temperatures above about 900°C is predominantly carried by self-interstitials (I) and below that temperature mainly by vacancies (V). We mention just a few of the difficulties which prevented a clear-cut verification or falsification of the Seeger-Chik suggestion: i) Thermal equilibrium concentrations of point defects at the melting point are orders of magnitude lower in silicon than in metals. Therefore a direct determination of their nature (I or V) by Simmons-Balluffi type experiments [3] has not been possible, see e.g. ref [4]. (ii) The accuracy of calculated point-defect formation and migration enthalpies appears to be so doubtful that they also do not help in distinguishing between the predominant presence of I or V at high temperatures; for specific values and references, see [1,4,5]. iii) The interpretation of low-temperature experiments on the migration of irradiation-induced point defects is complicated by the occurrence of radiation-induced migration of self-interstitials[6,7,8]. In addition there are indications that the structure and properties may change considerably by going from low to high temperatures [9]. iv) The observation of interstitial-type dislocation loops ("A-swirl defects") in macroscopically dislocation-free float-zone silicon showed that self-interstitials must have been present in appreciable concentrations at high temperatures after crystal growth [10,11]. However, it remains disputed whether these self-interstitials were present in thermal equilibrium [10,12] or were introduced during crystal growth by non-equilibrium processes [11].

Mat. Res. Soc. Symp. Proc. Vol. 14 (1983) ©Elsevier Science Publishing Co., Inc.

Fortunately, in the last two years progress in two areas helped to answer the question concerning the nature of thermal equilibrium point defects in silicon. The first area is associated with the diffusion of gold in silicon. It was shown that an essential part of silicon self-diffusion is carried by self-interstitials [13-15], but it could not be decided whether besides self-interstitials also vacancies are present under thermal equilibrium conditions. A decision in favor of the presence of both self-interstitials *and* vacancies came from the second area, namely that of oxidation-enhanced or -retarded diffusion of group III and group V dopants [16-19]. In the present paper we will concentrate on these two areas.

We will start with basic considerations on self- and dopant diffusion taking into account both self-interstitials and vacancies. Then we will describe how surface oxidation may influence point defect concentrations and how these concentration changes may be measured by the growth of stacking faults. Knowing these concentration changes we then proceed to determine the diffusion components of group III and group V dopants via self-interstitials and via vacancies. After discussing the significance of a reaction barrier against self-interstitial - vacancy recombination we will finally turn to the diffusion of gold and nickel in silicon and to the resulting implication for the nature of point defects and for the mechanism of self-diffusion at high temperatures.

BASIC CONSIDERATION

Self-diffusion. In the following we take into account that both self-interstitials (I) and vacancies (V) may be present in silicon under thermal equilibrium conditions. Their respective equilibrium concentration (in dimensionless atomic fractions) is given by [1]

$$C_X^{eq} = \exp(S_X^F/k) \exp(-H_X^F/kT),$$ (1)

where X stands for I or V, and H_X^F and S_X^F are the enthalpy and entropy of formation, respectively (k: Boltzmann's constant, T: absolute temperature). Material transport properties are not only determined by the concentrations C_I^{eq} and C_V^{eq} but also by the diffusion coefficients D_I and D_V which may be written in the form

$$D_X = D_X^o \exp(-H_X^M/kT),$$ (2)

where H_X^M is the respective enthalpy of migration and the pre-exponential factor D_X^o contains besides the attempt frequency also the lattice parameter and the entropy of migration S_X^M.

Point defects in semiconductors may occur in various charge states, the individual concentration of which depends on the position of the Fermi level and of their energy level in the band gap [1,4,20,21]. We will not deal with specific charge-state effects in this paper and will therefore not comment any further on this topic.

In self-diffusion experiments the measured quantity usually is the tracer self-diffusion coefficient [1,4,20,21]

$$D^T = f_I D_I C_I^{eq} + f_V D_V C_V^{eq},$$ (3)

where $D_I C_I^{eq}$ and $D_V C_V^{eq}$ are the uncorrelated self-diffusion coefficients via I and V, respectively. The temperature independent correlation factors f_I and f_V can reasonably assumed to be between about 0.5 and 1. Since the diffusion mechanism via self-interstitials (commonly termed interstitialcy mechanism) appears to be much less known than the vacancy mechanism we have schematically indicated the diffusion of a bonded interstitial through a diamond type lattice in Fig. 1.

Some experimentally determined results on D^T [22-25] are compiled in Table 1 in the form

$$D^T = D_o^T \exp(-H^{SD}/kT).$$ (4)

The observation that D^T may reasonably well be described by a single activation enthalpy H^{SD} over a wide temperature range (at least within a given set of experimental results) indicates either that one diffusion process dominates in this temperature range or that $f_I D_I C_I^{eq}$ and $f_V D_V C_V^{eq}$ are similarly thermally activated and accidentally are in the same order of magnitude within, say, a factor of ten.

In order to explain the high pre-exponential factor D_o^T (containing the entropies of formation and migration) Seeger and Chik [1] suggested that point defects, and especially self-interstitials in Si possess an *extended* configuration at high temperatures. This extension, which corresponds to a high entropy of formation is supposed to be a high temperature effect and to decrease with decreasing temperature resulting in usual "point-like" defect configurations at sufficiently low temperatures, e.g. at room temperature or below [9,26]. Bourgoin and Lannoo [27] have argued that the high D_o^T value may also be understood in terms of relaxation effects around vacancies.

If both I and V are present simultaneously under thermal equilibrium conditions, naturally the question arises whether they can co-exist complete independently of each other or whether they react with each other according to

$$V + I \rightleftharpoons O, \qquad (5)$$

where O denotes the undisturbed lattice. Reaction (5) implies local dynamical equilibrium between recombination and spontaneous thermal bulk creation of I and V described by the law of mass action according to [28,29]

$$C_I C_V = C_I^{eq} C_V^{eq}. \qquad (6)$$

For sufficiently long times and high temperatures eq. (6) turns out to be fulfilled [17,18,30] but in the course of this paper we will also discuss situations in which dynamical equilibrium between I and V has not yet been established.

Substitutional dopant diffusion. The diffusion coefficients D^S of substitutional group III and group V dopants in Si are larger than the tracer self-diffusion coefficient D^T. This is most likely due to the formation of highly mobile point-defect dopant complexes the stability and mobility of which will depend on the specific point-defect dopant interaction. Following Hu [16] we assume that under thermal equilibrium conditions the diffusivity D^S is composed of a diffusion component D_I^S involving silicon self-interstitials and of a diffusion component D_V^S via vacancies:

$$D^S = D_I^S + D_V^S. \qquad (7)$$

The normalized diffusion components G_I^S and G_V^S, defined by

$$G_I^S = 1 - G_V^S = D_I^S/D^S, \qquad (8)$$

are expected to depend on temperature and to be different for different dopants because of different elastic and electrostatic point-defect dopant interactions. The experimental observation that D^S may reasonably well be described in terms of a single activation enthalpy H_s^M again indicates as in the case of D^T that either only one point defect species dominates the dopant diffusion process or that self-interstitials and vacancies contribute to a similar extent to D^S over a wide temperature range (For P and B there are indications of a change of H_s^M to lower values for temperatures below about 900°C [31]).

For the determination of $G_I^S = 1 - G_V^S$ it is necessary to measure D^S for a case in which the concentrations C_I and C_V differ from their thermal equilibrium concentration C_I^{eq} and C_V^{eq}, a situation which may be created during surface oxidation of silicon. The diffusivity D_{ox}^S under these conditions is described by

$$D_{ox}^S = D_I^S(C_I/C_I^{eq}) + D_V^S(C_V/C_V^{eq}). \qquad (9)$$

With the self-interstitial and vacancy supersaturation ratios

$$S_I = (C_I - C_I^{eq})/C_I^{eq}, \quad S_V = (C_V - C_V^{eq})/C_V^{eq} \qquad (10a,b)$$

respectively, we may rewrite eq. (9) in terms of the normalized diffusion-enhancement Δ_{ox}^S as

$$\Delta_{ox}^S = (D_{ox}^S - D^S)/D^S = G_I^S S_I + G_V^S S_V. \qquad (11)$$

For times which are short against the average time τ_{eq} to establish dynamical equilibrium between self-interstitials and vacancies S_I and S_V may be treated as being independent of each other. For longer times $t > \tau_{eq}$ eq. (6) holds and S_I and S_V are coupled according to

$$S_V = -S_I/(1 + S_I), \qquad (12)$$

leading to [18]

$$\Delta_{ox}^S = (2G_I^S + G_I^S S_I - 1)S_I/(1 + S_I). \qquad (13)$$

In Fig. 2 Δ_{ox}^s is plotted as a function of the self-interstitial supersaturation ratio S_I for three different values of G_I^s. For $G_I^s = G_V^s = 0.5$ both $S_I>0$ as well as $S_I<0$ result in a diffusion enhancement. For dopants with $G_I^s>0.5$ enhanced diffusion is expected for $S_I>0$. In the region $S_I <0$, Δ_{ox}^s passes through a minimum

$$\Delta_{ox}^s(\min) = 2(G_I^s G_V^s)^{1/2} - 1, \tag{14}$$

which corresponds to the maximally attainable diffusion retardation for a given value of G_I^s. Analogously for dopants with $G_I^s<0.5$ enhanced diffusion is expected for $S_I<0$. In the region $S_I >0$, Δ_{ox}^s passes through a minimum also given by eq. (14). Before analyzing data on oxidation-enhanced and -retarded diffusion we will make plausible in the following section that surface oxidation usually leads to a self-interstitial supersaturation $S_I>0$, but that for sufficiently high temperatures and long times also $S_I <0$ may be realized which corresponds to a vacancy supersaturation.

POINT DEFECT GENERATION BY SURFACE OXIDATION

During oxidation a thin film of amorphous SiO_2 forms on the silicon surface. Oxygen from the gas phase diffuses (primarily in the form of O_2 molecules [32]) via holes in the SiO_2 network towards the SiO_2-Si interface to form new SiO_2 material. This reaction is associated with a large ($\sim100\%$) volume increase which at sufficiently high temperatures (say above about $950°C$ [33]) is almost entirely accommodated by visco-elastic flow of the oxide towards the surface of the SiO_2 film [34,35]. A small percentage of the volume increase, however, is accommodated by the injection of excess Si self-interstitials from the SiO_2-Si interface into the Si crystal (Fig. 3a) as has been suggested first by Dobson [36] and later on more specifically by Hu [16].

The self-interstitial supersaturation $S_I >0$ created and maintained *during* oxidation is highest for (100), less for (110), and lowest for (111) surface orientation [37]. In principle, absorption of silicon vacancies at the SiO_2-Si interface could also accommodate part of the volume increase associated with SiO_2 formation (Fig. 3a), but recent experimental observations [30] indicate that I-injection is the more efficient process, at least before dynamical equilibrium between I and V according to eq. (6) is established. After dynamical equilibrium has been established there is no measurable physical difference any more between I-injection and V-absorption. The oxidation-induced S_I decreases with increasing temperature because the viscous flow of the SiO_2 is more easily accomplished at higher temperatures.

Experimentally it turned out that S_I is non-linearly coupled with the oxide growth rate via

$$S_I \propto (dx_{ox}/dt)^m, \tag{15}$$

where x_{ox} is the oxide thickness and m is close to 0.5 [30,35,37-40] which leads in the diffusion-controlled growth regime of the oxide film to S_I prop. to $t^{-0.25}$. For suggested explanations for this non-linear behavior ($m \neq 1$) the reader is referred to the original literature [35,37-39,41].

Let us now discuss how the picture of point-defect generation by surface oxidation changes for high temperatures and thick oxides. The reaction zone in which SiO_2 formation occurs shifts with increasing thickness of the oxide film from the SiO_2-Si interface to the interior of the SiO_2 film (Fig. 3b). For the diffusion of silicon from the interface to the reaction zone ordinary Si self-diffusion in SiO_2 [42] or diffusion via SiO molecules [18] has been proposed. The diffusion of Si into the SiO_2 film leads to the injection of vacancies into (and/or the absorption of Si self-interstitials from) the silicon crystal resulting in a vacancy supersaturation $S_V>0$ associated with a self-interstitial undersaturation $S_I<0$. This effect may be best observed during the oxidation of (111) oriented silicon, since for this orientation the I-injection rate, which has to be overcompensated by the V-injection rate, is comparatively low from the beginning.

Since the oxidation process itself slows down with increasing oxide thickness and time, we expect that for a given temperature $S_I(t)$ starts with a positive value, decreases with time and may become even negative for sufficiently thick oxides. The change-over to negative S_I values is expected to occur after a shorter oxidation time for (111) than for (100) oriented crystals.

A chlorine compound in the atmosphere leads to a vacancy injection rate due to a chemical reaction between Si and Cl at the SiO_2-Si interface. This vacancy injection rate is to a good approximation uncorrelated to the oxidation-induced supersaturation $S_I(O_2)$ so that altogether S_I may be written as

$$S_I \sim S_I(O_2) + S_I(Cl), \tag{16}$$

where the chlorine-induced $S_I(Cl)$ is negative. For details and references see [43].

OSF AS A PROBE FOR POINT DEFECT SUPERSATURATIONS

In the present section we will deal with the possibility to measure point defect supersaturations quantitatively via the growth or shrinkage rate of oxidation-induced stacking faults (OSF). OSF consist of agglomerated self-interstitials on (111) planes surrounded by a partial dislocation which may climb by the absorption or emission of point defects. In the following we will calculate the growth or shrinkage kinetics of circular OSF with radius r_{SF} in the bulk of a Si crystal based on the treatment given in ref. [18,35,43-45]. With minor modifications the results are also applicable to (often non-circular) surface OSF [46]. Assuming that both I and V are present with supersaturation ratios S_I and S_V related via (12) we obtain [18,43,45]

$$dr_{SF}/dt = -[(D_I C_I^{eq} + D_V C_V^{eq})(\gamma/kT) - D_I C_I^{eq}S_I + D_V C_V^{eq}S_V](A\alpha_{eff}/\Omega), \tag{17}$$

where A is the area per atom in the stacking fault ($6.38 \times 10^{-16}cm^2$), Ω the atomic volume ($2 \times 10^{-23}cm^3$), and γ the stacking fault energy of about 0.026eV per atom [47]. The dimensionless quantity α_{eff} depends on the interaction between the point defects and the partial dislocation as well as on possible reaction barriers opposing the climb process [44].

For $S_I=0$ and $S_V=0$, i.e. for an inert atmosphere and for negligible I generation by oxygen precipitation, eq. (17) reduces to

$$(dr_{SF}/dt)_{in} = -A\alpha_{eff}\gamma(D_I C_I^{eq} + D_V C_V^{eq})/(kT\Omega), \tag{18}$$

which describes the experimentally observed time-independent shrinkage rate of OSF in an inert atmosphere; for references, see [44]. The length l_{SF} of almost semicircular surface OSF (as, e.g., revealed by etching) should be identified with $2r_{SF}$. A quantitative comparison of eq. (18) with experimental results indicates that the shrinkage of OSF (as well as their growth) is diffusion-controlled rather than controlled by a reaction barrier [45,46].

Following Antoniadis [19] we may determine $S_I(t,T)$ without knowing α_{eff}, $D_I C_I^{eq}$ and $D_V C_V^{eq}$ explicitly by the expression

$$S_I(t,T) \sim [1 - (dr_{SF}/dt)_{ox}/(dr_{SF}/dt)_{in}](\gamma/kT), \tag{19}$$

where $(dr_{SF}/dt)_{ox}$ is the OSF growth rate for the oxidation conditions for which we would like to know $S_I(t,T)$ and $(dr_{SF}/dt)_{in}$ is the (negative) growth rate in an inert atmosphere at the same temperature. Eq. (19) holds exactly for small S_I and is a good approximation for high S_I provided $D_I C_I^{eq} > D_V C_V^{eq}$. For dry oxidation of (100) oriented Si and for diffusion-controlled oxide growth eq. (19) leads approximately to [18]

$$S_I(t,T) \sim 6.6 \times 10^{-9} \exp(2.52eV/kT)t^{-1/4}, \tag{20}$$

which in the following section we will use to analyze data on oxidation-enhanced diffusion (OED) and on oxidation-retarded diffusion (ORD).

OXIDATION-ENHANCED AND OXIDATION-RETARDED DIFFUSION

In actual experiments on OED or ORD a time-averaged value

$$\overline{\Delta}_{ox}^s = \int_0^t \Delta_{ox}^s(t')dt'/t \tag{21}$$

is measured. Eq. (13) remains a good approximation for all practical cases, provided $S_I(t)$ is replaced by the corresponding time-averaged value $\overline{S}_I(t)$, which means, e.g., that the pre-exponential factor 6.6×10^{-9} in eq. (20) has to be replaced by 8.8×10^{-9}. We will discuss the results for $S_I > 0$ and for $S_I < 0$ separately.

Self-interstitial supersaturation ($\overline{S}_I > 0$). All group III dopants investigated so far (B, Al, Ga, In [17,30,40,48-50]) show OED for $\overline{S}_I > 0$. In the case of group V dopants the influence of surface oxidation appear to depend on their size. The small phosphorus atoms show a large diffusion-enhancement (similar as in the case of B [17,30,51]) and the big antimony atoms clearly show a *diffusion-retardation* [17,30,53]. ORD of Sb as first observed by Mizuo and Higuchi [17] could quantitatively be explained by the assumption that Sb diffuses almost exclusively via vacancies ($G_V^s = 1 - G_I^s \sim 0.98$) and that the *I-supersaturation* $\overline{S}_I$ is associated with a corresponding vacancy *undersaturation* $\overline{S}_V$ according to eq. (12); see Fig. 4 [18]. This result as well as others [30,53] showed that for the times used in the diffusion experiments of Mizuo and Higuchi (e.g., $t > 3$ hours at 1100°C) dynamical equilibrium between I and V is actually established. The fact that for the same oxidation conditions Sb shows ORD whereas P and B show OED also rules out the possibility that under thermal equilibrium conditions either I only or V only are present [54]. In Table 2 we have compiled calculated G_I^s values for various group III and group V dopants at 1100°C based on experimentally observed values of $\overline{\Delta}_{ox}^s$. The analysis of OED is made difficult by the considerable scatter in the experimental diffusion data. Estimation of G_I^s values at other temperatures indicate a tendency for an increase of G_I^s with increasing temperature [19].

Self-interstitial undersaturation ($\overline{S}_I < 0$). A self-interstitial undersaturation $\overline{S}_I < 0$ associated with a corresponding vacancy supersaturation $\overline{S}_V > 0$ according to (12) may be realized in practice by oxidizing (111) surface oriented silicon for sufficiently long times at high temperature. The condition $\overline{S}_I < 0$ is expected to cause ORD for dopants with $G_I^s > 0.5$ (provided the corresponding $\overline{S}_V$ is not too large, see Fig. 2) and OED for dopants with $G_I^s < 0.5$. The first observation of ORD of a dopant with $G_I^s > 0.5$, namely of P, was reported by Francis and Dobson [42] for dry oxidation of (111) silicon at 1160°C for 17.5 hours. Similar results of ORD have been reported by Hill [31] for B and by Mizuo and Higuchi [48,49] for Al and Ga.

The quantitative analysis of negative $\overline{\Delta}_{ox}^s$ values via eq. (13), although possible in principle also for dopants with $G_I^s > 0.5$, cannot be performed in practice since hardly any reliable stacking fault data for the determination of $\overline{S}_I$ are available. This is so, because at the high temperatures required for the observation of ORD, OSF shrink away rapidly. Therefore, we use a different approach to estimate G_I^s for the $G_I^s > 0.5$ dopants P, Al, Ga, and B. If we knew the $\overline{\Delta}_{ox}^s(\min)$ (which is negative, corresponding to the most pronounced diffusion retardation), we could calculate G_I^s via

$$G_I^s \sim 0.5 + 0.5[1 - (1 + \overline{\Delta}_{ox}^s(\min))^2]^{1/2}, \tag{22}$$

which follows from (14) for $G_I^s \geq 0.5$. Since $\overline{\Delta}_{ox}^s(\min)$ may even be lower than the lowest $\overline{\Delta}_{ox}^s$ value actually observed, we obtain a lower bound to G_I^s by using this $\overline{\Delta}_{ox}^s$ value in eq. (22). Estimates of G_I^s obtained in this way are compiled in Table 3 together with the corresponding temperatures. This procedure for determining G_I^s is independent of the way in which $\overline{S}_I < 0$ has been generated. Therefore we give in Table 3 also G_I^s values derived from ORD of P at 1150°C [55] and of B at 1200°C [31] in an oxidizing atmosphere containing several percent of HCl.

For a dopant with $G_I^s < 0.5$, such as antimony, $\overline{S}_I < 0$ should lead to oxidation-enhanced diffusion. OED of Sb has actually been observed for a 18.5 hour dry oxidation of (111) surface oriented silicon at 1160°C; for details see the contribution of Tan and Ginsberg in these proceedings [53,56].

Self-interstitial diffusion across silicon wafers. As yet we have implicitly assumed that surface oxidation leads to *depth-independent* I and V concentration changes in a silicon crystal. How fast the changes in point defect concentrations can actually spread into the Si bulk depends on D_I and D_V. In addition, it has to be taken into account whether there are sinks of and sources for point defects present in the bulk. The concentration changes of C_I and C_V depending on time and on the vertical distance x to the surface being oxidized may be described by the following two modified diffusion equations

$$\partial C_I / \partial t = D_I \partial^2 C_I / \partial x^2 + k_{rec}(C_I^{eq}C_V^{eq} - C_I C_V) - k_I(\text{sink}) + k_I(\text{source}), \tag{23}$$

$$\partial C_V / \partial t = D_V \partial^2 C_V / \partial x^2 + k_{rec}(C_I^{eq}C_V^{eq} - C_I C_V) - k_V(\text{sink}) + k_V(\text{source}), \tag{24}$$

where k_{rec} denotes the reaction constant governing vacancy-interstitial recombination. The term $k_I(\text{sink})$ takes care of possible sinks for I, e.g. dislocation loops or bulk OSF, and $k_I(\text{source})$ accounts for possible sources of I such as oxygen precipitates in Czochralski-grown Si during the precipitation process. Both k_I(sink) and k_I (source) may be time- and material-dependent. The analogous considerations hold for k_V(sink) and k_V(source). Bulk sources for vacancies could be, e.g. dissolving oxygen precipitates.

Sinks for I and V induced by oxygen precipitation will prevent that a noticeable S_I or S_V will extend too far below the oxidized silicon surface. In fact, in oxygen containing Czochralski-grown silicon S_I decreases exponentially with increasing distance x from the oxidized surface with a typical decay length in the order of 20-30μm [37,57]. The situation is different for *float-zone* grown silicon as has recently been demonstrated by Mizuo and Higuchi [58-60]. They measured the influence of *back-side oxidation* on the diffusivity of P,B, or Sb implanted into the silicon from the *frontside* of the wafer. The front side is covered by a Si_3N_4 film in order to prevent point-defect generation by front-side oxidation. The practically coinciding diffusion enhancements of P and B as well as the diffusion retardation of Sb in dependence on the wafer thickness d are shown in Fig. 5 for 1100°C and two oxidation times. The effect of back-side oxidation on front-side dopant diffusion increases with increasing time and decreases with increasing wafer thickness for a given time t. These features show that it takes a certain time until the point-defect concentration changes induced at the wafer back-side arrive at its front-side where they are measured by the corresponding changes in the dopant diffusivity. The opposite behavior of Sb compared to that of P and B indicates that local dynamical equilibrium between I and V according to eq. (6) with $C_I = C_I(x,t)$ and $C_V = C_V(x,t)$ is established. In this case, for float-zone grown silicon in which the two k_I and the two k_V terms may be neglected a simple diffusion equation for S_I or S_V with an effective diffusivity

$$D_{eff}^{I,V} \sim (D_I C_I^{eq} + D_V C_V^{eq})/(C_I^{eq} + C_V^{eq}) \qquad (25)$$

may be derived from eq. (23,24), provided the oxidation-induced concentration changes are so small that $S_I \sim -S_V$ is a reasonable approximation.

For a proper analysis of the data of Mizuo and Higuchi [58-60] the boundary conditions for S_I have to be known. In the case of an *oxide without oxidation* or of an oxide below an Si_3N_4 layer the SiO_2-Si interface maintains $S_I \sim 0$. In the case of an oxide during the process of oxidation the $S_I(t)$ at the interface is that which we discussed in previous sections, e.g., for dry oxidation of (100) silicon the S_I given by eq. (20) should be used. The case of an Si_3N_4-Si interface is difficult to handle since it is neither perfectly absorbing (S_I=0) nor perfectly reflecting ($\partial S_I/\partial x = 0$) for point defects but has a certain "leakage rate", as is, e.g., known from the work of Hu [61]. With no reliable quantitative information available on this leakage rate at least a lower limit of D_I may be determined based on reflecting boundary conditions. This leads to $D_I > 3 \times 10^{-9} cm^2 s^{-1}$ [58-60].

Influence of oxidation on self-diffusion. In principle self-diffusion may be influenced by surface oxidation in a similar way as the diffusion of group III and group V dopants leading to

$$\Delta_{ox}^T = (D_{ox}^T - D^T)/D^T = (2G_I^T + G_I^T S_I - 1)S_I/(1 + S_I) \qquad (26)$$

with $G_I^T = f_I D_I C_I^{eq}/D^T$, provided dynamical equilibrium between I and V is established. Experimental data on Δ_{ox}^T would enable a determination of G_I^T which represents the relative tracer self-diffusion component via self-interstitials. Presently, such data are not available.

A barrier against vacancy-interstitial recombination. In 1978 Hu [62] suggested that an energy barrier of "a few to several eV" practically prevents vacancy-interstitial recombination, so that I and V may be regarded as not influencing each other. In contrast, the experiments on OED and ORD especially those of Mizuo and Higuchi [17,59,60] at 1100°C showed that for the oxidation times used (t>3 hours) dynamical equilibrium between I and V is established. By the same type of Sb diffusion experiments as performed by Mizuo and Higuchi [17] but for shorter oxidation times Antoniadis and Moskowitz obtained a small diffusion *enhancement* ($\Delta_{ox}^s \sim 0.1$) after 5 minutes at 1100°C, which gives way to a diffusion retardation after about 10 minutes. The diffusion retardation approaches the value expected from the dynamical equilibrium condition for longest oxidation time of about 60 min. The results may be rationalized as follows: During oxidation self-interstitlas are injected from the SiO_2-Si interface i nto the silicon but it takes about an hour until I-V recombination processes have established dynamical equilibrium. Therefore, at the very beginning a vacancy undersaturation is not yet present ($S_V \sim 0$), so that the self-interstitial supersaturation $S_I > 0$ has a chance to enhance the Sb diffusion via the small interstitialcy diffusion component of $G_I^s \sim 0.02$ as expressed in eq. (11). The analysis of the Antoniadis-Moskowitz data showed that the time τ_{eq} to reach dynamical equilibrium I and V is about

10^5 times longer than the time $\tau_d \sim \Omega/(4\pi D^T r_o)$ expected for a diffusion-controlled I-V recombination process, where r_o is the I-V recombination radius [30,63]. Therefore, it is concluded that I-V recombination is controlled by the overcoming of a recombination barrier that exceeds the Gibbs free energy of diffusion by ΔG_b. In this picture the factor 10^5 arises from the Boltzmann factor in

$$\tau_{eq} \sim \tau_d \exp\ (\Delta G_b/kT) \tag{27}$$

by which the barrier influences the recombination process, with $\Delta G_B = 1.4eV$. Antoniadis and Moskowitz [30] attributed the l.4eV to the enthalpy contribution ΔH_b in $\Delta G_b \sim \Delta H_b - T\Delta S_b$, whereas Gösele et al. [63] attribute this value mainly to the term $-T\Delta S_b$, where ΔS_b is negative, i.e., to an *entropy barrier* against I-V recombination. Gösele et al [63] followed the arguments of Seeger and co-workers [1,9,26] that self-interstitials and to a certain degree also vacancies possess "extended" configurations which have to be contracted to "point-like" defect configurations before recombination can take place. This contraction corresponds to a decrease in entropy, i.e. to an entropy barrier $\Delta S_b < 0$.

Based on the strong temperature dependence of D^T in τ_d of eq. (27) we expect, that independent of the detailed nature of the recombination barrier, the time τ_{eq} to reach dynamical equilibrium will strongly increase with decreasing temperature and will be in the order of days for 900°C. If besides reaction (5) reactions influencing the vacancy self-interstitial balance may occur then τ_{eq} may be considerably shorter. Such additional reactions will be discussed in the following section on the diffusion of gold in silicon.

DIFFUSION OF GOLD IN SILICON

In the last two years it has been shown that most features of the diffusion behavior of Au in Si can be understood by assuming that self-interstitials are the dominant point defects in Si under thermal equilibrium conditions [13-15,26,64-66]. It could also be excluded that vacancies are the only equilibrium point defects in silicon. In the following we will describe how, in addition to $D_I C_I^{eq}$, also an estimate of the ratio $D_V C_V^{eq}/D_I C_I^{eq}$ and of $D_V C_V^{eq}$ may be obtained from a more detailed investigation of the diffusion of Au in Si.

Au in silicon my occupy substitutional (Au_s) as well as interstitial (Au_i) sites. Whereas the solubility C_s^{eq} of Au_s is much larger than the solubility C_i^{eq} of Au_i, the reverse is true for the diffusivity. The diffusion of Au in Si involves the interchange between Au_s and Au_i positions. We consider two mechanisms for such an interchange, namely the *Frank-Turnbull mechanism* [67, 68]

$$Au_i + V \rightleftharpoons Au_s \tag{28}$$

and the *kick-out mechanism* [13,14]

$$Au_i \rightleftharpoons Au_s + I. \tag{29}$$

Besides (28) and (29) we take into account the reaction (5)

$$V + I \rightleftharpoons 0. \tag{5}$$

We now assume that local dynamical equilibrium between I, V, Au_i, and Au_s is established for which it is required that at least two of these processes (5,28,29) are sufficiently fast. Under these circumstances the normalized Au_s concentration $C = C_s/C_s^{eq}$ to be measured in an experiment on the diffusion of Au into dislocation-free silicon may approximately be described by the diffusion equation

$$\partial C/\partial t = \partial[D_{eff}\partial C/\partial x]/\partial x \tag{30}$$

with the effective diffusion coefficient [45,69]

$$D_{eff} = \left(C^{-2} + D_V C_V^{eq}/D_I C_I^{eq}\right)D_I C_I^{eq}/C_s^{eq}. \tag{31}$$

For the derivation of the approximate expression (31) it has been assumed that $C_i = C_i^{eq}$, $C_V^{eq} << C_s^{eq}$, and $C_s < (C_s^{eq} C_I^{eq})$

In the limiting case that only vacancies and no self-interstitials are present ($D_I C_I^{eq} = 0$) eq. (31) yields a constant effective Au_s diffusivity [67,68]

$$D_{eff}^V = D_V C_V^{eq}/C_s^{eq}.$$ (32)

The same result is obtained if $D_I C_I^{eq} \neq 0$ but only reaction (28) operates, whereas reactions (5) and (29) are so slow that they may be neglected. In the other limiting case that only I and no V are present ($D_V C_V^{eq} = 0$) eq. (31) predicts a strong concentration dependent diffusivity [13,14]

$$D_{eff}^I = \left(D_I C_I^{eq}/C_s^{eq}\right)C^{-2}.$$ (33)

The same result holds if from the reactions (5,28,29) only (29) operates efficiently.

Experimental concentration profiles after diffusion of gold into dislocation-free silicon at and above 800°C could not be described in terms of constant diffusivity as predicted by (32) whereas they could satisfactorily be fitted by using the concentration-dependent D_{eff}^I. In Fig. 6 a typical experimental diffusion profile of Au in Si as obtained by Stolwijk and co-workers [66] by means of neutron activation analysis is shown together with a fit based on D_{eff}^I. A complementary error function profile as expected for D_{eff}^V is seen to be *not* in accordance with the experimental profile as had already been noticed by Wilcox et al. [70] in the case of their own diffusion data. In Fig. 6 the measured total gold concentrations C_{Au} are given which practically coincide with the corresponding Au_s concentrations C_s because of the negligible solubility C_i^{eq} of Au_i. Stolwijk et al. [66] also determined the solubility $C_{Au}^{eq} \sim C_s^{eq}$ and were thus able to extract the quantity $D_I C_I^{eq}$ from their data, ranging from 800-1100°C, via a profile analysis based on D_{eff}^I of eq. (33). The result is

$$D_I C_I^{eq} \sim 914 \exp\left(-4.84eV/kT\right) cm^2 s^{-1}.$$ (34)

In Fig. 7 these data in the form $f_I D_I C_I^{eq}$ (with f_I chosen as 0.5) are compared to two sets of tracer self-diffusion data. The differences between these two sets reflect the uncertainty of more than a factor of two in measured tracer data. From the correspondence between tracer data which comprise the sum $f_I D_I C_I^{eq} + f_V D_V C_V^{eq}$ as given in eq. (3) and the self-interstitial component $f_I D_I C_I^{eq}$ as determined from the diffusion of Au in Si we conclude that at least above 860°C

$$D_V C_V^{eq} \lesssim D_I C_I^{eq}$$ (35)

holds and that an essential part of silicon self-diffusion is carried by self-interstitials.

Yet let us now look at whether a non-negligible contribution of $D_V C_V^{eq}/D_I C_I^{eq}$ in D_{eff} of eq. (31) in experimental Au diffusion profiles may be detected. Since we deal with in-diffusion experiments in which over a wide part of the profiles analyzed $C^{-2} > 1$ hold we have to keep in mind that the influence of the self-interstitial component $D_I C_I^{eq}$ is enhanced by this factor compared to the influence of $D_V C_V^{eq}$. The fact that Au in-diffusion profiles may reasonably well be described by D_{eff}^I simply means that $D_V C_V^{eq}$ is either smaller or in the same order of magnitude as $D_I C_I^{eq}$. This is basically the same condition for $D_V C_V^{eq}/D_I C_I^{eq}$ as we derived above in eq.(35) from a comparison of $D_I C_I^{eq}$ with tracer self-diffusion data. Presently, a numerical analysis based on eq. (30,31) is carried out for a series of the Au diffusion profiles at a given temperature [69]. First preliminary results indicate that for short diffusion times and $C^{-2} > 1$ the experimental Au diffusion profiles are described by D_{eff}^I within experimental error, whereas for longer times and C approaching 1, small but noticeable deviations occur when D_{eff}^I is used. By minimizing these deviations Morehead et al. [69] obtain $D_V C_V^{eq} \sim D_I C_I^{eq}$ within about 20% for a temperature of 1000°C. In Fig. 8 this value has been used as an estimate for $D_V C_V^{eq}$ at 1000°C.

In order to obtain the temperature dependence of $D_V C_V^{eq}$ we need a value of $D_V C_V^{eq}$ at another temperature, which we get in the following way: It can be expected that at temperatures below about 900°C the recombination reaction (5) is so slow that it may be neglected for typical times of gold diffusion experiments. At even lower temperatures a simple estimate indicates that the kick-out mechanism should become "frozen-in" (i.e., so slow that it may be neglected for typical diffusion times) before this happens to the Frank-Turnbull mechanism. With only the Frank-Turnbull mechanism operating the self-interstitial concentration C_I remains unaffected by the gold diffusion and we expect Au to show a complementary error function profile as predicted for a constant effective diffusivity, such as D_{eff}^V. Wilcox and co-workers [70] found such diffusion behavior at 700°C, which indicates that at this temperature the kick-out mechanism is actually "frozen-in". Using $D_{eff}^V = 8.4 \times 10^{-14}$ cm^{2}s^{-1} from Wilcox et al. [70] and an extrapolated value of $C_s^{eq} \sim 1.05 \times 10^{-8}$ from the data of Stolwijk et al. [66] we obtain $D_V C_V^{eq} \sim 8.8 \times 10^{-22}$ cm^{2}s^{-1} at 700°C. From this value at

700°C and the value at 1000°C given above we get the rough estimate

$$D_V C_V^{eq} \sim 0.57 \exp(-4.03 eV/kT) cm^2 s^{-1}. \tag{36}$$

considering that $D_V C_V^{eq}$ is based on two points only it should be regarded as preliminary until more experimental data have been obtained and analyzed. In Fig. 8 we have plotted $D_I C_I^{eq}$ from eq. (34) and $D_V C_V^{eq}$ from eq. (36) as a function of reciprocal temperature. Self-interstitials contribute more to self-diffusion above about 1000°C and vacancies contribute more below that temperature. Such a behavior has already been suggested by Seeger and Chik [1]. For the sake of thoroughness, we mention that the gold diffusion also allows us to obtain information holding at least between 800 and 1100°C:

$$C_I^{eq} \ll C_s^{eq} \sim 18 \exp(-1.98 eV/kT), \tag{37}$$

$$D_I \gg 51 \exp(-2.86 eV/kT) cm^2 s^{-1}. \tag{38}$$

PRECIPITATION AND DIFFUSION OF NICKEL IN SILICON

Nickel, similarly as gold, migrates in silicon via a substitutional-interstitial diffusion mechanism. Experiments on the precipitation and diffusion of substitutional Ni in dislocated silicon indicate that Ni diffuses in this temperature regime via the Frank-Turnbull mechanism [71,72]. The experiments do not show a contribution from Si self-interstitials. We suggest the following explanation to this apparent contradiction to the results obtained from the diffusion of Au into silicon: The Ni experiments have been performed below about 900°C for typical precipitation or diffusion times of hours. Therefore, the I-V reaction (5) is much too slow to establish dynamical equilibrium between I and V. We now assume, similarly as in the case of Au, that the kick-out mechanism for Ni becomes "frozen-in" at a higher temperature T_f than the Frank-Turnbull mechanism would. We expect that this temperature T_f to be higher than in the case of Au, since, because of the about 100-1000 times smaller solubility C_s^{eq} of substitutional Ni, D_{eff}^V of Ni is larger by such a factor and therefore typical precipitation or diffusion times are correspondingly shorter. Treating the case of Ni in Si as the case of gold diffusion at 700°C completely in terms of the Frank-Turnbull mechanism we have estimated the vacancy contribution $D_V C_V^{eq}$ to Si self-diffusion at 900°C from the data of Kitigawa et al. [72]. Within experimental error the result shown in Fig. 8 is in agreement with the information derived on $D_V C_V^{eq}$ via the diffusion behavior of Au in silicon.

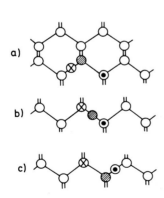

Fig. 1. Interstitialcy diffusion mechanism, schematically, a),b) and c) represent a sequence in time.

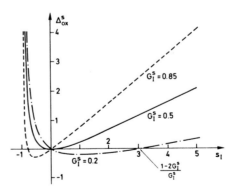

Fig. 2 Normalized diffusion enhancement or retardation Δ_{ox}^s versus self-interstitial supersaturation S_I for various relative interstitialcy diffusion components $G_I^s = D_I^s / D^s$ [18].

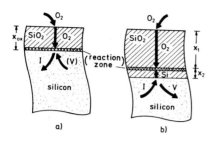

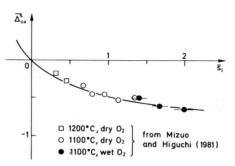

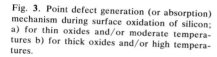

Fig. 3. Point defect generation (or absorption) mechanism during surface oxidation of silicon; a) for thin oxides and/or moderate temperatures b) for thick oxides and/or high temperatures.

Fig. 4. Theoretical fitting of the Sb ORD data of Mizuo and Higuchi [17] to eq. (13) with $G_I^S = 0.02$[18].

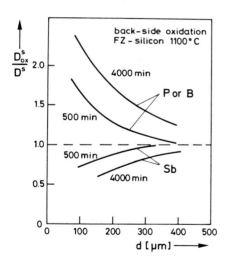

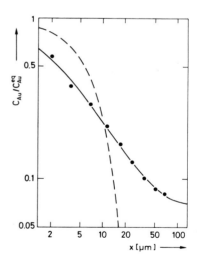

Fig. 5. Diffusion-enhancement (P,B) or retardation (Sb) of dopants at the front side of a Si wafer oxidized at the back-side according to Mizuo and Higuchi [59,60].

Fig. 6. Double-logarithmic plot of the Au penetration profile into dislocation-free silicon after a 1 hour anneal at 900°C. Solid line: fit based on D_{eff}^I of eq. (33). Dashed line: fit based on D_{eff}^V of eq. (32). [66].

56

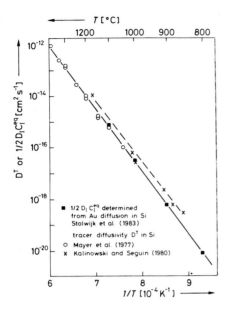

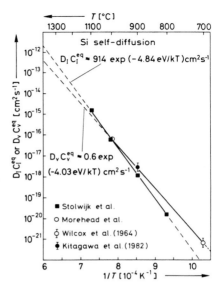

Fig. 7. Tracer self-diffusion coefficient D^T in Si as a function of reciprocal temperature. o and x: direct tracer data measuring $f_I D_I C_I^{eq} + f_V D_V C_V^{eq}$ [23,25]. ■ : $f_I D_I C_I^{eq}$ based on Au diffusion in Si, with f_I chosen as 0.5 according to Stolwijk et al. [66].

Fig. 8. Components $D_I C_I^{eq}$ and $D_V C_V^{eq}$ of Si self-diffusion versus reciprocal temperature calculated from the diffusion of Au into dislocation-free Si [66,69,70] and $D_V C_V^{eq}$ calculated from the diffusion and precipitation of Ni in dislocated Si.

$D_0[cm^2s^{-1}]$	$H^{SD}[eV]$	ref.
900	5.13	[22]
1460	5.02	[23]
8	4.1	[24]
154	4.65	[25]

Table 1: Data on tracer self-diffusion in silicon fitted to eq. (4).

	group III dopants			group V dopants		
	B	Ga	Al	P	As	Sb
r_s/r_{Si}	0.75	1.08	1.08	0.94	1.01	1.16
G_I	0.8-1.0	0.6-0.7	0.6-0.7	0.5-1.0	0.2-0.5	0.02
ref.	[17]	[49]	[48]	[17,19,51]	[17,30,51]	[17,30]

Table 2: Fractional diffusivity G_I^s via self-interstitials for some substitutional group-III and group-V dopants at 1100°C. The ratio of the radii r_s of the substituional dopant and r_{Si} of a silicon atom is also given. [45]

dopant	T [°C]	lowest $\overline{\Delta}_{ox}^s$ observed	G_I	ref.
P	1150	-0.13	>0.75	[54]
P	1267	-0.58	>0.95	[42]
Ga	1150	-0.36	>0.88	[49]
Al	1150	-0.29	>0.85	[48]
B	1200	-0.27	>0.83	[31]

Table 3: Fractional interstitialcy diffusion components G_I^s of various substitutional dopants calculated from oxidation-retarded diffusion and $S_I<0$.

ACKNOWLEDGMENT

We thank our colleagues for numerous discussions on the subject of this paper. We are especially grateful to Dr. Stolwijk and co-workers and to Dr. Morehead and co-workers for allowing us to discuss their results prior to final publication.

58

REFERENCES

1. A. Seeger and K.P. Chik, Phys. Stat. Sol. 29, 455 (1968).
2. A. Seeger, Rad. Effects 9, 15 (1971).
3. R.O. Simmons and R.W. Balluffi, Phys. Rev. 125, 862 (1962).
4. S.M. Hu in : Diffusion in Semiconductors, D. Shaw ed. (Plenum Press, London, 1973) p.217.
5. T.V. Mashovets, Sov. Phys. Semiconductors 16, 1 (1982).
6. L.C. Kimerling and D.V. Lang in Lattice Defects in Semiconductors 1974 (Inst. Phys. Conf. Ser. 23, 1975) p. 589.
7. G.D. Watkins, J.R. Troxell, and A.P. Chatterjee in: Defects and Radiation Effects in Semiconductors 1978 (Inst. Phys. Conf. Ser. 46 1979). p.16.
8. L.C. Kimerling in ref. [7], p.56.
9. A. Seeger, W. Frank, and U. Gösele, in ref. [7], p. 148.
10. H. Föll and B.O. Kolbesen, Appl. Phys. 8, 117 (1975).
11. P.M. Petroff and A.J.R. de Kock, J. Cryst. Growth 30, 117 (1975).
12. H. Föll, B.O. Kolbesen, U. Gösele, J. Cryst. Growth 40, 90 (1977); 52, 907 (1981).
13. U. Gösele, W. Frank, and A. Seeger, Appl. Phys. 23, 361 (1980).
14. U. Gösele, F. Morehead, W. Frank, and A. Seeger, Appl. Phys. Lett. 38, 157 (1981).
15. A. Seeger, Phys. Stat. Sol. (a) 61, 521 (1980).
16. S.M. Hu, J. Appl. Phys. 45, 1567 (1974).
17. S. Mizuo and H. Higuchi, Jap. J. Appl. Phys. 20, 739 (1981).
18. T.Y. Tan and U. Gösele, Appl. Phys. Lett. 40, 616 (1982).
19. D.A. Antoniadis, J. electrochem. Soc. 129, 1093 (1982).
20. D. Shaw, Phys. Stat. Sol. (b) 72, 11 (1975).
21. A.F.W. Willoughby, Rep. Progr. Phys. 41, 1665 (1978).
22. J.M. Fairfield and B.J. Masters, J. Appl. Phys. 38, 3148 (1967).
23. H.J. Mayer, H. Mehrer, and K. Maier in: Lattice Defects in Semiconductors 1976 (inst. Phys. Conf. Ser. 31, 1977) p. 186.
24. J. Hirvonen and A. Anttila, Appl. Phys. Lett. 35, 703 (1979).
25. L. Kalinowski and R. Sequin, Appl. Phys. Lett. 35, 211 (1979); 36, 171 (1980).
26. W. Frank in: Festköperprobleme (Advances in Solid State Physics), Vol. 21, J. Treusch ed. (Vieweg, Braunschweig 1981) p.221.
27. J.C. Bourgoin and M. Lannoo, Rad. Effects 6, 157 (1980).
28. S. Prussin, J. Appl. Phys. 43, 2850 (1972).
29. E. Sirtl in: Semiconductor Silicon 1977, H.R. Huff and E. Sirtl eds. (The Electrochem. Soc., Princeton 1977) p.968.
30. D.A. Antoniadis and I. Moskowitz, J. Appl. Phys. Oct. issue (1982).
31. C. Hill in: Semiconductor Silicon 1981, H.R. Huff, R.J. Kriegler, and Y. Takeishi eds. (The Electrochem. Soc., Pennignton 1981) p.988.
32. E. Rosencher, A. Straboni, S. Rigo, and G. Amsel, Appl. Phys. Lett. 34, 254 (1979).
33. E.P. EerNisse, Appl. Phys. Lett. 30, 290 (1977).
34. W.A. Tiller, J. Electrochem. Soc. 127, 619, 625 (1980).
35. T.Y. Tan and U. Gösele, Appl. Phys. Lett. 39, 86 (1981).
36. P.S. Dobson, Phil. Mag. 24, 567 (1971); 26, 1301 (1972).
37. S.M. Hu in: Defects in Semiconductors, J. Narayan and T.Y. Tan eds. (North-Holland, New York 1981) p.333.
38. B. Leroy, J. Appl. Phys. 50, 7996 (1979).
39. A.M.-R. Lin, R.W. Dutton, D.A. Antoniadis, And W.A. Tiller, J. Electrochem. Soc. 128, 1121 (1981).
40. A.M.-R. Lin, D.A. Antoniadis, and R.W. Dutton, J. Electrochem. Soc. 128, 1131 (1981).
41. R.B. Fair, J. Electrochem. Soc. 128, 1361 (1982).
42. R. Francis and P.S. Dobson, J. Appl. Phys. 50, 280 (1979).
43. T.Y. Tan and U. Gösele, J. Appl. Phys. 53, 4767 (1982).
44. U. Gösele and W. Frank in ref. [37], p.55.

45. U. Gösele and T.Y. Tan in *Aggregation Phenomena of Point Defects in Silicon*, E. Sirtl ed. (The Electrochem. Soc., Pennington 1982). in press.
46. B. Leroy, J. Appl. Phys. *53*, 4779 (1982).
47. H. Alexander, H. Eppenstein, H. Gottschalk, and S. Wendler, J. Microscopy *118*, 1 (1980).
48. S. Mizuo and H. Higuchi, Jap. J. Appl. Phys. *21*, 56 (1982).
49. S. Mizuo and H. Higuchi, Denki Kagaku (J. Jap. Electrochem. Soc.) *50*, No. 4 (1982).
50. D.A. Antoniadis and I. Moskowitz, J. Appl. Phys. Dec. issue (1982).
51. Y. Ishikawa, Y. Sakino, H. Tanaka, S. Matsumoto, and T. Niimi, J. Electrochem. Soc. *129*, 644 (1982).
52. D.A. Antoniadis, A.M. Lin, and R.W. Dutton, Appl. Phys. Lett. *33*, 1030 (1978).
53. T.Y. Tan and B.J. Ginsberg, submitted to Appl. Phys. Lett..

54. T.Y. Tan, U. Gösele, and F. Morehead, to be published.
55. Y. Nabeta, T. Uno, S. Kubo, and H. Tsukamoto, J. Electrochem. Soc. *123*, 1416 (1976).
56. T.Y. Tan and B.J. Ginsberg, these proceedings.
57. K. Taniguchi, K. Karasawa, and M. Kashiwagi, J. electrochem. Soc. *127*, 2243 (1980).
58. S. Mizuo and H. Higuchi, Jap. J. Appl. Phys. *21*, 272 (1982).
59. S. Mizuo and H. Higuchi, J. Electrochem. Soc. (in press).
60. S. Mizuo and H. Higuchi, Jap. J. Appl. Phys. (in press).
61. S.M. Hu, J. Appl. Phys. *51*, 3666 (1980).
62. S.M. Hu, J. Vac. Science and Technol. *14*, 17 (1977).
63. U. Gösele, W. Frank and A. Seeger, Solid State Comm., in press.
64. W. Frank, A. Seeger, and U. Gösele in ref. [37], p.31.
65. M. Hill, M. Lietz, and R. Sittig, J. electrochem. Soc. *129*, 1579 (1982).
66. N.A. Stolwijk, B. Schuster, J. Hölzl, H. Mehrer, and W. Frank in: *Proc. 12th Int. Conf. Defects in Semiconductors*, Amsterdam Aug./Sept. (1982), to be published.
67. F.C. Frank and D. Turnbull, Phys. Rev. *104*, 617 (1956).
68. W.R. Wilcox and T.J. La Chapelle, J. Appl. Phys. *35*, 240 (1964).
69. F. Morehead, W. Stolwijk, W. Meyberg, and U. Gösele, to be published.
70. W.R. Wilcox, T.J. LaChapelle, and D.H. Forbes, J. Electrochem. Soc. *111* 1377 (1964).
71. M. Yoshida and K. Saito, Jap. J. Appl. Phys. *6*, 573 (1967).
72. H. Kitagawa, K. Hashimoto, and M. Yoshida, Jap. J. Appl. Phys. *21*, 276 (1982).

MODELING OF DOPANT DIFFUSION AND ASSOCIATED EFFECTS IN SILICON

RICHARD B. FAIR
Microelectronics Center of North Carolina, Post Office Box 12889
Research Triangle Park, North Carolina 27709

ABSTRACT

Research in the area of dopant diffusion in Si has focused on identifying the specific mechanisms and point defects involved. Recent approaches include observing the effects of diffusion and doping on oxygen precipitation, stacking fault growth or shrinkage, enhanced/retarded diffusion of one dopant in the presence of another. Very few of these studies have yielded unambiguous interpretations as a result of the indirect nature of the experiments. However, taken together we can infer the relative importance of vacancies versus Si self-interstitials in the diffusion of each dopant species.

INTRODUCTION

The purpose of this paper is to review the progress that has been made in modeling dopant diffusion in silicon. Since diffusion processes are known to influence the growth of structural imperfections and oxygen precipitates, studies in these areas will be cited to help elucidate diffusion mechanisms. Oxidation of the Si surface also can enhance or retard the diffusion of dopants. Studies in this area have provided new insight into the relative importance of vacancies versus silicon self-interstitials in diffusion, and into point defect recombination kinetics.

Point Defect Models of Diffusion

The results of diffusion studies in metals and ionic crystals have led to the establishment of several basic atomic diffusion mechanisms. These mechanisms dominate the interpretation of silicon diffusion experiments with the exception that in silicon there is a wide energy range available to the Fermi level. This leads to a point defect appearing in a variety of ionized states.

Under thermal equilibrium conditions a silicon crystal will contain a certain equilibrium concentration of vacancies, C_V^O and a certain equilibrium concentration of silicon self-interstitials, C_I^O. In diffusion models based on the vacancy, $C_V^O >> C_I^O$, and dopant as well as self diffusion can be explained as[1]

$$D_i = D_i^X + D_i^- + D_i^= + D_i^+ \tag{1}$$

where D_i is the measured diffusivity and D_i^X, D_i^-, $D_i^=$ and D_i^+ are the intrinsic diffusivities of the species through interactions with vacancies in the neutral, single acceptor, double acceptor or donor charge states respectively.

Analogous to the vacancy models, silicon self-interstitials can be assumed to be dominant such that $C_I^O >> C_V^O$. For such a model, dopant and self diffusion are assumed to occur via an interstitialcy mechanism,[2] shown schematically in Fig. 1. Highly mobile complexes consisting of self-interstitials in various charge states and impurities are assumed to exist.

In principle, both vacancies and self-interstitials may occur simultaneously, and somewhat independently. Indeed, any relationship that may exist between

C_V^O and C_I^O may be dominated by the silicon surface which can act as a source or sink for either species. On the other hand it has been suggested that an "energy" barrier may exist that makes vacancy/self-interstitial recombination an activated process[3]. Thus, if both types of point defects are important, diffusion processes may involve both types[4,5].

$$D_i = D_i^I + D_i^V \qquad (2)$$

where D_i^I is the interstitialcy contribution and D_i^V is the vacancy contribution to the total measured diffusivity, D_i. One way in which vacancies and self-interstitials could cooperate in affecting impurity diffusion is the Watkins replacement mechanism[6] shown in Fig. 2. Interstitial dopant impurities can be created by the exchange between a self-interstitial and a substitutional dopant atom. The newly created interstitial impurity would migrate until it finds a vacancy and then it is free to diffuse again as a substitutional impurity.

In order to determine which of the dopant defect models apply to the various dopant diffusion processes in Si, several experimental/analytical approaches have been tried. We begin by describing oxidation-related experiments.

Effects of Oxidation on Dopant Diffusion
Oxidation Enhanced Diffusion (OED)
Thermal oxidation of silicon is known to promote the growth of intrinsic stacking faults and dislocations and to enhance the diffusion of B, P and As. It has been pointed out many times that these two phenomena can be understood only if the oxidizing Si/SiO_2 interface acts as a source of self-interstitials[5,7-11]. If the surface generates a local steady-state concentration of self-interstitials, C_I, then the supersaturation ratio is C_I/C_I^O. The continuity equation for the total dopant concentration C_T, in terms of the substitutional impurity concentration C_{SI} is[11]

$$\frac{\partial C_T}{\partial t} = D_i \left(1 + f_{II} \frac{C_I}{C_I^O}\right) \frac{\partial^2 C_{SI}}{\partial x^2} \qquad (3)$$

where f_{II} is the fractional partial-interstitialcy contribution to impurity diffusion. Fair[11] has determined that 10-20% of the dopant impurity flux during oxidation is due to interstitialcy diffusion. The fractional interstitialcy factor, f_{II}, correlates with the energy required to remove an impurity from a substitutional lattice position. Antoniadis[12] has established f_{II} to be ≈ 0.37 for phosphorus at 1000°C.

Oxidation Retarded Diffusion (ORD)
Mizuo and Higuchi[13] found that the diffusion of antimony was retarded during oxidation of the silicon surface. This was explained by assuming that Sb diffuses only by a vacancy mechanism, and the self-interstitials generated at the oxidizing surface combine with vacancies to reduce their concentration. This effect was further studied by Antoniadis and Moskowitz[15] whose results are shown in Fig. 3 along with data of Mizuo and Higuchi[13]. Antomony diffusion is initially enhanced slightly during oxidation, but steadily decreases, reaching 40% of its nominal value in a non-oxidizing situation after ~100 minutes at 1100°C. Assuming that an energy barrier exists to the annihilation of vacancies with interstitials, Antoniadis[14] proposed that the vacancy lifetime, τ_V is given by

$$\tau_V = (K \, c_I^O)^{-1} \tag{4}$$

where the reaction rate is given by

$$K = \frac{4\pi r_{I-V}}{\Omega n_H} (D_I + D_V) \exp\left(-\frac{E_B}{kT}\right) \tag{5}$$

In eq. (5),

D_I = diffusivity of silicon self-interstitials

D_V = diffusivity of vacancies

r_{I-V} = 2a, a=5.43Å, the cubic lattice constant

Ω = $a^3/8$, the atomic volume

n_H = $5 \times 10^{22} cm^{-3}$, the atomic concentration

E_B = recombination barrier energy

By fitting Sb diffusivity data with calculations, a value of E_B= 1.4eV was obtained. This result implies that the gradual reduction of Sb diffusivity in Fig. 3 is caused by a slow decay of vacancies through bimolecular recombination with self-interstitials. As the oxidation rate decreases over time, the supply of generated self-interstitials also decreases. Thus, the vacancy supply gradually increases (after ~100 min. in Fig. 3) which causes the Sb diffusivity to increase.

The idea of an energy or enthalpy barrier against vacancy-interstitial recombination in Si is not consistent with room temperature electron irradiation experiments where generated vacancies and interstitials readily recombine[15]. Göselle et al[16] have shown that a 1.4eV barrier at high temperatures would lead to a recombination probability of 10^{-24} at room temperature. Thus, vacancy-interstitial recombination would not take place, contrary to experimental evidence. Instead, they propose that an entropy barrier exists to recombination as a result of the "extended" nature of self-interstitials and vacancies at high temperatures. Recombination can only take place after a simultaneous contraction of both defects to about one atomic volume. This contraction corresponds to a decrease of entropy, i.e. an entropy barrier. The settlement of this question is far from being over. Additional experiments are needed to clarify the vacancy-interstitial recombination issue.

Backside Oxidation

In a study aimed at determining the diffusion length of point defects in FZ and CZ silicon wafers, Mizuo and Higuchi[17] prepared samples as described in Fig. 4. They found that the diffusion of B and P in the front surface of FZ Si wafers covered by Si_3N_4 was enhanced by oxidation of the back surface. In addition, they found that the front side diffusion of Sb was retarded by backside oxidation of FZ wafers covered with Si_3N_4, but that no effect was observed in CZ wafers. These results for FZ silicon backside-oxidized at 1100°C in dry O_2 for various times are summarized in Fig. 5. The ordinate on the graph is the front side junction depth ratio, where X_{JBO} and X_{JBN} are described in Fig. 4. The abscissa is the distance between the oxidized Si/SiO_2 backside interface and the front surface covered with Si_3N_4. The results of this study show:

1. Oxidation-enhanced diffusion (OED) of B and P and oxidation-retarded diffusion (ORD) of Sb involve the same point defects generated by an oxidizing Si surface.
2. The diffusion length of these point defects increases with diffusion time in FZ silicon.

3. Long range oxidation enhanced/retarded diffusion does not occur in CZ silicon.

4. A Si_3N_4 covered Si surface is not a sink for point defects.

Doping Dependence of OED

Taniguchi, et al[18] found that OED decreases as the concentration of the diffusing impurity increases beyond the point where concentration-dependent diffusion occurs. This effect was explained in terms of the reduction of oxidation-produced self-interstitials by recombination with the increasing supply of vacancies. Fair[19] assumed that the equilibrium vacancy concentration is unaffected initially by the self-interstitials generated at the oxidizing surface. But, the quasi-steady-state value of C_I/C_I^o is inversely proportional to the vacancy concentration which increases with doping above n_i. The oxidation-enhanced dopant diffusivity is then

$$D_e = D_n + \Delta D_o$$
$$= D_i \left(\frac{C_v}{C_v^o}\right) + D_i \; f_{II} \; \left(\frac{C_I}{C_I^o}\right)_i \left(\frac{C_v^o}{C_v}\right) \tag{6}$$

where $(C_I/C_I^o)_i$ is the self-interstitial supersaturation under intrinsic doping conditions, and C_v/C_v^o is the vacancy enhancement that occurs when doping exceeds n_i. Equation 6 is divided into the contributions to diffusion under non-oxidizing conditions, D_n, and the OED contribution ΔD_o.

The data of Taniguchi, et al[18] are shown in Fig. 6 for OED of P and B vs. the total number of dopant impurities/cm^2, Q_T. The calculated values of D_n and ΔD_o[19] are shown in comparison with the data. Reasonable agreement was obtained. Thus Taniguchi's model[18] of self-interstitial recombination with vacancies is consistent with the high concentration diffusion models[20,21] of B and P used by Fair[19] in his calculations.

Effect of Chlorine on OED

If chlorine or chlorine-bearing compounds are added to oxygen in sufficient concentrations in the furnace that stacking fault retrogrowth occurs in Si, OED will become neglible[22]. This result is believed to be due to the generation of vacancies at the Si/SiO_2 interface when chlorine reacts with silicon atoms on lattice sites to produce SiCl by the reaction[1,23]

$$Si + \tfrac{1}{2} \; Cl_2 \rightarrow SiCl + V_{Si} \tag{7}$$

The vacancy generated is then available to recombine with a silicon self-interstitial produced due to oxidation:

$$I_{Si} + V_{Si} \rightarrow Si \tag{8}$$

As a result, the supersaturation of self-interstitials in the Si surface and bulk is reduced or eliminated, inhibiting stacking fault growth and OED. The concentration of generated self-interstitials during oxidation in a HCl atmosphere depends upon $(p_{HCl})^{-\frac{1}{2}}$ where p_{HCl} is the partial pressure of HCl in the gas stream. Thus, the effective impurity diffusivity is[19]

$$D_e = D_i \left(1 + K_H \; (p_{HCl})^{-\frac{1}{2}}\right) \tag{9}$$

where K_H is temperature and orientation dependent.

EFFECT OF DOPING AND DIFFUSION ON STRUCTURAL DEFECTS
Stacking Fault Shrinkage
The growth of oxidation-induced stacking faults (OSF) occurs by a climb mechanism involving either the emission of vacancies or the absorption of silicon self-interstitials[9]. The extent to which doping levels and diffusion processes disturb the equilibrium concentrations of vacancies and self-interstitials will also affect the growth or shrinkage of OSF's. Previous studies of the role of doping levels on OSF shrinkage rates have produced conflicting results[24-27]. However, in a recent study Fair and Carim[28] showed that OSF shrinkage can be enhanced during N_2 annealing by the presence of shallow P-implanted layers. Shrinkage rate data are shown in Fig. 7 for the P-implant doses shown. The Si samples had pre-grown surface OSF's. After implantation the wafers received 1150°C anneals in N_2 + 0.5%O_2 for 30-120 min. It can be seen that OSF shrinkage rate increases with increasing P dose. Table I shows OSF shrinkage rates versus the average P surface doping. Measured values are compared with calculated rates based upon vacancy absorption by the fault or self-interstitial emission from the fault. Considerably better agreement with the data is obtained from the model that vacancy absorption increases with doping. However, at intrinsic doping concentrations the rate-limiting step in OSF shrinkage is believed to be self-interstitial emission.

Oxygen Precipitation
The role of surface processing and dopant diffusion on oxygen precipitation in CZ silicon was studied to correlate point defect models of dopant diffusion with models of oxygen precipitation[29]. It was found that surface treatments can affect both the nucleation and the growth of oxygen precipitates through the generation of point defects. For oxygen supersaturation ratios s>5 at 1000°C it was found that P diffusion near the Si surface largely inhibited oxygen precipitation during 20 hour anneals in N_2 or O_2 ambients. A similar result was achieved if the wafers were annealed in dry O_2 without the P diffusion. The observed small decrease in interstitial oxygen concentration, ΔC_{oi}, as measured by the 1107cm^{-1} absorption band was equal to the amount of oxygen going into SiO_2 precipitates, estimated from the 1230cm^{-1} absorption band. However, for s<5 the P-diffusion samples that were similarly annealed showed no inhibition in ΔC_{oi} as compared to undiffused samples. However, the amount of oxygen tied up in SiO_2 precipitates, (1230cm^{-1} band) was 2.5 times less than ΔC_{oi}. For As-diffused samples ΔC_{oi} was equal to the oxygen concentration in SiO_2 precipitates. A similar result was obtained for undiffused samples annealed in both N_2 and O_2.

By analyzing these results in the light of various models of nucleation and growth of oxygen precipitates, it was concluded that the type of point defects generated by the phosphorus diffusion were different from those generated by an oxidizing surface. This result was further supported by observing defects in cleaved wafers that were Secco etched[29]. Only in the P-diffused wafers with s>5 were many bulk stacking faults observed, whereas few faults were found in undiffused wafers that were annealed in O_2 or N_2. It is believed that vacancies generated by the P diffusion recombine with silicon self-interstitials and, thus, lower the concentration of self-interstitials in the wafer. Through mass action this would tend to enhance the precipitate growth rate of already existing oxygen precipitates of critical size. With an enhanced local concentration of self-interstitials around the precipitate, stacking fault growth would be encouraged in the precipitate's vicinity.

NEW OBSERVATIONS IN DIFFUSION
Codiffusion of As and P
Fair and Meyer[20] have shown that P diffusion in Si is greatly enhanced in the presence of a highly doped As layer. By implanting 5×10^{14} P/cm^2 into a

pre-diffused As layer doped >10^{20}cm-3 and codiffusing the two dopants, P can overtake the As and diffuse through the As junction. The resulting P tail that is formed also shows enhanced diffusion. This is particularly interesting since As alone does not produce a tail region. An example of this phenomenon is shown in Fig. 8. The pre-diffused As layer (2×10^{16}As/cm^2, 150Kev at 1050°C for ½ hour) was implanted with P and codiffused at 900°C for 6 hours. The diffusion of P under the As as well as in the tail region is significantly enhanced. In effect, similar results are obtained if the highly doped layer is P.

These results provide strong support for a Fermi-level controlled point defect model of P diffusion with impurity-point defect pair dissociation. Under the highly doped As background P diffusivity varies as $(n/n_i)^2$, just as if the background doping were P^{31}. Also, the enhanced P tail diffusivity can be calculated from the Fair-Tsai P model[31].

Codiffusion of As with lightly-doped P also enhances As diffusion (as shown in Fig. 9) provided that P penetrates the As junction. Thus, the same mechanism that enhances the P tail diffusivity also enhances As diffusion. This coincident enhanced As diffusion in the presence of a P tail can be calculated by estimating the point defect supersaturation created by dissociating P-defect pairs. Since As diffusion is only slightly affected by self-interstitial supersaturation (small OED effect), the point-defects involved here are most likely vacancies.

Diffusion Under Si$_3$N$_4$ Films

The diffusion of dopants in Si with a deposited Si$_3$N$_4$ film directly on the surface has been investigated recently[17,32,33]. A typical experimental structure[33] is shown in Fig. 10 in which diffusion in Si capped with Si$_3$N$_4$, SiO$_2$ under Si$_3$N$_4$ and a free oxidizing surface can be studied simultaneously. Substrates used were both CZ and FZ Si. The results from this study are shown in Fig. 11. Mizuo and Higuchi found that in FZ Si, B and P diffusion (low concentration) under the Si$_3$N$_4$ film, X_{JN}, is retarded relative to diffusion under the SiO$_2$ capped with Si$_3$N$_4$, X_{JON}. However, in CZ Si $X_{JN}/X_{JON} > 1$ for both dopants. It is believed that the compressive stress introduced into the Si by the tensile Si$_3$N$_4$ film causes an increase in the Si bandgap through a volume change:

$$\Delta Eg = -K' d\ln(V) \tag{10}$$

where K' is a constant. The intrinsic concentration of donor-type vacancies in stressed Si relative to unstressed Si is[32]

$$[V^+]_i^S = [V^+]_i \exp\left(\frac{-\Delta Eg}{2kT}\right) \tag{11}$$

Thus, a $+\Delta Eg$ causes $[V^+]_i^S/[V^+]_i < 1$. As a result, the diffusion coefficient of B would be reduced under compressive stress. A similar argument can be made for P with V$^-$ vacancies.

For the CZ Si another phenomenon is operating that overrides the stress effect - Si self-interstitial generation due to oxygen precipitation. Hu[34] has shown that OSF's actually grow in Si during argon annealing at 1150°C. He proposed that a self-interstitial supersaturation occurs in the Si as a result of self-interstitial emission from growing precipitates. At the surface the interstitials can participate in surface regrowth processes as they do during oxidation. However, with a Si$_3$N$_4$ film on the Si surface, surface bond rearrangements are made more difficult, and the surface is no longer a self-interstitial sink. The resulting supersaturation causes enhanced diffusion of dopants in the surface region.

For high concentration Sb diffusion under a Si_3N_4 film, $X_{JN}/X_{JON}>1$ for FZ Si, just the opposite of low concentration B and P diffusion[17]. A similar result was obtained for high concentration As[32]. Since the high concentration diffusivity is $D_i(n/n_i)$, the stress effect on n_i must be included such that under extrinsic conditions[32],

$$[V^-]^S = [V^-] \exp \left(\frac{+\Delta Eg}{2kT} \right) \tag{12}$$

Thus, a compressive stress producing $a+\Delta Eg$ causes $[V^-]^S/[V^-]>1$. In CZ Si, the supersaturation of self-interstitials under the Si_3N_4-capped Si surface reduces the concentration of vacancies available for Sb diffusion. Thus X_{JN}/X_{JON} is smaller for CZ Si than for FZ Si[17], as shown in Fig. 12.

Phosphorus Diffusion

The previously described effects of P diffusion on oxygen precipitation/ stacking fault growth and As codiffusion on P profiles support a P-vacancy pair model of P diffusion. Using Fair and Tsai's model[31], Mathiot and Pfister[35] obtained a rigorous numerical solution for the coupled P and vacancy diffusion equations. They found that the supersaturation of vacancies due to E-center dissociation can account for the formation of a tail region in high concentration P profiles, however, they maintain that this supersaturation causes an important flux of vacancies toward the surface, which draws the P atoms back towards the surface. As a result, their calculated P diffusivity is reduced above $C_p/n_i=10$, and the flat region characteristics of P profiles could not be reproduced. Such a decrease in P diffusivity has been observed for $40 \leq C_p/n_i \leq 100$[36], and this was attributed to misfit stain effects.

Other workers have also addressed the high concentration plateau region of P profiles, trying to account for the non-electrically active P in this region. Nobili, et al[37] performed TEM examinations of annealed, ion-implanted P samples using a weak beam technique. They detected a high density of silicon phosphide precipitates which could account for the difference between total P and the free-carrier concentration. The coherent structure of the precipitates puts previous determinations of precipitated P concentrations using RBS in question[38]. Nobili, et al concluded that the precipitation is associated with a large enhancement of P diffusivity, which is responsible for the flat surface region. The results of their experiments are shown in Fig. 13. The carrier density profiles of two ion-implanted and laser annealed speciments are indicated, along with the results following an 850°C, 30 minute anneal. The amount of enhanced diffusion in the tail regions of both profiles relative to the "kink" in the profiles is the same. Thus, instead of proving that enhanced tail diffusivity depends upon the amount of P precipitation, they have shown that the source of point defects responsible is independent of precipitation after a short, initial period of time.

Contrary to the work of Mathiot and Pfister[35], Yoshida[39] was able to rigorously calculate P profiles by assuming equilibrium was maintained between the various charge states of the E centers. In order to explain the "kink" in the P profile, he incorporated a mechanism that assumes that the vacancy formation energy decreases with increasing P concentration. Thus, the equilibrium concentration of V^- vacancies and hence P diffusivity in the tail region is increased sufficiently to reproduce the "kink". However, the codiffusion experiment of Fair and Meyer[30] in which As is substituted for P and a similar result obtained invalidates Yoshida's model.

68

Arsenic Diffusion

The study of As diffusion in Si has centered around the clustering of As atoms at high concentrations. The most recent contribution by Guerrero, et al[40] suggests that clustering involves As atoms along with one negative charge (electron or V^- vacancy). This result was obtained by fitting various models to room temperature free electron concentration versus total As data. Other authors have tried to fit some of these same data and have arrived at different conclusions[41,42]. The scatter in the data makes any interpretation feasible. The problem may be in using data obtained at room temperature. It has been reported that the free electron concentration in highly doped As layers is higher at diffusion temperatures than at room temperature[43,44]. If the total As concentration is assumed to be the sum of monatomic As^+ and various As-vacancy pairs (V^-As^+, $V^=As^+$ or $V^=As_2^+$), then the sum of the concentrations of these species at 700°C and 1000°C is shown in Fig. 14[45]. At 700°C the contribution of $V^=As^+$ becomes significant. This also corresponds to the temperature at which As diffusion becomes greatly enhanced, much like P diffusion at higher temperatures ($V^=P^+$ pair analogy). Thus the wide spread in n vs C_{As} data taken at room temperature may be due to a rapidly equilibrating $V^=As^+$ complexes whose concentration depends upon wafer cooling rate. The kinetics of $V^=As_2^+$ clustering are assumed to determine carrier concentrations in samples subjected to extended annealing following high temperature diffusion. However, additional studies are needed to support these models.

CONCLUSION

A major thrust in modeling dopant diffusion in Si has been to study the effects of diffusion on microstructural defect growth and precipitation. While many details need to be resolved regarding individual dopant models, it is clear that both vacancies and self-interstitials play an important role in affecting the diffusion of dopants both directly and indirectly through biomolecular reactions of the point defects.

REFERENCES

1. R. B. Fair in Applied Solid State Science, Supplement 2B, D. Kahng ed. (Academic Press, New York, 1981) pp. 1-103.

2. U. Gosele and H. Strunk, Appl. Phys., 20, 265 (1979).

3. S. M. Hu, J. Vac. Sci, Technol., 14, 17 (1977).

4. A. Seeger and K. P. Chick, Phys. Stat. Sol, 29, 455 (1968).

5. S. M. Hu, J. Appl. Phys., 45, 1567 (1974).

6. G. D. Watkins, in Effects Des Rayonnements Sur Les Semiconductors (Dunod, Paris 1965).

7. G. N. Wills, Solis-State Electron., 12, 133 (1969).

8. A. M. Lin, D. A. Antoniadis and R. W. Dutton, J. Electrochem. Soc., 128, 1131 (1981).

9. G. R. Booker and W. J. Tunstall, Philos. Mag., 13, 71 (1966).

10. U. Gosele and W. Frank in Defects in Semiconductors, J. Narayan and T. Y. Tan, eds. (North Holland, New York, 1981) p. 55.

11. R. B. Fair, J. Appl. Phys., 51, 5828 (1980).

12. D. A. Antoniadis in Semiconductor Silicon 1981, H. R. Huff, R. J. Kriegler, Y. Takeishi, eds. (The Electrochemical Society, Inc., Pennington, NJ, 1981) pp. 947-962.

13. S. Mizuo and H. Higuchi, Japan J. Appl. Phys., 20, 739 (1981).

14. D. A. Antoniadis and I. Moskowitz, to be published in J. Appl. Phys., Oct. 1982.

15. M. D. Matthews and S. J. Ashby, Phil. Mag., 27, 1313 (1973).

16. U. Gosele, W. Frank and A. Seeger, unpublished.

17. Mizuo and Higuchi, to be published.

18. K. Taniguchi, K. Kurosawa, and M. Kashiwagi, J. Electrochem. Soc., 127, 2243 (1980).

19. R. B. Fair, J. Electrochem. Soc., 128, 1360 (1981).

20. R. B. Fair, J. Electrochem. Soc., 122, 800 (1975).

21. R. B. Fair, ibid, 125, 323 (1978).

22. Y. Nabeta, T. Uni, S. Kubo and H. Taukamoto, ibid, 123, 1416 (1976).

23. S. P. Murarka, Phys. Rev. B., 21, 692 (1980).

24. H. Hashimoto, H. Shibayama, H. Masaki, and H. Ishikawa, J. Electrochem. Soc., 123, 1899 (1976(.

25. Y. Enomoto, T. Ishioka, S. Matsumoto and T. Miimi, unpublished.

26. C. L. Claeys, G. J. Declerck and R. J. Van Overstraeten in Semiconductor Characterization Techniques, P. A. Barnes and G. A. Rozgonyi, edc. (The Electrochemical Society, Pennington, NJ, 1978) p. 366.

27. C. L. Claeys, G. J. Delcerck, and R. J. Van Overstraeten, Rev. Phys Appl., 13, 797 (1978).

28. R. B. Fair and A. Carim, J. Electrochem. Soc., 129, 2319 (1982).

29. R. B. Fair, Meeting of the Electrochemical Society, Montreal, Spring, 1982.

30. R. B. Fair and W. G. Meyer, Silicon Processing Symposium, San Jose, Jan. 1982, to be published.

31. R. B. Fair and J. C. C. Tsai, J. Electrochem. Soc., 124, 1107 (1977).

32. R. B. Fair in Impurity Doping Processes in Silicon, F. F. Y. Wang, ed. (North Holland Press, Amsterdam, 1981) pp. 317-442.

33. S. Mizuo and H. Higuchi, Japan J. Appl. Phys., 21, 231 (1982).

34. S. M. Hu, J. Appl. Phys., 51, 3666 (1980).

35. D. Mathiot and J. C. Pfister, J. Appl. Phys., 53, 3053 (1982).

36. R. B. Fair, J. Appl. Phys., 50, 860 (1979).

37. D. Nobili, A. Armigliato, M. Finetti, and S. Solmi, J. Appl. Phys., 53, 1484 (1982).

38. R. Fogarrassy, R. Stuck, J. C. Muller, A. Grob, J. J. Grob, and P. Siffert, J. Electron Mater., 9, 1977 (1980).

39. M. Yoshida, Japan J. Appl. Phys., 19, 2427 (1980).

40. E. Guerrero, H. Potzl, R. Tielert, M. Grasserbauer, and G. Stingeder, J. Electrochem. Soc., 129, 1826 (1982).

41. R. B. Fair and G. R. Weber, J. Appl. Phys., 44, 273 (1973).

42. Y. Wada and N. Hashimoto, J. Electrochem. Soc., 172, 461 (1980).

43. J. Murota, E. Arai, K. Kobayashi, and K. Kudo, J. Appl. Phys., 50, 804 (1979).

44. S. Matsumoto, T. Miimi, J. Murota, and E. Arai, J. Electrochem. Soc., 127, 1650 (1980).

45. R. B. Fair in Semiconductor Silicon 1981, H. R. Huff, R. J. Kriegler, Y. Takeishi, eds. (The Electrochemical Society, Pennington, NJ, 1981) pp. 963-978.

71

TABLE I
Doping Concentrations and Stacking Fault Shrinkage Rates After 30 Minutes in N_2 at 1150°C

Implant Dose	Ave. Surface Doping	OSF Shrinkage Rate		
		Meas. (Ave.)	Cal. (V Absorp.)	Cal. (I Emission)
$0\ cm^{-2}$	$2\times10^{15}\ cm^{-3}$	2×10^{-8} cm/sec	$\cong 0$ cm/sec	4×10^{-8} cm/sec
1×10^{14}	2.5×10^{18}	3×10^{-8}	$\cong 0$	4×10^{-8}
5×10^{14}	1.5×10^{19}	4.5×10^{-7}	1.2×10^{-7}	5.7×10^{-8}
1×10^{15}	2.5×10^{19}	5.6×10^{-7}	3.7×10^{-7}	1.0×10^{-7}
5×10^{15}	7×10^{20}	-------------	1.5×10^{-6}	1.35×10^{-7}
1×10^{16}	1×10^{20}	1.2×10^{-6}	2.4×10^{-6}	1.42×10^{-7}
5×10^{16}	1.9×10^{20}	$>2.5\times10^{-6}$	4.4×10^{-6}	1.46×10^{-7}

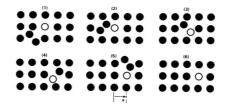

Fig. 1. Schematic of one diffusional jump of a substitutional atom (open circle) via an interstitialcy mechanism.

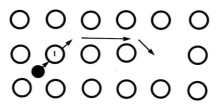

Fig. 2. Schematic of the Watkin's replacement mechanism.

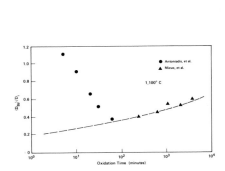

Fig. 3. Oxidation-retarded diffusion of Sb in Si.

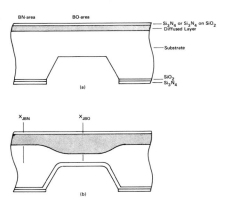

Fig. 4. Backside oxidation experiment of Mizuo.

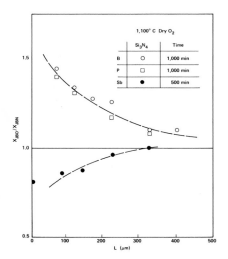

Fig. 5. X_{JBO}/X_{JBN} vs. distance from backside oxidizing surface for B, P and Sb diffusion (Mizuo).

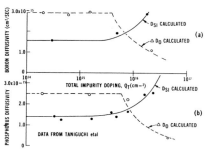

Fig. 6. Diffusivities of (a)B and (b)P during wet O_2 oxidation at 1100°C vs. implant dose (Fair).

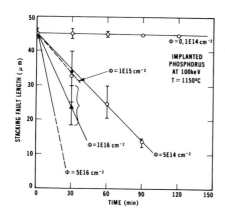

Fig. 7. Oxidation-induced stacking fault shrinkage in N$_2$ at 1150°C vs. P implant dose (Fair and Carim).

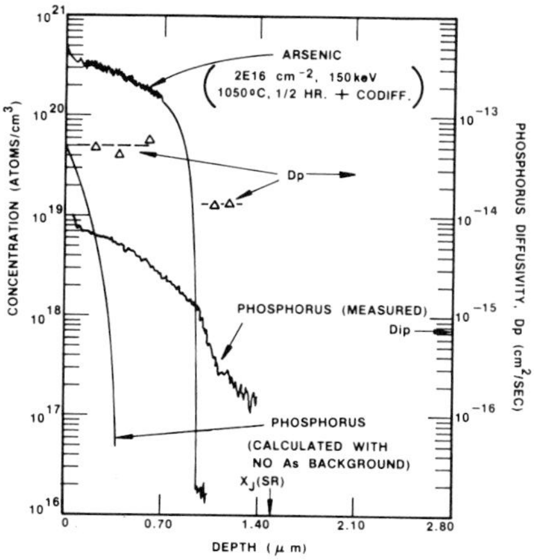

Fig. 8. P codiffusion with As at 900°C for 6 hours (Fair and Meyer).

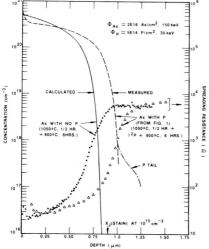

Fig. 9. As concentration and spreading resistance profiles with and without P diffusion through the As junction (Fair and Meyer).

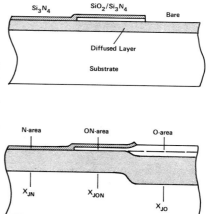

Fig. 10. Diffusion under a Si$_3$N$_4$ mask (Mizuo).

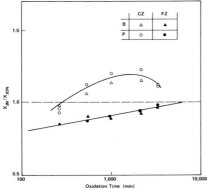

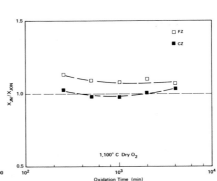

Fig. 11. X_{JN}/X_{JON} vs. oxidation time for Si_3N_4 covered samples diffused with B and P (Mizuo).

Fig. 12. X_{JN}/X_{JON} vs. oxidation time for Sb diffusion.

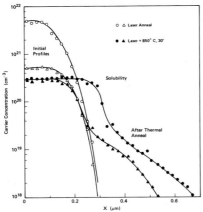

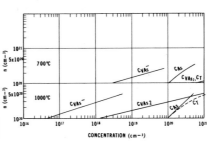

Fig. 13. Carrier density profiles for two ion implanted/annealed P doses (Nobili).

Fig. 14. Calculations of As complexes vs. electron concentration at 700°C and 1000°C (Fair).

DEFECTS IN AMORPHOUS SILICON

D. K. BIEGELSEN
Xerox Palo Alto Research Centers, Palo Alto, CA   94304

ABSTRACT

In this paper we argue that amorphous silicon can be treated as a relaxed random continuous random network. The optical and electronic properties are controlled by localized gap states which arise from characteristic features of a disordered tetrahedrally–bonded covalent network. Experimental results are reviewed which indicate that the dominant (perhaps only) electrically–active defect in hydrogenated amorphous silicon is the topologically distinct, silicon dangling bond. Finally, we suggest that the same, disorder–related characteristics might also typify the electronic properties of some macroscopic crystalline silicon defects.

INTRODUCTION

In a symposium on defects in crystalline materials, a paper treating defects in amorphous silicon may seem anomalous. Defects are conceptualized as deviations from perfection. For many years amorphous silicon was widely considered to be intractably defective crystalline silicon. The theme we wish to develop here is that in fact amorphous silicon can be treated as a well–characterized phase, and that, therefore, the notion of a topologically distinct, electrically active defect is sensible, useful and, hopefully, correct! We then suggest that a variety of defects associated with crystalline silicon, in which the effects of disorder can dominate, might manifest electronic properties similar to amorphous silicon.

NETWORK STRUCTURE

Pure amorphous silicon (a–Si) is – most likely at all temperatures – a metastable state. Whether deposited by evaporation or sputtering or created by bombarding crystalline silicon with ionic species, the material has a local structure characterized by covalently–bonded tetrahedral units. The nearest neighbor spacing is almost identical to that in crystalline silicon. In Figure 1, we show the radial distribution functions (RDF) for amorphous and crystalline germanium (a very similar system) [1]. The corresponding position and width of the first and second neighbor peaks (strong short–range order) are a result of the dominant chemistry of the covalent bonds. The strong potential minimum at the optimal bond length [2] produces a system in which the total energy of the network can be minimized by localizing strains (e.g. in weak or broken bonds) [3]. The rapid decay of the radial distribution function at longer distances (short correlation length) arises primarily from bond angle and dihedral angle variations which incur only small energy increases. Because of the over–constrained nature of tetrahedrally coordinated amorphous semiconductors [4], the relatively low energy bond angle variations are

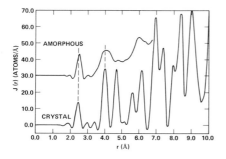

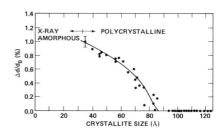

Fig. 1. Radial distribution functions for crystalline and amorphous germanium.

Fig. 2. Lattice expansion versus crystallite size for deposited polycrystalline silicon films.

not sufficient to accommodate all network strains. One thus finds that pure a–Si is highly defective, having ~ 10^{20} paramagnetic states per cm^3 [5] and an abundance of internal surfaces.

We first consider then what is known about the a–Si network. In the context of this forum it is particularly appropriate to consider the distinction between a–Si and defective or microcrystalline silicon. A nearly discontinuous transition between the ordered and disordered phases has been observed by many authors. In microcrystalline material [6] it has been shown that the coherence length, as measured by x–ray diffraction, drops suddenly for crystallite size $\leq$ 30 Å. In particular, the intensity of the third neighbor peak in the RDF drops, implying a loss in dihedral angle definition.

Veprek et al. [7] have controllably produced plasma deposited films with varying crystallite size by varying the substrate temperature and thereby the degree of deviation from equilibrium growth. They have shown that there exists an inverse correlation between crystallite size and lattice expansion (Figure 2) (due to an increasing relative contribution of the surface energy to the total energy of crystallites with decreasing dimensions – i.e., decreasing average bond strength). The resulting increase in elastic energy apparently makes the crystalline structure unstable relative to the amorphous phase. The second method which allows one to follow the crystalline to amorphous transition entails high energy ion implantation. Figure 3 is a composite of data from similar studies for varying energy deposition into atomic displacements. Brower and Beezhold [8] measured electron spin resonance (ESR) in oxygen–bombarded silicon. At low fluence they found an increase in vacancy–related centers. These centers evidence crystalline anisotropy. Above a deposited energy of ~ 10^{24} eV cm^{-3}, the crystalline defects disappear and a macroscopically isotropic line with a characteristic resonance g–value of 2.0055 appears. Mayer et al. [9] studied Rutherford backscattering (RBS) on phosphorus and silicon–implanted material. The loss of lattice order correlates well with the onset of the a–Si ESR line. Conductivity measurements in heavily–doped silicon by Müller and Kalbitzer [10] show a strong drop in the same damage regime most likely due to relaxation of dopant coordination from 4–fold (doping) to 3–fold (electrically inactive). Similarly, Raman scattering spectra of the 525 cm^{-1} Si–Si vibrational mode by Bourgoin et al. [11] convert rather sharply from a narrow crystalline matrix line to a broad amorphous network line in the same dose range.

Fig. 3.   Order–disorder   transition
            versus energy deposition as
            observed by (A) resistivity,
            (B) ESR, (C) RBS, (D) Raman
            scattering.

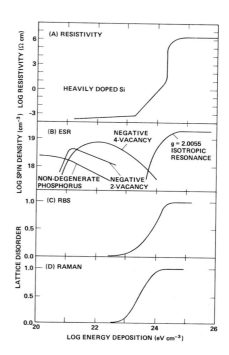

Brower and Beezhold interpreted their measurements to indicate that damage regions ~ 35 Å in diameter were produced in the undamaged crystalline matrix. The resulting picture is that when damaged regions overlap, the crystalline elastic constraints are lost and the lattice relaxes to a lower energy amorphous phase. Whether or not the transformation is truly a first order transition is not important here. Rather, the notion that a–Si can be described as having a characteristic relaxed tetrahedral network represents a great simplification in its understanding. Many of the important electronic properties of a–Si will be shown to follow then from general features of the a–Si network.

ELECTRONIC STATES AND DEFECTS

The valence and conduction bands in tetrahedrally–bonded amorphous semiconductors are derived from bonding and anti–bonding states of the continuous random network. The band gap and electronic state distribution are fixed by the local covalent chemistry. Small variations in bond angles and lengths and ring statistics have been shown by calculations to affect only the electronic structure within the bands and to introduce tails of localized states at the band edges [12]. On the other hand, coordination defects or large fluctuations in bond length can produce states deep in the gap. The simplest defect in a–Si is a dangling silicon bond – a 3–fold coordinated silicon atom. Nevertheless, one might expect that an overwhelming range of possible network defects and void structures would make hopelessly complex any understanding or use of a–Si as a viable semiconductor. Indeed, a–Si has a very high density of states throughout the "gap" with ~ 10^{20} states cm^{-3} eV^{-1}.

Before proceeding to consider the electrically–active defects which do occur in amorphous silicon, we introduce here an additional ingredient which in fact results in further conceptual and practical simplification. That is, alloying with atomic hydrogen – a–Si:H. It is now widely accepted that, because the Si–H bond is stronger than the Si–Si bond, hydrogen reduces the density of states in the gap by passivating weak bonds and dangling bonds, i.e., removing those states from the band gap. The rest of this paper will deal then only with a–Si:H.

ESR was one of the earliest experimental tools applied to the study of defects in a–Si [5]. A characteristic signature, the inhomogeneously broadened resonance with a g-value of 2.0055, is the only signal seen in undoped amorphous silicon independent of preparation conditions. The signal has plausibly, but without direct microscopic evidence, been ascribed to silicon dangling bonds. The density of these centers depends strongly on preparation conditions, varying from ~ 10^{20} cm^{-3} for pure a–Si to $\leq 10^{15}$ cm^{-3} for "good" a–Si:H. When material is doped n– or p–type, the dangling bond resonance disappears. At high doping levels, when E_F enters the band tails, new ESR signals arise – one each from singly occupied conduction and valence band tail states. Figure 4 shows resonances [13] and Figure 5 shows an example of the low temperature spin densities in moderately

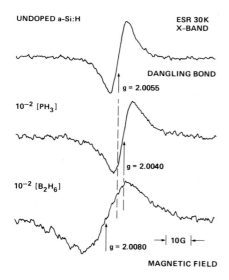

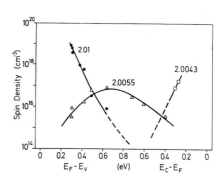

Fig. 4. Characteristic ESR spectra for undoped and doped a–Si:H corresponding to dangling bonds, and band tail electrons and holes.

Fig. 5. Spin densities of the three paramagnetic gap states versus Fermi energy in the gap.

defective samples as a function of Fermi energy in the gap [14]. The peak in the defect spin density just below midgap implies that the spin-½ center is paramagnetic when neutral and has a positive effective correlation energy, (the extra energy to add a second electron to the singly-occupied site) $\tilde{U} \sim 0.4$ eV. In Figure 6 we have sketched a schematic density of states, and shown the two defect energy levels corresponding to single and double occupancy. From the invariance of the spin signal for gross changes in material parameters (e.g. microstructure, presence of hydrogen, etc.) it can be deduced that the defect electron wave function is strongly localized. The dangling bond is the simplest defect to satisfy these requirements. Other consistencies will be described below, but until hyperfine studies are completed on Si^{29} enriched samples, positive identification cannot be made.

When light with energy greater than the band gap (~ 2 eV) is absorbed in a–Si:H, electron–hole pairs are generated. Because the scattering length, even in the bands, is approximately an interatomic distance, the carriers quickly thermalize down into the localized band tail states. Each charged carrier is paramagnetic and has the characteristic ESR signature shown in Figure 4. At temperatures below ~ 100 K, the carriers recombine either (a) radiatively by tunneling, emitting a photon

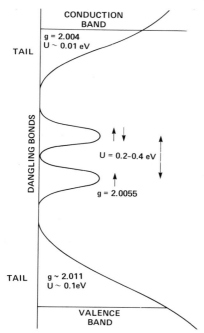

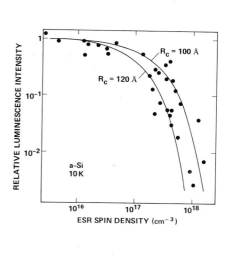

Fig. 6.   Density of states model including singly and doubly occupied dangling bond levels.

Fig. 7.   Luminescence efficiency versus dangling bond spin density.

with energy ~ 1.4 eV (band–edge luminescence), or (b) non–radiatively after the electron tunnels to a neutral dangling bond defect (see Figure 6). Step (b) is also weakly radiative with an energy ~ 0.9 eV (defect luminescence) [15]. The role of the dangling bond as a non–radiative recombination center is demonstrated in Figure 7. Here we plot the luminescence efficiency, η, versus spin density, N_s, for a wide range of sample preparation conditions [16]. The solid curves are theoretical estimates of η based on the results of time–resolved luminescence studies, and assuming a random distribution of isolated dangling bonds.

In recent photoconductivity measurements [17] quenching of the photo–current and range limitation of drifting carriers have been correlated with N_s. The results again indicate that dangling bonds are the dominant (if not only) recombination center in a–Si:H. Deep level transient spectroscopies (DLTS) have also been extended to amorphous silicon. Two results are of particular interest here. In the density of gap states derived for doped samples [18], a broad, often double–peaked band is found near mid–gap. In a related study, modulation of the dangling bond ESR was observed by varying the depletion width in a p$^+$–n diode [19]. The ESR increased as E_F was moved from near the conduction band towards midgap. From the temperature and frequency dependence of the induced ESR, the identity of the negative dangling bond state and the upper defect band in the DLTS density of states was demonstrated. The results are shown in Figure 8.

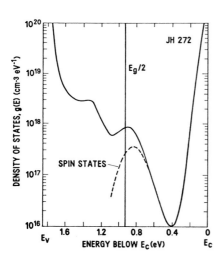

Fig. 8. Density of states derived from DLTS measurements for a phosphorus doped sample along with the doubly occupied dangling bond density of states for the same sample, ascertained by field modulated ESR.

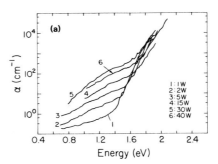

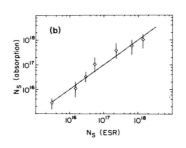

Fig. 9. (A) PDS spectra for undoped samples with varying N_s; (B) integrated extrinsic absorption plotted versus N_s.

Another manifestation of the defect states in a–Si:H is extrinsic optical absorption. Using a novel technique (photothermal deflection spectroscopy, or PDS) which is capable of measuring $\alpha d \geq 10^{-5}$ in thin films, Jackson and Amer [20] have measured a broad shoulder in undoped a–Si:H extending down from the absorption edge to ~ 0.5eV. Figure 9 (a) shows spectra for samples deposited under various deposition conditions; Figure 9 (b) demonstrates the linear correlation between the integrated absorption and N_s. The authors showed [21] that the absorption excites an electron from a neutral dangling bond ~ 1 eV below the conduction band, to an empty conduction band tail state. Furthermore, by comparing the absorption shoulders in undoped and n–type a–Si:H [22], they concluded that the dangling bond $\bar{U}$ is 0.25–0.45 eV.

Undoped a–Si:H grown under optimal conditions can be an excellent semiconductor – and, indeed, it is being developed rapidly for technological applications. Doping of a–Si:H is accomplished by adding, for example, PH_3 or B_2H_6 to the gas mixture during deposition [23]. However, unlike crystalline silicon, doping in a–Si:H has been found to introduce dangling bond defects [24]. Figure 10 shows estimates of defect densities as a function of dopant concentration. Experiments used here are light–induced ESR (LESR, which produces a non–equilibrium dangling bond resonance when the minority carrier is trapped by an oppositely charged dangling bond) and band–edge luminescence (which is quenched by the competitive channel afforded by the presence of dangling bonds).

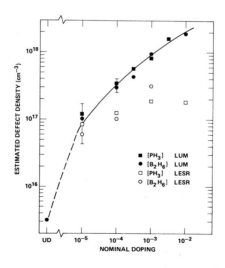

Fig. 10. Dangling bond defect density versus dopant concentration . (LESR values are lower bounds because the non–equilibrium signal depends on light intensity.)

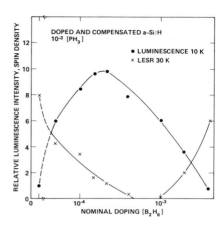

Fig. 11. LESR dangling bond densities and luminescence efficiencies for co-doped samples.

PDS and DLTS measurements are in excellent agreement with the luminescence data [21]. Clues to the origin of doping-related defect incorporation come from experiments in co-doped samples. Conductivity and thermopower measurements demonstrate that E_F moves through midgap at the nominal compensation level [24]. Figure 11 shows LESR dangling bond densities and luminescence efficiencies for samples having fixed phosphorus concentration and varying boron levels.

The light-induced N_s disappears near compensation even though twice as many impurities are present as in singly-doped samples. Similarly, the luminescence efficiency peaks at nominal compensation. PDS data on the same samples also imply the reduction in dangling bond density. (The three techniques also indicate an increase in a new valence band tail density of states. These states may be due to larger disorder potential fluctuations, or, more likely, to boron/phosphorus complexes which act as hole traps.)

The strong implication of these data is that the defect creation is related solely to the positions of E_F. An autocompensation mechanism has been proposed to explain the results [24, 25]. During deposition defects are created concomitant with four-fold dopant incorporation to minimize the net energy. The electronic energy increase is greatest when $|E_d-E_F|$ is greatest. (Here E_d is the energy of the ionized dopant.) This trend agrees with the data in Figure 10 which shows a slightly sub-linear increase in defect density when $|E_d-E_F|$ is large, and a strongly sub-linear behavior as $|E_d-E_F|$ decreases.

Many similar studies have now been performed on amorphous germanium, a-Ge:H. The similarities between the two systems are dramatic. In both systems, three characteristic ESR lines are found [26], corresponding to localized band tail carriers and dangling bonds (Figure 12). In Figure 13 we see the g-shifts (where g_0 is the free electron g-value of 2.0023) and line widths of the three centers in a-Ge:H plotted against the values for similarly doped a-Si:H samples. If the two ordinates are scaled by the ratio of atomic spin-orbit coupling factors ($\lambda_{Ge}/\lambda_{Si} \sim 7$), the resulting identity of the three ESR signals is striking. Thus, most of the important characteristics of the tetrahedrally-bonded amorphous semiconductors are intrinsic to the local chemistry and disorder in the network and the effects of disorder are similar in the two systems.

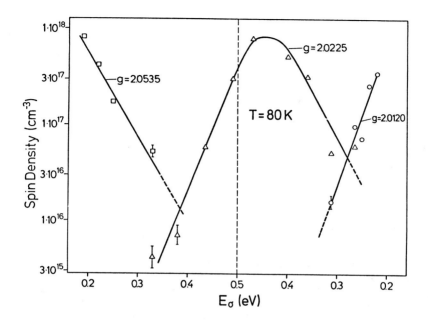

Fig. 12. Band tail carrier and
dangling bond spin densities
versus Fermi level position in
a–Ge:H.

To summarize, a model which is consistent with experimental measurements
and theoretical calculations [27] is that the electronic properties of a–Si:H can be
described by a continuous random network of tetrahedrally–coordinated silicon
atoms. As is even the case with vacancy–related defects embedded in a crystalline
silicon matrix, Jahn–Teller energies are sufficient to cause weak bond
reconstruction in a–Si [28]. This leads to band tail states. (The possibility of band
tail states involving charge transfer pairs of 3–coordinated silicon defects [29] –
$(T_3{}^+, T_3{}^-)$ – cannot be ruled out, but seems unlikely in hydrogenated material.)
The deep gap states, which are the non–radiative recombination centers, are
dominated by the dangling bond coordination defects. The decoupling of the
electrical properties from details of actual microstructure are then easily explained,
in that internal surface bonds either (1) reconstruct, becoming band tail states, (2)
become hydrogenated and drop into the valence band, or (3) relatively infrequently
remain as dangling bonds.

84

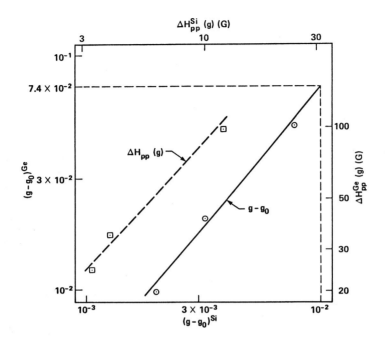

Fig. 13. Defect and band tail resonance g–values and linewidths plotted against each other for a–Ge:H and a–Si:H. The lines show a linear relationship.

DISORDERED SILICON SYSTEMS

In the little space which remains we wish to suggest a possible connection between the characteristics of a–Si and several macroscopic defects in crystalline silicon. These are defects in which the electrical properties may be dominated by effects of disorder – in particular, the silicon/silicon dioxide interface and grain boundaries. Caplan et al. [30] have shown that the dominant defect at the silicon {111} – silicon dioxide interface is the silicon dangling bond. (A broadened powder pattern of the axially–symmetric g–tensor is remarkably similar to the dangling bond signature in a–Si) [13]. The defects are isolated, positive Ũ centers which can be passivated by hydrogenation and which apparently have three charge states in the gap [31]. Electrical measurements find band tails which also can be greatly reduced by hydrogenation [32]. Although much experimental work still must be carried out, it is very tempting to apply the model of a–Si with its localized band tail states derived from fluctuations in interfacial bonding configurations and a single amphoteric dangling bond defect.

In a similar way, recent photocapacitance measurements by Werner et al. [33] find an electronic density of states for a single [110] tilt grain boundary comprised of band tail states and one or two deep levels near midgap. Previously groups studying single grain boundaries [34] and multiple grain boundaries [35] have concluded that two levels (an electron and hole trap) of approximately equal magnitude spanning midgap and separated by ~ 0.3 eV is needed to explain transport measurements. ESR measurements studies on polycrystalline silicon [36] have shown the dominance of the amorphous silicon dangling bond defect signature and its passivation by hydrogenation.

We do not wish to imply here that these defects have properties identical to those of a–Si. Rather we suggest that, in systems in which the relaxation of constraints is sufficient, many of the simple features which become dominant in a–Si can also be important.

REFERENCES

1. G. A. N. Connell and R. A. Street, Handbook on Semiconductors, Vol. 3, S. P. Keller, ed. (North Holland, New York, 1980) p. 689.

2. L. Pauling, The Nature of the Chemical Bond (Cornell Univ. Press, New York, 1960).

3. J. Robertson, Phys. Chem. Glasses 23, 1 (1982).

4. J. C. Phillips, J. Non–Cryst. Solids 34, 153 (1979).

5. M. H. Brodsky and R. S. Title, Phys. Rev. Letters 23, 581 (1969).

6. M. H. Brodsky, Light Scattering in Solids, M. Cardona, ed. (Springer–Verlag, New York, 1975), p. 208.

7. S. Veprek, Z. Iqbal and F.–A. Sarott, Phil. Mag. B 45, 137 (1982).

8. K. L. Brower and W. Beezhold, J. Appl. Phys. 43, 3499 (1972).

9. J. W. Mayer, L. Eriksson, S. T. Picraux and J. A. Davies, Can J. Phys. 46, 663 (1968).

10. G. Müller and S. Kalbitzer, Phil. Mag. B 41, 307 (1981).

11. J. C. Bourgoin, J. F. Morhange and R. Beserman, Radiation Effects 22, 205 (1974).

12. N. F. Mott, J. Phys. C 13, 5433 (1980); M. L. Thorpe, D. Weaire and R. Alban, Phys. Rev. B 7, 3777 (1973); J. D. Joannopoulos, Phys. Rev. B 16, 2764 (1977).

13. D. K. Biegelsen, Nuclear and Electron Resonance Spectroscopy, E. N. Kaufmann and G. K. Shenoy, eds. (North Holland, New York, 1981), p. 85.

14. H. Dersch, J. Stuke and J. Beichler, Phys. Stat. Sol. B 105, 265 (1981).

15. R. A. Street, Phys. Rev. B 21, 5775 (1980).

16. R. A. Street, J. C. Knights and D. K. Biegelsen, Phys. Rev. B 18, 1880 (1978).

17. R. A. Street, Phil. Mag. B 46, 273 (1982); and R. A. Street (to be published).

18. J. D. Cohen and D. V. Lang, Phys. Rev. B 25, 5321 (1982).

19. J. D. Cohen, J. P. Harbison and K. W. Wecht, Phys. Rev. Letters 48, 109 (1982).

20. W. B. Jackson and N. M. Amer, Phys. Rev. B 25, 5559 (1982).

21. W. B. Jackson and N. M. Amer, Phys. Rev. B (to be published).

22. W. B. Jackson, Solid State Comm. (in press).

23. W. E. Spear and P. G. LeComber, Phil. Mag. 33, 935 (1976).

24. R. A. Street, D. K. Biegelsen and J. C. Knights, Phys. Rev. B 24, 969 (1981).

25. D. K. Biegelsen, R. A. Street and J. C. Knights, AIP Conf. Proc. 73, 166 (1981).

26. M. Stutzmann and J. Stuke (to be published).

27. J. D. Joannopoulos, J. Non-Cryst. Sol. 35–36, 781 (1979).

28. G. D. Watkins and J. W. Corbett, Phys. Rev. 138, A543 (1965).

29. S. R. Elliott, Phil. Mag. B 38, 325 (1978).

30. P. J. Caplan, E. H. Poindexter, B. E. Deal and R. R. Razouk, J. Appl. Phys. 50, 5847 (1979).

31. C. Bronström and C. Svensson, Solid State Comm. 37, 399 (1981).

32. N. M. Johnson, D. K. Biegelsen and M. D. Moyer, in Physics of MOS Insulators, ed. by G. Lucovsky, S. T. Pantelides and F. L. Galeener (Pergamon, New York, 1980) p. 311.

33. J. Werner, W. Jantsch and H. J. Queisser, Sol. State Comm. 42, 415 (1982).

34. G. E. Pike and C. H. Seager, J. Appl. Phys. 50, 3414 (1979); J. P. Colinge (to be published).

35. S. W. Depp, B. G. Huth, A. Juliana and R. W. Koepcke, Grain Boundaries in Semiconductors, G. E. Pike, C. H. Seager and G. Leamy, eds. (North Holland, New York, 1982) p. 297.

36. N. M. Johnson, D. K. Biegelsen and M. D. Moyer, Appl. Phys. Letters 40, 882 (1982).

DIRECT EVIDENCE OF DIFFUSION OF SELF-INTERSTITIALS IN SILICON

GOBINDA DAS
IBM General Technology Division, East Fishkill, New York 12533

ABSTRACT

High temperature (1200°C) HCl oxidation treatment has been employed to float-zone (FZ) silicon wafers (625μm thick) containing swirl defects in order to study their diffusion characteristics. In treated wafers, swirl defects can be eliminated from both surfaces up to a depth of ~30μm. In the bulk of the wafers, however, large swirl defects (A-swirls) rearrange themselves into many small defects. The untreated portions of wafers contain large swirl defects (A-swirls) that extend up to both surfaces. Since swirl defects are primarily clusters of silicon self-interstitials, their rearrangement in the bulk and elimination from the surfaces demonstrate that migration of interstitials takes place on a large scale and is not confined to SiO_2/silicon interface only. The above observations appear to provide direct[2] evidence for the dominant role of self interstitials for diffusion mechanism in silicon at high temperature and can be rationalized in terms of an interstitialcy mechanism. Alternatively, however, dominance of interstitials can be related to a higher migration energy of vacancies proposed in a model where both species coexist at high temperature. The preference of one model over another must await theoretical calculations of diffusion energetics derived from both models.

INTRODUCTION

One of the fundamental problems in the field of diffusion of silicon is understanding of diffusing species during diffusion. In the sixties, vacancy mechanism enjoyed popularity [1] while in the past decade, both experimental evidence [2-4] and theoretical models [5,6] strongly favored interstitial or intersitialcy mechanism. In particular, the observations of extrinsic prismatic dislocation loops [2,3] associated with swirl defects in the float-zone silicon crystals (FZ) were interpreted in terms of clusters of self-interstitials. In Czochralski silicon (CZ), many investigators [7,8] observed oxidation-induced extrinsic stacking faults. In bipolar devices built on CZ substrate, extrinsic prismatic dislocation loops [4] were observed in electrically active areas. Additionally, oxidation enhanced diffusion of B, P [9] and oxygen precipitation induced defects in (CZ) silicon [10] have been also associated with silicon self-interstitials. The role of interstitials also gained technological interest when it was discovered that stackings faults (which are clusters of self-interstitials) are sensitive to device leakage [11]. Elimination of stacking faults from device regions has become an essential goal for obtaining high yield of devices with low leakage. Hu [12] and Shiraki [13], reported that dry oxidation at high temperatures (in excess of 1100°C) resulted in the elimination of bulk stacking faults, (i.e., self-interstitials) from CZ silicon. Shiraki also demonstrated that the addition of HCl in dry oxygen accelerated removal of bulk stacking faults.

Although these observations qualitatively show that at high temperature interstitials are quite mobile, they can annihilate by recombining with

Mat. Res. Soc. Symp. Proc. Vol. 14 (1983) © Elsevier Science Publishing Co., Inc.

as-grown vacancies or their clusters in CZ silicon as in the case of retardation of oxygen precipitation [14]. Similarly swirl patterns associated with
CZ silicon have been attributed to oxygen precipitation [15]. Therefore, in
order to avoid any possible contribution from oxygen clusters or precipitates,
the behavior of interstitials in silicon should be studied with FZ silicon
crystals. From the work of Secco D'Aragona [16] it can be established that
swirl defects can be eliminated from FZ silicon by annealing at a temperature
between 1200°C and 1250°C in argon.

In the present investigation we have performed HCl oxidation at high
temperature on FZ silicon to eliminate swirl defects and thus study diffusion
characteristics of self-interstitials. Oxidation was chosen to determine the
effect of an oxide layer on the surface of silicon. HCl was added to dry
oxygen to promote efficiency of sink for interstitials at silicon/silicon
dioxide interface [13]. The principal objective of the present investigation
has been to provide evidence for migration of interstitials in order to
emphasize their dominant role in diffusion mechanisms in silicon. No attempt
will be made to identify interstitials to be solely responsible for self
diffusion in silicon. Indeed, if vacancies and interstitials coexist at high
temperature, our observations here would clearly establish sluggishness of
vacancy migration as compared to the mobility of the self-interstitials in
silicon at high temperatures.

EXPERIMENTAL

N-type float-zone silicon crystals containing swirls which were used in
these experiments were grown in argon ambient with [111] growth axis. The
diameter of the crystals is 45mm. The resistivity of silicon is 140-160Ω cm
with oxygen and carbon content of 10^{15}/cm^3 and 10^{16}/cm^3 respectively. Wafers
were polished on both sides down to 625µm in thickness. Each wafer was broken
into two parts; one part was oxidized at 1200°C for two hours in dry oxygen
and 1% HCl and cooled slowly while the other part was not subjected to any
heat treatment. Both parts of the wafers were etched in Sirtl etch for five
minutes in order to develop the swirl defect pattern. X-ray scanning topographs and section topographs were obtained from each part of a wafer. In
order to facilitate the observation of swirl defects by x-ray topography, both
parts of the wafers were copper decorated at 950°C.

RESULTS AND DISCUSSION

The effect of high temperature HCl oxidation is illustrated in Fig. 1. It
can be noticed that the untreated part (A) exhibited typical swirl defects
upon etching with Sirtl etch. In contrast, the adjacent part of the wafer
which was oxidized does not contain any swirl defects on its surface region.
The absence of swirl defects from the surface of the oxidized wafers demonstrates removal of interstitials [2,3] during thermal treatment. Figures 2(a)
and 2(b) are section topographs representing untreated and treated halves of a
wafer. Only Pendellosung fringes indicative of lack of strain fields associated with swirl defects, are visible. Figures 3(a) and 3(b) are corresponding
topographs after copper decoration at 950°C. In Fig. 3(a) characteristic
A-swirl defects of approximately 10-30µm in diameter (actual size is expected
to be smaller), extending to both surfaces, are seen. No B-swirl defects are
seen. In Fig. 3(b) which is obtained from the high temperature treated part,
swirl defect-free zones of approximately 30µm are noticed on both surfaces.
In the mid-section of the wafers small defects are randomly distributed. Some
large A-swirls are also seen in Fig. 3(b), but they are mixed with small

defects and not arranged in rows, characteristic of A-swirl. This is a significant observation reported for the first time. It appears that many A-swirl defects have redissolved during the high temperature oxidation (1200°C). On cooling super-saturated interstitials out-diffuse from surface regions to create the observed swirl-defect free zone. In the bulk, however, they nucleate and foster growth of smaller defect clusters. The nature of the small defects (A or B defects) cannot be established now in view of lack of TEM evidence. It should be noted, however, that if small defects observed here are B-swirl defects, they would be difficult to be resolved even by TEM [2,3]. Furthermore, if these are of vacancy type, they have to be formed as a result of thermal treatment since they were not present in the as-grown condition. One would require to supply (a) formation energy of both interstitials and vacancies and (b) their respective migration energies which are larger than energy required to cluster pre-existing interstitials. Finally, one should examine the possibility that some small vacancy clusters (so small that they are not detected by x-ray even after copper decoration) are always present in the as-grown crystals [17]. Can they grow and form the small defects after heat treatment? This is very unlikely as excess interstitials should recombine with pre-existing vacancy clusters making the formation of small defects in Fig. 3(b) very difficult. On the basis of these arguments, the small defects observed in Fig. 3(b) are thought to be of interstitial in origin.

The observed DFZ is variable, approximately 15-30µm (Fig. 3(b)) at 1200°C for two hours. The estimated diffusion length for the self-interstitials based on an interstitialcy model of Seeger et al. [6], is approximately $2\sqrt{D_I t}$ where $D_{self} = f_I C_I D_I$. Noting that D_{self} at 1200°C is 1.03×10^{-14} cm^{-2} sec^{-1} [18], $f_I = .72$, and taking $C_I \sim 2 \times 10^{-7}$ as a reasonable value at 1200°C, the diffusion length for the interstitials is 450µm which is somewhat larger than the DFZ observed here. This discrepancy appears to indicate that removal of interstitial at the Si/SiO$_2$ interface is dependent on the thickness of the oxide grown. During dry oxidation at 1200°C for two hours, 3700°A of SiO$_2$ is grown. If the migration of interstitials across the interface is limited by the diffusion of oxygen through the oxide to form SiO as suggested by Tan and Gosele [19], it is reasonable to believe that thinner oxide layers act as more effective sink for interstitials. Perhaps, free surface of silicon is even more efficient in absorbing the out-diffused interstitials at 1200°C [16]. Finally, the dominant nature of interstitials observed here compels one to look for the fate of vacancies if both were present in thermal equilibrium. One possible explanation can be that equilibrium concentration of vacancies is much smaller than that for interstitials. This implies vacancies have higher activation energy for diffusion. Assuming similar formation energy for both, one concludes a rather high value for migration energy for vacancies. Validity of these ideas is illustrated in Fig. 4 which shows that for the migration of a vacancy at least two bonds [20] have to be broken which certainly requires quite high energy (3.66 eV [1] or 11.8 eV [5]). Thus, once formed, vacancies are fairly immobile. Similarly, condensation of vacancies in a row can be a very high energy configuration. On the other hand, split interstitials can be energistically favorable [21,22] and condense into a plane to reduce strain energy associated with them.

SUMMARY AND CONCLUSION

We have clearly established that out-diffusion of self-interstitials in silicon takes place at high temperature. During dry HCl oxidation at 1200°C for two hours, a defect-free-zone of 30µm can be achieved. This diffusion of

90

interstitials is a bulk phenomenon and is not confined to the silicon/silicon dioxide interface only. In bulk, rearrangements of interstitials into small defects take place. These observations have been explained in terms of the dominant role that interstitials play in determining diffusion mechanisms in silicon. No attempts have been made, however, to identify interstitials as solely responsible for self-diffusion in silicon. A new model based on coexistence of interstitials and vacancies has been proposed where sluggishness of vacancy migration is attributed to its high migration energy.

ACKNOWLEDGMENTS

The author wishes to acknowledge his sincere thanks to B. O. Kolbesen (Siemens AG, Munich) for supplying FZ silicon wafers, used in the present investigation. He also thanks Drs. S. M. Hu and W. A. Westdorp for valuable discussions. The experimental assistance of C. Hoogendoorn and J. T. Versusky is also gratefully acknowledged.

REFERENCES

1. R. A. Swalin, J. Phys. Chem. Solids 18 290 (1961).

2. P. M. Petroff, A. J. R. DeKock, J. Crys. Growth 30 117 (1975).

3. H. Foll, B. O. Kolbesen, Appl. Phy. 8 319 (1975).

4. A. Seeger, M. L. Swanson, In Lattice Defects in Semiconductors, R. R. Hasiguti, Editor, University of Tokyo Press, Tokyo, 1968, p. 93.

5. G. Das, In High Voltage Electron Microscopy, P. R. Swann, C. J. Humphreys, M. J. Goringe, Editors, Academic Press London and New York, 1974, p. 277.

6. A. Seeger, H. Foll, W. Frank, Inst. Phys. Conf. Ser. No. 31, p. 12 (1977).

7. G. R. Booker, R. Stickler, Phil. Mag. 11 1303 (1965).

8. C. M. Hsieh and D. M. Maher, J. Appl. Phys. 44 1302 (1973).

9. S. M. Hu, J. Appl. Phys. 45 1567 (1974).

10. D. M. Maher, A. Staudinger, J. R. Patel, J. Appl. Phys. 47, 3813 (1976).

11. C. J. Varker, K. V. Ravi, J. Appl. Phys. 45 272 (1974).

12. S. M. Hu, J. Appl Phys. 4 165 (1975).

13. H. Shiraki, Jap. J. Appl. Phys. 15 83 (1976).

14. S. M. Hu, Appl. Phys. Lett. 36 561 (1980).

15. T. Abe, K. Kikuchi, S. Shirai, Semiconductor Silicon, 1977, H.R. Huff, E. Sirtl, Editors, The Electrochemical Society, Princeton, NJ, p. 95.

16. F. Secco D'Aragona, Phys. Stat. Sol. (a) 7 557 (1971).

17. S. M. Hu, J. Vac. Sci. Technol. $\underline{14}$ 17 (1977).

18. H. J. Mayer, H. Mehrer, K. Maier, Inst. Phys. Conf. Ser. No. 31, 1977, p. 186.

19. T. Y. Tan, U. Gosele, Appl. Phy. Lett. $\underline{40}$ 616, (1982).

20. J. A. Van Vechten, Phys. Rev. B4, 1482, (1974).

21. G. D. Watkins, R. P. Messmer, C. Weigel, D. Peak, J. W. Corbett, Phys. Rev. Lett. $\underline{27}$ 1573, (1971).

22. S. P. Singhal, Phys. Rev. $\underline{B5}$ 4203, (1972).

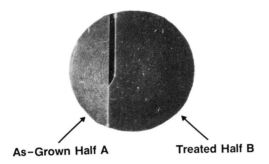

As–Grown Half A   **Treated Half B**

Fig. 1.   Elimination of swirl defects after 1200 °C Dry HCl
            oxidation for two hours.

            As-grown half of the wafer contains characteristic
            swirl pattern while swirl defects are eliminated
            from the surfaces of the treated half.

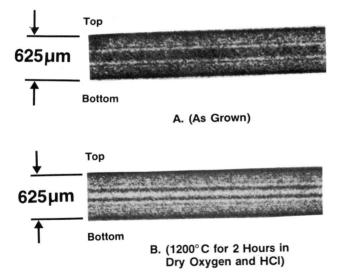

Top

625µm

Bottom

A. (As Grown)

Top

625µm

Bottom

**B. (1200°C for 2 Hours in
Dry Oxygen and HCl)**

Fig. 2.   X-Ray section-topographs of swirl-rich float zone
            silicon; (A) as-grown and (B) after HCl oxidation.

            Note Pendellosung fringes in both micrographs
            indicative of lack of strain in swirl defects. See
            text for details.

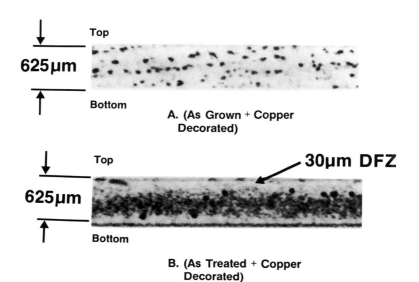

↓
625μm
↑

Top

Bottom

A. (As Grown + Copper Decorated)

↓
625μm
↑

Top

30μm DFZ

Bottom

B. (As Treated + Copper Decorated)

Fig. 3. X-Ray section-topographs of copper-decorated (A) as-grown half (B) HCl-oxidized half.

Note swirl defect free zone (DFZ) of 30 μm at both surfaces of the treated half of the wafer due to out-diffusion of interstitials. A-swirls are visible in (A) while "small defects" and A-swirls are seen in (B). See text for details.

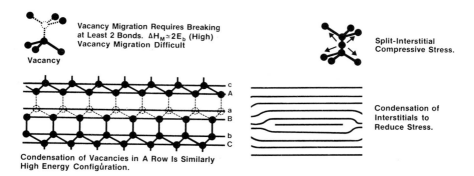

Vacancy

Vacancy Migration Requires Breaking at Least 2 Bonds. $\Delta H_M \simeq 2E_b$ (High) Vacancy Migration Difficult

Split-Interstitial Compressive Stress.

Condensation of Vacancies in A Row Is Similarly High Energy Configuration.

Condensation of Interstitials to Reduce Stress.

Fig. 4. Condensation of interstitials lowers defect energy while vacancies agglomerate into a high energy state.

METASTABLE DEFECT CONFIGURATIONS IN SEMICONDUCTORS

J. L. BENTON AND M. LEVINSON
Bell Laboratories, Murray Hill, New Jersey 07974

ABSTRACT

Deep Level Transient Spectroscopy is used to examine the stability of defect configurations in Si and InP. A systematic approach has been developed to study alternate structures of metastable defects through their representative electronic states. Regulation of defect charge state prior to analysis reveals dramatic transformations in the resulting spectra. These defect states are metastable and can be controlled with thermal or electronic energy. Model studies in electron irradiated InP and Si are presented. The barriers to configurational change are determined from the reaction kinetics. The roles of electric field, minority carrier injection and charge state are explored through junction bias techniques. It is believed that the observed behaviors represent a new class of defect reactions. The systematics of detecting and studying these reactions are presented. Model defect structures are discussed.

INTRODUCTION

Deep Level Transient Spectroscopy,[1] which has been used extensively to identify and characterize electrically active point defects in semiconductors,[2] is extended to examine the stability of defect configurations in Si and InP. A systematic method to observe alternate structures of a metastable defect and to determine the reaction kinetics for the configurational changes is presented. Recognition of this new class of defect reactions is important both in interpreting DLTS results in general as well as in specifically investigating and controlling the defect phenomenon.

In this work we report the observation of two metastable defect phenomena, one in Si and one in InP. The configurational changes are charge state controlled. These configurational changes in defect structures are believed to be a frequent occurrence in semiconductors. The properties of the two systems and the systematics of observation are presented.

EXPERIMENTAL PROCEDURE

The capacitance transient associated with the thermal emission of carriers from a defect in the semiconductor junction depletion region, is monitored by the DLTS measurement. The energy position of the defect level in the forbidden gap of the semiconductor is related to the temperature position of the corresponding peak in a DLTS spectrum. Configurational changes of a defect which result in altered electronic states can thus be identified by transformations in the resulting spectrum.

Figure 1 is a schematic representation of a defect which has two charge state controlled configurations. If the sample junction has an applied reverse bias, the Fermi level is "pinned" at mid gap and the defect state in the upper half of the gap of a n-type semiconductor is empty of electrons, (configuration B). However, if no bias is applied to the sample, the Fermi level lies above the defect level, which produces a filled state and configuration A, of a more negative character. In p-type material, a similar effect occurs which changes the charge state of hole traps in the lower half of the energy gap. A different DLTS spectrum is observed for each configuration. Minority carrier injection can also transform a defect to configuration B. If the defect state is set in one configuration, the sample is cooled, and then the charge state is changed; the defect will become metastable. The reaction mechanisms, potential barriers, and electronic properties are all subject to study.

The systematics of the procedure are simple. For example, the reaction from the metastable configuration A to configuration B is observed as follows:

1. Cool the sample at zero bias to a temperature T.

2. Apply a reverse bias and anneal for a time t.

3. Fast cool the sample ($>1°$ K sec^{-1}) to low temperature ($\sim 50°$ K).

4. Record DLTS or TSCAP spectrum.

A series of isochronal anneals reveals the transformation temperatures and the defect states involved. Isothermal anneals quantitatively probe the reaction kinetics. A similar procedure is used to examine the reverse (B $\rightarrow$ A) transformation. Configuration B is set under a reverse bias and anneals are performed at zero bias.

Metastable defect states can be observed using this capacitance transient technique if the carriers are emitted from the defect at a temperature lower than the transformation temperature. If the carrier emission occurs at a temperature equal to or greater than that of the transformation the phenomenon may proceed undetected. It is believed that similar reactions were observed in the work of Jellison[4] and Sibille and Mircea.[3] The experimental procedure outlined above, however, permits proper analysis of the reaction kinetics.[5]

Application of this method to metastable defects in InP and Si systems is explained in the following sections.

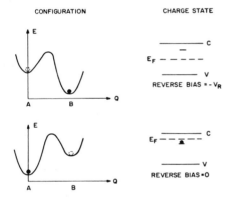

Figure 1. Schematic drawing of a defect having two charge state controlled configurations. If the reverse bias is changed at low temperature, the defect becomes metastable.

MODEL STUDIES

Indium Phosphide

Mesa diodes consisting of liquid phase epitaxial p$^+$InP layers grown on nominally undoped (n = 5×10^{15} cm^{-3}), liquid encapsulated, crucible grown n-InP substrates were irradiated with 1 MeV electrons to a dose of 10^{16} e$^-$/cm^2. The samples were subsequently annealed at 200°C for 1 hr. DLTS spectra were taken after cooling the sample with the bias off (configuration A) and after cooling with 5 volts reverse bias (configuration B) as shown in Figure 2. Because the DLTS measurement is made with an applied reverse bias, the spectrum corresponding to configuration B is normally observed. The corresponding defects labeled E1, E3, and E7 with respective electron emission activation energies of 0.09 eV,

0.14 eV and 0.37 eV were previously characterized.[6] The carrier emission activation energy of the defect state, E7 in configuration A is 0.15 eV.

The reaction A → B was examined by a series of isochronal anneals, (Fig. 3). For each point, the sample was cooled from above 200° K with the bias off and annealed with the bias on for 1 min at the indicated temperature. Changes in peak heights which are proportional to changes in defect state concentrations were monitored. The relation $\Delta EA = -(\Delta E1 + \Delta E3) \approx -\Delta E7$ was found. Since the temperature of peak from E7 is near that of the transformation A → B, kinetic studies were limited to E1, E3, and EA. The data in Figure 3 show that the transition A → B occurs in two stages at ∼110° K and ∼160° K. Conversely, the B → A transition occurs in one stage at ∼140° K.

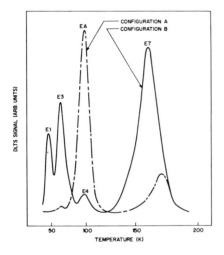

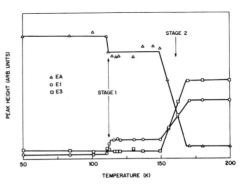

Figure 2. DLTS spectra ($\tau = 6\ ms$) of 1 MeV e^- irradiated n-InP. Configuration A was observed after cooling with the bias off, configuration B was observed after cooling with an applied reverse bias.

Figure 3. Isochronal annealing data for the transformation A → B in n-InP. For each point, the sample was cooled at zero bias from above 200°K, annealed for 1 min under reverse bias, and peak heights of E1, E3, and EA monitored.

Following the same annealing procedure, isothermal anneals furnished data to determine the time constant for transformation at each temperature. The kinetics for all reactions were then calculated from Arrhenius plots.

$$A \rightarrow B \text{ stage 1: } R = 10^{18}\exp[-.40\ eV/kT]$$
$$\text{stage 2: } R = 10^{11}\exp[-.42\ eV/kT]$$
$$B \rightarrow A \qquad R = 10^{7}\exp[-.24\ eV/kT]$$

The electrical and structural properties of the metastable defects and the transformation mechanism were probed with auxiliary tests. Forward bias minority carrier injection promoted the reaction A → B at temperatures much lower than the above reactions. Since no electric field is present in this experiment, the configurational change is directly related to the charge state change. Additionally, electron emission rates increased with increasing junction electric field at a rate consistent with a Poole-Frenkel mechanism.[7] Thus, states E1, E3 and EA were identified as singly charged donors. Finally, thermally stimulated capacitance measurements (TSCAP) were recorded with the defect in each configuration. The results show that there is more positive charge in configuration B than in configuration A by an amount equal to the concentration of [EA]. A complete analysis of these results suggests a model for the metastable configurational interaction which involves a short range pairing of a defect associate.[8]

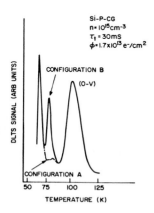

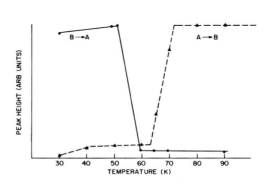

Figure 4. DLTS spectra of 1 MeV e⁻ irradiated, phosphorus doped silicon. Configuration A was observed after cooling with zero bias, configuration B was observed after cooling with reverse bias on.

Figure 5. Peak height of E(0.1 eV) versus temperature of isochronal anneals (1 min) for the transformations A → B and B → A in n-type Si.

Silicon

The silicon material used for these experiments was Czochralski-grown, phosphorus doped to a carrier concentration of 10^{15} cm^{-3}. The junction region was defined by evaporated Au-Pd Schottky barrier contacts. The samples received a 1 MeV electron dose of 1.7×10^{13} e⁻/cm^2.

The resulting DLTS spectra are shown in Figure 4. The peak labeled (O-V) is the previously identified A-center or oxygen-vacancy complex with an activation energy for electron emission of 0.18 eV.[9] The measured activation energy for the (O-V) defect in these experiments is .20 eV which may indicate the presence of another overlapping peak.

The defect E(0.1 eV), is shown in each of its two metastable states. The sample, cooled under bias to carrier freezeout and then scanned by DLTS, gives configuration B. The sample, shorted during cool down, produces configuration A. This peak is observed only in silicon of high carbon concentration.

The defect configuration was regulated before cooling, and then isochronal anneals were performed. The results (Figure 5) reveal that the transformation B → A occurs at a lower temperature than the transition A → B. Isothermal anneals were performed by following the experimental procedure outlined above. The results are shown on the Arrhenius plot Figure 6. The kinetics for the metastable defect transitions are

$$A \rightarrow B \quad R = 10^6 \exp[-.11 \ eV/kT]$$

$$B \rightarrow A \quad R = 10^{16} \exp[-.20 \ eV/kT]$$

Formulation of a configurational model based on these results is in progress.

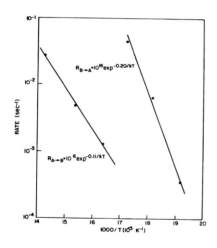

Figure 6. The reaction kinetics for E(0.1 eV) defect transformations A → B and B → A in silicon.

CONCLUSIONS

A systematic method to identify and analyze changes in defect configurations in semiconductors was described. It was developed to isolate the phenomenon responsible for irreproducible spectral changes observed in routine DLTS scans.

The metastable defect state E(0.1 eV) present in Si is the same phenomenon reported by Jellison as confirmed by the defect spectral peak position. However, two differences in results should be noted. First, no complementary behavior of the oxygen-vacancy, (O-V), peak was found in this work because this peak appears at a higher temperature than either transformation. But an increase in the height of (O-V) peak is observed with increasing forward bias pulse width similar to the effect reported by Jellison. This effect is not consistent with the electron capture cross section of 10^{-14} cm^2 for (O-V), but may relate to a signal overlapping that of (O-V). Second, expressions were presented by Jellison to describe the variation of E(0.1 eV) peak height with pulse width and pulse interval. These expressions do not agree with the kinetic equations above. The differences require study, as no simple explanation for the discrepancies is apparent.

Changing the junction bias condition during cooldown is a simple test for defect metastability. Since carrier emission is the process probed by DLTS, it is necessary to apply the described techniques to observe charge state controlled defect structures. Extension of the method to other semiconductor systems should uncover defect states hidden during standard DLTS analysis. These experiments point out dramatically the interaction between defect configurations and their electronic properties.

ACKNOWLEDGMENTS

The assistance of H. Temkin in providing the InP samples is appreciated. We wish to thank L. C. Kimerling for his incisive suggestions and probing discussions.

References

(1) D. V. Lang, J. App. Phys. *45*, 3023 (1974).

(2) G. L. Miller, D. V. Lang, and L. C. Kimerling, Ann. Rev. Mater. Sci., *7*, 377 (1977).

(3) A. Sibille and A. Mircea, Phys. Rev. Lett., *47*, 142 (1981).

(4) G. E. Jellison, J. Appl. Phys. *53*, 5715 (1982).

(5) Pulsing methods introduce an interplay between capture and emission processes. One process may limit the other, while both "annealing" treatments modify the measurement conditions.

(6) M. Levinson, J. L. Benton, H. Temkin and L. C. Kimerling, Appl. Phys. Lett., *40*, 990 (1982).

(7) L. C.Kimerling and J. L. Benton, Appl. Phys. Lett., *39*, 410 (1981).

(8) M. Levinson, J. L. Benton, and L. C. Kimerling, to be published.

(9) L. C. Kimerling, "Radiation Effects in Semiconductors 1976" (Inst. Phys. Conf. Ser. 310 p. 221.

FORMATION OF INTERSTITIAL DEFECTS IN HIGH CONCENTRATION SHALLOW PHOSPHOROUS DIFFUSIONS IN Si.

RALPH JACCODINE
Sherman Fairchild Laboratory, Lehigh University, Bethlehem, PA 18015

ABSTRACT

An experimental study of high concentration (10^{19}-10^{21}/cm^3) shallow diffusion was undertaken using TEM to investigate the nature of these diffusion-related defects. Phosphorous from a wide variety of sources (PBr$_3$, POCl$_3$, etc.) other than ion implantation was used in temperature range from 950°- 1100°C and for times of 15 minutes to 1 hour. Care was taken in surface preparation and material selection to avoid extraneous defects from sources other than the diffusions. A high concentration of small dislocation loops (10^{12}/ cm^2) was present in the top few microns of the wafers. Diffraction contrast on the loops revealed that they are of edge character with Burger's vector b = ½<110>. Identification of the nature of these loops by tilting method [(g·b)·s] and anomalous dark-faced black-white lobes show they are of the interstitial type, i.e. the observed defects place the surrounding matrix in a compressive state.

INTRODUCTION

The basic mechanism for diffusions in Si, both for silicon itself and impurities, has been extensively investigated, and a clear consensus has not yet developed on whether vacancies or interstitials are the dominant defects.

Hettrick [1] et al. agree with the early work of Seeger and Chik [2] that the high temperature processes are dominated by silicon self-interstitials and the low temperature processes by acceptor-like vacancies. Evidence of the role of interstitials in Si comes from work on swirl defects in Si [3]. Here a positive identification of swirl defects was made and loops were shown to result from silicon self-interstitial condensation. These workers also propose in their papers reactions between negatively-charged interstitials (I⁻) and the n-type doping (Sb⁺) forming (Sb⁺I⁻) complex to account for their observations of the influence of doping on these defects.

On the other hand, from diffusion studies Schwettman and Kendall [4] suggest that the n-type impurity pairs with a vacancy (E center) move together as a fast-diffusing specie and thus account for the "kink" in phosphorous profile.

Kandall and Carpio [5] suggest a model in which the Si surface is the source of excess vacancies which in the form of neutral and negatively charged E centers P⁺V⁻ and P⁺V⁼ pairs) are maintained throughout the depth of the diffusion zone. This process has been elaborated by the Fair-Tsai model [6]. In this model, it is proposed that the majority atom (phosphorous) enters the surface and some fractions become paired with a double charged vacancy V⁼ according to the reaction P⁺+V⁼ ⇌ P⁺V⁼. This reaction is assumed to occur throughout the highly-doped surface region. These pairs then diffuse down the doping gradient and do not dissociate until the electron concentration drops below concentration at the kink.

From this brief review, one can see that phosphorous diffusion is postulated to occur by either interstitials or vacancies and that defect phosphorous pairs

Mat. Res. Soc. Symp. Proc. Vol. 14 (1983) ©Elsevier Science Publishing Co., Inc.

are also expected in the diffusion zone. It is reasonable to expect that, for high concentrations of n-doping, high concentrations of these point defects should manifest themselves in the defects formed in the region of a shallow high concentration diffusion. This motivated a study of high concentration ($C_s \geq 10^{20}$/cm^3) shallow ($\sim 2\mu$) phosphorous diffusions at 900-1050°C in which specimens were rapidly cooled in order to study defects that formed in these shallow layers. It was the purpose to infer from the resultant defects which of these several mechanisms was operative.

EXPERIMENTAL DETAILS

Specimens were prepared from a wide variety of materials, dislocation count, resistivity of {111} orientation from float zone (Fz) and Czochralski (Cz) material prepared in wafer form by both mechanical and chemical polishing techniques. Other specimens in this study made use of Si ribbon crystals with natural crystal faces that required no additional preparation prior to diffusion. Care was exerted in preparation and handling to eliminate outside sources of defect introduction.

The phosphorous diffusions were accomplished by various chemical vapor source processes (P_2O_5, PBr_3, $POCl_3$ and PH_3). Ion implantation processes were purposely not used in the experiment. The diffusions were carried out at temperatures from 950°C to 1100°C in various nitrogen or argon ambient which sometimes included O_2 for times ranging from 15 minutes up to one hour. Some selected samples were given subsequent lower temperature anneals in the 600-700°C region for various times extending to 100 hours to enhance their size. All specimens were carefully stripped of the phosphorous glass prior to being cleaned and chemically thinned for TEM investigation. Specimens so thinned were examined in a Phillips TEM under light and dark field conditions and detailed defect identifications using several contrast techniques were carried out and cross-checked.

RESULTS

Figure 1 shows the defects that originate in a thin topmost region of a Cz wafer that has been phosphorous diffused to 10^{20}/cm^3 at 950°C for 30 min. using a PBr_3 source. A high concentration of small spot defects associated with small precipitates and/or small dislocation loops can be observed. Contrast of these defects could be influenced by specimen tilting indicating they were small dislocation loops with small precipitates present.

Similar defects could be observed even when material was varied from Cz, Fz to ribbon, both n and p type, and for the various sources used in this study. When diffusions were extended in time or in temperature so as to extend to greater depths (several microns) then these loops disappeared in favor of dense dislocation networks. As previously noted, defects could be more conveniently studied after low temperature (600-700°C) anneals which were carried out in He$_2$ or N$_2$ atmospheres.

The results of these treatments shown in Fig. 2,3 show an enhanced defect size with the double-lobed contrast expected of dislocation loops. From the black-white vector direction change with respect to change in operating diffraction vector $\vec{g}$, it appeared that these loops had a spherically symmetric strain field. In addition, contrast effects from loops near the surface of the foil, observed in both light and dark field, were used in loop identification. From Ashby and Brown's study [7] it is known that "interstitial" defects taken on a positive print of a dark field have an "extra" wide image with its dark side toward positive $\vec{g}$. Wilkens [8] also discusses contrast techniques to be used in determining the nature of small (compared to "ξ_g" defects.

These techniques indicated that the loops in this study were of an interstitial nature.

Contrast experiments on the larger loops formed after the low temperature anneal allowed more detailed study of the loops' properties. The Burger's vector, $\vec{b}$, of the loops was determined using the usual extinction criterion, i.e. $\vec{g}\cdot\vec{b} = 0$; although this is usually applied to straight dislocations, it can be used for loops where $\vec{g}\cdot\vec{b}\times\vec{n} \neq 0$ [9].

The Burger's vector was determined to be $\frac{a}{2}<110>$ lying on the (111) foil plane (i.e. the plane of the diffusion).

Knowing the Burger's vector, $\vec{b}$, and the normal of the loop plane $\vec{n}$, the dislocation strain field, i.e. whether interstitial type or vacancy type, was determined [10]. This method uses the fact that for a crystal oriented off from the exact Bragg condition, the dislocation image lies inside or outside the trace loop position depending on the type of loop, the operating diffraction vector, and the sign of the deviation parameter. See example Fig. 2, reference 10. This is the basis of the $(g\cdot b)\cdot s \gtrless 0$ technique. From application of this criterion, the loops were identified as being of "interstitial" character, i.e. a defect whose strain causes matrix atoms to be displaced away from its center. This is more often pictured as an extra plane of atoms inserted into the matrix.

DISCUSSION

The result of this study eliminates the presence of a high concentration of vacancy-like defects, existing after the insertion into the Si lattice of a high concentration ($\sim 10^{20} cm^3$) of phosphorous atoms. The loop identification fits into the idea that these loops with Burger's vector = $1/2<110>$ result in condensation of point defects forming an extra half plane {111} inserted in a manner to relieve the strain resulting from the diffusion mismatch. This configuration is probably a forerunner to the formation of dislocation networks, and observations have been made of a loose network of dislocations with loops in adjacent regions. It should be noted that the loops would be perfect prismatic loops if they stabilized on the {110} plane normal to diffusion surface.

In a related study, Das [11] identified several similar type dislocations ($\vec{b} = \pm\frac{a}{2}<110>$) of interstitial type resulting from arsenic diffusions which he concluded occurred in the presence of excess interstitials.

These findings fit closely into the framework proposed by Seeger and Chik [2] that diffusions (at temperatures >800°C) take place as "extended interstitials." This idea can be extended, as proposed by DeKock [3], to include an "interstitial-N dopant" complex.

A criticism of this interpretation has been proposed by Mathews and Van Vechten [12] who point out that the presence of interstitial loops can be explained by several other alternatives in addition to the obvious condensation of self-interstitials and prismatic punching of loops. Since the loops in this study were all nearly perfect single prismatic loops with no evidence of stacking fault contrast or climb processes, most of their proposed explanations could be discounted.

In silicon (diamond cubic lattice) there is room to accommodate interstitial atoms with very little resulting strain. Compared to f.c.c. lattice (74% filled) the diamond structure (34% filled) is very open and each interstitial site is large enough to hold a displaced or impurity atom (.88 R_{Si}). Therefore a mechnism that involves the supersaturation of interstitials or interstitial-dopant complexes is not hindered. H. Strunk et al.[13] have isolated dislocation helices in device structures occurring along with "emitter-push." They asserted that this was indicative of supersaturation of interstitials in the base region and an interstitialcy diffusion mechanism. In a more detailed article [14], the same authors propose that a phosphorous point defect complex may be used to explain the self-interstitial supersaturation and a mechanism related to the "Kirkendall effect."

This study is consonant with the above proposed mechanisms. The presence of interstitial loops may be interpreted in terms of supersaturation of self-interstitials or prismatic punched loops resulting from a Si-P precipitate. Since no clear evidence existed for concluding that every loop did or did not have evidence of a precipitate both mechanisms must be considered as accounting for the results.

REFERENCES

1. G. Hettick, H. Mehrer and K. Maier, Int. Conf. on Defects and Radiation Effects in Semiconductors, Nice, France (1978); H. J. Mayer, H. Mehrer and K. Maier, Radiation Effects in Semiconductors (1976) IPO Conf. Ser. #31, Inst. of Physics, Bristol (1977) p. 196.

2. A. Seeger and K. P Chik, Phys. Stat. Sol. 29, 455 (1968).

3. A. J. R. deKock and W. Van de Wijgert, J. Crystal Growth 49, 718 (1980). P. M. Petroff and A. J. R. de Kock, J. Crystal Growth 35, 4 (1976).

4. F. Schwettmann and D. Kendall, Appl. Phys. Lett. 21, 3 (1972).

5. D. L. Kendall and R. Carpio, Fall Mtg., ECS Pittsburgh (1978) RNP #396.

6. R. Fair, "Impurity Doping Processes in Silicon," Vol. 2, Material Processing Theory and Practices, F. F. Y. Wang, Chapter 7 (North Holland, New York) p. 374.

7. M. F. Ashby and L. Brown, Phil. Mag. 8, 1083 (1963).

8. M. Wilkens, Modern Diffraction and Imaging Techniques in Material Science (North Holland, New York, 1970) p. 233.

9. Howie, A., Whelan, M. J., Proc. R. Soc. A263 (1961) p. 217; Proc. R. Soc. A267 (1962) p. 206.

10. Maher, D., and Eyre, B., Phil. Mag. 23, 409 (1971).

11. Das, G., High Voltage Electron Microscopy, P. R. Swann (Academic Press, 1974) p. 277.

12. J. Mathews and J. A. Van Vechten, J. Crystal Growth 35, 343 (1976).

13. H. Strunk, U. Gösele and B. O. Kolbesen, Appl. Phys. Lett. 34(8), 530 (1979).

14. U. Gösele and H. Strunk, Appl. Phys. 20, 265 (1979).

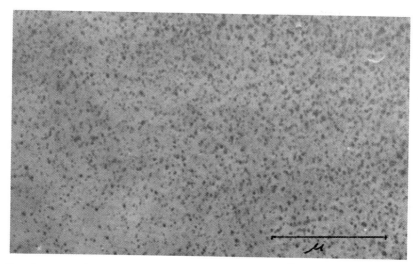

Fig. 1. Cz wafer diffused from a PBr source at 950°C for 30 min. to concentration of ~10^{20}/cm^3.

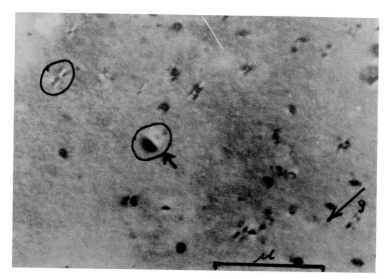

Fig. 2. Heat treated wafer (650°C, 60 hrs forming gas). Arrow points to circle with anomalous black-white lobes.

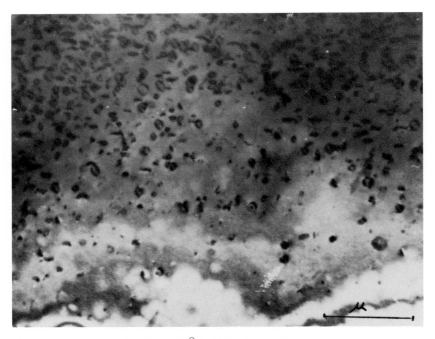

Fig. 3.   Heat treated wafer (650°C) 24 hrs).   Evidence of a precipitate
          can be seen at center of many of the loops.

II
DEFECTS IN SILICON: OXYGEN RELATED

EARLY STAGES OF OXYGEN CLUSTERING AND ITS INFLUENCE ON ELECTRICAL BEHAVIOR OF SILICON

Gottlieb S. Oehrlein
IBM Thomas J. Watson Research Center, Yorktown Heights, N.Y., USA

James W. Corbett
Institute for the Study of Defects in Solids, Physics Department, State University of New York at Albany, Albany, N.Y., USA

ABSTRACT

Our knowledge of phenomena connected to the early stages of oxygen clusters, especially their electrical activity is reviewed. In addition to the well-known 450 °C thermal donors, 'new oxygen donors', which occur in ca. 650 °C annealing, have emerged in conjunction with low temperature processing for VLSI and are discussed. The existing models of thermal donors are reviewed. In a new model of thermal donor formation, thermal donors are viewed as metastable oxygen clusters which lower the compressive strain of the surrounding silicon matrix via bonding of 2p oxygen lone pair orbitals. It is shown how this bonding can result in shallow double donor states.

INTRODUCTION

Oxygen is incorporated into the lattice of a Czochralski silicon crystal (Cz Si) during crystal growth at a concentration of typically 10^{18} at/cm³ (approximately the solubility of oxygen in silicon near the melting point) [1,2]. The oxygen originates from the quartz crucible which is partly dissolved by the molten Si [3]. The solubility of oxygen in silicon decreases exponentially with temperature [4]. This means that at room temperature or at temperatures which would be used during device manufacturing the solid solution is highly supersaturated. At temperatures higher than about 300 °C oxygen becomes mobile and reduces this supersaturation via outdiffusion or formation of SiO_x - clusters (precipitates) [1,2]. The clustering of oxygen leads to a large variety of observable phenomena which have received a great deal of attention, especially during the last years (see e.g. [5,6]).

It has been ascertained that oxygen can influence the properties of a silicon wafer in a detrimental way (such as a change in resistivity because of the formation of oxygen related donors [1], or the introduction of precipitate related microdefects within the depletion layer of a device [7]). More recently beneficial effects like the mechanical hardening effect [8] and heavy metal gettering at precipitate related dislocations (intrinsic gettering [9,10]) were discovered and extensively studied. At present it is not yet possible to predict precisely the behavior of oxygen in a given wafer during thermal processing. In particular the formation and nature of oxygen related donors, like the 'thermal donors' which are produced at temperatures

Mat. Res. Soc. Symp. Proc. Vol. 14 (1983) © Elsevier Science Publishing Co., Inc.

from 300-525 °C [11-13] or the 'new donor(s)' (observed after heat treatment from 550-900 °C) [14-16], is still not clear. In addition there is not a general consensus why the defect density after a given heat treatment at higher temperature can vary from wafer to wafer despite a constant oxygen concentration (i. e. the question of what initially nucleates oxygen precipitates) [17,18].

It is the purpose of the present paper to review critically our knowledge of phenomena connected to these early stages of oxygen clusters. In section 2 some important properties of oxygen in silicon are mentioned. We then summarize in section 3 experimental results on oxygen related donors. We also examine critically models of thermal donor formation which have been suggested. We then propose a new model of oxygen related donor formation. The paper concludes with some remarks on the nucleation of oxygen precipitates at higher heat treatment temperatures (900-1000 °C).

SOME PROPERTIES OF OXYGEN IN SILICON

This section gives a summary of properties of oxygen in silicon which are important for an understanding of the early stages of oxygen precipitates after heat treatment (i.e. configuration of isolated oxygen, oxygen content of a Si wafer, solubility, diffusivity, microdistribution). No attempt has been made to cover the whole available information in detail (for such reviews see [1,2,12,21,22]).

Isolated oxygen exists in silicon in a puckered Si_2O configuration. The oxygen is located off the <111>-axis at an interstitial site [19-21]. The oxygen content (interstitial) of a Si wafer is usually determined by measuring the strength of the oxygen associated 9 μm infrared line (for a critical review see [22]). Interstitial oxygen concentration and IR absorption are related by a calibration coefficient. A recent determination gave [23]

$$[O_I] = 2.45x10^{17} \, \alpha_{O_I} \, at/cm^2 \tag{1}$$

($[O_I]$ in at/cm^3 and α_{O_I} is the absorption coefficient of the 9 μm line).

A number of investigators found that oxygen is not only present as isolated oxygen O_I but also exists in the form of SiO_x - precipitates in as-grown Si crystals [24-28]. Therefore, it escapes detection by the above method. The effect can be quite important and Jastrzebski et al. found in recent measurements in which they subjected their samples to 1350 °C in order to redissolve grown-in oxygen precipitates that in their samples about 20% of the total oxygen concentration was precipitated while about 80% was in interstitial form [28].

There is a large number of oxygen diffusivity measurements which have been recently reviewed [12]. Even more recently Stavola and co-workers [13] extended the low temperature diffusivity measurements of Corbett et al. [19] based on the recovery of stress-induced dichroism of the 9 μm band. The dichroism data, together with Mikkelsen's [30] mass transport data and the often used Takano and Maki measurements [31] are displayed in figure 1. Excellent agreement exists between all sets of data over the temperature range 1250 °C to 330 °C [13]. A best fit to all these data is shown in the figure and gives for the temperature dependence of the oxygen diffusivity

$$D_{O_I}(T) = 0.16 \, \exp \, (- 2.53/kT) \tag{2}$$

A critical discussion of oxygen diffusion measurements based on the method of thermal conversion [32] is given in the section on oxygen-related donors.

The segregation coefficient of oxygen was determined to be 1.25 [33]. Other investigators found their results consistent with a segregation coefficient greater than 1 [34] while in more recent measurements Abe et al. found that the segregation coefficient should be smaller than one [35]. It is clear that more investigations are required on this point.

The problem of the microdistribution of oxygen, i.e. the question if oxygen is distributed homogeneously or nonhomogeneously (e.g. oxygen striations) is related to the question of the value of the segregation coefficient. There are some measurement results indicating that oxygen is indeed distributed homogeneously [17,36], which would be consistent with a distribution coefficient close to 1 [37]. These measurements were based on 450 °C thermal conversion (see section 3) and spreading resistance measurements, i.e. an indirect method. The majority of investigators found that oxygen was distributed nonhomogeneously [34,38-47]. Various methods were employed in these studies and most of them were ultimately based on thermal conversion [34,38-42]. However, more recently Ohsawa et al. and Rava et al. [43-47] measured the oxygen microdistribution directly by scanning a collimated beam of 9 μm laser light (~ 200 μm diamter) across a wafer cut either perpendicular or parallel to the growth axis. They found that $[O_I]$ varies locally and that the fluctuations corresponded to oxygen striations in the silicon crystal [43-45].

We conclude this section by pointing out that as regards to oxygen diffusivity and solubility there is good agreement between different sets of experimental data, provided the data were obtained by direct methods. However, two assumptions which were often used when analyzing experiments on the clustering of oxygen, namely (1) that oxygen is distributed randomly and (2) that oxygen is initially present only interstitially contradict the experimental evidence. We also note that Stavola and co-workers [13] have observed an anomalously fast diffusion of oxygen in "as-received" samples (i. e. "as-grown" + 2 hours 900 °C [13]). Whether the non-randomness of the oxygen, or strain interaction between oxygens can account for this anomaly remains to be seen.

OXYGEN RELATED DONORS IN SILICON

Thermal Donors

The thermal donor problem in Si is one of the most studied of all defect problems in semiconductors [1,2,11-14,16,32,36,39-41,46-88]. Despite the large number of investigations, no real progress has been made in terms of a defect model which could explain all experimental observations consistently. Because of the number of studies, some of which are contradictory to each other, it has become difficult to know what exactly a defect model of the thermal donor should explain. We shall initially summarize experimental results which seem to be significant for an understanding of the thermal donor and which have to be accounted for by any defect model.

Kinetic information. (1) The initial rate of donor formation is proportional to the fourth power of $[O_I]$ [11,75].

(2) The maximum donor concentration is approximately proportional to the third power of $[O_I]$ [11,72].

(3) The logarithm of the difference - maximum donor concentration minus the donor concentration at a given time - is linear versus time prior to the appearance of the maximum [11].

(4) Donors formed by heat treatment below 550 °C can be destroyed by heating above 550 °C [11]. Recently however, a slower reduction of thermal donors at 550 °C has been observed [68,82]. In one study [82] three different kinds of thermal donors were distinguished by their thermal stability.

(5) The activation energy for the forward reaction is equal to the the activation energy of the reverse reaction and about 2.5-2.8 eV [11,15,51,72].

Reported energy levels. (1) Hall measurements give two energy levels at 0.13 eV and 0.06 eV below the conduction band edge [51,70]. These shift with longer heat treatment time to lower energy [70].

(2) Deep level transient spectroscopy (DLTS) measurements give energy levels at 0.15 eV and 0.07 eV which shift to lower energy after prolonged heat treatment [13,74]. Both defect states exhibit strong electric field enhanced emission in accordance with the Poole-Frenkel mechanism, which verifies their donor character [13,74].

(3) IR measurements show many electronic transitions between 300-550 cm^{-1} and 700-1200 cm^{-1} [52,62,64,71,87,88]. They are explained as ground state - excited state transitions of double donors [71,87,88]. In a recent study [87] nine distinct double donors were found having ionization energies in the range 69.1-52.9 meV and 156.3-118 meV. With increasing annealing time new donors with lower ionization energies are formed at the expense of preceding species with higher ionization energies [71,87].

(4) In some investigations an oxygen-related acceptor level was reported at about 0.35 eV above the valence band [58,88].

Influence of dopants. (1) High acceptor concentrations (B, Al and Ga) increase the rate of donor formation and the maximum donor concentration [53,54,85]. The donors formed in Al doped Si show a very high stability upon heating above 500 °C. While the donor activity disappears slowly after heat treatment in the range 700-900 °C, the Al acceptors are only recovered upon heating higher than 1000 °C [53,85].

(2) The presence of large concentrations of n-type dopants does not affect the maximum concentration of donors attained after heat treatment [62,85].

(3) Carbon concentrations in excess of 1.0×10^{17} at/cm^3 suppress donor formation [62,65,16,75]. However, a small number of donors is formed even in carbon rich material. The donor density increases monotonically even for very long times (in excess of 800 h)[65].

Resistivity stabilization. (1) Heating Si crystals for times of 5 minutes to several hours at a temperature of about 1000 °C reduces resistivity changes on subsequent heating at 450 °C [51,65]. However, in [65] no clear pattern in 450 °C donor generation following heat treatment for various times at 1050 °C was observed. In a recent study [13] it was found that the initial rate of thermal donor formation at 450 °C was about 60 times faster in "as-received" samples (i.e. "as-grown" + 2 hours 900 °C) as compared to samples where the oxygen had been dispersed by a 1350 °C anneal.

(2) Thermal cycles 450 °C - 560 °C reduce the rate of donor formation on subsequent heating at 450 °C [51].

(3) Stabilization reduces the rate of subsequent donor formation, but not the number of donors that can ultimately be formed [51,65].

(4) A 1300-1400 °C anneal erases the effects of stabilization [51].

Some miscellaneous results. (1) 450 °C heat treatments increase the minority carrier lifetime of Cz Si. The increase in lifetime and the number of donors formed are related [39,64]. This indicates that heavy metal impurities might be involved in the thermal donor (however, see the discussion on oxygen related trapping levels [11]).

(2) In EPR measurements about 10 different centers were observed after 450 °C heat treatment (depending on heat treatment time) [69]. Some are of <110> axial symmetry [69] (however, [39] observed <111> symmetry). It was not possible to establish the presence of

oxygen in any of these centers by O^{17} doping and observation of hyperfine interactions in the EPR spectra [67].

(3) In spreading resistance experiments it was found that the donor generation during 450 °C heat treatment is not spatially homogeneous [17,36,40,41,46,47]. In one investigation it was found that in regions of high swirl density [89] the concentration of introduced donors was reduced [36].

(4) It was observed that on a microscale (~ 30 μm spatial resolution) $[O_I]$ maxima do not necessarily coincide with thermal donor density maxima [46,47].

(5) Two investigators [61,64] found that for short annealing times the strength of an electronic IR band at about 9.6 μm is proportional to the concentration of introduced donors.

(6) Fuller and Logan [51] reported that the donor concentration produced in nonrotated silicon can be 2 orders of magnitude smaller than in rotated crystals, although the interstitial oxygen concentration differs only by a factor of two for the two kinds of crystals (see however [11]).

(7) Lattice defects introduced by 1 MeV electron-irradiation did not change donor formation in subsequent 400 °C heat treatments [13]. However, in [59] an enhanced formation rate of thermal donors was observed for a specimen which was heated (300 °C) and gamma-ray irradiated at the same time.

(8) Thermal donors give rise to a number of photoluminescence spectra [76-79].

New Oxygen Donors

Kanamori et al. [14] found in 1979 that new oxygen related donors can be created in the temperature range 550-800 °C after prolonged heating. This is a temperature range which is of great interest for VLSI low temperature processing and a large number of studies have already been devoted to this new oxygen donor problem [15,16,72,75,76,79,82,90-92]. We shall summarize the salient facts here:

(1) New donor generation is enhanced by low temperature (450-550 °C) preannealing. Thermal donor generation is not necessary for the enhancement [14,92,76].

(2) A high substitutional carbon concentration promotes new donor generation [14,76,90,91]. In some studies a relationship between density of new donors and reduction in substitutional carbon concentration $[C_s]$ was observed [16,91] However, no critical $[C_s]$ for new donor generation was found [91].

(3) High temperature preannealing (~800-1050 °C) adversely affects subsequent new donor formation at around 650 °C [14,16].

(4) The new donor can be annihilated by annealing at over 1000 °C [14].

(5) An anneal at 800 °C does not in itself produce donors. However, after preannealing at 470 °C, new donors are formed in subsequent 800 °C heat treatments [14].

(6) New donor formation is more pronounced in p-type wafers than in n-type. The maximum new donor concentration is about 10^{16} donors/cm^3 [14].

(7) Ionization energies in the range 30-120 meV were measured for the new donor [15]. In contrast to the thermal donor the ionization energy increases with length of heat treatment (at 650 °C) [15]. For higher temperatures (700-800 °C) the ionization energy goes through a maximum between 100-120 meV and decreases again for larger times [15].

(8) The activation energies for the different processes of new donor formation were found to depend on the temperature of the heat treatment [15,72]:

T=450-525 °C : E ~ 2.5 eV
T=525-600 °C : E ~ 0.7 eV
T=650-900 °C : E ~ 1.9 eV

In another study it was concluded that the reaction which creates the new donor is the oxygen diffusion limited growth of a precipitate [91].

Models of Thermal Donor Formation

The Kaiser, Frisch and Reiss (KFR) model. The kinetic information provided in the section on thermal donors was combined by Kaiser et al. in 1958 in a kinetic model [11]. They considered the formation of Si_mO_n - complexes (n,m integers) from initially dispersed oxygen. The clustering of diffusing oxygen interstitials O_I is represented by:

$$A_1 + O_I \overset{k_2}{\rightarrow} A_2 \tag{3}$$

$$A_2 + O_I \overset{k_3}{\rightarrow} A_3 \tag{4}$$

$$A_3 + O_I \overset{k_4}{\rightarrow} A_4 \tag{5}$$

$$A_4 + O_I \overset{k_p}{\rightarrow} P_5 \tag{6}$$

$$P_5 + O_I \overset{k_p}{\rightarrow} P_6 \tag{7}$$

The A_n and P_n are complexes containing n oxygen atoms. It is assumed that A_1 and the P_n are electrically inactive, while A_2, A_3 and A_4 are electrically active. In particular A_4 acts as a donor. The reaction constants k_i are diffusion limited. Their relative magnitude is such, that A_4 would have a high concentration as compared to A_2 and A_3. Therefore most of the observed electrical activity would be due to A_4. Under these conditions the model gives [11]:

$$d[A_4]/dt_{Initial} \; \alpha \; [O_I]^4 \tag{8}$$

$$[A_4]_{Max} \; \alpha \; [O_I]^3 \tag{9}$$

$$\ln ([A_4]_{Max} - [A_4(t)]) = ln([A_4]_{Max}) - kt \tag{10}$$

in agreement with the experimental observations. Kaiser et al. did not discuss the physical origin of the donor activity of their A_4 complex.

The Helmreich and Sirtl model. Helmreich and Sirtl [66] argued that the diffusivity of interstitial oxygen is too low to allow for the formation of Si_mO_n - complexes at around 450 °C. They therefore discarded the SiO_4 KFR model and proposed instead a donor consisting of a substitutional oxygen-vacancy complex. The O_sV - complex would be the observed donor and would be formed due to the interaction of thermal vacancies with interstitial oxygen during heat treatment.

The Gosele and Tan model. This model [12] is based on two arguments: (1) The effective diffusivity of oxygen around 450 °C is 4 orders of magnitude higher than extrapolated high-temperature oxygen diffusion data [32]. (2) The KFR model is internally inconsistent.

Gosele and Tan invoke then a gas-like oxygen molecule O_2, which would have no bonds to the Si lattice. A dynamic equilibrium between interstitial oxygen and molecular oxygen would exist according to

$$O_I + O_I \underset{\leftarrow}{\overset{\rightarrow}{}} O_2 \qquad (11)$$

The diffusivity of O_2 would be about 9 orders of magnitude higher than the diffusivity of $O_I(at\ 450\ °C)$. A molecular complex O_4 would be formed according to

$$O_2 + O_2 \underset{\leftarrow}{\overset{\rightarrow}{}} O_4 \qquad (13)$$

and O_4 would be a donor. Diffusion of additional O_I to O_4 would transform O_4 into an electrically inactive complex.

Comments on Thermal Donor Models

Prior to commenting on the KFR model, we would like to make a few remarks as to the validity of the other two models.

Helmreich and Sirtl model. The Helmreich and Sirtl model can account for the electrical activity of the oxygen-complex formed, but not only does it not explain any other experimentally observed features (including the kinetics of the donor formation), it is contradictory to some experimental information [2,12,71,93,94]. The model is based on a critical VO-complex, which Helmreich and Sirtl consider a substitutional oxygen. A VO-center is electrically active, but it acts as an acceptor, not as a donor [95,96]. Furthermore, the VO-center gives rise to a well known EPR-spectrum [95] and has a vibrational mode at about 835 cm^{-1} in the infrared [96]. Neither the EPR-spectrum [93] nor the IR absorption line [94] have been observed in 450 °C heat treated Cz Si. The multivacancy-oxygen complexes which were proposed by Helmreich and Sirtl introduce electronic levels into the Si bandgap [102]. They are known from EPR studies of electron-irradiated silicon [102]. However, they are different from the ones observed in 450 °C heat treated Cz silicon, since they are much deeper in the gap [102]. The Helmreich and Sirtl model clearly does not offer a satisfactory explanation of the donor phenomenon.

Gosele and Tan model. The first assumption of the Gosele and Tan model is that the 'effective diffusivity' of oxygen at temperatures around 450 °C is much higher (~ 4 orders of magnitude as was observed in [32]) than the extrapolation of high-temperature diffusion data. It is clear from figure 1 that diffusivity measurements of O_I around 450 °C coincide with the extrapolation of measurements at higher temperatures which raises some doubt as to the validity of the assumption. Nevertheless the 'effective diffusivity' of O_I could be higher, e.g. based an a dynamic equilibrium of O_I with fast diffusing O_2 as suggested.

The concept of a fast diffusing O_2 molecule which would control interstitial oxygen diffusion at temperatures below 700 °C leads to contradictions with experience. The effective oxygen diffusivity in terms of interstitial oxygen and O_2 diffusion is given by [12]:

$$D_{eff} = (D_{O_I}[O_I] + 2D_{O_2}[O_2])/([O_I] + 2[O_2]) \qquad (14)$$

Since at all temperatures $[O_I] >> [O_2]$ and using $k_{diss}[O_2] = 8\pi D_{O_I} R_{O_I}[O_I]^2$ we can write D_{eff} as:

$$D_{eff} = D_{O_I}(1 + 16\pi R_{O_I}[O_I]D_{O_2}/k_{diss}) \qquad (15)$$

At 450 °C molecular oxygen would control oxygen diffusion [12] which requires $16\pi R_{O_I}[O_I]D_{O_2}/k_{diss} >> 1$. At temperatures higher than 700 °C the contribution of O_2 to oxygen diffusion is negligible [12] and therefore $16\pi R_{O_I}[O_I]D_{O_2}/k_{diss} << 1$.

The activation energy of the forward reaction (donor formation) has been measured and is between 2.5-2.8 eV [11,15,51,72]. Exploiting this one can show within the framework of the Gosele and Tan model that $16\pi R_{O_I}[O_I]D_{O_2}/k_{diss} = C \exp (\delta/T)$ where $\delta > 0$ (C=constant, T=absolute temperature), i. e. an effective negative activation energy. Applying the above boundary conditions one can calculate the temperature dependence of D_{eff}.

We show in Fig. 2 the diffusion data of Fig. 1 together with the experimental value of D_{eff} of 2.7×10^{-14} cm^2/sec at 450 °C. One can notice that this value is higher than Mikkelsen's value for D_{O_I} at around 700 °C. We also show in the figure the temperature dependence of D_{eff} which we calculated requiring that $16\pi R_{O_I}[O_I]D_{O_2}/k_{diss}$ at 700 °C was equal to 10^{-5} or 0.1 and at 450 °C would equal the experimental value. It is clear that the behavior of D_{eff} contradicts experience since it requires an unreasonable value for D_{eff} at temperatures around and slightly above room temperature.

The actual experiment on which the 'effective diffusivity' concept of Gosele and Tan is based [32] seems to be inaccurate and inferior to the method of stress-induced dichroism [13,19] by which the other low temperature data were taken. It is an oxygen outdiffusion method (at 450 °C) in which a depth profile of $[O_I]$ is obtained by the spreading resistance method [32,36]. Measurement of $[O_I]$ via the thermal conversion method (i.e. indirectly) is not very precise, since the standard deviation of donor density versus $[O_I]$ is large [75]. This might explain the experimental result [32].

The inconsistencies which Gosele and Tan observe in the KFR model are based on the assumption that the capture radii of the reactions $A_3 + O_I \rightarrow A_4$ and $A_4 + O_I \rightarrow P_5$ are the same. Mathematically only $R_4[A_3] = R_5[A_4]$ is required. This is different from their assumption, since according to Kaiser et al. [A_4] should be larger than [A_3].

We conclude the discussion of this model by noting that since the O_4-complex (donor) is presumably not bonded to the lattice and may be mobile, which was considered by Fuller and Logan but was not observed [51]. It is furthermore inconsistent with thermal cycling experiments.

Kaiser, Frisch and Reiss (KFR) model. This model can reproduce the overall kinetic features of thermal donor formation. There is some disagreement between measured oxygen diffusivity and diffusivity based on thermal donor data [11,12]. It is possible that this discrepancy can be reconciled by considering that O_I is actually not randomly distributed and that oxygen is initially not only present as O_I (see section 2). Also the anomalous interstitial oxygen diffusivity results of Stavola et al. [13] have to be considered in this context.

The KFR model however provides no explanation for the majority of observations. A new model should maintain and extend (to account for the fine kinetic details of thermal donor formation, e.g. the evolution of many slightly different donors from each other during heat treatment, the required oxygen diffusivity etc.) the kinetic properties of the KFR model. In addition it should offer explanations for the other observations.

An Improved Thermal Donor Model (An extension of the KFR model)

It is now generally accepted that the supersaturation of interstitial oxygen in Cz Si crystals is reduced by formation of SiO_x - precipitates (x ~ 2) during heat treatment at high temperatures (600-1100 °C) [2]. The growth of oxygen clusters at high temperatures is controlled by the diffusion of interstitial oxygen [5,98]. There is a large volume misfit between an SiO_2 - precipitate and the silicon lattice (V_{SiO_2}/Si ~ 2) which causes the generation of silicon self-interstitials at the precipitate-silicon interface during growth of the oxygen aggregate [5].

We shall now propose a model of thermal donor formation where we view donor generation as the beginning of the oxygen precipitation process. Oxygen donors are suggested to be metastable oxygen precipitates occuring at an early stage of the clustering process. The donor activity of these small clusters provides strain relief as ejected Si self-interstitials provide strain relief for the larger clusters. The model can explain a large amount of the data provided in the first section of this chapter and it seems plausible.

However, only future experimental tests and theoretical work can show if it corresponds to the actual microscopic situation.

Electrical activity. According to kinetics given by the KFR model we assume that Si_3O_2, Si_4O_3 and Si_5O_4 centers are present after a certain length of 450 °C heat treatment time. Because of the large volume misfit as compared to silicon these clusters are compressed in the Si lattice. The strain energy with 2, 3 or 4 oxygen atoms in a complex is not yet large enough to break a bond (i.e. we have a coherent interface). However, the strain is large enough to distort the Si-O-Si bond angles severely, causing the 2p lone pair orbitals of two oxygen atoms to interact. This is shown schematically in figure 3. The two interacting lone pair orbitals result in bonding σ and antibonding σ^* orbitals. The perpendicular lone pair orbitals interact only weakly and the energy of the resulting π orbitals is therefore not changed by much as compared to the separated 2p orbitals. An energy level diagram corresponding to this situation is shown in figure 4.

If the energy of the σ^* orbital is sufficiently above the Si conduction band edge this state would be doubly ionized. Consequently the Si_mO_n - complex would be a double donor. The simple positive or double positive charge left on the oxygen atoms would give a set of Rydberg states.

The lone pair bonding between the oxygen atoms of the Si_3O_2, Si_4O_3 and Si_5O_4 centers presumably results in increased density of the complexes which in turn would lower the volume misfit strain energy. The complexes would then be stabilized.

The suggested mechanism is applicable to strained Si_mO_n - centers with any number of oxygen atoms as long as n $\geq$ 2 and the lone pair bonding oxygen atoms have the required positions with respect to each other. The Si_5O_4 complex which according to the KFR model is the main donor does not necessarily has to have a tetrahedral coordination. A number of different donor structures differing in coordination and in ionization energy are possible.

We should point out that lone pair bonding of proximate oxygen atoms in a Si_mO_n cluster is only one possible way of relieving strain via electrical activity. Alternatively e.g. a strained Si_mO_n - cluster could give rise to a truly substitutional oxygen atom (i.e., tetrahedral bonding). According to recent first principles calculations such a structure also could result in shallow double donor activity [101].

Disappearance of electrical activity. During prolonged 450 °C heat treatment the Si_mO_n aggregates (n $\leq$ 4) transform into small Si_mO_n - precipitates with n>4. This growth strains the SiO bonds even further and eventually has to result in the rupture of bonds. Upon reconstruction of the bonds the small oxygen precipitate should be relaxed. One Si atom has to be displaced away from the precipitate into interstitial position to make this possible. Because of the ejected silicon atom the compression of the oxygen containing complex is reduced. The 2p lone pair orbitals of the oxygen atoms no longer interact and consequently the donor activity ceases.

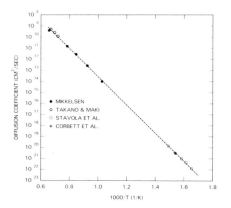

Fig. 1: Temperature dependence of the diffusivity of interstitial oxygen in silicon. The data are from the following references: Mikkelsen [30], Takano & Maki [31], Stavola et al. [13] and Corbett et al. [19]. A least-squares fit to all data points is also shown.

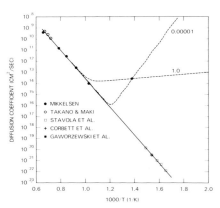

Fig. 2: The diffusivity data of Fig. 1 together with the data point by Gaworzewski et al. [32]. Note that the diffusivity according to Gaworzewski et al. is higher at 450 °C than according to Mikkelsen at about 700 °C. The dashed curves display the temperature dependence of the 'effective diffusivity' after Gosele and Tan if it is subjected to the required boundary conditions (see text).

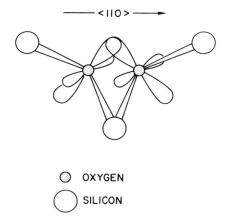

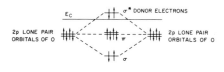

Fig. 3: Schematic illustration of the overlapping of the 2p lone pair orbitals of two oxygen atoms in a distorted Si_mO_n - complex.

Fig. 4: Energy level diagram which corresponds to the oxygen-oxygen bond formation. E_c denotes the energy position of the Si conduction band.

The above mechanism appears to be reasonable, since we know from oxygen precipitation studies at higher temperatures that large concentrations of Si self-interstitials are eventually produced. They can be observed at temperatures as low as 650 °C after condensation into rod-like defects [97]. There has to be a critical size of an oxygen aggregate (which could depend on the spatial configuration) where this mechansim starts to provide strain relief. In order to be consistent with the KFR model we propose that for Si_mO_n - complexes with $n \geq 5$ this happens.

The neutralization of thermal donors by growth into small oxygen precipitates occurs very slowly at 450 °C. That it occurs is demonstrated by the appearance of a maximum in thermal donor concentration versus time. At 550 °C the oxygen cluster growth rate is much higher because of the higher oxygen diffusivity (donor annihilation). For annealing temperatures higher than ~ 650 °C we expect that dissolution of some of the thermal donors takes place.

<u>Comparison with experiment.</u> The above model of thermal donor formation and annihilation can explain a great deal of the experimental observations. With the incorporation of the KFR model it provides the correct kinetic features of donor formation. In our model the annihilation of thermal donors after 550 °C heat treatment occurs by growth of thermal donors into small oxygen precipitates and not by dissolution of thermal donor complexes. This agrees with the measured activation energy of the reverse reaction (~ activation energy of oxygen diffusion) [11,51]. It also explains why thermal donor formation and annihilation is of great importance for the nucleation of oxygen precipitates at higher temperatures [88] (see also the following section).

The double donor character of thermal donors and the spectrum of excited states is explained. A multiplicity of thermal donor species is expected within the framework of the model.

The model can also explain why a large concentration of carbon delays donor formation. Because of the small size of carbon, carbon-oxygen complexes are not very compressed in the silicon matrix. Since the 2p lone pair orbitals of the oxygen atoms do not interact no electrical activity results.

Since carbon-oxygen interaction is appreciable at 450 °C [62], a large concentration of substitutional carbon provides clustering sites for a significant fraction of the available interstitial oxygen without formation of donor centers. The 'effective' interstitial oxygen concentration available for thermal donor formation is reduced. This together with the $[O_I]$ power dependence of thermal donor formation results in a significant reduction in donor concentration. That the lowering of the 'effective' oxygen concentration is the reason for the effect of carbon on donor formation can be seen from a different result. The KFR model predicts that lowering the oxygen concentration available for donor formation results in a time delay in the occurence of the donor maximum [11]. This effect can be observed in the data of Capper et al. [65] who compared the kinetics of thermal donor formation in seedend and tailend sections of Si crystals.

That carbon-oxygen complexes which are formed at 450 °C are not active as donors is shown in figures 5 and 6. In figure 5 we compare the vibrational characteristics of carbon-oxygen complexes observed after 450 °C heat treatment (i.e. thermal donors present) with characteristics of carbon-oxygen complexes after 900 °C heat treatment (no donors). The IR lines in the lower part of Fig. 5 are due mainly to carbon-oxygen complexes [99]. The 1087 and 1079 cm^{-1} lines are exceptions, which are due to a vibrational mode of O^{18} [100]. We note from the figure that the two spectra are almost identical. The one additional band in the upper part of the figure is an electronic transition of a thermal donor, since it can only be observed after 450 °C heat treatment. This fact is supported by the temperature dependence of the IR bands in Fig. 6. The electronic transition shows the temperature dependence which has

118

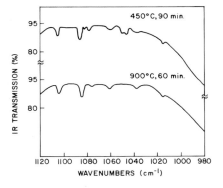

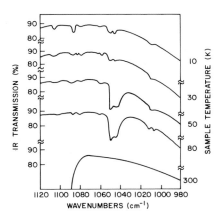

Fig 5: Comparison of IR spectra between 1120 and 980 cm^{-1} in B-doped Cz Si which had been annealed at 450 °C for 90 min. or at 900 °C for 60 min. (measured at a sample temperature of 15 K).

Fig. 6: Temperature dependence of IR bands observed in heat treated B-doped Cz Si after short annealing time (90 min.) at 450 °C.

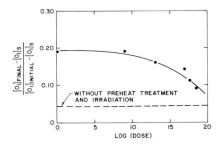

Fig. 7: Fraction of interstitial oxygen concentration which did not precipitate after 68 hours at 900 °C as a function of electron-dose which the Si samples had received before heat treatment at 900 °C. Initially (before irradiation) all samples had received a high-temperature preanneal at 1150 °C for 26 hours to dissolve grown-in precipitate nuclei. $[O_I]_S$ denotes the oxygen solubility at 900 °C.

been observed for thermal donors [62,64,71], but not the carbon related bands. Therefore, carbon-oxygen complexes formed at 450 °C are not electrically active, which is in agreement with our model.

Application to new donors. The formation of new donors can be understood to a certain extent within the framework of the above model. We expect that carbon-oxygen complexes which are not compressed much during 450 °C heat treatment reach a high volume misfit strain level after continued growth at higher temperatures. Electrical activity would come about in the same way as for the thermal donor via bond angle distortions and oxygen lone pair interaction. It appears that the effect of carbon on new donor formation can be explained by such a mechanism.

The fact that new donor formation is enhanced after 470 °C preannealing suggests that new donors evolve from centers formed at 470 °C. The new donors should be centers with many more oxygen atoms than the thermal donor, and consequently should have an incoherent interface (differing from the carbon containing new donors). It seems plausible that (in the absence of a high carbon concentration) the observed electrical activity then would originate at the interface. However, before we can discuss microscopic models, more detailed investigations are needed.

REMARKS ON THE NUCLEATION OF HIGH TEMPERATURE OXYGEN PRECIPITATES (~ 900-1000 °C)

According to the above model the formation of thermal donors is the beginning of the oxygen precipitation process. The precipitate density after 900-1000 °C annealing is much smaller than the thermal donor density at 450 °C on the other hand [18,19], which indicates that most of the donor complexes should dissolve at 900-1000 °C. The question arises which of the complexes formed during thermal donor formation are nuclei. According to Schaake et al. [88] the latter should be thermal donors characterized by an incoherent interface between the Si matrix and the thermal donor. The following two observations show that the existence of an incoherent interface can indeed be crucial for the stability of nuclei.

Leroueille [16] found that during a 750 °C heat treatment the substitutional carbon concentration decays to zero. The variation in $[C_S]$ was related to new donor generation and the precipitation of $[O_I]$. The new donor complexes however were dissolved by a 900 °C anneal and most of the carbon returned to its substitutional position [16]. If we explain these findings in terms of the concept suggested by Schaake et al. [88] we must conclude that the carbon containing complexes of Leroueille were still characterized by a coherent interface and nucleation was not completed (see also the discussion on electrical activity of carbon containing new donors). Characteristic for an incoherent interface in our model are emitted Si self-interstitials which would condense into rod-like defects at temperatures around 650-750 °C [97]. The complete absence of such defects in carbon-rich material which had been annealed at 650-750 °C [90], despite high new donor density and strong reduction in $[O_I]$ strongly supports the notion of a coherent interface for carbon containing new donor complexes. In conjunction with the finding that high temperature annealing can dissolve carbon-containing complexes the previous observation supports also the notion of Schaake et al..

That lattice defects (which would be present at an incoherent interface) can produce enhanced nucleation at 900 ° can be shown by the following experiment. Similar wafers are heat treated initially at 1150 °C to dissolve grown-in nuclei. They are then irradiated to various doses with 2 MeV electrons, thus creating displacement damage. All specimens are then heat-treated at 900 °C for 68 hours. The reduction in $[O_I]$ is measured as a function of electron dose. The results are shown in Fig. 7 and demonstrate clearly that lattice defects can be important for nucleation of oxygen precipitates at temperatures as high as 900 °C.

If we combine the view of Schaake et al. with our model we can conclude the following: A large number of stable nuclei for oxygen precipitates at high temperatures is introduced by a 450 °C anneal once the maximum in donor concentration has been reached, since then a substantial number of donors has been neutralized. According to our model these centers are characterized by an incoherent microprecipitate-silicon matrix interface. We could alternatively form thermal donors at 450 °C and subsequently annihilate them at 550 °C which would complete nucleation.

SUMMARY AND CONCLUSIONS

A critical survey of the available literature on oxygen related donor formation in silicon is presented. We conclude that the donor formation models of Helmreich and Sirtl [66] and Gosele and Tan [12] are not consistent with the experimental observations. We suggest a model for the formation of thermal donors which extends the ideas of Kaiser, Frisch and Reiss [11]. Thermal donors are viewed as metastable oxygen clusters which lower the compressive strain of the surrounding silicon matrix via bonding of 2p oxygen lone pair orbitals. The bonding results in shallow double donor states. Our model can account for a significant number of the experimental observations which could not be explained by the model of Kaiser, Frisch and Reiss.

ACKNOWLEDGMENTS

The idea that antibonding combinations of lone pair orbitals on proximate oxygen atoms in Si_mO_n - clusters are plausible candidates for the thermal donor state was suggested by Professor Lawrence Snyder of the State University of New York at Albany Chemistry Department. We would like to thank J. L. Lindstrom, S. N. Sahu and V. A. Singh for helpful discussions. B. Pajot is gratefully acknowledged for pointing out the origin of the 1087 and 1079 cm^{-1} IR lines of Figs. 5 and 6.

REFERENCES

[1]   J. R. Patel in "Semiconductor Silicon 1977", eds. H. R. Huff and E. Sirtl (Electrochem. Soc., Princeton, 1977), 521.
[2]   J. R. Patel in "Semiconductor Silicon 1981", eds. H. R. Huff, R. J. Kriegler, Y. Takeishi (Electrochem. Soc., Pennington, 1981), 189.
[3]   A. Murgai, W. J. Patrick, J. Combronde, J. C. Felix, IBM J. Res. Develop. 26, 546 (1982).
[4]   R. A. Craven, Op. Cit., Ref. 2, 254.
[5]   Op. cit., Ref. 2.
[6]   "Defects in Semiconductors", eds. J. Narayan and T. Y. Tan (North-Holland, New York 1981).

121

[7]   G. H. Schwuttke, Microelectron. Reliab. 9, 397 (1970).
[8]   D. Thebault, L. Jastrzebski, RCA Review 41, 592 (1980).
[9]   W. K. Tice, T. Y. Tan, Appl. Phys. Lett. 28, 564 (1976).
[10] W. K. Tice, T. Y. Tan in Ref. [6], 367.
[11] W. Kaiser, H. L. Frisch, H. Reiss, Phys. Rev. 112, 1546 (1958).
[12] U. Gosele, T. Y. Tan, Appl. Phys. A 28, 79 (1982).
[13] M. Stavola, J. R. Patel, L. C. Kimerling, and P. E. Freeland, Appl. Phys. Lett. 42, 73 (1983).
[14] A. Kanamori, M. Kanamori, J. Appl. Phys. 50, 8095 (1979).
[15] K. Schmalz, P. Gaworzewski, phys. stat. sol. (a) 64, 151 (1981).
[16] J. Leroueille, phys. stat. sol. (a) 67, 177 (1981).
[17] S. M. Hu, J. Appl. Phys. 52, 3974 (1981).
[18] S. Kishino, Y. Matsushita, M. Kanamori. T. Iizuka, Jap. J. Appl. Phys. 21, 1 (1982).
[19] J. W. Corbett, R. S. McDonald and G. D. Watkins, J. Phys. Chem. Solids 25, 873 (1964).
[20] D. R. Bosomworth, W. Hayes, A. R. L. Spray and G. D. Watkins, Proc. Roy. Soc. A317, 133 (1970).
[21] R. C. Newman, "Infra-Red Studies of Crystal Defects", (Barnes & Noble, London New York 1973).
[22] B. Pajot, Analysis 5, 293 (1977).
[23] K. Graff, E. Grallath, S. Ades, G. Goldbach, and J. Toelg, Solid-State Electr. 16, 887 (1973).
[24] V. A. Malyshev, Sov. Phys. Semicond. 8, 92 (1974).
[25] E. M. Ryzhkova, I. I. Traveznikova, V. E. Chelnokov, and A. A. Yakovenko, Sov. Phys. Semicond. 11, 628 (1977).
[26] F. Shimura, H. Tsuya, and T. Kanamura, Appl. Phys. Lett. 37, 483 (1980).
[27] F. Shimura, Y. Ohnishi, and H. Tsuya, Appl. Phys. Lett. 38, 867 (1981).
[28] L. Jastrzebski, P. Zanzucchi, D. Thebault, and J. Lagowski, J. Electrochem. Soc. 129, 1638 (1982).
[29] H. J. Hrostowski and R. H. Kaiser, J. Phys. Chem. Solids 9, 214 (1959).
[30] J. C. Mikkelsen, Appl. Phys. Lett. 40, 336 (1982).
[31] Y. Takano and M. Maki in "Semiconductor Silicon 1973", eds. H. R. Huff and R. R. Burgess (Electrochem. Soc., Princeton, 1973), 469.
[32] P. Gaworzewski and G. Ritter, phys. stat. solidi A 67, 511 (1981).
[33] Y. Yatsurugi, N. Akiyama, Y. Endo, and T. Nozaki, J. Electrochem. Soc. 120, 975 (1973).
[34] A. Murgai, H. C. Gatos and W. A. Westdorp, J. Electrochem. Soc. 126, 2240 (1979).
[35] T. Abe, H. Harada, and J. Chikawa, Op. cit. Ref. [13] (to be published).
[36] F. G. Vieweg-Gutberlet in "Spreading Resistance Symposium", Nat. Bur. Stand. Spec. Publ. 400-10 (1974), 185.
[37] S. M. Hu, J. Vac. Sci. Technol. 14, 17 (1977).
[38] T. Abe, K. Kikuchi, and S. Shirai, Op. cit., Ref. [1], 95.
[39] K. Graff, J. Hilgarth, and H. Neubrand, Op. cit., Ref. [1], 575.
[40] P. Gaworzewski and H. Riemann, Kristall u. Technik 12, 189 (1977).
[41] P. Gaworzewski, S. Hahle and H. Riemann, Kristall u. Technik 12, 871 (1977).
[42] A. Murgai, J. Y. Chi, and H. C. Gatos, J. Electrochem. Soc. 127, 1182 (1980).
[43] A. Ohsawa, K. Honda, S. Ohkawa, and R. Ueda, Appl. Phys. Lett. 36, 147 (1980).
[44] A. Ohsawa, K. Honda, S. Ohkawa, and K. Shinohara, Appl. Phys. Lett. 37, 157 (1980).
[45] A. Ohsawa, H. Honda, S. Shibatomi, and S. Ohkawa, Appl. Phys. Lett. 38, 787 (1981).
[46] P. Rava, H. C. Gatos, and J. Lagowski, Appl. Phys. Lett. 38, 274 (1981).

122

[47]   P. Rava, H. C. Gatos, and J. Lagowski, Op. cit., Ref. [2], 232.
[48]   C. S. Fuller, N. B. Ditzenberger, N. B. Hannay, E. Buehler, Phys. Rev. 96, 833 (1954).
[49]   W. Kaiser, Phys. Rev. 105, 1751 (1957).
[50]   R. A. Logan, J. Appl. Phys. 28, 819 (1957).
[51]   C. S. Fuller, R. A. Logan, J. Appl. Phys. 28, 1427 (1957).
[52]   H. J. Hrostowski, R. H. Kaiser, Phys. Rev. Lett. 1, 199 (1958).
[53]   C. S. Fuller, F. H. Doleiden, J. Appl. Phys. 29, 1264 (1958).
[54]   C. S. Fuller, F. H. Doleiden, K. Wolfstirn, J. Phys. Chem. Solids 13, 187 (1960).
[55]   Y. Matukura, J. Phys. Soc. Japan 14, 918 (1959).
[56]   T. Arai, J. Phys. Soc. Japan 17, 246 (1962).
[57]   V. N. Mordkovich, Sov. Phys.-Solid State 4, 2662 (1963), 6, 1716 (1965).
[58]   V. N. Mordkovich, Sov. Phys.-Solid State 6, 654 (1964).
[59]   M. I. Starchik, Fiz. Tekh. Poluprovodnikov 3, 153 (1969).
[60]   P. M. Kurilo, E. Seitov, and M. I. Khitren', Soviet Phys. - Semicond. 4, 1953 (1971).
[61]   Y. P. Koval, V. N. Mordkovich, E. M. Temper, Soviet Physics - Semicond. 5, 1076
        (1971).
[62]   A. R. Bean, R. C. Newman, J. Phys. Chem. Solids 33, 255 (1972).
[63]   F. W. Voltmer, T. G. Digges, Jr., J. Crystal Growth 19, 215 (1973).
[64]   K. Graff, H. Pieper, J. Electronic Mat. 4, 281 (1975).
[65]   P. Capper, A. W. Jones, E. J. Wallhouse, and J. G. Wilkes, J. Appl. Phys. 48, 1646
        (1977).
[66]   D. Helmreich, E. Sirtl, Op. cit., Ref. [1], 626.
[67]   H. Kolker, Electrochem. Soc. Spring Meet. 1977, Abstract No. 117.
[68]   A. Kanamori, Appl. Phys. Lett. 34, 287 (1979).
[69]   S. H. Muller, M. Sprenger, E. G. Sieverts, and C. A. J. Ammerlaan, Solid State Com-
        mun. 25, 987 (1978).
[70]   P. Gaworzewski, K. Schmalz, phys. stat. sol. (a) 55, 699 (1979).
[71]   D. Wruck, P. Gaworzewski, phys. stat. sol. (a) 56, 557 (1979).
[72]   P. Gaworzewski, K. Schmalz, phys. stat. sol. (a) 58, K223 (1980).
[73]   L. C. Kimerling, Op. cit., Ref. [6], 21.
[74]   L. C. Kimerling, J. L. Benton, Appl. Phys. Lett. 39, 410 (1981).
[75]   V. Cazcarra, P. Zunino, J. Appl. Phys. 51, 4206 (1980).
[76]   M. Tajima, A. Kanamori, S. Kishino, T. Iizuka, Jap. J. Appl. Phys. 19, L755 (1980).
[77]   M. Tajima, S. Kishino, M. Kanamori, T. Iizuka, J. Appl. Phys. 51, 2247 (1980).
[78]   M. Tajima, A. Kanamori, T. Iizuka, Jap. J. Appl. Phys. 18, 1401 (1979).
[79]   M. Tajima, T. Masui, T. Abe, T. Iizuka, Op. cit., Ref. [2], 72.
[80]   H. Nakayama, J. Katsura, T. Nishino, Y. Hamakawa, Jap. J. Appl. Phys. 19, L547
        (1980).
[81]   H. Nakayama, T. Nishino, Y. Hamakawa, Appl. Phys. Lett. 38, 623 (1981).
[82]   J. Reichel, phys. stat. sol. (a) 66, 277 (1981).
[83]   K. D. Glinchuk, N. M. Litovchenko, phys. stat. sol. (a) 58, 549 (1980).
[84]   K. D. Glinchuk, N. M. Litovchenko, Z. A. Salnik, phys. stat. sol. (a) 71, 83 (1982).
[85]   J. W. Cleland, J. Electrochem. Soc. 129, 2127 (1982).
[86]   T. V. Mashovets, Sov. Phys. Semicond. 16, 1 (1982).
[87]   B. Pajot, H. Compain, J. Lerouille, B. Clerjand, "16th Int. Conf. Phys. Semicond.",
        Montpellier 1982 (to be published).
[88]   H.F. Schaake, S.C. Baber, R.F. Pinizzotto, Op. cit., Ref. [2],273.
[89]   A. J. R. DeKock, Op. cit., Ref. [6], 309.
[90]   K. Yasutake, M. Umeno, H. Kawabe, H. Nakayama, T. Nishino, Y. Hamakawa, Jap. J.
        Appl. Phys. 21, 28 (1982).

[91]   A. Ohsawa, R. Takizawa, K. Honda, A. Shibatomi, and S. Okhawa, J. Appl. Phys. 53, 5733 (1982).

[92]   P. M. Grinshtein, G. V. Lazareva, E. V. Orlova, Z. A. Sol'nik, V. I. Fistul', Sov. Phys. Semicond. 12, 68 (1978).

[93]   S. Muller, Ph.D. Thesis, Amsterdam 1981.

[94]   G. S. Oehrlein, Ph.D. Thesis, Albany 1981.

[95]   G. D. Watkins, J. W. Corbett, Phys. Rev. 121, 1001 (1961).

[96]   J. W. Corbett, G. D. Watkins, R. M. Chrenko, R. S. McDonald, Phys. Rev. 121, 1015 (1961).

[97]   K. Tempelhoff, F. Spiegelberg, R. Gleichmann, D. Wruck, phys. stat. sol. (a) 56, 213 (1979).

[98]   K. Wada. N. Inoue, K. Kohra, J. Crystal Growth 49, 749 (1980).

[99]   G.S. Oehrlein, J.L. Lindstrom, J.W. Corbett, Appl. Phys. Lett. 40, 241 (1982).

[100]   B. Pajot, J. Phys. Chem. Solids 28,73 (1967).

[101]   V. A. Singh, A. Zunger, U. Lindefelt, Phys. Rev. B (accepted for publication).

[102]   Y. H. Lee, J. W. Corbett, Phys. Rev. B 13, 2653 (1976).

PRECIPITATION OF OXYGEN AND MECHANISM OF STACKING FAULT FORMATION IN CZOCHRALSKI
SILICON BULK CRYSTALS

KAZUMI WADA, NAOHISA INOUE AND JIRO OSAKA
Musashino Electrical Communication Laboratory, Nippon Telegraph and Telephone
Public Corporation, Musashino, Tokyo, 180 JAPAN

ABSTRACT

This paper describes recent progress on nucleation and
growth of oxide precipitates and stacking faults in
Czochralski silicon.   Conclusions on the growth kinetics
of oxide precipitates are drawn from the experiments and
analysis of growth kinetics of two-dimensional precipitates:
The experimentally obtained growth kinetics, three-quarter
power law is theoretically derived and the precipitate
growth is demonstrated to be diffusion-limited by oxygen
interstitials.   The formation mechanism of stacking faults
is the Bardeen-Herring mechanism.   Based on diffusional
growth model, the growth kinetics of stacking faults are
analyzed, assuming a coexistence of self-interstitial
supersaturation and vacancy undersaturation.   It is found
that the growth is driven by vacancies in undersaturation.
Vacancy component of self-diffusion has been determined and
found to be predominant at low temperature.   The
possibility of growth model proposed for increase of oxide
precipitate density during annealing has been excluded.
Both processes, homogeneous and heterogeneous nucleation,
have been taking place during annealing.

INTRODUCTION

It is generally accepted that oxygen interstitials in Czochralski silicon
precipitates during annealing [1] and the resulting oxide precipitates induce
dislocation loops, i.e., stacking faults and/or perfect dislocaton loops [2].
However, nucleation and growth kinetics of the microdefects has not been fully
understood. Recent review by Patel has clearly shown gaps in our understanding
of nucleation and growth kinetics [3]. The present paper intends to add some
fundamental aspects of recent results.

MICRODEFECTS AND CONTROLLED INTRODUCTION

Various kinds of microdefects are induced in Czochralski silicon during
annealing. Figure 1 shows transmission electron microscopic images of the
defects. At lower temperature around $600^{\circ}C$ rod-like defects and small sphere-
like defects are the predominant defects. The nature of the rod-like defects is
as yet unknown. The small sphere-like defects are presumably oxide precipitates
of early growth stage. At temperature around $750^{\circ}C$ square-shaped platelet
precipitates are the predominant defects, whose habit planes and peripheral edge
directions are {100} and <110>, respectively. At temperature above $900^{\circ}C$
the dislocation loops and the oxide precipitates are predominant. Since the
density of observed dislocation loops were too low and the morphology was too
complex to study the nucleation and growth kinetics, we developed a two-step
annealing technique consisting of a long term low temperature first annealing,
e.g. at $750^{\circ}C$ for more than one hundred hours and a high temperature second

Mat. Res. Soc. Symp. Proc. Vol. 14 (1983) ©Elsevier Science Publishing Co., Inc.

annealing around 1000°C. The purposes of the 1st annealing are
(1) to make accurate measurements by increase of the detection efficiency due to introducton of high density of the defects,
(2) to generate self-interstitials in supersaturation (and vacancies in undersaturation) due to volume expansion associated with the precipitate growth,
(3) to form large precipitates to reveal the stage of the interaction between the precipitate and the stacking fault in initial growth stage,
(4) to eliminate formation of the prismatic punching out loops and/or complex dislocation loops during a high temperature second annealing by saturating the precipitate growth during first annealing (suppressing the rapid growth of the precipitates during the second annealing).
The purpose of the second annealing is to nucleate and grow the stacking faults and dislocation loops at the precipitates. Thus, a high density of simple-shaped dislocation loops can be formed, as shown in Fig.1.
   In this paper we focus on nucleation and growth kinetics of oxide precipitates and stacking faults. The proportionality coefficients for oxygen interstitital and carbon substitutional concentration used in the present paper are 3×10^{17} at./cm^3 and 0.8×10^{16} at./cm^3 in the infrared absorption measurement method [4], respectively (the former will be the Japanese standard).

GROWTH OF MICRODEFECTS

   Stacking faults are two-dimensional defects and oxide precipitates are almost constant in thickness during annealing [5]. In other words, the growth front of these defects is just the peripheral edge. Therefore, the growth kinetics can be analyzed by a toroidal approximation method proposed by Seidman and Balluffi [6], which is briefly outlined below and is applied to analyze the experimentally obtained growth kinetics.

Theory
   The growth kinetics of two-dimensional defect (ideally stacking fault) have been expressed by approximating the peripheral edge (dislocation loop) to be in a shape of torus with the same large radius r as the defect and with a circular cross section of atomic jumping distance (a dislocaton core radius r_c). In this approximation the volume increase of the defect becomes equal to solute flux across the surface of torus multiplying atomic volume of the solute. The quasi-steady state diffusion flux across the surface is given for diffusion-limited shrinkage by [7]

$$F = 4 \pi \, r^{eff} \, D \, (C_\infty - C_0), \qquad (1)$$

where F is the flux, r^{eff} is the effective radius given by analogy of electrostatic capacity, D is the diffusion coefficient and C_∞ and C_0 is the concentration of solute at large distance and at the surface of the torus, respectively. The effective radius is expressed by an infinite series of Legendre polynomials. For $r > 3 r_c$, r^{eff} is approximated to be [8]

$$r^{eff} = (r^2 - r_c^2)^{1/2} \, \pi \, / \ln(8r/r_c). \qquad (2)$$

The growth rate of two-dimensional defects is given by

$$d(\pi \, r^2 \, d)/dt = v \, F, \qquad (3)$$

where d is the thickness of the defect and v is an atomic volume of the solute. From Eqs.(1),(2) and (3), the basic formula of the growth kinetics in the diffusion-limited case is given by

$$dr/dt = 2 \pi \ v/d \ (1-(r_c/r)^2)^{1/2} (\ln(8r/r_c))^{-1} D \ (C_\omega - C_0). \tag{4}$$

The equation has been also given by Hu [9]. Figure 2 shows the numerical calculation result of Eq.(4). Here we substitute half thickness of the stacking faults and the precipitates into r_c. Clearly, the resulting growth law depends on the defect size and can not be expressed by a unique exponent. For the precipitates r_c = 20 Å, the experimentally obtained growth kinetics, three-quarter power law, shown later, is approximately derived in the observed range of the precipitate size, a few hundred Å to sub μm. For stacking faults r_c = 1.57 Å, the derived growth kinetics is about nine tenth power law. However, the stacking faults observed have an initial radius about 500 Å, as shown later. Taking into account it, three-quarter power law can be also obtained in the observed range, 500 Å to several thousand Å.

Goesele et al. claimed that a derivation for two dimensional precipitates gives r to be propotional to t [10]. He started with the similar equation to Eq.(4). However, he regarded the term $\ln(8r/r_c)$ as constant and the term $(1 - (r_c/r)^2)^{1/2}$ as 1 to get r is propotional to t. His approximation is not valid whenever the aspect ratio r_c/r can not be ignored.

Growth of oxide precipitates.

Diffusion-limited growth.
Yang et al.[11] have studied the growth kinetics of octahedral oxide precipitates at $1100^{\circ}C$, using transmission electron microscopy. They adapted a model developed by Kahlweit [12] and suggested that the oxide precipitate growth is limited by diffusion of oxygen interstitials. Wada et al. [5] have also directly measured the growth of square-shaped platelet precipitates at 750° to $1050^{\circ}C$. They found that the precipitate growth follows three-quarter power law and the thickness keeps almost constant. In this paper we have employed the toroidal approximation method noted above to conclude the growth kinetics to be diffusion-limited by oxygen interstitials.

Figure 3 shows annealing time dependence of the half diagonal length L of the precipitates (C_∞ = C_i = 1.1 x 10^{18} at./cm^3). For the square shaped platelet precipitate growth, r in Eq.(4) is expressed by

$$r = (2 \ L^2/\pi)^{1/2} \fallingdotseq 0.8 \ L. \tag{5}$$

Substituting v = 4.8 x 10^{-23}cm^3 which is volume of $Si_{1/2}O$ in cristobalite, d = 40 Å, r_c = 20 Å, C_0 = C_e = 2.0 x 10^{21} exp(-1.03/kT) [13]. Best fitting shown by the solid lines in Fig.3 is obtained when the diffusion coefficient D is given by

$$D = 0.033 \exp(-2.43/kT). \tag{6}$$

The diffusion coefficient fits well into the range of the reported diffusion coefficient of oxygen interstitials. Thus, the growth of the square-shaped platelet oxide precipitates is quantitatively understood to be diffusion-limited by oxygen interstitials. Furthermore, the conclusion clearly shows that the slowest step of the precipitate growth is diffusion of oxygen interstitials and the other steps such as silicon interstitial emission (and vacancy absorption) by the precipitate growth are not the rate-determining process. We will come back the consideration later on.

Thickness of oxide precipitates.
There have been no direct measuements on the thickness of the platelet precipitates. Wada et al. have estimated the thickness from the weak bean images to be 40 Å [5]. Recently, Matsushita has reported the lattice image of the oxide precipitates grown during $800^{\circ}C$ annealing for 64 hrs [14]. The thickness was 12 Å which is three times smaller than the estimated value by Wada et al. It is to be noted that the length of the lattice image is about 100 Å whereas the half diagonal length is to be about 500 Å from Fig.3. This suggests that the very edge (thin

part) of the precipitate might be taken as the lattice image. Further
experimental information is necessary to clarify the detailed kinetics of
thickening of the precipitates.

Growth of stacking faults.

Extensive research has been performed on growth of oxidation-induced surface
stacking faults (OISF) in conjunction with oxidation-induced anomalous dopant
diffusion [15]. However, the growth kinetics have only been empirically
analyzed because concentration of point defects C_{∞} in Eq.(4) has not been
analytically determined. In contrast, the growth kinetics of bulk stacking
faults (BSF) formed by precipitation are much simpler and fit to quantitative
analysis, since the phenomenon occurs in a closed system. In this section we
concentrate on the growth of BSF.

Bardeen-Herring Mechanism. BSF is formed by the operation of the Bardeen-
Herring source in the metallogical view point, as demonstrated by Wada et al
[16]. Iizuka has firstly speculated the mechanism from observation of
concentric BSF with a precipitate at their center [17]. Direct TEM observation
on the birth stage of BSF became possible for two-step annealed specimens [16].
The first annealing was performed at 750°C for 16 hrs to introduce high
density of large oxide precipitates ($2 \times 10^{12} cm^{-3}$, L = 600 Å). The
second annealings were at 1050°C for 10 to 70 minutes to nucleate BSF and
initiate the growth. Figure 4 shows a succesive growth process of BSF. Although
these images of BSF were taken from the specimens 2nd-annealed for various
durations, they show the actual growth stage of BSF. The readers easily
understand that the change in shape of dislocation loop corresponds to the
operation of the Bardeen-Herring source. The specimen tilting showed that the
precipitate is on BSF without intersecting the fault plane. Figure 5 is
schematics of the growth mechanism of BSF. Figure 5(a) shows spacial
configuration between the precipitate ABCE and faulted plane ABH. It is readily
found that BSF is nulceated at a peripheral edge of the precipitate and that the
precipitate acts as the dislocation pinning points. Figure 5(b) depicts the
growth (climbing) process of BSF. The faulted half loop is nucleated at an edge
of the precipitate and grows by steps F_1 to F_4 and dislocation segments of
F_4 rejoin close to the point P. Thus a circular BSF is nucleated. Figure 6
shows a kidney-shaped perfect dislocation loop. The specimen was second-
annealed at 950°C for 1000 min. It is concluded that full loops such as
perfect dislocation loops (by climb) and BSF are nucleated by the Bardeen-
Herring mechanism. Criterion on nucleation of full loops from half loops is
derived from this finding. We will not describe it here. This consideration
based on the dislocation theory does not give any information on the driving
force of dislocation loop growth, self-interstitials in supersaturation or
vacancies in undersaturation. We will clarify that either point defect reaction
is responsible for the growth (climb) of the dislocation loops in the following.

Growth curves of bulk stacking faults. Figure 7 shows annealing time
dependence of radius of BSF by the 2-step annealing. Let us explain the typical
growth stage of BSF, using the data at 1000°C. BSF grows suddenly upto about
500 Å (r_0) in radius and the growth halts (growth halt stage). Later the
growth begins again (steady growth stage). The numerical integration of Eq.(4)
shown in Fig.2 fits the growth curve very well. Finally the growth saturates at
radius of about 6000 Å (growth saturation stage). The similar features are
observed in this wide temperature range.

Figure 8 shows the growth data of BSF by Patel et al. [18] and in Fig.7. The
numerical calculation result of Eq.(4) fits these growth data very well except
the growth saturation stage. However, these initial conditions appear to be
quite different, as pointed out by Patel [3]: BSF grows simultaneously with the
precipitate growth in the experiment by Patel et al., whereas BSF grows after
the growth saturation of the oxide precipitates in the present case. However,
the boundary conditions in these two cases become quite similar after complete

consumption of excess oxygen interstitials. Since BSF is nucleated at a few percent of precipitates [19], the precipitation should be finished sooner than, at least, growth saturation of BSF. In fact, the growth saturation has been observed in the experiment by Patel et al. Therefore, the boundary conditions has been similar in these two cases.

Rate-determining process. It is found that isolated BSF are slightly larger than BSF accompanied with a dislocation loop on one precipitate [20], even if they were observed in the same sample, which suggests that the rate determining process is diffusion of point defects. In contrast, in the growth of OISF the growth rate is independent of distances between them [21]. This finding suggested that the growth of OISF is limited by reaction between dislocation loop and point defects. Hu elaborated his model based on the reaction-limited growth [9]. On the other hand, the shrinkage kinetics of OISF in inert ambient is understood to be diffusion-limited [10]. It is not clear why the rate determining process is different between growth of BSF and OISF and even between growth and shrinkage of OISF. It is worth mentioning that in the case of OISF the bounding partial dislocation reaches the Si-SiO$_2$ interface, i.e., the source (sink) of point defects. This suggests that the transportation of point defects to or from the bounding partial occurs by pipe-diffusion through the dislocation core rather than by volume diffusion. The distance-independent growth of OISF might be explained in such a way. In the following we will analyze the growth data of BSF mainly in terms of the diffusion-limited growth model, but occasionally comment on the reation-limited growth model.

Driving force of stacking fault growth. BSF can grow by self-interstitial (I) supersaturation or vacancy (V) undersaturation. It has been recently reported that oxygen precipitation enhances dopant diffusion [22]. This finding clearly verifies our previous model that supersaturated I is generated by the oxide precipitate growth [23]. On the other hand, it was suggested from the phenomenon of oxidation-retarded diffusion of Sb that V undersaturation simultaneously occurs during oxidation [24]. Therefore, we will further take into account contribution of V undersaturation to BSF growth in the present paper. Let us assume for the moment that I in supersaturation and/or V in undersaturation homogeneously distributed in the entire specimen. Thus we can use the same boundary conditions as used in the oxide precipitate growth. In Eqs.(4') and (7) we have considered the contribution of V undersaturation. It can be seen from Fig.2 that 2π v/b CD t is 1.6×10^{-4} sec when r = 2000 Å in case of BSF (r$_c$ = 1.57 Å).

$$dr/dt = 2\pi v/d \ (1-(r_c/r)^2)^{1/2}(\ln(8r/r_c))^{-1} \ CD, \qquad (4')$$

$$CD = g_I \ (C_I - C_{IE}')D_I + g_V \ (C_{VE}' - C_V)D_V \qquad (7a)$$

$$= 1.6 \times 10^{-4} \ b/ \ 2 \pi \ v \ 1/t, \qquad (7b)$$

where g denotes a contribution factor lying between 0 and 1, the subindex I (V) self-interstitials (vacancies), the subindex E the equilibrium, the dash on the dislocation loop and b Burgers vector. Here, C_{IE}' is given by C_{IE} exp(- Δ G / kT) $\doteq$ C_{IE}. ΔG is the stacking fault energy, 0.026 eV/atom [10]. Similarly, C_{VE}' is by C_{VE} exp(Δ G / kT) $\doteq$ C_{VE}. Substituting annealing time t required for reaching 2000 Å in radius at 850° to 1200°C, b = 3.14 Å and atomic volume of lattice silicon v = 2 x 10^{-23} cm^3, CD is obtained, as shown in Table I. On the other hand, the product of self-diffusion coefficient and lattice concentration is given by

$$C_S D_S = f_I C_{IE} D_I + f_V C_{VE} D_V \ , \qquad (8)$$

where f_I and f_V are correlation factors which we regard as 0.5. Based on the assumption that $C_I \gg C_{IE}$ and $C_{VE} \gg C_V$, Eq.(7a) is written by

$$CD \doteq g_I C_I D_I + g_V C_{VE} D_V \qquad (7c)$$

It is readily understood that comparison between CD and $C_S D_S$ gives us the information on I supersaturation and/or V undersaturation. We have selected the self-diffusion coefficients reported by Mayer et al [25] and Kalinowski and Seguin [26].

Table I
Comparison between CD and $C_S D_S$

Annealing Temp.($^\circ$C)	850	886	950	1000	1050	1100	1150	1200
CD (at.cm^{-1} s^{-1})	2.2e4	8.8e4	1.0e6	5.5e6	2.2e7	6.4e7	1.7e8	4.6e8
CD/$C_S D_{S1}$ = n1	(9.7)	(7.8)	(6.4)	(5.4)	3.8	2.2	1.4	.92
CD/$C_S D_{S2}$ = n2	(2.0)	1.82	1.81	1.77	1.42	0.94	0.63	(0.47)

$C_S = 5 \times 10^{22}$ (at./cm^3), $D_{S1} = 1460 \exp(-5.02/kT)$ ($1047^\circ C < T < 1387^\circ C$) [25], $D_{S2} = 154 \exp(-4.65/kT)$ ($855^\circ C < T < 1175^\circ C$) [26]. The parentheses indicate values obtained by extrapolation.

Figure 9 shows the temperature dependence of CD and $C_S D_S$. In the case that the driving force is I in supersaturation ($g_V=0$), CD is always larger than $C_S D_S$, unless I component of the self-diffusion is negligibly smaller than V component, as deduced from Eqs.(7c) and (8). The calculation result n1 (based on the self-diffusion coefficient by Mayer et al.) is larger than 1 below $1200^\circ C$, which means that the driving force is I in supersaturation below $1200^\circ C$. In the case of V undersaturation ($g_I=0$), CD is always smaller than or equal to $C_S D_S$. The calculation result n2 (based on the self-diffusion coefficient by Kalinowski and Seguin) shows that CD is about two times larger than $C_S D_S$ below $1050^\circ C$ and is smaller than $C_S D_S$ above $1100^\circ C$. This implies that the driving force is I in supersaturation below $1050^\circ C$ and V in undersaturation above $1100^\circ C$. However, the following consideration of the temperature dependence of CD and $C_S D_S$ shows that the driving force is V in undersaturation in the entire temperature range in case of Kalinowski and Seguin. The activation energy of CD is calculated to be 4.4 eV in low temperature region where degree of supersaturation and undersaturation is high enough. As deduced from Eq.(7c), the activation energy is equal to that of D_I in the case of I supersaturation, whereas it is equal to that of $C_{VE} D_V$ in the case of V undersaturaton. In the case of Kalinowski and Seguin the temperature dependence of CD is equal to $C_S D_S$ below $1000^\circ C$, since CD/$C_S D_S$ is almost constant, about two, in the temperature range. Therefore the calculation (based on the self-diffusion coefficient by Kalinowski and Seguin) strongly suggests that the driving force is V undersaturation in the entire temperature range. These results are summarized in Table II.

TABLE II
Driving force of BSF growth

Self-difusion coefficient	Result	Driving force	CD =
Mayer et al.	CD>$C_S D_S$	I super	$C_I D_I$
Kalinowski et al.	CD<$C_S D_S$	V under	$C_{VE} D_V$

Therefore, in the following we will discuss on two models for the growth kinetics of BSF, I in supersaturation and V in undersaturation, and suggest that BSF grows by V in undersaturation.

I in supersaturation or V in undersaturation. We will first calculate the diffusion coefficient of I using the relation CD = $C_I D_I$ in Table II. C_I

has been determined in the following [23]. Let us recall that I generation from the interface by excess volume accomodation is not a rate-determining process since the growth of the oxide precipitates is purely diffusion-limited by oxygen interstitials, not generation-limited by I. This means that I in supersaturation has been generated during the first annealing at 750°C for 161 hrs. Assuming complete accomodation of excess volume by the precipitate formation, the number of I, n_{emi}, to be generated from one precipitate is given by

$$n_{emi} = B \, V_{OP} \, C_z, \tag{9}$$

where, $B = (C_S - C_y)/C_z = 0.67$, V_{OP} is volume of the precipitate, C_y and C_z are concentrations of silicon and oxygen atoms in cristobalite equal to 2.1×10^{22} and 4.2×10^{22} at./cm^3, respectively. Since the precipitates formed during the first annealing are 600 Å in half diagonal length and 40 Å in thickness, the number of I, n_{emi}, is estimated to be 6×10^5 atoms/precipitate. On the other hand, the number of I, n_{agg}, agglomerated into one BSF (r=500 Å) at the growth halt stage shown in Fig.7 is calculated to be 1×10^5 atoms/BSF in such a way, which is about six times smaller than n_{emi}. The difference in number is probably due to incomplete accomodation of strain around the precipitate, over-estimation of the precipitate thickness and so on. From the difference the range of B is given by

$$0.67/6 = 0.11 < B < 0.67 \tag{10}$$

Therefore, the initial concentration of I is expressed by

$$C_I = n_{emi} \, N_{OP} = B \, (C_i - C_e), \tag{11}$$

where $(C_i - C_e)$ is the concentration of excess oxygen interstitials.

The diffusion coefficient of I can be calculated by substituting Eq.(11) into Eq.(7c) and is shown in Fig.10. The highest value in B=0.67 corresponds to the lowest value in the diffusion coefficient. The temperature dependence of the diffusion coefficient is given by

$$D_I = D_0 \exp(-4.4/kT). \tag{12}$$

$$1.9 \times 10^6 < D_0 < 1.1 \times 10^7. \tag{13}$$

The diffusion coefficient of I obtained here (especially $D_0 = 1.1 \times 10^7$) agrees well with the previously estimated values; 1×10^{-6} cm^{2}s^{-1} at melting point by Seeger et al. [27] and 1.1×10^{-9} cm^{2}s^{-1} at 1100°C by Mizuo and Higuchi [28]. However, we have to point out two difficulties with the diffusion coefficient. Since the oxide precipitate growth is diffusion-limited by oxygen interstitials above 750°C, diffuse away of self-interstitials formed by the precipitate growth should be much faster than duffuse in of oxygen interstitials above the temperature. However, the diffusion coefficient of I becomes smaller than that of oxygen interstitials below about 900°C. This is a difficulty arising from the activation energy, 4.4 eV. Another difficulty is arising from the pre-exponential factor [29]. The pre-exponential factor of D_I and D_S is expressed by

$$D_0 \propto \exp(S_m / k) \tag{14}$$

$$D_{SO} \propto \exp((S_f + S_m)/k) \tag{15}$$

Since D_{SO} ranged from 10^2 to 10^4, D_{SO}/D_0 is calculated to be 10^{-2} to 10^{-5}. This results in negative S_f which does not make sense.

Therefore, it is not likely that the driving force is I in supersaturation. In the following , we will examine model based on V in undersaturation. Following the model, the vacancy component of self-diffusion coefficient can be obtained, using the relation $CD = C_{VE}D_V$ in Table II .

Since f_V and g_V are regarded as 0.5 and 1, respectively, the vacancy component of self-diffusion coefficient is given by $1/2 \; CD/C_S$, as deduced from Eq.(8). Figure 11 shows the temperature dependence of the vacancy component calculated. The vacancy component is given by

$$1/2 \; C_{VE}D_V = 110 \; \exp(-4.4 \; / \; kT) \tag{16}$$

Quite recently, Kitagawa et al. [30] have reported on dissociative diffusion of Ni at 900°C. The vacancy component of self-diffusion coefficient calculated by Goesele and Tan [31] from the result agrees well with the present result. The result by Kalinowski and Seguin is in very good agreement with the present result below 1000°C. This indicates that V contributes more to self-diffusion at least below 1000°C. The result by Mayer et al. (measured between 1047° and 1387°C) is also in good agreement with the present result, but the activation energy, 5.02 eV, is slightly different from that of the present result. This difference might be explained by self-interstitial component to be predominant at higher temperature, as suggested by Seeger and Chik [32] or by change of diffusion mechanism in terms of a relaxation effect of V at higher temperature.

The bending of the present data at higher temperature is presumably due to reduction of degree of V undersaturation with elevating temperature, since shrinkage of precipitates has been observed above 1100°C [23]. We can calculate C_{VE} from this result.

We conclude that the driving force of BSF growth can be quantitatively understood to be V in undersaturation. The present result indicates a possibility that V is also responsible point defects for growth and shrinkage of OISF.

NUCLEATION OF OXIDE PRECIPITATES

The nucleation kinetics is less understood than the growth kinetics. One of the reasons is lying on the variety of starting specimens. We will remark a possibility that initial states of silicon specimens used in experiments would be very different: As-grown or as-received , crystal cooling history and so on. For this reason we will first consider briefly what has happened during cooling of crystal growth. Then we will discuss on the phenomena during annealing.

Phenomena during cooling of crystal growth.   Oxygen interstitials are supersaturated during cooling of crystal growth. They tend to cluster at some growth irregularities, if there has been, to form large precipitates of several hundred to a few thousand $\overset{\circ}{A}$ in as-grown crystals [33]. However, most of them are to form small clusters, embryos, at sites of oxygen interstitials (homogeneous process) or at the other sites (heterogeneous process). Formation of the steady state embryo distribtuion requires a certain length of time, i.e., induction time (short at high temperature and long at low temperature). Therefore, in cooling process of crystal growth the embryo distribution is considered to be easily formed at higher temperature. Piling up of embryo distribution will be finished at a certain temperature due to increasing of induction time. Thus, two-types of frozen-in distributions of embryos have been formed by homogeneous and heterogeneous process and coexist in as-grown crystals. Oxide precipitates will be simultaneously nucleated from these embryo distributions during cooling. Therefore, the frozen-in distributions of embryos from homogeneous and heterogeneous process, resulting microprecipitates from these embryo distributions, and large precipitates formed at growth irregularities would be contained in as-grown crystals. What we should note here is that the shape of these embryo distributions and size and density of micro- and large precipitates strongly depend on concentration of oxygen interstitials and heterogeneous nucleation sites, growth condition and cooling history. Furthermore, additional heat treatments such as donor killer annealing around 650°C and homogenization at high temperature completely change the initial conditions of starting wafer. Thus, people have used various wafers with very different initial conditions. We should establish a certain procedure to set the initial conditions to clarify the nucleation kinetics; e.g. a set of high temperature homogenization and gradual cooling at a certain rate.

Nucleation during annealing. The precipitate density increases during annealing. Three models have been proposed for the increase: Growth [34], homogeneous nucleation [35] and heterogeneous nucleation models [36].

  Growth model. This model claims that the increase of the precipitate density is due to only growth of pre-existing microprecipitates noted above [34]. It has been reported that high temperature annealings above 1050°C suppress the precipitation in subsequent lower temperature annealings around 800°C [34]. Following this model, high temperature annealings have redissolved the preexisting microprecipitates, which results in no precipitation observed in subsequent low temperature annealing. However, it has been recently found that the nucleation in high temperature preannealed specimens was not suppressed but was retarded for a long duration, as shown in Fig.12 [37]. Hu has also observed the precipitation retardation after high temperature annealing [9]. This phenomenon can not be explained by the growth model. We consider this phenomenon as time-lag in nucleation [37], which is expressed to be $t_i > t$ in the following equation (16).

  Nucleation model. Nucleation of the oxide precipitates has occured during annealing, as demonstrated above. Next problem is to clarify that either process is responsible for the observed precipitate nucleation. General equation for nucleation rate J is expressed by

$$J = Z \, w^* \, N \, \exp(-\Delta G/kT) \, \exp(-t_i/t), \qquad (16)$$

$$w^* = 4\pi(r^*)^2 \, n \, D_i/a, \qquad (17)$$

$$J_{total} = J_{homo} + J_{hetero}, \qquad (18)$$

where Z is Zeldovich non-equilibrium factor, w^* frequency factor, on the rate at which single atom joins the critical nucleus, N atomic nucleation site density, ΔG free energy for formation of the critical nucleus, t_i induction time, r^* critical radius, n concentration of single atoms and a jumping distance. Differences between homogeneous and heterogeneous nucleation are briefly summarized in Table III.

TABLE III
Main differences between homogeneous and heterogeneous nucleation

	Homogeneous nucleation	Heterogenous nucleation
N	oxygen concentration	impurity (or complex) concentration
ΔG	$\Delta G(\Delta G_v, \sigma, W)$	$\Delta G(\Delta G_v, \sigma, W, E_b, ...)$
r^*	$r^*(\Delta G_v, \sigma, W)$	$r^*(\Delta G_v, \sigma, W, E_b, ...)$

Here, ΔG_v is volume free energy change, σ interfacial energy, W volume strain energy, E_b binding energy between heterogeneous site and oxygen. Based on the assumption that J_{hetero} and W is negligible, Freeland et al. [35] have proposed that the precipitates are nucleated through homogeneous nucleation and Osaka et al. [38] have quantitatively analyzed the nucleation kinetics by

134

Eq.(16), as shown in Fig.13. However, some important findings have recently been reported: (1) presence of oxygen related precipitation and carbon related precipitation [39], (2) reduction of carbon concentration accompanied with annealing [40], (3) new donor concentration increase related to carbon reduction [41], (4) carbon/oxygen complex presence [42], (5) oxidation-retarded precipitation [43]. Pinizzotto et al. have reported that in specimens with high concentration of oxygen interstitials the oxygen interstitials have precipitated independently of carbon whereas in specimens with low concentration of oxygen interstitials carbon effects precipitation [39]. In their experiment all specimens have been preannealed at $1350^{\circ}C$ and gradually cooled to room temperature. Therefore, differences in shape of the embryo distribution and in size and density of micro- and large precipitates is eliminated by this high temperature annealing. Their result first suggests qualitatively that both homogeneous nucleation and heterogeneous nucleation occur simultaneously. In order to discuss on the nucleation kinetics we have to clarify that either process is responsible for the experiment reported so far (e.g. Fig. 13). On the other hand, there is a implication that formation of embryo distributions for homogeneous and heterogeneous nucleation might not be independent events. Thermal donor, which is produced by $450^{\circ}C$ annealing and are proposed to be SiO_4 [44], is a candidate of embryos for homogenous nucleation. It has been clearly shown that the concentration of thermal donor decreases about one order when carbon concentration becomes one order high [40]. This suggests that the total nucleation rate is not expressed by linear combination like Eq.(18). Further studies are necessary to elaborate the coexistence model.

Behaviors of carbon as heterogeneous sites are rather complicated. As deduced from Eq.(16) and table III, in zero order approximation, the nucleation rate is proportional to N, carbon concentration. Therefore, amount of oxygen precipitation should be proportional to carbon concentration. However, correlation between oxygen reduction ratio and total concentration of carbon is very weak [13]. Similar result is obtained in relation between new donor density and carbon concentration [41]. This suggests that single carbon atom does not act as nucleation sites. The findings (2), (3) and (4) imply the possibility that carbon complex are heterogeneous nucleation sites.

Based on his finding (5), Hu proposed a model that V clusters play as nucleation site of the precipitates to accomodate strain energy due to volume expansion by embryo formation. Following his model, I injected by oxidation eliminates V cluster via pair annihilation to retard the precipitation. However, his model can not explain time-lag in nucleation shown in Fig.12. A different viewpoint is given by Schaake et al. [45]. Assuming that composition of embryo is SiO_2, reaction for embryo formation is accompanied with emission of I.

$$Si + 2 O_I \rightarrow SiO_2 + B I. \tag{19}$$

In this chemical reaction, SiO_2 favors to decomposite into Si and O_I when concentration of I increases. This indicates that embryo distribution shrinks by oxidation and the nucleation of the precipitates becomes retarded. This idea is interesting.

In summary, it is likely that homogeneous nucleation and heterogeneos nucleation occur simultaneously. Homogeneous and heterogeneous nucleations might not be independent events. Nature of heterogeneous nucleation sites is related to carbon atoms but is not likely single atoms. The nucleation mechanism is beginning to be understood. Further experimental information on specimens with well specified initial conditions and theoretical effort is necessary to understand quantitatively the nucleation kinetics.

Acknowledgment

The authors would like to thank A. Osawa, Y. Matsushita and H. Tsuya for their helpful discussions. One of the authers (K. Wada) is grateful to J. R. Patel for valuable suggestions on the theories of growth kinetics.

References

1. K. Tempelhoff, F. Spieglberg and R. Gleichmann in: Semiconductor Silicon 1977, H. R. Huff and E. Sirtl, eds. (The Electrochemical Society, Princeton, 1977), p.585.
2. D. M. Maher, A. Staudinger and J. R. Patel, J. Appl. Phys. 47, 3813 (1976).
3. J. R. Patel in: Semiconductor Silicon 1981, H. R. Huff, R. J. Kriegler and Y. Takeishi, eds. (The Electrochemical Society, Pennington, 1981), p.189.
4. Y. Endo, Y. Yatsurugi, N. Akiyama and T. Nozaki, Analytical Chemistry 44, 2258 (1972).
5. K. Wada, N. Inoue and K. Kohra, J. Cryst. Growth 49, 749 (1980).
6. D. N. Seidman and R. W. Balluffi, Philos. Mag. 13, 649 (1964).
7. C. P. Flynn, Phys. Rev. 134, A241 (1964); ibid., 133, A587 (1964).
8. H. Buchholz, Electrische und Magnetische Potentialfelder (Springer-Verlag, Berlin, 1957), p.233.
9. S. M. Hu in: Defects in Semiconductors, J. Narayan and T. Y. Tan eds. (North Holland, New York, 1981), p.333.
10. U. Goesele and W. Frank in: Defects in Semiconductors, J. Narayan and T. Y. Tan eds. (North Holland, New York, 1981), p.55.
11. K. H. Yang, H. F. Kappert and G. H. Schwuttke, Phys. Stat. Sol. A50, 221 (1978).
12. M. Kahlweit in: Progre in Solid State Chemictry, H. Reiss eds. (Pergamon Press, New York, 1965), p.134.
13. R. A. Craven in: Semiconductor Silicon 1981, H. R. Huff, R. J. Kriegler and Y. Takeishi, eds. (The Electrochemical Society, Pennington, 1981), p.254.
14. Y. Matsushita, J. Cryst. Growth 56, 516 (1982).
15. S. M. Hu, J. Appl. Phys. 45, 1567 (1974).
16. K. Wada, H. Takaoka, N. Inoue and K. Kohra, Jpn. J. Appl. Phys. 18, 1629 (1979).
17. T. Iizuka, Jpn. J. Appl. Phys. 4, 1018 (1966).
18. J. R. Patel, K. A. Jackson and H. Reiss, J. Appl. Phys. 48, 5279 (1977).
19. H. Takaoka, J. Osaka, and N. Inoue, Jpn. J. Appl. Phys. 18, Suppl. 18-1, 179 (1979).
20. K.Wada, unpublished.
21. B. Leroy, J. Appl. Phys. 50, 1567 (1979).
22. S. Mizuo and H. Higuchi, Jpn. J. Appl. Phys. 21, 281 (1982).
23. K. Wada and N. Inoue in: Defects and Radiation Effects in Semiconductors 1980, (Inst. Phys. Conf. Ser. 59) p.461 (1981).
24. S. Mizuo and H. Higuchi, Jpn. J. Appl. Phys. 20, 739 (1981).
25. H. J. Mayer, H. Mehrer and K. Maier in: Lattice Defects in Semiconductors 1976, (Inst. Phys. Conf. Ser. 31, 1977), p.186.
26. L. Kalinowski and R. Seguin, Appl. Phys. Lett. 35, 211 (1979).
27. A. Seeger, W. Frank and H. Foell in: Lattice Defects in Semiconductors 1976, (Inst. Phys. Conf. Ser. 31, 1977), p.12.
28. S. Mizuo and H. Higuchi, Jpn. J. Appl. Phys. 21, 272 (1982).
29. T. Y. Tan, IBM Thomas J. Watson Research Center, Yorktown Heights, N. Y. private communication.
30. H. Kitagawa, K. Hashimoto and M. Yoshida, Jpn. J. Appl. Phys. 21, 276 (1982).
31. U. Goesele and T. Y. Tan, these proceedings.
32. A. Seeger and K. P. Chik, Phys. Status Solidi 29, 455 (1968).
33. K. Wada, H. Nakanishi, H. Takaoka and N. Inoue, J. Cryst. Growth 57, 535 (1982).
34. S. Kishino, Y. Matsushita, M. Kanamori and T. Iizuka, Jpn. J. Appl. Phys.

21, 1 (1982).

35. P. E. Freeland, K. A. Jackson, C. W. Lowe and J. R. Patel, Appl. Phys. Lett. 30, 31 (1977).

36. A. J. R. Dekock and W. M. van de Wijgert, J. Cryst. Growth 49, 718 (1980).

37. N.Inoue, K. Wada and J. Osaka in: Semiconductor Silicon 1981, H. R. Huff, R. J. Kriegler and Y. Takeishi, eds. (The Electrochemical Society, Pennington, 1981), p.282.

38. J. Osaka, N. Inoue and K.Wada, Appl. Phys. Lett. 36, 288 (1980).

39. R. F. Pinizzotto and S. Marks, these proceedings.

40. J. Leroueille, Phys. Status Solidi (a) 67, 177 (1981).

41. A. Osawa, R. Takizawa, K. Honda, A. Shibatomi and S. Ohkawa, J. Appl. Phys. 53, 5733 (1982).

42. G. S. Oehrlein, J. L. Lindstroem and J. W. Corbett, Appl. Phys. Lett. 40, 241 (1982).

43. S. M. Hu, Appl. Phys. Lett. 36, 561 (1980).

44. W. Kaiser and P. H. Keck, J. Appl. Phys. 28, 882 (1957).

45. H.F.Schaake, S. C. Baber and R. F. Pinizzotto in: Semiconductor Silicon 1981, H. R. Huff, R. J. Kriegler and Y. Takeishi eds. (The Electrochemical Society, Pennington, 1981), p.273.

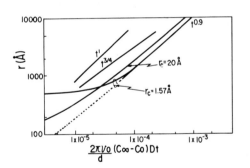

Fig. 2 Growth kinetics of two-dimensional defects by toroidal approximation.

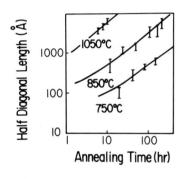

Fig. 3 Growth of oxide precipitates.

137

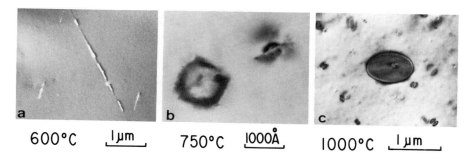

600°C ___1µm___   750°C ___1000Å___   1000°C ___1µm___

Fig. 1 Microdefects induced by annealing. a) Rod-like defects
b) Oxide precipitates c) Stacking fault

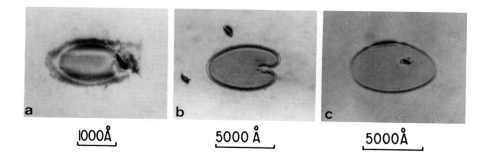

___1000Å___   ___5000 Å___   ___5000Å___

Fig. 4 Succesive growth process of stacking faults.

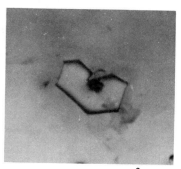

Fig. 6 Kidney-shaped
perfect dislocation loop.

___5000Å___

138

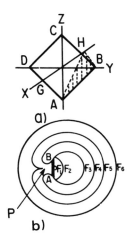

a)

b)

P

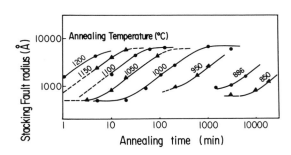

Fig. 7 Growth of bulk stacking faults.

Fig. 5 Schematics of growth mechanism of stacking fault. a) spacial configuration b) growth process

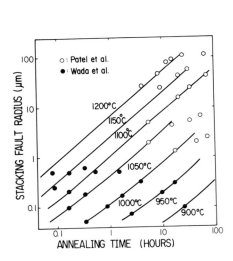

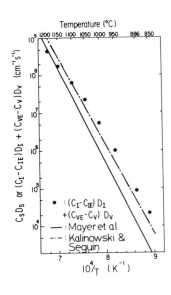

Fig. 8 Growth of bulk stacking faults.

Fig. 9 Comparison between CD and $C_S D_S$.

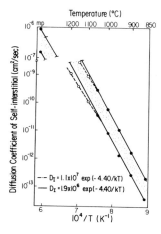

Fig. 10 Result based on
assumption that driving force
of bulk stacking faults is
self-interstitials in
supersaturation.

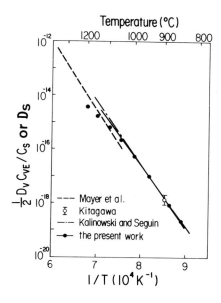

Fig. 11 Result based on vacancies
undersaturation.

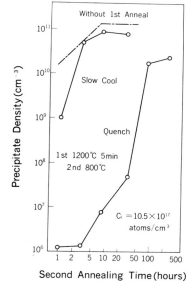

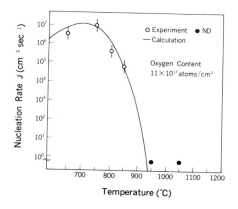

Fig. 13 Nucleation rate vs.
Temperature.

Fig. 12 Time-lag in nucleation.

OBSERVATION OF OXIDATION-ENHANCED AND -RETARDED DIFFUSION
OF ANTIMONY IN SILICON: THE BEHAVIOR OF (111) WAFERS

T. Y. TAN AND B. J. GINSBERG
IBM Thomas J. Watson research Center, Yorktown Heights, New York, 10598

ABSTRACT

An experiment was performed to study the oxidation-enhanced and -retarded diffusion (OED and ORD) of Sb in silicon wafers oxidized in dry O_2 at 1160°C. The ORD data of (100) wafers agree well with the prediction of a model assuming that Si self-interstitials (I) and vacancies (V) coexist in thermal equilibrium at high temperatures. An adjustment of the I supersaturation values is needed to bring the ORD/OED data of (111) wafers to fit with the model satisfactorily. This indicates the existence of a mechanism which injects V into (111) wafers in addition to the normal mechanism of I injection due to SiO_2 growth.

INTRODUCTION

By analogy to metals it was believed for a long time that vacancies (V) is the only thermal equilibrium point defect species in Si at high temperatures. Coexistence of V and Si self-interstitials (I) was suggested by Seeger and Chik [1] and by Hu [2] but was questioned by others [3-5]. Recent experimental and theoretical progresses, however, showed that coexistence of I and V is most likely. The analysis of Au diffusion [6] led to the conclusion that I is present, and the analysis of oxidation-enhanced and -retarded diffusion (OED and ORD) of Sb and other substitutional dopants [7-10] led to the conclusion that I and V coexist. For the analysis of OED/ORD phenomena, it was shown that qualitative discrepancies exist between experimental results and predictions of models based on the assumption that either I or V is the only point defect species [8,10]. On the other hand, all available experimental results are in qualitative agreement with a model assuming that I and V coexist, and the ORD results of Sb [9,11] agree well with the model on quantitative basis [7-10]. The analysis of the substitutional dopant OED/ORD phenomena led to an expression for the time averaged incremental diffusivity of the dopant given by [7]:

$$\delta \bar{D}_s^{ox} = \bar{D}_s^{ox}/D_s - 1 = (2G_I + G_I \bar{s}_I - 1)\bar{s}_I/(\bar{s}_I + 1). \qquad (1)$$

In Eq. (1) $\bar{s}_I = (\bar{C}_I - C_I^{eq})/C_I^{eq}$ is the time averaged supersaturation ratio of I caused by oxidation, with $\bar{C}_I$ and C_I^{eq} being the (time averaged) actual and thermal equilibrium I concentration respectively, G_I is the fractional I-component of the dopant diffusivity in thermal equilibrium, i.e., $G_I = D_s^I/D_s$ with $D_s = D_s^I + D_s^V$. In deriving Eq. (1) the local equilibrium condition that $C_I C_V = C_I^{eq} C_V^{eq}$ [12] was used. Equation (1) is plotted in Fig. 1 for three different G_I values: 0.7, 0.5 and 0.02. Briefly, the main features are that: for $G_I=0.5$, it is improbable to observe ORD; for $G_I>0.5$, ORD is observable for a range of $\bar{s}_I$ values smaller than zero, i.e., with a net I undersaturation; for $G_I<0.5$, ORD is observable for a range of $\bar{s}_I$ values larger than zero, i.e., with a net I supersaturation. In order to use Eq. (1), $\bar{s}_I$ values must be

Mat. Res. Soc. Symp. Proc. Vol. 14 (1983) © Elsevier Science Publishing Co., Inc.

known. For oxidation of (100) Si wafers in dry O_2 for a few hrs the $\bar{s}_I$ values may be obtained from the growth/shrinkage kinetics of the oxidation-induced stacking faults (OSF) [7]:

$$\bar{s}_I^{ox}=8.8 \times 10^{-9} t^{-1/4} \exp(2.52/kT),\tag{2}$$

where kT is the thermal energy in eV and t the oxidation time in sec. Changing the ambient conditions and wafer orientations affects the $\bar{s}_I$ values. For (111) wafers oxidized in dry O_2, Eq. (2) needs to be multiplied by a factor of 0.6 to 0.7; for (100) wafers oxidized in wet O_2, Eq. (2) needs to be multiplied by a factor of , e.g., 1.7. These factors are Leroy's OSF size multiplication factors [13]. For more general cases for which two or more chemical reactions may have occurred independently at the growing SiO_2-Si interface, the use of a multiplication factor fails. Instead, separate terms must be added to Eq. (2) to obtain the total $\bar{s}_I$ value. We have found that for oxidation of (100) wafers in dry O_2 containing a chlorine species, $\bar{s}_I=\bar{s}_I^{ox}-\bar{s}_I^{Cl}$ holds, with $\bar{s}_I^{Cl}=at^{m-1}$ represents the effect of Cl (the constants a and m can be determined from experimental results)[14].

Also shown in Fig. 1 are: (i) The Sb ORD data of Mizuo and Higuchi [11]; (ii) The P OED data of Mizuo and Higuchi [11], Antoniadis et al. [15] and Lin et al. [16] for T≥1100°C; (iii) The P ORD data of Francis and Dobson [17]. Except for the P ORD data of Francis and Dobson for which the $\bar{s}_I$ and $\bar{D}_S^{ox}$ values are schematical, for all other data points the $\bar{s}_I$ values were calculated from Eq. (2) and the $\delta\bar{D}_S^{ox}$ values from the original author's data. It can be seen that the G_I=0.02 curve fit the Sb ORD data well and we therefore regard the model as satisfactory. While no curve with a single G_I value can seem perfectly fit the few available P OED data, it nevertheless can be seen that G_I≥0.5 fop P with T≥1100°C. This means for observing P ORD, there has to be a net V supersaturation. The P ORD data was obtained from a (111) wafer oxidized in dry O_2 for 17.5 hrs at 1160°C, and Francis and Dobson suggested that the V injection resulted from solid state diffusion of Si atoms at the SiO_2-Si interface into the growing SiO_2. In view of the very low diffusivity of Si in SiO_2, we have suggested the formation of gaseous SiO molecules and their diffusion into the SiO_2 as a more likely alternative [7]. Clearly, an experiment yielding some additional Sb ORD data would now help to provide a further check of the correctness of the model. Moreover, under similar experimental conditions for observing P ORD [17] Sb should show OED, since the model predicts opposite behaviors between these two elements on qualitative basis, see Fig. 1. This occurs for $\bar{s}_I$<0, and judging from the conditions for observing P ORD, it is fulfilled only for oxidizing (111) wafers at a very high temperature for very long times. Knowledge derived from OSF studies, i.e., Eq. (2) multiplied by a factor of 0.6 to 0.7, does not readily provide this information. This may be simply due to that there seems to exist only one set of usable OSF size data for (111) wafers with a three hr oxidation time [18] and Leroy's theoretical fitting of the data [13] did not yield results that are reliable enough for extrapolations. Hence, such an experiment would also provided an evidence that a net V injection can occur for (111) wafers oxidized in dry O_2. The purpose of this paper is to report that we have carried out an experimental study of Sb diffusion in dry O_2 oxidation and observed both ORD and OED. In particular, the Sb OED is observed under conditions ORD was observed for P.

EXPERIMENTAL

Our experiment consisted of p-type Czochralski Si with a resistivity of 20 Ω -cm with both (100) and (111) orientatiions. A dose of $7 \times 10^{13} cm^{-2}$ of Sb^+ ions were implanted into the wafers at 400 kev and subsequently annealed at 1000°C for 20 min. in N_2 to remove damages. A 1.5μm thick undoped epitaxial Si layer was then grown by CVD deposition at 1050°C. A 50nm thick SiO_2 layer was thermaly grown on the wafers and followed by CVD deposition of a 200nm thick Si_3N_4 layer. The Si_3N_4/SiO_2 layers were selectively removed by photo-etching with the use of a conveniently available photo-mask. The wafers were oxidized at 1160°C in dry O_2 for 0.67, 1.5, 3.33, 9 and 18.5 hrs. The oxidized wafers were angle lapped and stained for observing the junction depth using an optical microscope.

RESULTS AND DISCUSSION

As a result of this study, both ORD and OED of Sb were observed. ORD was observed for (100) wafers for all five oxidation times and for (111) wafers oxidized up to 9 hrs. The 18.5 hr (111) wafers, however, showed OED. In Fig.2 we show the observed junction depth pictures of two (111) wafers, one oxidized for 3.33 hrs and the other for 18.5 hrs. It is seen that for the 3.33 hr sample the junction depth of the oxidized portion is shallower than that of the unox-idized portion while the reverse is true for the 18.5 hr sample, i.e., the 3.33 hr sample showed ORD while the 18.5 hr sample showed OED.

Data obtained from all cases were plotted in Fig. 3. It is seen that the five data points of (100) wafers fit the $G_I=0.02$ curve almost exactly. Thus, our results agree well with the prediction of the model and the results of Mizuo and Higuchi. This provides a substantial support to the conclusion that our model is essentially correct [7]. Data from the (111) wafers, however, exhibited a quite different behavior. With the $\bar{s}_I$ values for (111) wafers taken as $0.65\bar{s}_I^{ox}$, all five data points lie above the curve and at best only the three data points with shorter oxidation times may be regarded as fitting satsi-factorily with the curve. Deviations of the two longer time data from the curve are quite large, and, in particular, the 18.5 hr data lies on the OED side of the plot which is a qualitative discrepancy. The use of just another multiplication factor to obtain the $\bar{s}_I$ values, e.g., 0.3 instead of 0.6-0.7, has failed to resolve this discrepancy, and we interpret this anomalous (111) wafer behavior as due to the existence of another point defect generating mechanism operating at the SiO_2-Si interface when oxidized at very high temper-atures. In analogy to the effect of Cl on V injection, this mechanism apparent-ly also injects V and it acts independent of the I injection caused by SiO_2 growth. This means the effect is additive to the $\bar{s}_I^{ox}$ values rather than multiplicative. In order to explain the present results, we choose

$$\bar{s}_I = 0.65\bar{s}_I^{ox} + \delta\bar{s}_I \qquad (3)$$

with $\qquad \delta\bar{s}_I = -6\times10^{-6}t. \qquad (4)$

The (111) wafer data corrected by the use of Eqs. (3) and (4) were also shown in Fig. 3 and it is now seen that all five data fit the $G_I=0.02$ curve quite nicely, and, in particular, the 18.5 hr data lies on the $\bar{s}_I < 0$ side of the plot and hence the qualitative discrepancy removed.

CONCLUSION

We consider that the existence of a V injection mechanism for (111) Si wafers oxidized in dry O_2 as established, since only then can our present Sb ORD/OED data be explained consistently with the (100) wafer data. The V injection is characterized by a constant rate and is independent of that of I injection due to SiO_2 growth. This is reasonable since the V injection should be due to diff-usion of Si into SiO_2 in the form of atoms or SiO molecules [7,17].

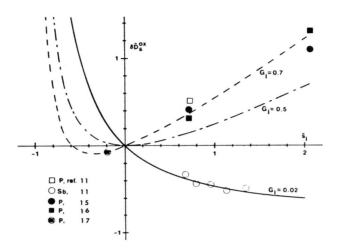

Fig. 1. Calculated curves per Eq. (1) for G_I=0.7, 0.5 and 0.02 cases. Also shown are available ORD data of Sb and OED and ORD data of P.

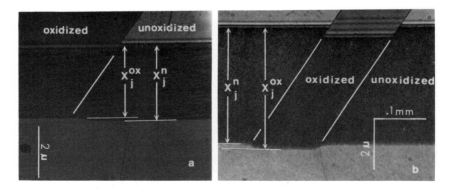

Fig. 2. Typical appearences of the diffused Sb junctions of (111) samples. (a) Oxidized for 3.33 hrs; (b) Oxidized for 18.5 hrs. Notice the reversal from ORD shown in (a) to OED shown in (b).

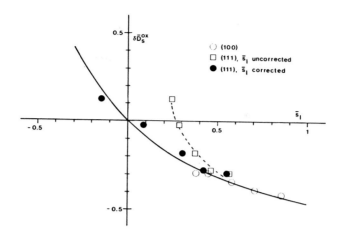

Fig. 3. The present ORD and OED data fitted to the $G_I=0.02$ curve. A dotted line was schematically fitted to (111) wafer data before a correction in $\bar{s}_I$ values was made. See text for other details.

REFERENCES

1. A. Seeger and K. P. Chik, Phys. Stat. Sol. 29, 455 (1968).
2. S. M. Hu, J. Appl. Phys. 45, 1567 (1974).
3. D. Shaw, Phys. Stat. Sol. b72, 11 (1975).
4. J. A. van Vechten, Phys. Rev. B17, 3197 (1978).
5. J. C. Bourgoin and M. Lannoo, Rad. Effects 46, 157 (1980).
6. U. Gosele, W. Frank and A. Seeger, Appl. Phys. 23, 361 (1980).
7. T. Y. Tan and U. Gosele, Appl. Phys. Lett. 40, 616 (1982).
8. T. Y. Tan and U. Gosele, in Extended Abstracts 82-1 The Electrochemical Society (Electrochemical Society, New Jersey, 1982) p. 314.
9. D. A. Antoniadis and I. Moskowitz, J. Appl. Phys. in press.
10. T. Y. Tan, U. Gosele and F. F. Morehead, to be published.
11. S. Mizuo and H. Higuchi, Jpn. J. Appl. Phys. 20, 739 (1981).
12. E. Sirtl, in Semiconductor Silicon 1977 H. R. Huff and E. Sirtl eds (Electrochemical Society, New Jersey, 1977) p.4.
13. B. Leroy, J. Appl. Phys. 50, 7996 (1979).
14. T. Y. Tan and U. Gosele, J. Appl. Phys. 53, 4767 (1982).
15. D. A. Antoniadis, A. M. Lin and R. W. Dutton, Appl. Phys. Lett. 33, 1030 (1978).
16. A. M. Lin, D. A. Antoniadis and R. W. Dutton, J. Electrochem. Soc. 128, 1131 (1981).
17. R. Francis and P. S. Dobson, J. Appl. Phys. 50, 280 (1979).
18. S. M. Hu, Appl. Phys. Lett. 27, 165 (1975).

CARBON AND THE KINETICS OF OXYGEN PRECIPITATION IN SILICON

R.F. Pinizzotto and S. Marks
Central Research Laboratories, Texas Instruments Incorporated
P.O. Box 225936, Dallas, TX 75265

ABSTRACT

Oxygen precipitation in Czochralski silicon has been studied as a function of anneal time, oxygen concentration and carbon concentration using FTIR. It was found that the oxygen supersaturation controls the precipitation kinetics in high oxygen content samples, whereas the carbon concentration is of prime importance in low oxygen content samples. The decrease in sustitutional carbon concentration after nucleation and its subsequent increase with extended growth anneals supports the view that carbon affects precipitate nucleation, but not precipitate growth. The measured oxygen solubility at 1000°C was found to depend on both the initial oxygen concentration and the initial carbon concentration.

INTRODUCTION

Interstitial oxygen in the Czochralski grown silicon used for the manufacture of integrated circuits can cause beneficial and detrimental effects simultaneously. Interstitial oxygen and/or small oxygen clusters [1,2] strengthen silicon wafers and inhibit process induced warpage. Intrinsic gettering can be used to remove unwanted impurities from the active device regions [3,4]. Oxygen donors formed at various temperatures can significantly alter the electrical properties of the material [5]. As a result, the behavior of oxygen in silicon has received intense study in the past several years [6,7]. In addition, the effects of carbon on both oxygen precipitation and oxygen donor formation have been investigated in detail [3-12]. However, these studies were done at low temperatures [8,11], with defect centers generated by irradiation [9], or with a small number of samples [10]. The goal of the work reported here was to study a series of samples with various oxygen and carbon concentrations in an attempt to unravel the causes of some of the synergistic effects demonstrated by the two impurities.

EXPERIMENTAL DETAILS

The samples were silicon disks 12 mm in diameter and 2.3 mm thick. The disks were cut from the center of thick slices of <100> boules grown by the Czochralski technique. The silicon was p-type with a resistivity of 8-10 ohm-cm. The disks were metallurgically polished to a mirror finish on both sides to minimize scattering during the concentration measurements. The carbon and oxygen concentrations were measured with an IBM Model 98 Fourier Transform Infrared Spectrometer (FTIR). This technique measures the concentrations of interstitial oxygen and substitutional carbon, not the total bulk concentrations. Oxygen and carbon in other forms cannot be quantified. The measurements were made in vacuum using a mid-IR DTGS detector with a KBr window. The first version of the "Carbox" software supplied by IBM was used for data reduction. The sample thicknesses were measured with a micrometer to

Mat. Res. Soc. Symp. Proc. Vol. 14 (1983) © Elsevier Science Publishing Co., Inc.

an accuracy of 0.1%. The 95% confidence detection limits for this particular experimental apparatus (using the methods of Currie [13]) are 1.4 x 10^{16} atoms/cm^3 for carbon and less than 0.5 x 10^{16} atoms/cm^3 for oxygen. The calibration of Kaiser and Keck [14] was used to convert the measured absorbances to atomic concentrations of oxygen and that of Newman and Willis [15] was used for carbon. The IBM instrument yields oxygen concentration values that are 16% lower than those measured with a Nicolet FTIR and carbon concentrations that are 74% lower. The discrepancy is caused by the different techniques used for obtaining the sample thickness. The Nicolet instrument uses the thickness determined from the absorbancy of the sample, whereas the actual mechanically measured thickness was used as an input to the IBM software. The samples were loaded into an automatic sample changer. Position 0 was left empty and a sample of silicon grown by the float zone method was placed in position 1. FTIR spectra were obtained for the blank and for the FZ sample. The ratio of the FZ Si to the blank was stored on disk as the reference spectrum for the Carbox program. For each of the samples, the blank was remeasured and two separate spectra were obtained. The ratios of the experimental sample spectra to the new blank spectrum were used as inputs to the Carbox program. The FZ spectra were used to check instrumental reproducibility from run to run. The calculated oxygen and carbon concentrations were always less than the stated detectability limits when two different FZ spectra were used for the reference and the unknown.

The oxygen and carbon concentrations were measured after crystal growth and the resistivity stabilization anneal, but before additional processing. The samples were then annealed for 4 hours at 1350°C in a N_2 + 3% O_2 ambient. The concentrations were then remeasured. This high temperature solution anneal was designed to dissolve precipitate nuclei formed during the crystal growth and the resistivity stabilization anneal. A proprietary oxygen precipitation nucleation anneal was done in Ar and the oxygen and carbon concentrations were measured again. The samples were subsequently annealed in Ar at 1000°C to promote precipitate growth. The growth anneal was intermittently stopped to monitor the concentrations. The times chosen were 0.25, 0.5, 1.0, 1.5, 2.0, 3.0, 4.5, 6, 12, 105 and 205 hrs. The samples were cleaned using standard production processes both before and after each furnace cycle.

RESULTS AND DISCUSSION

The 13 samples were separated by oxygen concentration into three groups; the high oxygen concentration samples had approximately 1.5 x 10^{18} O/cm^3 (30ppm); the medium concentration samples had 1.4 x 10^{18} O/cm^3 (28ppm); the low oxygen concentration samples had 1.25 x 10^{18} O/cm^3 (25 ppm). These concentrations are within the specification range for wafers used in MOS processes that explicitly exploit intrinsic gettering.

The graphs in Fig. 1 show the behavior of both the interstitial oxygen and substitutional carbon concentrations as a function of time for all three sets of samples. The initial points are the concentrations measured for the as-received samples, that is after crystal growth and a resistivity stabilization anneal. The second set of points are the concentrations after the solution anneal. For all 13 samples, the oxygen concentration increased and the carbon concentration decreased after the solution anneal. The increase in oxygen content is due both to the dissolution of oxygen precipitate nuclei that were formed during crystal growth and the resistivity stabilization anneal, plus a small amount of diffusion of oxygen into the samples, since the oxygen solubility at 1350°C is 2 x 10^{18} O/cm^3. The decrease in carbon concentration is harder to explain. The solubility at 1350°C is 2.3 x 10^{17} C/cm^3, which is larger than the concentration in any of the samples. It was noticed, however, that a scale had formed on the samples during the anneal. The amount of scale was propor-

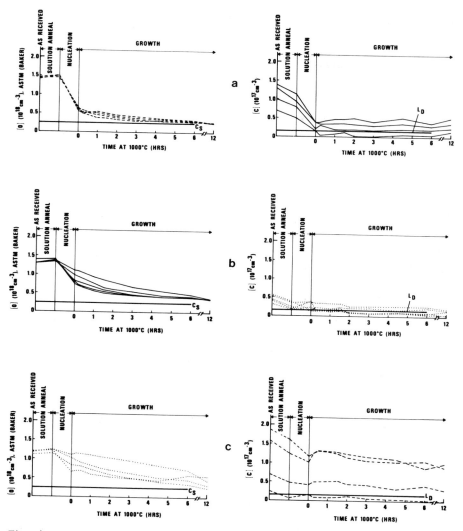

Fig. 1 Oxygen and carbon concentration as a function of time. The plots in Fig. 1a are for the high oxygen concentration samples. Fig. 1b and 1c are for the medium and low oxygen concentration samples, respectively.

tional to the initial carbon concentration in the sample. It is possible that a mixed oxide, nitride and carbide layer was formed that preferentially removed carbon from the sample.

The third points in Fig. 1 correspond to the concentrations measured after the nucleation anneal. The measured concentrations of both oxygen and carbon are dramatically reduced due to the formation of oxygen precipitate nuclei. The interdependence of the two species is complex. Figure 3a shows the dependence of the oxygen reduction on the initial oxygen concentration and Figure 3b relates the oxygen reduction to the initial carbon concentration. From Fig. 3a, it is seen that the oxygen reduction depends primarily on the initial oxygen concentration for the high and medium oxygen content samples. However, there is no obvious relationship between the oxygen reduction and the initial oxygen concentration for the low oxygen content samples. Conversely, there is no dependence of the oxygen reduction on initial carbon concentration for the high and medium content samples, but a strong dependence on carbon concentration for the low oxygen content samples (Fig. 3b). These observations can explain why various workers have reported conflicting effects of carbon on oxygen precipitation. A small change in oxygen concentration, for example from 1.25 to 1.4 x 10^{18} O/cm^3 (25 to 28 ppm) can change the experimentally determined dependence from that primarily controlled by carbon to a regime where the oxygen concentration is of paramount importance. For the high and medium oxygen concentration samples, the controlling factor is the oxygen supersaturation, whereas for the low oxygen content samples, the controlling factor is the effect of carbon on the formation of oxygen precipitate nuclei. The carbon may affect the nuclei in two ways. It can lower the interfacial free energy or it can lower the volume free energy. The result is the same in both cases: easier nucleation and faster precipitation kinetics.

After nucleation, the samples were annealed at 1000°C to promote precipitate growth. As shown in Fig. 1, the oxygen concentrations monotonically decrease with time. The carbon concentration is independent of time for the high and medium oxygen content samples, but decreases with time for the low oxygen content samples. There are two particular features that should be noted. The final oxygen concentration after 12 hours at 1000°C depends on both the initial oxygen concentration and the initial carbon concentration. This is particularly evident for the high and medium oxygen content samples. This behavior was found to persist even after 205 hours at 1000°C. The solubility values reported in the literature must, therefore, be used with caution, since these dependencies are not usually acknowledged. After 105 hours at 1000°C, the carbon concentrations in all the samples with carbon levels above the detectability limit were found to have increased over the values measured after 12 hrs at the same temperature. This data, along with the relatively time independent nature of the carbon concentration implies that carbon affects mainly the nucleation process and not the growth of previously formed nuclei. The rise in carbon concentration with extended anneal time is probably due to a reduction in the number of particles, the coarsening phenomenon. When a precipitate dissolves, it would release the carbon that it contained. This carbon would again become a substitutional impurity and be detected by FTIR. This is further illustrated in Fig. 4, which shows the time dependence of the carbon concentration in samples that received a slightly different nucleation anneal. During nucleation, no measures were taken to insure survivability of the nuclei at high temperatures. As a result, the carbon concentrations are found to immediately increase when the samples are annealed at 1000°C. The carbon concentrations continue to increase with time. This is due to the dissolution of nuclei that are too small to survive distinct the 1000°C process. All of these data are consistent with the existence of two nucleation mechanisms, one for low oxygen content samples and another for high oxygen content samples. The time independent nucleation rate of any solid state phase transformation can be found from [16]

$$J = N_0 bZ \exp [- \frac{16}{3} \pi \frac{v^2}{(kt)^3} \frac{\sigma^3}{(\ln S)^2}] \tag{1}$$

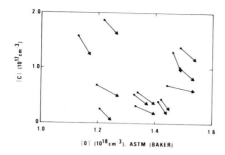

Fig. 2 Oxygen concentration versus carbon concentration. The starting points of the arrows are the concentrations in the as-received state. The arrows end at the values measured after the solution anneal.

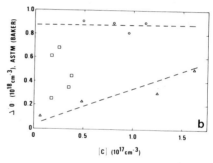

Fig. 4 The time dependence of the carbon concentration in samples where no measures were taken to insure the survivability of nuclei at high temperatures.

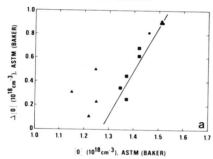

Fig. 3 The reduction in interstitial oxygen concentration as a function of: a. initial interstitial oxygen concentration, and b. initial carbon concentration.

where J = the nucleation rate, N_0 = the number of subcritical precipitate embryos, b = the capture rate, z = the Zeldovich factor, v = the atomic volume, T = the temperature, σ = the interfacial free energy and S = the oxygen supersaturation. Using the data of Fig. 1, a change in σ of 7% would be necessary to maintain a constant nucleation rate for the supersaturation difference of the high and low oxygen content samples. This value is several orders of magnitude greater than the estimated value based on the behavior of other materials [17] and agrees with the previous conclusion that the oxygen supersaturation must be the controlling factor in high oxygen content samples.

CONCLUSIONS

1. Carbon content does not strongly affect oxygen precipitation in high oxygen content samples (30 ppm), but does control the oxygen precipitation rate in low oxygen content samples (25 ppm).
2. Carbon is involved in precipitate nucleation, but not in precipitate growth.
3. The initial carbon and oxygen concentrations affect the final measured oxygen solubility.

152

ACKNOWLEDGEMENTS

It is our privilege to acknowledge our colleagues who made this work possible. In particular, G. Larrabee, H. Schaake, R. Hartzell, R. Massey and D. Heidt provided insightful critiques of our data interpretation. M. Jarvis contributed excellent technical support.

REFERENCES

1. S.M. Hu and W.J. Patrick, J. Appl. Phys., 46, 1869 (1975).

2. S.M. Hu, Appl. Phys. Letts., 31, 53 (1977).

3. R.A. Craven, Semiconductor Silicon/1981, edited by H.R. Huff, J. Kriegler and Y. Takeishi, The Electrochemical Society, Vol. 81-5, 1981, page 254.

4. S. Kishino, Y. Matsushita, M. Kanomori and T. Iizuka, Jap. J. Appl. Phys., 21, 1 (1982).

5. U. Gosele and T.Y. Tan, Appl. Phys. A, 28, 79 (1982).

6. Defects in Semiconductors, Proc. Mater. Res. Soc., Vol. 2, North-Holland, 1981.

7. Semiconductor Silicon/1981, The Electrochemical Society, Vol. 81-5, 1981.

8. J. Leroueille, phys. stat. sol. (a), 67, 177 (1981).

9. G.S. Oehrlein, D.J. Challou, A.E. Jaworowski and J. W. Corbett, Phys. Letts., 86A, 117 (1981).

10. G.S. Oehrlein, J.L. Lindstrom and J.W. Corbett, Appl. Phys. Letts., 40, 241 (1982).

11. A. Ohsawa, R. Takizawa, K. Honda, A. Shibatomi and S. Ohkawa, J. Appl. Phys., 53, 5733 (1982).

12. B.O. Kolbesen and A. Muhlbauer, Sol. St. Elec., 25, 759 (1982).

13. L.C. Currie, Anal. Chem., 40, 586 (1968).

14. W. Kaiser and P.H. Keck, J. Appl. Phys., 28, 882 (1957).

15. R.C. Newman and J.B. Willis, J. Phys. Chem. Sol., 26, 373 (1965).

16. H. Weidersich and J.L. Katz, Correleation of Neutron and Charged Particle Damage, National Technical Information Service, Springfield, VA, 1976, page 21.

17. L.E. Murr, Interfacial Phenomena in Metals and Alloys, Addison-Wesley, Reading, MA, 1975.

THERMAL DONOR FORMATION BY THE AGGLOMERATION OF OXYGEN IN SILICON

U. GÖSELE
Max-Planck-Institut für Metallforschung, Stuttgart, Fed. Rep. Germany

T. Y. TAN
IBM Th. J. Watson Research Center, Yorktown Heights, N. Y. 10598, USA

ABSTRACT

We suggest that thermal donor formation in silicon involves fast-diffusing, gas-like molecular oxygen in dynamical equilibrium with atomic oxygen in interstitial position. We will discuss still remaining difficulties in understanding thermal donor formation in the light of recent experimental observations by Stavola, Patel, Kimerling and Freeland, indicating that the diffusivity of interstitial oxygen apparently depends on the thermal history of the silicon sample.

INTRODUCTION

Most of the silicon used nowadays for the fabrication of electronic devices contains oxygen in the concentration range of about 5×10^{17} to 2×10^{18} cm^{-3}. Heating of such oxygen containing silicon between about 350-500°C for long times leads to the formation of so-called "thermal donors" which may convert lightly p-doped into n-type material [1]. Thermal donor formation has been investigated in detail with various experimental techniques such as electrical resistivity, electron paramagnetic resonance, Hall effect, infrared absorption, photo-luminescence and deep level transient spectroscopy, (for references, see [2]). In spite of these experimental efforts no agreement on the mechanism of thermal donor formation has been reached. This is partly so because thermal donor forma-tion is strongly influenced by a number of parameters such as the carbon con-centration [3] or the concentration of group III dopants [4]. In the following we will only deal with thermal donor formation in carbon-lean, near-intrinsic silicon.

DIFFUSION OF INTERSTITIAL OXYGEN

Oxygen is incorporated in silicon in slightly off-centered interstitial posi-tion between two neighbouring silicon atoms to which it is chemically bound as indicated in Fig.1 [5,6]. Oxygen in this configuration is further on denoted as O_i. The diffusivity D_i of O_i has been determined by various groups at tempera-tures ranging from the melting point down to 330°C; for references see [2]. The best fit is given by [7]

$$D_i = 0.17 \exp(-2.54 eV/kT) cm^2 s^{-1} , \qquad (1)$$

as also shown in Fig.2 as line 1. In Fig.2 we have also indicated as point 2 the oxygen diffusivity at 450°C measured by means of an out-diffusion experiment by Gaworzewski and Ritter [8]. This diffusivity is more than four orders of magnitude higher than predicted by eq. (1). Recently we suggested that this high diffusivity value is an effective oxygen diffusivity D_{eff} resulting from the influence of fast-diffusing, gas-like molecular oxygen O_2, which is in dynamical equilibrium with atomic oxygen O_i in interstitial position [2]. Before presenting a model of thermal donor formation which is based on the action of molecular oxygen we first discuss some essentials of the original thermal donor model of

154

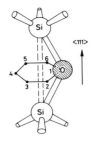

Fig. 1 (above):
Configuration of inter-
stitial oxygen O_i in
silicon [5,6].

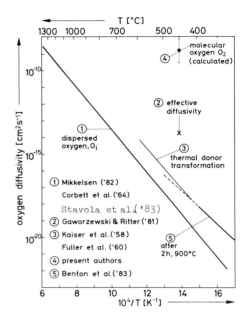

Fig. 2:
Oxygen diffusivity as a function
of reciprocal temperature based
on data of ref. [4,6-10,13,14,18].

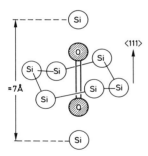

Fig. 3:
Proposed configuration of
O_2 molecules in crystalline
silicon; schematic and not
to scale [2].

Kaiser et al. [5,9].

THE MODEL OF KAISER ET AL.

Kaiser et al. [5,9] assume that interstitial oxygen O_i is in dynamical equilibrium with immobile and electrically inactive oxygen complexes A_2 and A_3 containing two and three oxygen interstitials, respectively. The addition of one oxygen interstitial O_i leads to the formation of a tetrahedral SiO_4 complex which is believed to be electrically active and to constitute the thermal donor. Addition of further O_i leads to the transformation of the SiO_4 unit into an electrically inactive complex. Within the model of Kaiser et al. this transformation reaction limits the attainable thermal donor concentration to a maximum value C_{TD}^* which turns out to be proportional to C_i^3, where C_i is the concentration of O_i. Within the Kaiser model it can also be explained why the initial donor formation rate k_{trans} is proportional to C_i^4. From the first order reaction rate constant for the transformation (induced by oxygen diffusion to the thermal donor),

$$k_{trans} = 4\pi D_i r_{TD} C_i \, , \qquad (2)$$

the oxygen diffusivity D_i may be determined. The result based on data of Kaiser et al. [9] and Fuller et al. [4] is shown in Fig.2 as line 3, where we have assumed a value of 5×10^{-8} cm for the thermal donor reaction radius r_{TD}.

As shown elsewhere in detail [2] the model of Kaiser et al. [9] has three difficulties: i) It is unclear why the SiO_4 tetrahedral unit should act as thermal donor. ii) The oxygen diffusivity derived from the thermal donor data based on eq.(2) although showing about the same activation energy as expected for O_i diffusion is about 40 times higher than the values measured for isolated O_i according to eq.(1). iii) In order to explain the fourth order dependence of the initial donor formation rate on C_i Kaiser et al. have to assume that the dynamical equilibrium concentrations C_2^* and C_3^* of the complexes A_2 and A_3 are reached orders of magnitude faster than the maximal donor concentration C_{TD}^*. A close inspection shows that this cannot be fulfilled if the same diffusion process is responsible for all three agglomeration processes leading to A_2, A_3, and the thermal donor SiO_4 [2].

Helmreich and Sirtl [3] suggested a thermal donor model based on comlexes of substitutional oxygen and vacancies. Whereas this model can reasonably explain the donor activity of such complexes it can not explain the kinetic features of thermal donor formation. Therefore and because irradiation-induced vacancies do not lead to enhanced thermal donor formation [7] we do not consider this model any further. We rather suggest a thermal donor model based on the action of oxygen molecules.

THERMAL DONOR FORMATION VIA MOLECULAR OXYGEN

In order to avoid the difficulty iii) in the Kaiser model on one side and to maintain its general features on the other side we introduce a second diffusivity in the thermal donor formation process by proposing a dynamical equilibrium between interstitial oxygen O_i and gas-like molecular oxygen O_2 according to

$$O_i + O_i \rightleftharpoons O_2 \, . \qquad (3)$$

This gas-like molecular oxygen possessing no Si-O bonds is supposed to be only weakly coupled to the silicon lattice and to show a high diffusivity $D_2 \gg D_i$ because no bonds have to be broken in the diffusion process. The encounter of two oxygen molecules O_2 is assumed to lead to a molecular complex O_4 which involves four oxygen atoms and acts as thermal donor:

$$O_2 + O_2 \rightarrow O_4 \quad \text{(thermal donor)}. \tag{4}$$

The observed fourth-order dependence of the initial donor formation rate follows naturally from (3) and (4) by applying the law of mass action. Similarly as in the original Kaiser model [9] the diffusion of one additional interstitial oxygen O_i to O_4 leads to the <u>transformation</u> of O_4 into an <u>electrically inactive</u> complex. Therefore eq. (2) for k_{trans} still holds and consequently the difficulty ii) of a two high diffusivity D_i determined from eq. (2) still remains in our model. We will comment on possible reasons for this discrepancy later on.

The difficulty iii) we have avoided by the introduction of a second diffusivity. As for the difficulty i) concerning the electrical activity we speculate that in the formation of the O_4 complex the two oxygen molecules interact without forming bonds with silicon atoms leading to a molecular configuration

$$\begin{array}{c} O\!-\!O \\ {}_{+}O^{\diagup} \quad {}^{\diagdown}O^{+} \end{array} \tag{5}$$

which in principle could be the ionized form of a double donor, a property which has experimentally be shown to hold for "thermal donors" [7,10]. The association of additional O_2 molecules to the O_4 complex resulting in a slight change of its electrical activity appears feasible and could explain the variety of electrically slightly different thermal donors formed successively during long time anneals [10].

The assumption of the existence of oxygen molecules O_2 in Si allows to interpret the high diffusivity found at low temperatures by Gaworzewski and Ritter [8] as well as by others [11,12] measuring the oxygen transport over macroscopical distances as an effective diffusivity

$$D_{eff} = (D_i C_i + 2D_2 C_2^*)/(C_i + 2C_2^*) \tag{6}$$

strongly dominated by the diffusivity D_2 of molecular oxygen. We mention explicitly that within this interpretation the result of Gaworzewski and Ritter [8] does not contradict the results on D_i by Corbett et al. [6] and Benton et al. [7] since in these latter cases single jumps of the configuration O_i rather than the oxygen transport over macroscopic distances has been determined. Combining $D_{eff} \approx 2.7 \times 10^{-14}$ cm^{2}s^{-1} [8] with data on thermal donor formation at 450°C [13,14] we obtain a rough estimate of 2×10^{-9} cm^{2}s^{-1} for the diffusivity D_2 of O_2 at 450°C, a value which is also shown in Fig.2 as point 4. D_2 is predicted to depend linearly on C_i [2]. As a possible configuration of O_2 in the silicon lattice we suggest the configuration schematically indicated in Fig.3. We assume that even for high O_i concentrations the influence of O_2 diffusion on D_{eff} becomes negligible at temperatures above about 700°C [2]. In the context of molecules in silicon we note that it has been suggested that nitrogen is incorporated in silicon predominantly in the form of N_2 molecules [15].

IMPLICATIONS OF THE OBSERVATIONS OF STAVOLA ET AL.

By means of stress-induced dichroism Benton et al. [7] measured the frequency of single diffusion jumps of oxygen in silicon in the temperature range of 330-440°C. Their results, which are in accordance with the previous result of Corbett et al. [6] at 377°C, are represented by D_i of eq.(1) provided the silicon sample was treated beforehand at 1350°C in order to guarantee dispersed oxygen atoms. Benton et al. [7] observed an about two orders of magnitude higher jump frequency for silicon samples which had been hold at 900°C for two hours. The diffusivities D_i calculated from these data are shown in Fig.2 as line 5 and are seen to be close to the oxygen diffusivity values calculated from

thermal donor formation, at least around 450°C. Only further experiments on
thermal donor formation after a high temperature treatment near the melting point
and analogous experiments after a 900°C heat treatment can answer the question
whether the transformation constant k_{trans} of eq.(2) depends on a preceding heat
treatment in a similar way as the oxygen diffusivity.

Benton et al. [7] speculate that either the local strain fields or the non-
equilibrium concentration of self-interstitials I generated by the oxygen precipitation at 900°C could cause the enhanced oxygen diffusivity they observed. We
consider the latter possibility as more likely, especially since there are a number
of indications that O_i and I form complexes [16,17] which might be more mobile
than O_i itself. Even if these complexes are not stable around 400°C they might
lead to a higher apparent oxygen diffusivity via a dynamical equilibrium between
O_i and I. Irradiating an oxygen-containing silicon sample after a 1350°C dispersion treatment at, say, 400°C with 1 MeV electrons up to a dose of 10^{18}cm^{-2}
or more should reveal whether or not the diffusivity of O_i can be influenced by
self-interstitials.

Finally we mention that we can not rigorously exclude the possibility that
the transformation rate k_{trans} is not given by the diffusion of O_i to the thermal donors as described by eq.(2) but rather by the break-up or dissociation of
the thermal donor complex. Then the comparative closeness of curve 3 to D_i and
the accordance of the activation energies were purely accidental.

CONCLUSIONS

The kinetics of thermal donor formation may be understood in terms of fast
diffusing, gas-like molecular oxygen. There remains a discrepancy of a factor
of 40 between the diffusivity of O_i calculated from thermal donor formation and
the diffusivity D_i of O_i extracted from more direct measurements. Recent observations of Benton et al. [7] indicate that the diffusivity of O_i strongly depends on the thermal history of the silicon sample and that the just mentioned
discrepancy is possibly caused by a difference in the preceding heat treatments.

REFERENCES

1. J.R. Patel in: Semiconductor Silicon 1981, H.R. Huff, R.J. Kriegler, and
   Y. Takeishi eds. (The Electrochem.Soc., Pennington 1981) p. 189.
2. U. Gösele and T.Y. Tan, Appl.Phys. A 28, 79 (1982).
3. D. Helmreich and E. Sirtl in Semiconductor Silicon 1977, H.R. Huff, and
   E. Sirtl eds. (The Electrochem. Soc., Princeton 1977) p. 626.
4. C.S. Fuller, F.H. Doleiden, and K. Wolfstirn, J.Phys.Chem.Solids 13,187(1960)
5. W. Kaiser, Phys. Rev. 105, 1751 (1957).
6. J.W. Corbett, R.S. McDonald, and G.D. Watkins, J.Phys.Chem.Solids 25,873(1964).
7. M. Stavola, J. R. Patel, L. C. Kimerling and P. E. Freeland, Appl. Phys.
   Lett. 42, 73 (1983).
8. P. Gaworzewski and G. Ritter, phys.stat.sol. (a) 67, 511 (1981).
9. W. Kaiser, H.L. Frisch, and H. Reiss, Phys.Rev. 112, 1546 (1958).
10. P. Gaworzewski and K. Schmalz, phys.stat.sol.(a) 55, 699 (1979).
11. T.J. Magee and B.K. Furmann, J. Appl. Phys. 53, 1227 (1982).
12. H.F. Schaake, J. Appl. Phys. 53, 1227 (1982).
13. P. Gaworzewski and H. Riemann, Kristall und Technik 12, 189 (1977).
14. D. Wruck and P. Gaworzewski, phys.stat.sol (a) 56, 557 (1979).
15. P.V. Pavlov, E.I. Zorin, D.I. Tetelbaum, and A.J. Khokholov, phys.stat.sol.
    (a) 35, 11 (1976).
16. A. Brelot in: Radiation Damage and Defects in Semiconductors, J.E. Whitehouse
    ed. (Inst.Phys.Conf.Ser. No 16, 1973) p. 191.
17. G.S. Oehrlein, J.L. Lindström, I. Krafcsik, A.E. Jaworowski, J.W.Corbett in[7].
18. J.C. Mikkelsen, Appl.Phys.Lett. 40, 336 (1982).

EFFECT OF OXYGEN ON RADIATION-ENHANCED DIFFUSION IN SILICON

V.E. BORISENKO
Minsk Radioengineering Institute, P. Browka 6, Minsk, USSR

ABSTRACT

Low-energy ion bombardment has been used to enhance diffusion of phosphorus and antimony atoms in silicon. Oxygen free silicon crystals both containing phosphorus and antimony doped surface layers and original crystals were bombarded at 400-700 $^\circ$C with 400 eV oxygen or argon ions. Impurity and electrical carrier profiles were measured to analyse the role of oxygen in the radiation-enhanced diffusion. The results obtained are explained on assuming complexes such as vacancy-oxygen and vacancy-substitutional impurity to be involved in the process.

INTRODUCTION

In the planar technology of semiconductor devices oxygen enters silicon lattice during crystal growth and during device fabrication steps involving heat treatment in an oxidizing ambient or ion implantation through SiO_2 masks. Oxygen contaminations inevitably influence physical properties of silicon [1]. Therefore oxygen-related heat induced and radiation generated states have become the subject of growing interest [2]. However, there have been no detailed investigations connected with the effect of oxygen on radiation-enhanced diffusion in silicon.

The fundamental idea behind the radiation enhancement of substitutional impurity migration is that their diffusion is caused by the local increase in vacancies and interstitial atoms due to radiation damage [3-8]. Moreover, defect interaction effects and interference of fluxes of various diffusive species such as vacancies, host atoms, impurity atoms, vacancy-impurity pairs have been shown to be able to play an important role in this process [9-12].

The role of oxygen-related defects in phosphorus and antimony diffusion enhanced by low-energy ion bombardment is experimentally analysed in the present study.

EXPERIMENTAL

Silicon oxygen free p-type (111) oriented Monsanto wafers with a resistivity of 10 Ohm.cm were used in the experiment. The samples were implanted with phosphorus (80 keV, $3,6\times10^{14}$ ion/cm^2 or antimony (60 keV, $6,0\times10^{14}$ ion/cm^2) ions and annealed at high temperature in dry nitrogen. The sample surfaces were chemically etched in order to eliminate residial radiation defects. Both original and doped silicon crystals were bombarded with 400 eV oxygen or argon ions for 15 min in the

ion-plasma generator described in [9]. Current density of
2,0 mA/cm² for oxygen ions and 0,75 mA/cm² for argon ions were
used to provide an equal sputtering yield which characterizes
the rate of defect generation. Sample temperature during ion
bombardment was fixed within the range from 400 to 700°C. Samples were cooled up to the room temperature at the rate of 80-
100 degree/min just after the bombardment was stopped.

Neutron activation analysis and spreading resistance technique
were employed to obtain impurity and electrical carrier profiles.

RESULTS

Low-energy ion bombardment of phosphorus and antimony doped
silicon crystals causes redistribution of these impurities in the
whole temperature range investigated. Fig.1 presents typical
phosphorus profiles in the samples irradiated with oxygen and
argon ions. Initial distribution of phosphorus atoms is also
plotted. The application of low-energy ions to enhance diffusion
process is followed by sputtering of surface layers. Thus, the
bombarded profiles are removed with respect to the initial surface position according to the thickness of sputtered layers (as
determined by weight loss measurements).

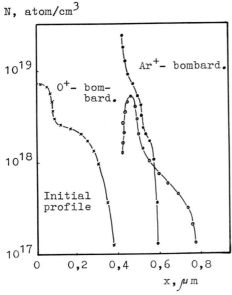

Fig.1. Redistribution of
phosphorus atoms due to oxygen and argon low-energy
bombardment at 600 °C.

Phosphorus diffusion enhanced by oxygen ion bombardment is
characterized with lowering of the impurity concentration at the
surface. An impurity peak appears at the distance of about 0,08-
0,09 μm from the surface. At that time argon irradiated samples
have an increased surface concentration of phosphorus atoms but
their redistribution has appeared to be deeper than that in the

oxygen bombarded ones.

Variation of antimony atom profiles in the samples subjected to the oxygen ion bombardment is identical to the phosphorus redistribution. As far as the argon ion bombardment is concerned, no detectable impurity concentration has been found in the irradiated samples.

Spreading resistance measurements display n-layer formation in the original silicon crystals irradiated with oxygen ions while argon bombarded crystals do not change the resistivity of the surface layer. Fig.2 shows temperature dependence of the n-layer depth. In the temperature range from 400 to 500 °C n-layer

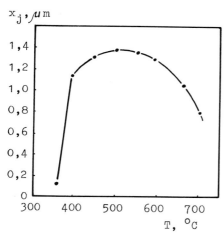

Fig.2. Depth of n-layer induced in p-silicon as a function of sample temperature during low-energy oxygen ion bombardment.

depth permanently increases and the maximum value of $1,4\,\mu$m is achieved. When a silicon crystal is heated above 500 °C the layer become thinner and its depth reduces to $\sim 0,9\,\mu$m at 700 °C.

Electron distributions in the n-layers are presented in Fig.3. The profiles are well approximated by the sum

$$n = n_1 \exp(-x/L_{d1}) + n_2 \exp(-x/L_{d2})$$

where $L_{d1} = 0,24\,\mu$m and $L_{d2} = 0,51\,\mu$m. Partial surface concentrations n_1 and n_2 are temperature dependent. They gradually increase when temperature rises in the range from 400 to 600 °C, and reach maximum value. At 700 °C n_1 decreases up to the zero while n_2 remains practically unchanged.

DISCUSSION AND INTERPRETATION

The main feature of low-energy ions interaction with solid is that they loss their energy in a very thin surface layer [13]. Interstitials are preferentially sputtered out of this layer or annihilated at the surface. Therefore vacancies become the defects preferably influencing diffusion processes.

The experiments using argon ion bombardment can be related to the "classical" case when phosphorus and antimony redistribution

162

n, e/cm^3

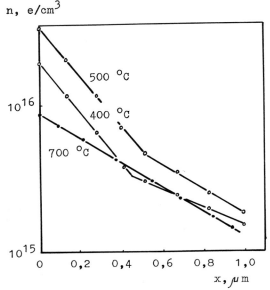

Fig.3. Electron profiles
generated in 10 Ohm.cm
p-silicon crystals by
400 eV oxygen ion bom-
bardment.

results from vacancy-exchange diffusion in the presence of localized vacancy source at the surface. Effective diffusivity of the
impurities is proportional to the equilibrium diffusivity and
local increase of vacancy concentration. It explains the increase of phosphorus concentration at the surface. The redistribution depth is controlled by vacancy diffusion length which is
about $0,086 \mu$m [3,9,11]. As for antimony, which has more than two
order lower equilibrium diffusivity [14] its enhanced diffusion
is suppressed by the sputtering process.
The use of oxygen low-energy ions to enhance the impurity
diffusion transforms the set of radiation defects involved in
the process. Vacancy-oxygen complexes are supposed to be generated at the surface and diffuse into the crystal. The complexes
getter vacancies introduced by ion bombardment. Surface impurity
concentration is reduced as it has been observed for both phosphorus and antimony.
Vacancy-oxygen complexes are quasi-stable at the elevated
temperature [15]. They dissociate in the crystal releasing free
vacancies and oxygen atoms. Appearence of n-layer in the oxygen
irradiated samples demonstrates introducing oxygen-related donor
states in silicon. Diffusion lengths of $0,24$ and $0,51 \mu$m show
that two kinds of vacancy- oxygen complexes take part in the
formation of n-layer. The shorter length one has been already
observed in the enhanced diffusion experiments [9,11]. It was
supposed to be a vacancy-oxygen pair. The longer length complex
becomes dominant at 700 °C. Hence, it must be a multipartical
one.

The diffusion process via vacancy-impurity pairs (vacancy-oxygen - V-O, vacancy-substitutional impurity - V-D) can be considered as an exchange by impurity atoms in the pairs [16]:

$$\overset{\frown}{D-V} + O-V \qquad V-D \overset{\frown}{+} O-V \qquad V-O + \overset{\frown}{D-V}$$

reorientation   exchange   reorientation

Since migration of the pairs is characterized with the longer diffusion length compared to the vacancies diffusion length, they provide the deeper redistribution of the substitutional impurities during oxygen ion bombardment.

CONCLUSION

This report is a part of a continuing effort to study the role of oxygen atoms in radiation induced processes and diffusion processes in silicon. The effect of oxygen on phosphorus and antimony diffusion enhanced by low-energy ion bombardment has been examined. Despite some experimental limitations of the present study the main features of the influence have been indicated and interpreted. Extended detail information about the particals involved in the enhanced diffusion would be of great interest.

ACKNOWLEDGMENT

The author would like to express his thanks to Professor V.A.Labunov for support and many helpful discussions.

REFERENCES

1. Semiconductor Silicon 1981, ed. by H. R. Huff, R. J. Kriegler and Y. Takeishi (Electrochemical Society, Pennington, NJ 1981).

2. Defects in Semiconductors, ed. by J. Narayan and T. Y. Tan (North-Holland, New York 1981).

3. R. L. Minear, D. C. Nelson and J. F. Gibbons, J. Appl. Phys. 43, 3468 (1972).

4. Y. Ohmura, S. Mimura, M. Kanazawa, T. Abe and M. Konaka, Rad. Eff. 15, 167 (1972).

5. E. W. Maby, J. Appl. Phys. 47, 830 (1976).

6. P. Baruch in: Radiation Effects in Semiconductors, ed. by N. B. Urli and J. W. Corbett (Institute of Physics, Bristol 1977) p. 126.

7. B.J. Masters and E. F. Gorey, J. Appl. Phys. 49, 2717 (1978).

8. Y. Morikawa, K. Yamamoto and K. Nagami, Appl. Phys. Lett. 36, 997 (1980).

9. V. A. Labunov, V. E. Borisenko and V. A. Ukhov, Elektron. Tekh. (Sov.) ser. 6, No 11, 72 (1977).

10. W. Akutagawa, H. L. Dunlap, R. Hart and O. J. Marsh, J. Appl. Phys. 50, 777 (1979).

11. V. E. Borisenko, L. D. Buyko, V. A. Labunov and V. A. Ukhov, Fiz. Tekh. Poluprov. (Sov.) 15, 3 (1981).

12. V. A. Labunov, V. E. Borisenko, Fiz. Tekh. Poluprov. (Sov.) 15, 1413 (1981).

13. G. Carter and D. G. Armous, Thin Solid Films 80, 13 (1981).

14. S. M. Hu in: Atomic Diffusion in Semiconductors, ed by D. Shaw (Plenum-Press, New York 1973) p. 217.

15. J. W. Corbett, J. P. Karins and T. Y. Tan, Nucl. Instr. Methods 182/183, 457 (1981).

16. V. E. Borisenko, L. F. Gorskaya, A. G. Dutov, V. A. Labunov and K. E. Lobanova, Fiz. Tekh. Poluprov. (Sov.) 16, 910 (1982).

PHOTOLUMINESCENCE STUDY OF THERMALLY TREATED SILICON CRYSTALS

J. WEBER AND R. SAUER
Physikalisches Institut (Teil 4), Universität Stuttgart
Pfaffenwaldring 57 - D 7000 Stuttgart 80, F.R. Germany

ABSTRACT

CZ silicon samples heated to $450^\circ C$ or higher temperatures for several hours exhibit many sharp photoluminescence lines well below the band edge. We study the production of the lines in terms of carbon and oxygen doping and of the heating temperature and duration. First results as to a level scheme of the P(0.7672 eV) line are reported.

Many new photoluminescence (PL) features have been observed in the last few years in thermally treated CZ silicon crystals and have been ascribed to defects incorporating oxygen and/or carbon. These observations can be roughly classified according to three basic types of spectra: (1) Broad PL bands at near-band edge energies arise in CZ wafers dependent on the annealing temperature [1,2,3]. Tajima and coworkers distinguished between four annealing stages characterized by A-, B-, C- and E-type spectra. The A-type spectrum consists of the familiar sharp bound exciton lines in as-received material. The B-type spectrum obtained after annealing at $450^\circ C$ for 64 hours shows a relatively broad PL band with various phonon replicas. It is substantially broadened and modified in shape upon a $600^\circ C$ anneal (C-type spectrum). Finally, annealing at $900^\circ C$ results in the E-type spectrum of very similar appearance to the original A-type spectrum. Referring to the annealing conditions, Tajima et al. ascribed the B- and C-type spectra to oxygen "thermal donors" or "new donors", respectively. (2) Sharp many-lines spectra emerge in oxygen rich, carbon rich silicon upon annealing at $500^\circ C$ for 100 hours [4,5]. At least 19 lines labelled S_j are observed extending from near-band gap energies over a spectral range of $\simeq 100$ meV. Their origin is not yet clear. A recent tentative suggestion ascribes the lines to donor-acceptor pair transitions [6]. (3) Several sharp lines at energies well below the band edge are reported by Minaev and Mudryi [7] in n-type CZ silicon after heating at $450^\circ C$ for 1 to 300 hours. Most prominent are the lines labelled P, H, T and 8,9 which were found to correlate either with the oxygen or the carbon concentration, and the line I which seemed to be dependent on the oxygen concentration as well as on the carbon concentration.

In the present report we study the last mentioned lines which - as we believe - are more characteristic for oxygen and carbon doped thermally treated silicon than the broad PL bands and the many-lines spectrum mentioned above under the headings (1) or (2).

We observe the P, H, T and I lines not only in n-type silicon as Minaev and Mudryi did but also in p-type material (Fig. 1). The positions of the lines (see Table 1) remain unaltered in p-type silicon. The P and H lines are dominant when the oxygen concentration is substantially larger than the carbon concentration, and are quenched upon increased carbon concentrations. Accounting for this influence of carbon we conclude from our data that the P and H line intensities essentially neither depend on the shallow impurity species nor on its doping level. This is in accord with the original suggestion by Minaev and Mudryi from the n-type samples alone. Contrary to these authors we find that both the T and the I line correlate with the carbon concentration and with the oxygen concentration.

Another set of PL lines appears in carbon rich CZ silicon which is annealed

Mat. Res. Soc. Symp. Proc. Vol. 14 (1983) ©Elsevier Science Publishing Co., Inc.

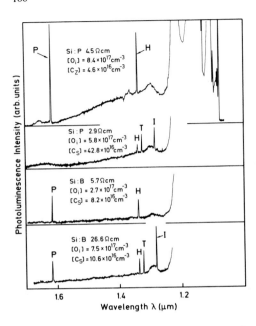

Fig. 1. PL spectra of n- and p-type CZ silicon crystals after annealing at 450°C for 100 hours. Notation of lines (P, H, T, I) after Minaev and Mudryi [7]. Oxygen and carbon concentrations were determined from infrared absorption data (O_i: Ref. 14, C_s: Ref. 15). Cut-off spectra on right-hand side are due to bound exciton (BE) and electron-hole drop (EHD) emission. Bath temperature 4.2 K.

at 650°C for about 100 hours after a 470°C preanneal (Fig. 2). These are the conditions under which the oxygen "new donors" form [8,9] . We observe these lines in the identical sample in which Tajima et al. [10] detected the C_{II}-spectrum at photon energies closer to the band edge than in our case. The latter authors did not observe the present lines near 1.4 µm as they used a S1-type cathode photomultiplier whose detectivity is exponentially cut-off for longer wavelengths in this spectral range whereas we recorded the spectra with a Ge-detector which is highly sensitive. The present lines are identical with the lines 8 and 9 in the work of Minaev and Mudryi recorded at sample temperatures

TABLE 1
Positions of various PL lines at 4.2 K and correlations with doping elements (upper part). P line associated satellites and positions (30 K, lower part). Labelling of line "9" (Fig. 2, T = 40 K) according to Minaev and Mudryi [7].

Line	Wavelength	Photon energy	Dependence on/Defect
P	1.6156 µm	0.7672 eV	Oxygen/Thermal donor
H	1.3388 µm	0.9258 eV	" " "
T	1.3248 µm	0.9356 eV	Oxygen and carbon
I	1.2840 µm	0.9653 eV	" " "
"9"	1.372 µm	0.9034 eV	Oxygen/New donor

P line satellites (30 K, Figs. 3,4):

1	1.5688 µm - 0.7901 eV		5	1.711 µm - 0.724 eV
2	1.6316 - 0.7597		6	1.757 - 0.705
3	1.654 - 0.7494		7	1.7650 - 0.7022
4	1.7036 - 0.7276		8	1.7820 - 0.6956

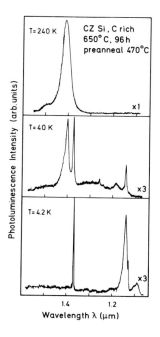

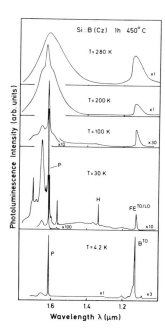

Fig. 2. (left) PL spectrum of CZ, carbon rich silicon after preannealing at 470°C and subsequent annealing at 650°C for ≃ 100 hours. The close lines at 1.37 μm and 1.40 μm (T=40 K) correspond to lines "9" and "8", respectively, in Ref. 7. Line shifts are due to the band gap shrinkage at increased temperatures.

Fig. 3. (right) P line spectrum at various temperatures. The high temperature broad band near the P line position is strongly observed up to appoximately 400°C.

of 15 K. At 4.2 K the line at 1.372 μm ("9") shows up alone and is at increasing temperatures accompanied by the 1.4 μm satellite which relatively grows and finally takes over the total PL intensity as a broad emission band. From the interrelation of the PL lines' occurrence and the new donor formation it is suggestive to ascribe the lines to these donor-like defects. This again supports the conclusions recently drawn by Minaev and Mudryi.

We investigated the P line in greater detail. Fig. 3 shows that the line grows from 4.2 K to 30 K and is in the latter temperature range accompanied both by low and high energy satellites which are depicted on an expanded scale in Fig. 4. The satellites disappear at higher temperatures and are gradually replaced by a broad band. Whereas the P line is no longer seen above ≃ 200 K, the band persists up to ≃ 600 K and begins only then to disappear. More quantitatively, the P line is activated by a factor of ≃ 300 up to 30 K possibly due to an effectively enhanced excitation from other photoexcited states which are emptied thermally in this temperature interval. The P line is then quenched with a thermal dissociation energy of ≃ 45 meV. This value places its transition initial and final states at energies of ≃ 45 meV or ≃ 357 meV below the bands in the forbidden gap where we have used the displacement of the P line (402 meV) from the low temperature band edge (1.1695 eV). The broad high temperature band whose maximum is at 280 K spaced from the band edge (at the same temperature) by ≃ 360 meV is then possibly a band-to-defect transition. This identification is consistent with the dependence of its intensity and halfwidth on temperature. The deep ≃ 360 meV level so determined corresponds to the binding energy of the charge carrier which is primarily localized at the defect and should be observable by DLTS or Hall effect measurements. Indeed, Mordkovich [11] reported pre-

168

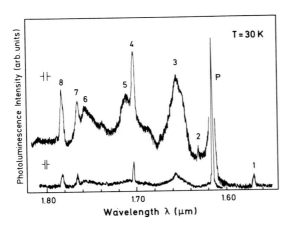

Fig. 4. Survey spectrum of P line and associated satellites at 30 K.

viously a level, E_v + 350 meV, which he obtained from Hall effect measurements on thermal oxygen donors in p-type, oxygen rich silicon, heated at 430°C for approximately 2 hours. It is possible that this level correlates with our present defect state. The band developing from the free exciton ($FE^{TO/LO}$) emission at higher temperatures in Fig. 3 is attributed to the conduction band-to-valence band transition. It is, however, much stronger than in intrinsic material at comparable experimental conditions. This effect may be due to a reduction of the carrier diffusion by the oxygen defect associates resulting in an electron-hole confinement near the excitation spot.

In Fig. 4 the vicinity of the P line is plotted on an expanded scale showing the satellites which are arbitrarily labelled 1 through 8. The spectral positions of the satellites are listed in Table 1. High-resolution sensitive recordings of the spectrum show that the satellites 1 and 2 possess themselves closely spaced, low intensity associated lines 1' and 2' at energies of 0.7895 eV and 0.7573 eV, respectively. In the temperature range around 30 K none of the satellites exhibits an intensity relative to the intensity of the principle P line consistent with a Boltzmann factor. In particular, satellite 1 is therefore not due to an excited state of the same optical center which gives rise to the P line. When the P line is assumed to be a ground state-to-ground state transition between the two electronic states involved this is also evident from the 22.9 meV displacement energy of line 1 from P which is too large for thermal excitation of line 1 from the upper P level. So the nature of the sharp satellites 1, 2, 4, 7 and 8 is not yet clear. On the other hand, we explain the broad satellites together with the structured background as due to lattice phonons coupled to the electronic transition. The broad satellites 3, 5 and 6 are displaced from the P line by 17.8 meV, 39.6 meV and 62.2 meV, respectively. These values are very close to the energies where the phonon density of states exhibits pronounced peaks [12] indicating that the optical center couples to phonons from the whole lattice spectrum. This is a luminescence feature which is well known from other optically active deep level defects in silicon [13]. The spectrum provides, however, no evidence of discrete satellites due to low energy quasi-localized in-band vibrational modes.

ACKNOWLEDGEMENTS

We are much obliged to M. Tajima for the loan of several samples and many conversations on his PL results on thermally treated silicon, and to T. Nishino and

H. Nakayama for the gift of oxygen and carbon rich silicon samples. The financial support of the Bundesministerium für Forschung und Technologie (BMFT) is grate-fully acknowledged.

REFERENCES

1. M. Tajima, A. Kanamori and T. Iizuka, Jpn. J. Appl. Phys. 18, 1401 (1979)

2. M. Tajima, S. Kishino and T. Iizuka, ibidem 18, 1403 (1979)

3. M. Tajima, A. Kanamori, S. Kishino and T. Iizuka, ibidem 19, L 755 (1980)

4. H. Nakayama, J. Katsura, T. Nishino and Y. Hamakawa, J. Lumin. 24/25, 35 (1981)

5. H. Nakayama, T. Nishino and Y. Hamakawa, Appl. Phys. Lett. 38, 623 (1981)

6. R. Sauer and J. Weber, in "Lecture Notes in Physics", ed. J. Giber et al. (Springer, Heidelberg), in press

7. N.S. Minaev and A.V. Mudryi, phys. stat. sol. (a) 68, 561 (1981)

8. A. Kanamori and M. Kanamori, J. Appl. Phys. 50, 8095 (1979)

9. K. Schmalz and P. Gaworzewski, phys. stat. sol (a) 64, 151 (1981)

10. M. Tajima, T. Masui, T. Abe and T. Iizuka, in "Semiconductor Silicon1981", ed. H.R. Huff and R.J. Kriegler (The Electrochemical Society, Pennington, N.J.), p. 72

11. V.N. Mordkovich, Fiz. tverd. Tela 6, 847 (1964); see also the listing of TD oxygen levels in P. Gaworzewski and K. Schmalz, phys. stat. sol. (a) 55, 699 (1979)

12. A.P.G. Hare, G. Davies and A.T. Collins, J. Phys. C 5, 1265 (1972)

13. See, e.g., the review by R. Sauer and J. Weber, Physica B (Proc. ICDS-12, Internat. Conf. Def. Semicond., Amsterdam 1982), in press

14. W. Kaiser and P.H. Keck, J. Appl. Phys. 28, 882 (1957)

15. R.C. Newman and J.B. Willis, J. Phys. Chem. Solids 26, 373 (1965)

INFRARED-ABSORPTION OF THERMAL DONORS IN SILICON

ROBERT OEDER[+], PETER WAGNER[*]
[+]Wacker-Chemie, [*]Heliotronic GmbH, D-8263 Burghausen, FRG

ABSTRACT

Thermal donors generated in CZ-silicon by annealing at 450°C are investigated by infrared absorption spectroscopy. Up to nine double-donors can be distinguished due to their line spectra characteristic for effective-mass-like donors. Their number and their concentrations depend on the duration of the thermal treatment. The formation of the different donors is ruled by complicated kinetics.

INTRODUCTION

Thermal donors (TDs) can be created in oxygen containing silicon by suitable thermal treatment in the temperature range 400-1000°C. Abundant literature exists already dealing especially with the TDs formed at about 450°C. In recent publications infrared-absorption lines due to electronic transitions of such TDs have been reported /1,2,3,4/. According to these publications a whole series of double-donors appears by annealing CZ-Si at temperatures around 450°C. In /3/ four, in /4/ six different TDs have been reported. New observations on larger numbers of different TDs generated at 450°C and their electronic structure are shown in the present paper on the basis of IR absorption spectroscopy*. In addition preliminary experiments on the kinetics of TD formation are presented.

EXPERIMENTAL DETAILS

The IR-spectra shown in the present paper have been obtained by a Nicolet MX-1 Fourier Transform Spectrometer with a resolution of 1 cm^{-1}. Samples with different dopants (P, Sb, B, Al, undoped) have been investigated at 8K by means of a closed-cycle refrigerator. The interstitial oxygen concentration of all samples was about 10^{18} cm^{-3} as determined by the 9 μm absorption line at room temperature /5,6/. The substitutional carbon concentration was below the detection limit of $5x10^{15}$ cm^{-3}.

RESULTS

The electronic level energies of ideal, hydrogen-like and helium-like donors have been calculated using the effective-mass-theory (EMT) by /7/. Characteristic absorption line series have been observed at low temperatures for shallow donors in accordance with EMT /8,9,10,11/ and it is through this pattern that the TDs have been identified as EMT-like double donors /3/. In Fig. 1 the IR-absorption spectra between 300 and 600 wavenumbers of an "as-grown" B-doped sample after 3 different annealing steps are shown. The absorption lines of the different TDs are specified corresponding to the transitions of an EMT-donor. The labelling of the TDs starts with 1 for the TD with the highest binding energy /4/. The observed absorption lines are listed in Tab. I together with the corresponding transitions as well as the derived binding energies of the ground

---

*During completion of the present article we have learned of similar results having been presented by B. Pajot et al. at the 16th Int. Conf. Physics of Semiconductors, Montpellier, 6.-10.9.82.

Mat. Res. Soc. Symp. Proc. Vol. 14 (1983) ©Elsevier Science Publishing Co., Inc.

172

states. The highest transitions observed have been used for the derivation of such binding energies. The correlation of transitions and absorption lines is no longer unambigious for some lines and long annealing times, because the spectra of different TDs show interference (compare Tab. I). Some lines in Fig. 1 are clipped due to instrumental reasons. The identification of D_8 and D_9 has to be considered preliminarily. The absorption lines in Fig. 1 are caused by electronic transitions on neutral TDs (D_i^o). The corresponding spectra of singly ionized TDs (D_i^+) in partially compensated Si are shown in Fig. 2 and can again be explained as transitions of EMT-donors, the suitable model now being a He^+-atom

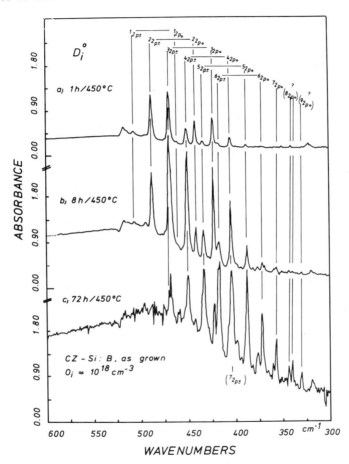

Fig. 1: Absorption spectra of the different neutral TDs D_i^o (i=1,2,...,9) of an as-grown sample, CZ-Si:B, after three annealing steps. a) 1h/450°C, b) 8h/450°C, c) 72h/450°C; $B = 6.10^{14} cm^{-3}$; T = 8K

instead of a H-atom. In contrast to the D_i^0-spectra the 2p± and 3p±-transitions are split in Fig. 2. The splitting of the 2p±-transitions increases from 5 cm^{-1} (D_1^+) to 16 cm^{-1} (D_5^+) with increasing donor number, the splitting of the 3+-transitions increases from 2 cm^{-1} to 3,5 cm^{-1}. The line energies of the D_i^+ transitions are also listed in Tab. I together with the binding energies of the ground states. The energetic positions of the absorption lines are independent of the dopants of the investigated samples as far as the lines can be observed. At 8K the D_i^+ can only be seen in B- or Al-doped samples because the TDs are double donors.

During accumulative annealing steps the amplitudes of the different absorption lines are changing drastically. This is shown in Fig. 1a)-1c) for a B-doped sample. The changing of the amplitudes of the D_i^0-2p$_o$-transitions as a function of annealing time is plotted in Fig. 3 for the same sample. In addition the corresponding curves for an undoped, pre-annealed sample without any TDs at the beginning are shown for short annealing times in broken lines. Tentatively these amplitudes are used as a relative measure for the concentration changes of the different TDs as long as the optical cross sections are not known. Variations of the amplitudes due to a changed occupancy of the levels induced by a shift of the Fermi-level are assumed to be small in the samples shown. Possible changes of the line widths are neglected. With these approximations in mind the curves of Fig.3 can be interpreted. TD D_1 and D_2 seem already to have had their maximal concentration in the "as-grown" sample. Their concentrations decrease in the course of the annealing procedure whereas the concentrations of TDs D_3 to D_6 increase, reach a maximum and seem to decrease thereafter. In a pre-annealed sample with no TDs at the beginning of the annealing process the initial increase also is observed for D_1 and D_2. During annealing the concentration of the interstitial oxygen decreases as monitored by the 9 µm band. The amplitude of the 320 cm^{-1} boron absorption line also decreases indicating that boron might be involved into the TD formation process /12/. This decrease of absorption line amplitudes after-heat treatment is observed for other dopants too.

TABLE I
IR-absorption-line energies and derived optical binding energies E_B at T = 8K

Donor-Nr.	1	2	3	4	5	6	7	8	9
D_i^0-transitions/cm^{-1}									
2p$_o$	461	442	423	405	388	372	357	343	330
2p±	507	488	470	451	434	417	404	385	376
3p±	533	514	496	477	-	443	-	-	-
E_B^0/meV	69.3	66.9	64.7	62.3	60.2	58.1	56.5	54.1	53.0
D_i^+-transitions/cm^{-1}									
2p$_o$	854	806	762	713	678	645	(614)	(584)	-
2p±	1044	991	945	889	846	(804)	(782)	-	-
	1048	998	951	904	862	825	(793)	-	-
3p±	1156	1105	1057	1011	-	-	-	-	-
	1160	1107	1059	1014	-	-	-	-	-
E_B^+/meV	156.3	149.7	143.8	138.2	132.5	127.9	-	-	-

174

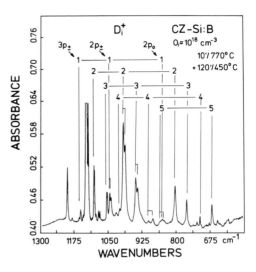

Fig.2: Absorption spectrum
of singly ionized TDs
D_i^+ in the same material as shown in Fig.1
after annealing steps
of $10'/770^\circ C + 120'/$
$450^\circ C$.

T = 8K

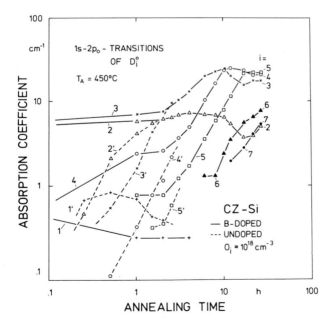

Fig.3:
Amplitudes of the
1s-2p-transitions
of D_i^{oo} as a function
of annealing time
at $T_A = 450^\circ C$. The
undoped sample was
pre-annealed, the
B-doped sample in
"as-grown"-state at
the beginning of
the annealing cycle.

DISCUSSION

One fact can be derived from Fig.3 very clearly: the different TDs come to existence one after another, starting with D_1^* (see Fig.1). They might tentatively be interpreted as different stages in a series of varying O-complexes that were studied already in a less detailed manner by /1,2/. Complicated kinetic seems to rule a regime of atomic aggregation that has to be investigated more thoroughly in the future. The splitting of the 2p+- and 3p+-transitions of the D_i^+ can be explained differently. One explanation is based on the assumption of a chemical splitting of the corresponding states (similar to interstitial Mg^+ /8/). This view is favoured by increased splitting with an assumed growth of the TD complexes. Another explanation for the splitting phenomenon might be an anisotropic, inhomogeneous potential, possibly with mechanical stress components of such complexes.

ACKNOWLEDGMENTS

The stimulating discussions with Prof. E. Sirtl, Heliotronic, Prof. J. Schneider, IAF Freiburg and Dr. W. Zulehner, Wacker-Chemitronic are greatly acknowledged by the authors. They also want to thank P. Stoeckl for the help with the experiments. This work was supported by the Bundesministerium für Forschung und Technologie (BMFT) of the Federal Republic of Germany under contract Nr. NT 0845/0846.

REFERENCES

1. K. Graff, H. Pieper, J. of Electronic Mat. 4, 281 (1975)

2. D. Helmreich, E. Sirtl in "Semiconductor Silicon 1977", Eds.: H.R. Huff, E. Sirtl, The Electrochemical Society, Inc., Princeton, p. 626

3. D. Wruck, P. Gaworzewski, phys.stat.sol.(a) 56, 557 (1979)

4. H.F. Schaake, S.C. Barber, R.F.Pinizotto in "Semiconductor Silicon 1981", Eds.: H.R. Huff, R.J. Kriegler, Y. Takeishi, The Electrochemical Society, Inc., Pennington, p. 273

5. W. Kaiser, P.H. Keck, C.F. Lange, Phys.Rev. 101, 1264 (1956)

6. K. Graff, E. Grallath, S. Ades, G. Goldbach, G. Toelg, Sol.State Electr.16, 887 (1973)

7. R.A. Faulkner, Phys.Rev. 184, 713 (1969)

8. A.K. Ramdas, S. Rodriguez, Rep.Progr.Phys. 44, 1297 (1981) and references therein

9. W.E. Krag, H.J. Zeiger, Phys.Rev.Letters 8, 485 (1962)

10. C.Swartz, D.H. Lemmon, R.N. Thomas, Sol.State Comm. 36, 331 (1980)

11. H.G. Grimmeiss, E. Janzen, H. Ennen, O. Schirmer, J. Schneider, R. Woerner, C. Holm, E. Sirtl, P. Wagner, Phys.Rev. B24, 4571 (1981)

12. C.S. Fuller, F.H. Doleiden, K. Wolfstirn, J.Phys.Chem.Sol. 13, 187 (1960)

---

* Large differences in the dipole strength might change the order

CORRELATION OF OXYGEN AND RECOMBINATION CENTERS ON A MICROSCALE IN AS-GROWN CZOCHRALSKI SILICON CRYSTALS

K. NAUKA, H. C. GATOS AND J. LAGOWSKI
Massachusetts Institute of Technology, Cambridge, Massachusetts 02139, USA

ABSTRACT

Quantitative microprofiles of the interstitial oxygen concentration and of the excess carrier lifetime, with a spacial resolution of about 20 μm, were obtained in as-grown dislocation-free CZ-Si crystals employing a double laser absorption technique. It was found that maxima (minima) in oxygen concentration along the crystal growth direction coincide with minima (maxima) of the lifetime. It was further found that the relation between changes in oxygen concentration and in lifetime varies in the radial direction indicating that "as-grown" oxygen precipitates are involved in lifetime limiting processes.

INTRODUCTION

It is now generally recognized that oxygen-induced defects in silicon and their interactions during processing at elevated temperature affect the electronic, structural, and mechanical properties and lead to phenomena which can be detrimental or beneficial to VLSI technology [1]. Extensive studies of oxygen-induced properties have been carried out, however, primarily on a macroscale [2]. Only recently, IR laser absorption scanning technique has provided a quantitative means for nondestructive, contactless determination of the oxygen concentration distribution in Si with a spacial resolution of about 20 μm [3,4]. We have applied this technique to the analysis of the formation and the annihilation of a thermal oxygen donor on a microscale; our findings demonstrated that an activation of the oxygen donor is not only a function of the oxygen concentration but also a function of microdefect distribution [4]. This result, unattainable from macroscale measurements, accounted for the quantitative differences in thermal donor microprofiles and the oxygen concentration microprofiles.

In the present study our IR laser scanning technique is extended to quantitative microprofiling of the excess carrier lifetime. This extension is achieved by the incorporation of the second laser (YAG laser), which is used to generate excess carriers. The concentration of excess carriers (and thus the lifetime) is determined from the increase in IR free carrier absorption [5].

Experimental

The present study was carried out on dislocation-free "as-grown" Czochralski Si crystals doped with boron (p-type, 300 K holes concentration 1 to 2 x 10^{15} cm^{-3}) or phosphorous (n-type, 300 K electron concentration 1 to 2 x 10^{15}). The average oxygen concentration was about 10^{18} cm^{-3}. Carbon concentration was about 10^{17} cm^{-3}.

Measurements were carried out parallel to the crystal growth direction at various distances from the periphery. The sample surfaces were covered with antireflecting coating (CdTe film) designed to reduce the reflection in the

Mat. Res. Soc. Symp. Proc. Vol. 14 (1983) © Elsevier Science Publishing Co., Inc.

employed IR region (9.1 - 10.3 μm) to a value below 3%. This antireflection
coating also effectively eliminates interfering effects caused by slight variations in sample thickness. The two lasers arrangement is shown in Fig. 1. The
CO_2 laser (tunable in the range 9-11 μm) is used for transmissivity profiles,
whereas the YAG laser radiation (λ = 1.06 μm) is used for excess carrier
generation. A germanium filter eliminates stray light and prevents YAG radiation from reaching the detector. Acquisition and processing of data is carried
out with a microcomputer.

Three types of measurements were carried out in order to determine microprofiles of the concentration of free carriers, the concentration of excess
carriers, and the concentration of interstitial oxygen. Free carrier concentration (n or p) profiles were obtained from the profiles of the absorption
coefficient, α, measured at λ = 10.3 μm (i.e., outside the oxygen absorption
band) as described in ref. (6). Then profiles of α were taken along the same
scan line employing both the CO_2 and YAG lasers. The difference between the
two types of profiles yields the concentration of excess carriers, Δn (or Δp)
generated by the YAG laser radiation. In the case of negligible trapping (assumed here) Δn = Δp and the excess carrier lifetime, τ, is given by the ratio
τ = G/Δn, where G is the optical generation rate G determined by the illumination density. Oxygen concentration microprofiles were obtained from absorption
measurements at λ = 9.17 μm as discussed elsewhere [3,4], employing the ASTM
standard [7]. The contribution from free carrier absorption was subtracted
from the absorption profile at λ = 9.17 μm.

Results and Discussion
    The procedure described above is determined in Fig. 2 a-d. Fig. 2a depicts
the absorption profile obtained at λ = 10.3. It reflects primarily free carriers and lattice contributions in IR absorption. As seen in Fig. 2b, YAG
excitation of excess carriers increases significantly the absorption coefficient at 10.3 μm. It is seen that this increase is not uniform. From the difference between microprofile a and b the spacial variation of the excess carrier lifetime is obtained (Fig. 2c). It is seen that the spacial variations
of the lifetime correlate well with variations in concentration of the interstitial oxygen, i.e., lifetime maxima (minima) occur where the oxygen concentration exhibit minima (maxima). Also for both microprofiles the amplitude
of spacial variations is similar (about 20%).

The correspondence between excess carrier lifetime and the oxygen concentration was observed for all as-grown n- and p-type CZ-Si crystals obtained from
different commercial sources. Statistical analysis showed that lifetime
minima were observed for more than 95% of oxygen concentration maxima. These
findings show that oxygen in CZ-Si as-grown crystals is involved in the formation of recombination centers. However, we have also found that these recombination centers cannot be directly assigned to interstitial oxygen. As shown
in Fig. 3, the central and the peripheral regions of Si crystals exhibit significant quantitative differences. Near the center of the crystal the lifetime
microprofile shows minima coinciding with oxygen concentration maxima, however,
the relative magnitude of lifetime variations ($\Delta\tau/\tau_{av} \simeq$ 10%) is relatively low
(much lower than the corresponding magnitude of oxygen concentration variations,
$\Delta[O_i]/[O_i]_{av} \simeq$ 20%).

The magnitude of lifetime variations increases drastically toward the periphery of the crystal. About 5 mm from the periphery of the crystal $\Delta\tau/\tau_{av}$
approaches a factor of 10, while $\Delta[O_i]/[O_i]_{av}$ does not exceed 1.5 (Fig. 3b).
The lifetime changes by as much as a factor of 20 0.6 mm from the periphery.
It is also of interest to note that the average value of τ decreased from the

center of the wafer to the edges by a factor of 2, while the average concentration of the interstitial oxygen decreased by 30%. In all instances, however, the lifetime minima (maxima) coincided with maxima (minima) of the oxygen concentration. The results of Figs. 2 and 3 indicate that lifetime limiting recombination centers in as-grown CZ-Si are due to complexes involving oxygen rather than to interstitial oxygen. The nature of these complexes is not known at present, however, the enhancement in magnitude of lifetime variations toward the periphery of the crystal can be accounted for by assuming that oxygen precipitates induce recombination centers.

Thus, it has been shown [8] that in as-grown Si a certain amount of oxygen is present in the form of precipitates. This amount increases toward the periphery of the crystal. It has also been suggested recently that certain types of oxygen precipitates in Si act as recombination centers [9].

The establishment of a direct relationship between excess carrier lifetime and oxygen precipitates (which is currently being pursued) should lead to a microscale electronic means for the determination of oxygen precipitates in as-grown dislocation-free CZ Si.

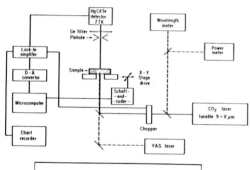

Figure 1. Experimental arrangement of double laser IR absorption scanning.

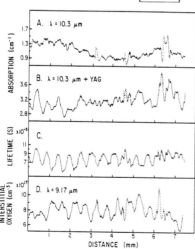

Figure 2. A. Absorption profile, at $\lambda = 1.03$ μm (CO_2 laser).

B. Absorption profile upon simultaneous irradiation with $\lambda = 1.03$ μm and $\lambda = 1.06$ μm (YAG laser).

C. Microprofile of excess carrier lifetime.

D. Microprofile of interstitial oxygen concentration.

180

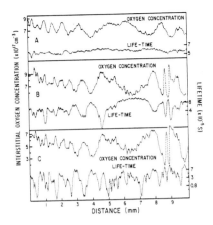

Figure 3. Microscale distribution of the interstitial oxygen concentration and excess carrier lifetime. Profiles obtained along crystal growth direction. A, B, and C correspond, respectively, to distances of 13 mm, 5 mm and 0.6 mm from the periphery of the crystal.

ACKNOWLEDGEMENTS

The authors are grateful to the National Aeronautics and Space Administration, Lewis Research Center, for financial support.

REFERENCES

1. L. Jastrzebski, Trans. IEEE ED-29, 475 (1982).

2. J.R. Patel, Semiconductor Silicon 1981, edited by H.R. Huff, R.J. Kriegler, and Y. Takeishi (Electrochemical Society, New York, 1981), p. 189.

3. A. Ohsawa, K. Honda, S. Okhawa, and R. Ueda, Appl. Phys. Lett. 36, 147 (1980).

4. P. Rava, H.C. Gatos, and J. Lagowski, Appl. Phys. Lett. 38, 274 (1981).

5. J.A. Mroczkowski, J.F. Stanley, M.B. Reine, P. LoVecchio and D.L. Polla, Appl. Phys. Lett. 38, 26 (1981).

6. L. Jastrzebski, J. Lagowski, and H.C. Gatos, J. Electrochem. Soc. 126, 260 (1979).

7. ASTM Standards, F 122-74 (Reappr. 1980).

8. L. Jastrzebski, P. Zounzucchi, D. Thebault and J. Lagowski, J. Electrochem. Soc. 129, 1638 (1982).

9. M. Miyagi, K. Wada, J. Osaka, and N. Inohue, Appl. Phys. Lett. 40, 719 (1982).

Oxygen Precipitation Effects in Degenerately - Doped Silicon

G. A. ROZGONYI, North Carolina State University, Raleigh, N.C.,
R. J. JACCODINE, Lehigh University, Bethlehem, PA. and C. W. PEARCE, Western
Electric, Allentown, PA.

Abstract

In this paper we report preliminary observations of oxygen precipitation in
degenerately-doped silicon using etching, optical microscopy and
transmission electron microscopy. It was found that n+ material was
resistant to precipitation, but p+ material precipitated readily. A multi-
step heat treatment starting with a low temperature step to achieve a high
supersaturation ratio was sucessfully used to induce precipitation in n+
material.

Introduction

For several years now an extensive literature has been developing on the
topic of oxygen precipitation in silicon. The reported research has dealt
almost exclusively with lightly-doped bulk material. The reasons for this
are twofold: (1)   There is a technological need to optimize intrinsic
gettering (IG) and denuded zone (DZ) formation in wafers used for integrated
circuit manufacture, and (2)   Lightly-doped materials offer experimental
simplicity. However, the trend to epitaxial wafers using degenerately-
doped substrates for MOS devices (1) has stimulated a need to study such
material. Initial work by deKock and van de Wijgert (2, 3) has revealed the
microdefect morphology of degenerately-doped silicon differs from the
lightly-doped case. There is an absence of microdefects in n+ silicon, when
the doping exceeds 5×10^{17} atoms/cm^3. However, microdefects (i.e. A type
defects etc.) do form in p+ material. Furthermore, deKock has proposed a
mechanism whereby silicon interstitials are involved in the heterogeneous
precipitation of oxygen at high temperatures. The interstitials are
postulated to be negatively charged. They pair with positively-charged
donor atoms in n+ material. This reaction inhibits precipitation in such
material. In p+ material the interstitials are not complexed and
precipitation is possible. Thus, degenerately-doped silicon offers an
interesting dimension to precipitation studies. We have found (4) n+
material to be resistant to precipitation in the temperature ranges (1000°C)
characteristic of device processing, but could be induced to precipitate
using the three step heat treatment described herein. However, p+ material
was observed to precipitate readily in all temperature ranges. This report
is concerned with a further comparison using TEM to identify the predominate
defect formed in each material following multi-step heat treatments.

Experimental

Sample were chosen at random from routine production material; they were of
<100> orientation, 100mm diameter, and from Czochralski-grown ingots. The
n+ wafers were antimony-doped to 5×10^{18} atoms/cm^3 and the p+ wafers to

Mat. Res. Soc. Symp. Proc. Vol. 14 (1983) Published by Elsevier Science Publishing Co., Inc.

1×10^{19} atoms/cm^3 with boron. To induce precipitation wafers were heated in the following sequence: 700°C, 24 hr, oxygen; 900°C, 24 hr, nitrogen; 1050°C, 24 hr, nitrogen. Samples were retained after each heat treatment electron for analysis by etching and transmission electron microscopy (TEM). The TEM studies were carried out in a Phillips model 300B microscope. Foils for TEM observation were obtained at varying distances from the wafer surface by etching the samples for various times in HNO$_3$-HF solutions. This assured a representative in-depth analysis of the bulk defect structure.

The TEM studies determined the densities and types of defects present at each stage of the three step treatment. Sirtl, Secco and Schimmel etch formulations were used as appropriate for each material type. Preferential etching was applied to wafer surfaces and cross-sections to determine defect densities and monitor the presence of denuded zones. This was assured by etching the samples for various times in HNO$_3$-HF solutions.

No direct measure of oxygen content was possible by IR analysis due to free carrier absorption in the highly doped substrates. However, the growing process used to produce the crystals typically contain from 10 to 20 ppma for p type crystals.

Additional samples were processed epitaxially after the treatment to determine if the surface was effectively denuded of defects which would otherwise nucleate epitaxial stacking faults etc.

Results

Figure 1a is a representative cleavage-face optical micrograph of an n+ sample heated for 24 hours in nitrogen at 1100°C. It is entirely void of any defects revealed by etching. Given similiar treatments, p+ material exhibits bulk-stacking fault densities of 10^5 to 10^6 faults/cm^2. Figure 1b is another micrograph of an n+ sample heated for the same time and in the same ambient as the first sample, but at 700°C. Note the presence of etch hillocks associated with the existance of SiO$_2$ precipitates in the material. Thus, the lack of nuclei for precipitation in n+ silicon at high temperature can be overcome by lowering the heat treatment temperature to the point where a substantial supersaturation condition exists. We estimate a supersaturation ratio of nominally 25 existed.

The TEM observations after the 700°C step revealed a similiar defect denisty to exist in each material type. The etch hillocks were correlated with the presence of small particles (precipitates) of undetermined composition. In n+ samples a density of 10^4-10^5/cm^2 was observed. Sizes were 50-500A. The p+ samples were of similiar density and 200-1000A in size. Figures 2 a and b compare typical results observed in n+ and p+ material after the 700°C step. The magnification is 50,000x. Following the 900°C step hillock etch features are still observed by etching, but the TEM analysis presented a more complex picture. In the p+ samples particles were still observed without any accompanying defect structure (see Fig. 2c), although coarsening was noted as the particles were now 1000-5000A in size. The n+ samples (Fig. 2d) showed evidence of long dipole dislocations and evidence of small loops within larger loops surrounding the particle consistent with prismatic punching about a precipitate nucleus. Finally, after the 1050°C processing extrinsic bulk stacking faults were observed in both material types (Fig. 2e and f) as the primary defect. However, the density of faults in the n+ samples (10^4-10^5/cm^2) was consistently lower than those obtained in p+ samples (10^6/cm^2). Prismatic loops were also observed at this point in the p+ samples along with square or hexangonal precipitates on or near the stacking fault plane. Prismatic loops were also seen in the n+ samples along with the small loops seen at 900°C.

Cross-section etching confirmed the presence of a denuded zone in the wafers of 8 to 20 microns in width following the three step anneal. This is consistent with the diffusion length of oxygen at 1050°C and the fact that precipitates formed at 700°C can be subcritical in size and therefore redissolve at 1050°C (see Fig. 3, ref. 5). Further proof of denuding was obtained by processing the samples epitaxially; high quality epitaxy was achieved on such wafers.

Conclusions

1. In outline form a three step heat treatment can be used to induce oxygen precipitation and associated defects in n+ and p+ silicon in a manner analogous to p- silicon.

2. Without such a treatment n+ silicon was found to be resistant to precipitation. A substantial supersaturation is evidently required to precipitate such material.

3. Differences in the defect formation were observed between n+ and p+ silicon suggesting a complex interaction exists between donor impurities, point defects and the nature of SiO_2 precipitation in bulk degenerately-doped material.

References

1. D. S. Yaney and C. W. Pearce, Proceedings - International Election Device Meeting 1981, p. 236.

2. A. J. R. deKock and W. M. van de Wijgert, Appl. Phys. Lett. 38 880 (1981).

3. A. J. R. deKock and W. M. van de Wijgert, J. Cryst. Growth, 49 718 (1981).

4. C. W. Pearce and G. A. Rozgonyi, "VLSI Science and Technology/1982", The Electrochemical Socity, Pennington N.J. p 53.

5. J. M. Andrews, S. Muller, G. A. Rozgonyi and C. A. Clark, ibid, p 43.

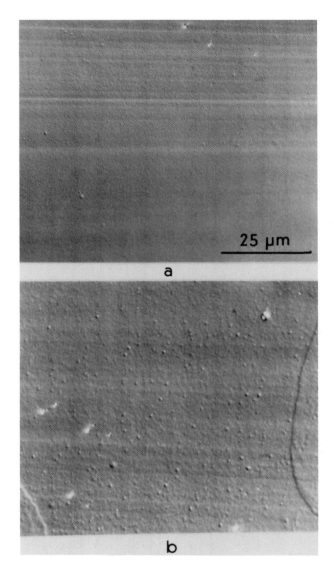

FIGURE 1.   Cleavage-face optical micrographs of n+ substrate
            material following 24 hour anneal in N_2 ambient at
            (a) 1100°C and (b) 700°C.

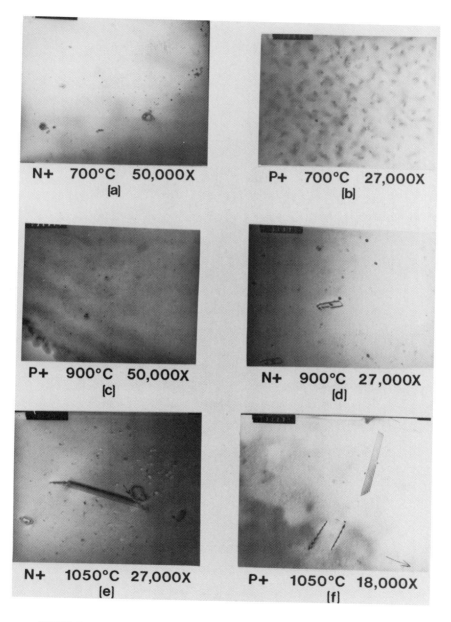

FIGURE 2.   Representative TEM micrographs of n+ and p+ silicon at various stages of the 700°C, 900°C and 1050°C treatment as noted.

OXYGEN PRECIPITATION IN SILICON - ITS EFFECTS ON MINORITY CARRIER RECOMBINATION AND GENERATION LIFETIME

C. J. VARKER, J. D. WHITFIELD AND P. L. FEJES
Semiconductor Research and Development Laboratories,
Motorola, Inc., 5005 E. McDowell Road, Phoenix, Arizona, USA

ABSTRACT

The effects of oxygen precipitation on the minority carrier recombination lifetime (τ_R) and the carrier generation lifetime (τ_G) have been characterized for a 'typical' silicon crystal grown with the Czochralski method. Infrared (IR) absorption measurements were obtained on polished wafers, before and after 2 step thermal anneals at 800°C and 1050°C to characterize the axial distribution of interstitial and precipitated oxygen in the ingot. Computerized measurements on NMOS diode and capacitor arrays were used to characterize the axial and radial distributions of carrier lifetime. The results indicate that oxygen precipitation is the dominant mechanism contributing to the degradation of both minority carrier recombination and generation lifetime.

INTRODUCTION

Oxygen and carbon are the most dominant unintentionally added impurities in silicon crystals grown with the Czochralski (CZ) method. The mechanism for the incorporation of these impurities has been the subject of many recent investigations [1,2,3]. The fundamental properties of oxygen in silicon such as its solubility, diffusion coefficient and precipitation kinetics have been reviewed by Patel [4]. The concentration of oxygen in typical CZ crystals is $1-2 \times 10^{18}$ cm^{-3} (ASTM F21-79) and is normally well above its solubility limit during processing. The axial and radial distribution of oxygen in particular, and carbon to a much lesser extent have been characterized [3,5,6] for silicon crystals grown with the Czochralski method.

The major problem arising from the high concentration of interstitial oxygen in dislocation free silicon crystals is its tendency to precipitate during thermal processing. Recent studies of oxygen precipitation in silicon using two step thermal anneals have produced evidence in support of both homogeneous and heterogeneous nucleation models [7,8]. Although the detrimental effects of oxygen precipitation in silicon are readily apparent from investigations of defects in silicon [9], recent claims regarding the beneficial effects of oxygen precipitation are not as obvious and are often difficult to reproduce.

In a recent study [10] the authors presented results which indicate that oxygen precipitation is the dominant mechanism degrading minority carrier lifetime in n$^+$p diode arrays for a 2 step thermal anneal designed to force oxygen precipitation. The objective of this investigation is to extend the work on oxygen precipitation to both carrier generation and recombination lifetime on the same dislocation free silicon crystal, using NMOS processing exclusively.

Mat. Res. Soc. Symp. Proc. Vol. 14 (1983) ©Elsevier Science Publishing Co., Inc.

EXPERIMENTAL PROCEDURE

Polished silicon wafers with a diameter of $\sim$75 mm were prepared from a boron doped ($N_A \sim 1 \times 10^{15}$ cm^{-3}) dislocation free crystal with a [100] orientation grown with the Czochralski method. The oxygen and carbon concentrations were obtained from infrared absorption (IR) measurements at 1106 cm^{-1} and 605 cm^{-1} respectively, with a Perkin Elmer Model 180, and Nicolet Model 7199 (170 SX) spectrometer on single side polished wafers before and after the 2 step anneal. Optical thickness corrections for the absorption peak at 1106 cm^{-1} which is attributed to interstitial atomic oxygen ([0]) in silicon [11-13], are intro-duced using the height of the silicon lattice absorption band at 615 cm^{-1}. The ASTM calibration factor of 4.81×10^{17} cm^{-3} (ASTM F121-79) was used to calculate the oxygen concentration. The substitutional carbon concentration was calculated using the calibration factor of 1.1×10^{17} cm^{-3} (ASTM F123-74).

A 2 step precipitation anneal was used consisting of; (1) 16 h at 800°C in argon and (2) 16 h at 1050°C in dry O_2 to force oxygen precipitation over an extended region of the crystal. The reverse sequence (dissolution anneal) was used to dissolve oxygen precipitates prior to device processing. Diode/capac-itor test arrays were fabricated on these annealed wafers using NMOS processing technology. The areas of the capacitor and diode in the NMOS test arrays are 5×10^{-4} cm^2 and 1.3×10^{-3} cm^2 respectively. Electrical measurements were obtained with a Tektronix S-3260 equipped with an automatic wafer prober or with an HP-85 computer and peripherals.

Average 'bulk' lifetime values (τ_R) for individual wafers were calculated from reverse recovery time (T_{RR}) measurements on diodes [14]. Generation lifetime values (τ_G) were calculated from capacitance relaxation time (T_S) measurements on capacitors adjacent to the diodes in the array. The calculated values of τ_G were obtained using the Zerbst method [15,16]. In addition to the lifetime measurements, the reverse current (I_R) was measured on all diodes using a test voltage (V_R) of 15 V at a temperature of 350°K. Comparisons between the near surface τ_G and 'bulk' τ_R lifetimes, were drawn for varying degrees of [0] precipitation. The experimental procedures described briefly above are covered in greater detail in a previous paper [10].

EXPERIMENTAL RESULTS

Figure 1 shows the initial concentration of interstitial oxygen ([0]$_i$) obtained from FTIR measurements on selected wafers plotted against their axial position measured from the seed end of the ingot. The values for [0]$_i$ show a linear decrease from [0]$_i \sim 2.1 \times 10^{18}$ cm^{-3} to a broad minimum at $\sim 1.3 \times 10^{18}$ cm^{-3}. In the tang-region (35 cm < X < 45 cm), [0]$_i$ increases to a value of 1.5×10^{18} cm^{-3}.

For the precipitation anneal the oxygen had precipitated to its solubility limit $\sim 5 \times 10^{17}$ cm^{-3} at the seed end of the crystal, for x <10 cm or [0]$_i$ > 1.8×10^{18} cm^{-3}. In the region corresponding to the oxygen minimum where [0]$_i \sim 1.3 \times 10^{18}$ cm^{-3} no significant precipitation is evident from the FTIR measurements. For the dissolution anneal little or no precipitation is observed over the major part of the ingot. Some precipitation is evident at the extreme seed and tang-ends. The concentration of substitutional carbon not shown in Figure 1 is below or near its detection limit $\sim 1 \times 10^{16}$ cm^{-3} throughout the ingot.

Figure 2 compares the axial distributions subsequent to the precipitation anneal for the generation and recombination lifetimes τ_G and τ_R respectively.

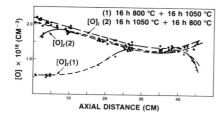

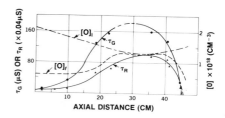

Fig. 1. Axial distribution profile of the initial $[0]_i$ and final $[0]_f$ concentration of interstitial oxygen in the crystal for a (1) precipitation and (2) dissolution anneal.

Fig. 2. Axial distribution profile of generation τ_G and recombination τ_R life times for the precipitation anneal and its relationship to the interstitial oxygen concentration measured before and after the annealing.

The dashed curves represent the initial and final oxygen distribution measured subsequent to the anneal. Both τ curves show a lifetime maximum at the 30-35 cm location in the ingot where $[0]_i$ is minimum. The maximum average values for τ_G and τ_R are approximately 180 µS and 4 µS respectively. Both curves show a decrease in lifetime at the seed and tang-ends of the ingot where oxygen precipitation is dominant. At the seed-end of the ingot, the minimum average values for τ_G and τ_R are 10 µS and 0.31 µS respectively. The values represented by the τ_G profiles are more than a factor of 10 greater than the corresponding values for the τ_R profile over the entire length of the ingot.

Figure 3 compares the axial distribution of τ_G for the wafer group shown in Figure 2 with a second group of wafers processed independently at a later date. The dashed curves represent $[0]_i$ and the concentration of precipitated oxygen $\Delta[0]$ measured subsequent to the 2 step anneal. The concentration of precipitated oxygen $\Delta[0]$ is calculated from the difference between the initial and the final concentration of interstitial oxygen. Both experimental groups show a projected maximum for the generation lifetime at $\sim$30 cm, of approximately 200 µS and 500 µS respectively. Although the data distribution did not permit an accurate determination of τ_G at the precise location of the oxygen minimum, the characteristic shape of the τ_G distribution curve is similar for both groups. On both sides of the oxygen minimum, where $[0]_i > 1.3 \times 10^{18}$ cm^{-3} the concentration of precipitated oxygen $\Delta[0]$ increases with $[0]_i$ with a corresponding decrease for the generation lifetime. The similarity between the axial profiles for τ_G and τ_R shown in Figures 2 and 3 suggest that a correlation can be expected.

In Figure 4 the reverse current I_R and τ_R are plotted against $\Delta[0]$ for the wafers represented in Figure 3. The I_R values which are measured at a reverse voltage $V_R = 15$ V and at a temperature of 350°K represent an average of $\sim$525 test diodes per wafer. The inverse relationship between τ_R and I_R and their strong dependence on $\Delta[0]$ provides direct evidence for the degrading effects of oxygen precipitation on the characteristics of the n^+p junction diodes. The current limit shown in Figure 4 at $\sim$400 pA represents the limiting diffusion current for $\Delta[0] < 10^{17}$ cm^{-3} where the corresponding values for τ_R approach a plateau. For $\Delta[0] > 10^{17}$ cm^{-3}, I_R increases sharply with $\Delta[0]$ corresponding to the regions shown in Figure 3 where the generation lifetime τ_G decreases with increased oxygen precipitation, for $[0]_i > 1.3 \times 10^{18}$ cm^{-3}.

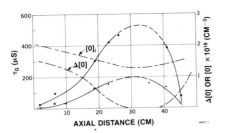

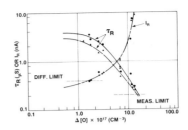

Fig. 3. Compares the axial distribution of generation lifetime τ_G for two processed lots showing their dependence on $[0]_i$ and the concentration of precipitated oxygen $\Delta[0]$.

Fig. 4. Shows the dependence of the reverse current I_R and the recombination lifetime τ_R on the concentration of precipitated oxygen $\Delta[0]$.

Figure 5 shows a distribution similar to Figure 3 for a 2 step dissolution anneal. The curves representing the initial and final concentration of soluble oxygen indicate that no significant oxygen precipitation is evident from the FTIR measurements subsequent to the anneals for the major part of the ingot. Whereas, FTIR measurements obtained on selected wafers subsequent to device fabrication (not shown in Fig. 5) indicate that significant precipitation had occurred during processing. Measurements on these wafers, indicate that the concentration of interstitial oxygen remains above the solubility limit at 1050°C ($[0]_f \approx 5 \times 10^{17}$ cm^{-3}) as shown in Figure 3 for x < 10 cm. Therefore the degree of pre-

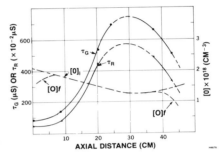

Fig. 5. Distribution profiles of τ_G and τ_R for the dissolution anneal and the interstitial oxygen concentration measured before and after annealing.

cipitation is less for the dissolution anneal. The axial distributions for τ_G and τ_R shown in Figure 5 reveal a profile similar to Figures 2 and 3, with significantly larger values for both τ_G and τ_R when compared with Figure 2.

The relationship between the generation and recombination lifetime is shown in Figure 6 for the data groups comprising the axial distribution plots in Figures 2, 3 and 4. The scatter plot is partitioned with diagonal lines to identify the trends indicated within each wafer group. Wafers within each group share common data point characters. Each point represents a pair of average values corresponding to a sample size of 12 measurements each for the 3 experimental groups. An overall correlation is indicated between τ_G and τ_R which covers a measurement range of ~ 2 decades for τ_G, 10 µS $\leq \tau_G \leq 1$ mS. The value of 30 ppm (1.5×10^{18} cm^{-3}) was selected to partition the scatter plot at the threshold concentration for the onset of oxygen precipitation as observed in the FTIR measurements. The data group to the left of this diagonal represents wafers having an initial oxygen concentration $[0]_i$ in the range 30 ppm < $[0]_i$ < 40 ppm. For these wafers measurable oxygen precipitation had occurred during the 2 step anneals or inadvertently during device processing. The group on the right side of the diagonal represents wafers with $[O]_i < 30$ ppm which show negligible precipitation as a result of thermal processing. This group consists of wafers on both sides of the oxygen minimum shown on the axial distribution plots in Figures 2, 3 and 4.

The circled characters in each diagonal sector shown in Fig. 6 correspond to adjacent wafers in the ingot having a similar oxygen concentration which were processed in two independent lots representing the precipitation anneal. The values of τ_G corresponding to group A are approximately a factor of 3 greater than those for group B. This can be attributed to process related fators, because no other accountable differences exist in the initial annealing procedures. All experiimental groups yield larger values for τ_G and τ_R when the initial oxygen concentration $[O]_i < 30$ ppm. Consequently, a consistent trend is observed which indicates that maximizing carrier lifetime requires minimizing oxygen precipitation.

Figure 7 presents a scatter plot for 2 groups of wafers which are not from the ingot discussed previously, but they represent randomly selected wafers typical of the same material type and quality. The data represents 12 measurement pairs per wafer for τ_G and τ_R for a total of 7 wafers. The 4 wafers represented by data group A contain an epitaxial film with a thickness (X_e) of ~ 5 μm.* The oxygen concentration in the epitaxial film is estimated to be 5×10^{17} (10 ppm). Data group B which consists of 3 wafers represents the same material type without an epitaxial film. The data for group A indicates that τ_G is relatively independent of τ_R. Whereas, for group B

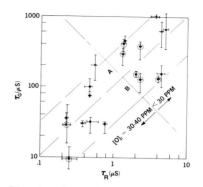

Fig. 6. Shows the relationship between τ_G and τ_R when oxygen precipitation is dominant, $[O]_i \approx 30$-40 ppm and when oxygen precipitation is minimal, $[O]_i < 30$ ppm.

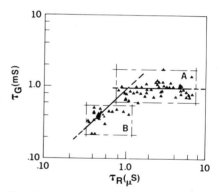

Fig. 7. Compares the relationship between τ_G and τ_R for wafers with a reduced oxygen concentration represented by a 5 μm epitaxial silicon film (A) and for typical CZ wafers (B).

an apparent correlation exits. Moreover, the wide range of values for τ_R which samples substrate lifetime, since the carrier diffusion length $L_N \approx 100$ μm, indicates that precipitation of oxygen in the substrate has probably occurred to some degree. The trend shown by the solid curve for group B shows the expected degradation of both τ_G and τ_R as shown in Figure 6. The limiting value for $\tau_G \approx 1$ mS shown by the solid horizontal curve for group A indicates that no oxygen precipitation has occurred in the epitaxial film. This is consistent with the previous data for the group to the right of the diagonal

---

*Data on epitaxial films was provided by P. J. Tobin of Advanced Product Research and Development Laboratory, Motorola, Inc., Mesa, Arizona.

where $[0]_i \leq 30$ ppm as shown in Figure 6 and is an expected result because the τ_G measurement samples primarily the depletion volume defined by the capacitor area and the depletion depth W. The test voltage used to measure τ_G is -10 V which satisfies the condition $W < X_e$, confining the measurement volume for τ_G to the epitaxial film. Analysis of the temperature dependence of the current transient indicates that the generation current is the dominant current component in the τ_G measurement.

DISCUSSION AND SUMMARY

When oxygen precipitation $\Delta[0]$ is evident in the FTIR measurements obtained subsequent to the thermal anneals carrier lifetime reduction is consistently observed for both τ_G and τ_R. The lifetime reduction is observed in wafers selected on either side of the oxygen minimum when $[0]_i \geq 1.5 \times 10^{18}$ cm^{-3}. The maximum lifetime reduction, ~ 2 decades (Fig. 6) occurs at the seed-end of the crystal where $[0]_i = 2.1 \times 10^{18}$ cm^{-3}. For wafers subjected to the dissolution anneal the thermal history of the fabrication process is a major factor. This effect is evident from the reduced lifetimes at the seed and tang-ends of the axial profiles for τ_G and τ_R in Fig. 5 which indicates that precipitation had occurred during processing. The larger values of τ_G and τ_R near the oxygen minimum in this group can be attributed to reduced oxygen precipitation resulting from the dissolution anneal. Wafers located near the oxygen minimum, $[0]_i \leq 1.5 \times 10^{18}$ cm^{-3} have a maximum value for τ_G of approximately 1 mS, as shown in Fig. 6. The results shown in Fig. 7 for the epitaxial film provide supporting evidence for the enhancement of generation lifetime when oxygen precipitation is minimized by reducing the concentration of oxygen near the surface. The value for τ_G in this figure agrees with the value shown in Fig. 6 for $[0]_i < 1.5 \times 10^{18}$ cm^{-3} (30 ppm) where oxygen precipitation is negligible.

The extended range of τ_R values, $0.8 \ \mu S \leq \tau_R \leq 8.0 \ \mu S$ shown in Fig. 7 for group A indicates that a varying degree of oxygen precipitation has occurred in these substrates. This is consistent with typical measurements on a random sample of wafers. The larger values of τ_R relative to those in Fig. 6 can be expected because these substrates were not annealed prior to processing. Moreover, the thermal cycle at 1150°C, associated with the epitaxial growth process would tend to function as a short dissolution anneal. The data presented here shows no direct evidence for generation lifetime improvements resulting from oxygen precipitation, as reported by other investigators [17]. Whereas, there is consistent experimental evidence for the detrimental effects of oxygen precipitation on both the generation and the recombination lifetime. Experimental evidence supporting the results of this investigation have recently been published by other investigators [18,19].

An important factor to be considered, is the 'cleanliness' of the wafer processing environment. If one encounters heavy metal contamination during processing, then near surface segregation of these impurities will probably set the upper limit on τ_G. For this case, oxygen precipitation in the interior of the wafer represents an alternative to back surface gettering, providing that the oxygen concentration in the near surface region is below the threshold for oxygen precipitation during processing. These competing factors require control of; the thermal history of the wafer before and during processing, the initial oxygen concentration in the wafer and the depth of the oxygen 'free' layer at the surface. These requirements are far more difficult to achieve than to reduce the oxygen concentration to an acceptable level during crystal growth and to maintain a level of cleanliness during wafer processing to achieve the optimum generation lifetime.

ACKNOWLEDGMENTS

The authors wish to thank Ross Boyle, Jerry Maki and Pat Huegle for their assistance with the experimental work, Marlene Scott for her help in preparing the manuscript, all with SRDL, Motorola, Inc., Phoenix, AZ., and to J. B. Price and P. Tobin for their help with the NMOS processing in APRDL, Motorola, Inc., Mesa, AZ. This work was supported in part by the National Science Foundation under its Industry/University Cooperative Program (IUC) Grant #DMR-79-18023.

REFERENCES

[1] K. E. Benson, W. Lin and E. P. Martin, Semiconductor Silicon 1981, Vol. 81-5, Electrochem. Soc. Inc., Pennington, N.J. (1981), p. 33.

[2] T. Abe, K. Kikuchi, S. Shirai, and S. Muraoka, ibid, p. 54.

[3] H. M. Liaw, Semiconductor Intl., Oct. (1979), p 71.

[4] J. R. Patel, ibid, p. 189.

[5] T. Abe, K. Kikichi and S. Shirai, Semiconductor Silicon 1977, Vol 71-1, Electrochem. Soc. Inc., Princeton, N.J., (1977) p. 95.

[6] A. Murgai, Semiconductor Silicon 1981, Vol 81-5, Electrochem. Soc. Inc., Electrochem. Soc. Inc., Pennington, N.J. (1981), p. 113.

[7] S. Kishino, J. Phys. Soc. Japan 49 Suppl. A, (1980), p 49.

[8] S. M. Hu, J. Appl. Phys. 52 (6), June (1981), p. 3974.

[9] 'Oxygen in Silicon,' Silicon Material Processing (II) Semiconductor Silicon 1981, Electrochem. Soc., Pennington, N.J. (1981), pp 208-303.

[10] C. J. Varker, J. D. Whitfield and P. Fejes, Proceeding of Symposium on Silicon Processing, San Jose, California, Jan. 19-22, 1982.

[11] W. Kaiser, P. H. Keck and Lange, Phys. Rev. 101, (1956), p. 1254.

[12] H. J. Hrostowski and R. H. Kaiser, Phys. Rev. 107, (1957), p. 966.

[13] J. W. Corbett, R. S. McDonald and G. C. Watkins, J. Phys. Chem. Solids, 25, (1964), p. 873

[14] H. J. Kuno, Trans. of IEEE ED-11, 8 (1964).

[15] F. P. Heiman, Trans. IEEE ED-14, (1967), p. 781.

[16] M. Zerbst, Z. Angew, Phys. Vol 22, (1966).

[17] K. Yamamoto, S. Kishino, Y. Matsushita and Tlizuka, Appl. Phys. Lett. 36, 3, 195 (1980).

[18] S. N. Chakravarti, P. L. Garbarino and K. Murty, Appl. Phys. Lett. 40 (7), 581 (1982).

[19] M. Miyagi, K. Wada, J. Osaka and N. Inoue, Appl. Phys. Lett. 40 (8), 719 (1982).

STRUCTURAL AND CHEMICAL MICROANALYSIS OF OXYGEN-BEARING PRECIPITATES IN SILICON

R.W.CARPENTER, I. CHAN, H.L. TSAI, C. VARKER* AND L.J. DEMER†
CENTER FOR SOLID STATE SCIENCE, ARIZONA STATE UNIVERSITY, TEMPE, AZ 85287
*Motorola Semiconductor Research Laboratories, Phoenix, AZ 85008
†Metallurgical Engineering Dept., University of Arizona, Tucson, AZ 85721

ABSTRACT

Precipitation in CZ-silicon during post-growth two-stage heat treatment has been examined using the methods of high resolution analytical electron microscopy. Electron transparent specimens prepared from these specimens, exhibited a low density of plate type precipitates on {100} planes. Microdiffraction experiments showed the precipitates to be consistently non-crystalline. Electron energy loss spectra showed that the precipitates contained oxygen, but carbon was not detected. It was found that carbon artifact absorption edges could be induced in spectra by specimen contamination in the microscope. The use of a low temperature stage eliminated this problem. Complementary characteristic x-ray microanalysis showed that metallic impurities had not segregated to these precipitates in this particular case, although this has been observed elsewhere.

INTRODUCTION

Many different morphological forms for precipitates in CZ-silicon have been reported [1, 2, 3] but rather little information concerning their crystallography or their composition has been given. The primary reason for the paucity of crystallographic and compositional data has been the lack of suitable microbeam instrumentation and the usually low volume concentration of precipitates. The latter circumstance usually results in low sensitivity of aperture-selected microdiffraction (SAD) to second-phase diffraction effects. We have used incident electron probes between 10 and ~3nm to examine the structure and composition of plate-type precipitates observed in silicon and some of our first results are given there.

MATERIALS AND EXPERIMENTAL METHODS

The CZ-silicon wafer used to prepare electron transparent specimens was cut with surface normal [100]. Infrared absorption spectroscopy indicated an as-grown interstitial oxygen concentration of ~2 X 10^{+18}/cm^3. The wafers were given a two-step anneal: 16 hrs. at 800°C in N_2 followed by 16 hrs. at 1050°C in dry O_2. Specimens were cut from the wafer (3mm max. dia.) and thinned by polishing both surfaces mechanically to remove several mils before final thinning to electron transparency. Final thinning was ordinarily done by Ar ion beam milling, but in a few cases final thinning was accomplished chemically, using HF + HNO_3 solutions.

A Philips EM400 analytical electron microscope was used for microstructural examination. This instrument is fitted with a field emission electron gun, which has well known advantages for small probe microanalysis [4, 5]. A hybrid x-ray spectrometer system with EDAX detector head interfaced to KEVEX electronics was used for EDS. A Gatan energy loss spectrometer interfaced to the same electronics was used for ELS. Spectral data were recorded on discs and transferred to an off-line computer system (PDP-1144 base) for analysis. A low temperature double-tilt low background specimen holder was used for some of the experiments. It has been our experience that many specimens undergo rapid contamination during micro-beam experiments, primarily resulting from specimen borne contamination in modern AEM's. The present experiments are no exception. The low temperature holder (from Gatan, Inc.) is cooled with liquid nitrogen and produced an indicated temperature of -182°C near the specimen. This was sufficient to reduce contamination in STEM to negligible levels on any particular specimen area for observational periods of several hours.

RESULTS AND DISCUSSION

The precipitates observed were either single plates on $\{100\}_{Si}$ or intersecting plates in cruciform with the same habit planes. Plate thickness was on the order of 10 nm and length 100 to 200 nm. Figure 1 shows a poorly developed cruciform precipitate viewed along the plate axis, so the long dimension runs top to bottom of the foil. The width of the plate perpendicular to the incident beam direction is ~10 nm. A CBED microdiffraction pattern from a similar plate is shown in Figure 2. Note the diffuse ring characteristic of an amorphous solid. The weak Bragg spots are from the surrounding matrix; the incident probe diameter was essentially equal to the precipitate width. Energy loss spectra taken from both the precipitate and the Si matrix adjacent of Fig. 1 are shown in Figs. 3 and 4. There are some distinct differences visible.

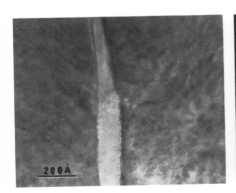

Fig. 1. Plate-type cruciform Si-O precipitate in Si. B≅[001]

Fig. 2. CBED microdiffraction pattern from precipitate. Note diffuse ring.

The most obvious one is the well defined oxygen K edge in the core loss region of the precipitate spectrum, and its absence in the matrix spectrum. There is also marked change in shape and chemical bond change induced shift in the silicon-L edge. The edge onset energy increased from about 99eV in Si to 108eV in SiOx.

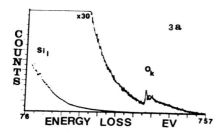

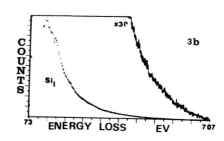

Fig. 3. Core loss regions of energy loss spectra from (a) the precipitate and (b) the adjacent matrix shown in Fig. 1.

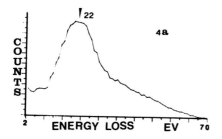

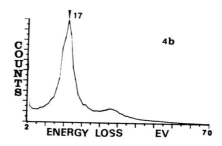

Fig. 4. Low loss regions of the spectra from the precipitate (a) and matrix (b) shown in Fig. 1. The low loss peak shifts from 17eV to 22eV in the precipitate.

The specimen thickness calculated from the zero loss to first plasmon intensity ratio was 49 nm, using Raether's eqn. [6]. The oxygen to silicon ratio in the precipitate was calculated from the net intensity ratio in the O-K and Si-L edges and was found to be ~1.2 using hydrogenic cross-sections and Egerton's formalism [7]. This composition ratio consistently reproduced with thin specimens. For thicker specimens the ratio becomes more silicon rich. Observations of the low loss region in the latter case show the precipitate plasmon peak to shift back toward the 17eV loss characteristic of Si and to become sharper. We conclude that the increase in silicon concentration occurs because precipitates did not extend all the way through the foil in that case, so the loss spectrum is that of a composite specimen. These spectra were obtained with the specimen normally at room temperature. The usual flooding procedure was used to suppress contamination, however, after sufficient examination time a small amount of contamination did occur, and was detectable as a local slight darkening of the image regions where the probe had been focused. The contamination also resulted in C-K edges in the loss spectra. Analysis of these spectra showed the Si:O:C atomic ratios to be in the range of 38:38:25 when the probe was on a precipitate and when the probe was on the adjacent silicon (t≃67nm) Si:C was about 40:60. The contamination layer responsible for these spectra was not heavy enough to cause serious degradation of strong beam BF TEM or STEM images. No C-K edges were observed in spectra obtained using the low temperature specimen holder. It appears that investigations of the effect of carbon on oxide precipitation in silicon require low specimen temperatures.

Matrix strain fields, easily visible under appropriate diffracting conditions, are associated with these precipitates, and often dislocations are attached to the precipitates. An example is shown in Fig. 5. This precipitate is also amorphous; the composition of the precipitate and the dislocations at each end were examined. No evidence for the segregation of metallic impurities to the precipitate or dislocations was obtained by EDS (Fig. 6). Energy loss spectra showed the precipitate itself (Fig. 7a) to contain oxygen, but when the probe was placed on either dislocation, no oxygen was detected (Fig. 7b). These dislocations are approximately in the foil plane, normal to the incident beam direction. Oxygen at their cores would occupy only a small fraction of the irradiated volume, resulting in a small signal.

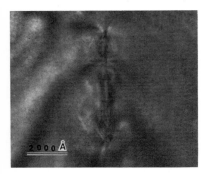

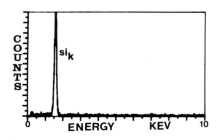

Fig. 5. Si-O precipitate in Si with dislocation segments extending from ends.

Fig. 6. STEM reduced frame scan EDX-ray spectrum from precipitate shown in Fig. 6.

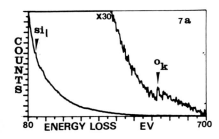

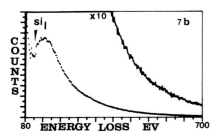

Fig. 7. Energy-loss spectra taken from various parts of the microstructure shown in Fig. 5. The spectrum in (a) is from the precipitate and in (b) from the dislocation on one end. No oxygen was detected on the dislocation. Note the similarity of (b) to Fig. 3(b). The plasmon loss energies for (a) and (b) were 22 and 17eV, respectively.

In an earlier, similar investigation Yang, et al. used ELS to examine etch pit residue in chemically thinned Si containing precipitates. Their precipitates were larger, ~1μm, and mostly etched out of the foil by the thinning process. They did succeed in observing O-K edges in spectra and chemical shifts in Si-L edges with which our results are in good agreement. The chemical shifts result from changes in outer electronic levels when Si-O bonds are formed. Bourret and Colliex [9] have reported oxygen segregation to large Burgers vector dislocations in germanium and silicon. In their case the dislocations cores were oriented end on to maximize the oxygen ELS signal. No quantitative analysis was made, but diffraction showed the precipitate to be crystalline.

In summary we note that direct observations of oxygen containing precipitates in silicon have been made using microanalysis, microdiffraction and imaging. Spatial resolution is adequate for both precipitates and segregation to dislocations when a field-emission gun is used. A cold stage is not absolutely necessary but it is certainly desirable. If carbon segregation is of interest the cold stage is necessary. We also note that transition metal impurity segregation to oxygen bearing precipitates in silicon has often been discussed. We have observed this in Si-epilayer material of considerable lower purity than the CZ-silicon used here [10].

This research was supported by the National Science Foundation, Division of Materials Research and the Chemistry Directorate at the Facility for High Resolution Electron Microscopy, Center for Solid State Science, ASU.

REFERENCES

1. D. M. Maher, A. Staudinger and J. R. Patel, Jour. App. Phys. 47, 3813(1976).

2. K. Tempelhoff, F. Spiegelberg and R. Gleichmann, p. 585 in Semiconductor Si 1977, Ed. by H. R. Huff and E. Sirtl, Proceed. Vol. 77-2, Electrochem. Soc.(1977).

3. J. R. Patel, p. 189 in Semiconductor Silicon 1981, Ed. by H. R. Huff and R. J. Kriegler, Proceed. Vol. 81-5, Electrochemical Soc.(1981).

4. R. W. Carpenter and J. Bentley, p. 153 in SEM/1979/I, Ed. by O. Johari, SEM, Inc., AMF O'Hare, USA.

5. J. Bentley, p. 73 in Proc. 38th Ann. Mtg. EMSA, Ed. by G. W. Bailey, Claitor's Pub. Div.(1980).

6. H. Raether, Excitation of Plasmons and Interband Transitions by Electrons, Springer-Verlag, Berlin(1980).

7. R. F. Egerton, Ultramicroscopy 3, 243(1978).

8. K. H. Yang, R. Anderson and H. F. Kappert, App. Phys. Lett. 33(3), 225 (1978).

9. A. Bourret and C. Colliex, p. 12 in Imaging and Microanalysis with High Spatial Resolution, Proc. Castle Hot Springs Conf. 1982, Ed. by O. Krivanek HREM Facility, ASU(1982).

10. I.Y.T. Chan, M. S. Abrahams and R. W. Carpenter, unpublished research, HREM Facility, Arizona State University, and R.C.A. Laboratories, Princeton, N.J.

CRYSTALLINE PARTICLES IN THERMALLY GROWN SILICON DIOXIDE

F. A. Ponce and T. Yamashita
Materials Research Laboratory, Hewlett-Packard Laboratories
1501 Page Mill Road, Palo Alto, California 94304, USA

ABSTRACT

Small crystalline particles in the vicinity of the Si/SiO_2 interface have been directly observed by high resolution transmission electron microscopy. These crystallites have typical diameters between 20 and 120 Å. Based on the observed interplanar spacings and angles in lattice images, the structure of these particles has been found to match those of cristobalite. Some orientation relationships also appear to exist between these particles and the silicon layer.

INTRODUCTION

Thin film dielectrics play a strong role in the operation of MOS devices in integrated circuit technologies. The structures of these dielectrics have been assumed to be amorphous. In particular, thermal oxide films grown on silicon have a structure similar to silica, which consists of SiO_4 tetrahedra linked in a three dimensional array and lacking long range order. The structural stability of the these dielectric films is an important requirement for reliable insulating and operating characteristics of MOS devices.

One of the main instabilities associated with the amorphous state is the process of devitrification. It is well known that under certain conditions fused silica tends to devitrify into several crystalline forms of SiO_2 (cristobalite, trydimite, quartz, etc.) [1]. It has been found that the devitrification process can be accelerated at elevated temperatures by the presence of impurities, especially under the influence of small amounts of alkali metals [2].

The stability of the structure of thermally grown silicon dioxide films has been the subject of several studies. Nagasima and Enari reported the observation of crystalline forms in SiO_2 films which had been heavily doped with alkali metal impurities and in films grown on rough silicon surfaces [3]. Aleksandrov and Edelman studied the crystallization of silica and silicon oxynitrides by thermal annealing and reported substrate orientation effects [4,5]. Sugano et al [6,7] reported the presence of crystalline forms in SiO_2 films, which were interpreted as silicon clusters that form at the Si/SiO_2 interface during the oxidation process.

In this paper we report on a study of the structure of the Si/SiO_2 interface region of thermally grown SiO_2 films, using high resolution transmission electron microscopy (TEM). These films had been produced using standard oxidation processes. During our investigations, crystalline particles have been observed on several occasions in the oxide near the silicon interface. The structure of these particles is found to be the same as that of cristobalite. In addition, orientation relationships between the crystallites and the silicon substrate have been observed which suggest some degree of epitaxy.

Mat. Res. Soc. Symp. Proc. Vol. 14 (1983) Published by Elsevier Science Publishing Co., Inc.

EXPERIMENTAL TECHNIQUES

The materials examined in this study were CZ silicon and epitaxial silicon on sapphire (SOS) grown by CVD [8]. The oxide was grown under standard steam oxidation procedures at 1000 °C, with flow rates of 3800 cc/min of O_2 and of 6700 cc/min of H_2 for 11 minutes. Prior to and after oxidation, the furnace was purged with 10 liters/min of N_2 for 10 and 30 minutes, respectively. All processes were carried out at 1000 °C. The wafers were introduced to and pulled-out from the furnace at a rate of 16 cm/min. TEM specimens were prepared in cross section, with a <110> direction of the silicon normal to the plane of the specimen. A JEOL 200CX transmission electron microscope equipped with a high resolution top entry stage was used, with an instrumental point resolution better than 2.5 Å. The lattice images were obtained under axial illumination using an objective lens aperture size of 0.55 $Å^{-1}$.

EXPERIMENTAL RESULTS

Several materials oxidized under similar conditions were examined, and in the large majority of cases homogeneous SiO_2 layers were observed, with no visible deviations from the expected amorphous structure. In some cases, however, dark spots with diameters of the order of a few nanometers were observed in the SiO_2 layer. These spots were mostly concentrated near the interface with the silicon substrate. Figure 1 is a high resolution lattice image of the Si/SiO_2 interface of a SOS crystal. The interface itself is planar and abrupt to within a monolayer of silicon. In the middle of the figure, a microtwin reaching the interface produces a hump at the interface. Dark spots in the SiO_2 layer are observed. These spots scatter more electrons than the surrounding amorphous regions, which suggests either some degree of crystallinity (diffraction) or a density difference.

Figure 1. A high resolution TEM micrograph of the Si/SiO_2 interface region. Note the presence of dark spots within the amorphous SiO_2 structure.

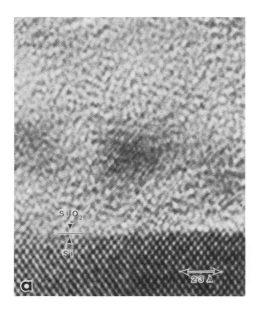 

Figure 2. (a) Lattice image of a particle near the Si/SiO$_2$ interface.
(b) Optical diffraction pattern of the SiO$_2$ particle in (a).

The dark spots have diameters that range from 20 to 40 Å, and they are located randomly in the amorphous SiO$_2$ matrix. Figure 2a shows a magnified view of one such spot in the SiO$_2$. The particle is about 25 Å in diameter and it is located ~40 Å from the silicon interface. The crystalline structure is evident from the presence of lattice fringes in the image. Two sets of fringes separated by 2.4 Å and making an angle of approximately 60° are observed in the image, and this can be determined accurately by laser diffractometry of the original negative. An optical diffractogram of the particle in figure 2a is shown in figure 2b. It confirms the observed 2.4 Å interplanar spacings and the ~60° angle between the two sets of planes. Similar characteristics were observed in other regions of the SiO$_2$ film.

In figure 3, a particle which is attached to the silicon substrate is shown. 2.4 Å interplanar separations are again observed and one set of planes in the particle is parallel to the {111} silicon planes parallel to the Si/SiO$_2$ interface. Some degree of epitaxy is therefore evident.

In order to interpret these observations, the observed interplanar spacings and angles were compared to those of known structures. When comparing these values with those of the cristobalite forms of SiO$_2$ we found excellent correspondence. We have not found any other known structure with reasonable chemical composition which can match the experimental observations.

The two possible forms of cristobalite are shown in figure 3b. The alpha-cristobalite structure is a tetragonal structure with lattice parameters $a_o = b_o$ = 4.97 Å and c_o = 6.93 Å [9]. When viewed in the <021> projection as in figure 3b, two {200} and a {112} planes with interplanar separations of 2.49 and 2.47 Å, respectively, are observed edge on. On the other hand, the beta-

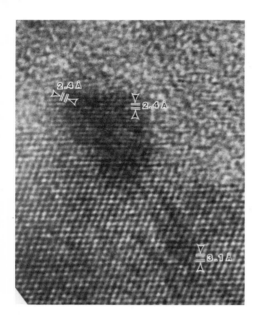

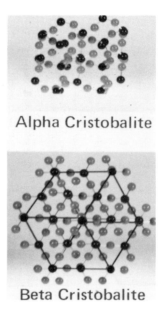

Alpha Cristobalite

Beta Cristobalite

Figure 3. Lattice image of a particle attached to the silicon substrate.

Figure 4. Molecular models of cristobalite structures.

cristobalite structure is cubic with a lattice parameter of 7.16 A. In the <111> projection (shown in figure 3b) three sets of {220} planes with interplanar separations of 2.53A are observed edge on. The alpha structure is a slight modification of the beta structure. The angles between the sets of planes observed in figure 3b is very close to 60 degrees. Multislice image simulations of these two structures using actual experimental parameters are shown in figure 5. Due to the finite, small dimensions of these particles, some slight deviations are to be expected from the bulk lattice structures. Also, the accuracy of the measurement of fringe spacings is about 0.1 A because of the small size of the particle. Hence we conclude that these structures are too close to be distinguished solely by lattice imaging.

DISCUSSION

It must be emphasized that the experimental observations reported here are not typical of thermally-grown silicon dioxide layers, and that they were observed at all was rather fortuitous. Crystalline particles of the type discussed in this paper were not observed in films with good dielectric properties. They appear to be present in materials that are not adequate for device fabrication. The possibility of directly imaging a particle of the sort reported here should be quite remote since several critical conditions must be met simultaneously. First, the imaging of the structure of cristobalite requires point resolutions of less than 2.5 A under axial illumination, which is at the limit of state-of-the-art instruments. Second, the particle must be oriented so that its projection gives an open structure which can be imaged in

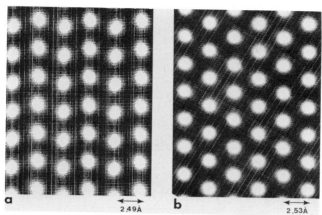

Figure 5. Calculated images of the cristobalite structures: (a) alpha, in the
<021> projection, (b) beta, in the <111> projection.

the TEM. Third, in addition to their typically very low numbers, the size of
these particles is quite small and only those in very thin regions (<80Å) of
the specimen can be imaged with good resolution. Regarding the origin of these
crystalline particles, we can only speculate at the present time.
Devitrification is known to be catalyzed by the presence of metallic impurities,
and it has been observed in bulk silica at about 1000°C [1], which is the
temperature at which thermal oxidation processes are carried out.
Irregularities on the silicon surface, as manifested in figure 1 by the presence
of a crystal fault at the surface, have also been thought to be a possible
source for devitrification [3]. We have observed higher incidence of
crystalline particles in SiO_2 films grown on SOS films than in those grown on CZ
silicon. This is consistent with the fact that SOS has a high density of
aluminum impurities and a high density of faults at the Si/SiO_2 interface.
    In summary, we have reported here the direct observation of small crystalline
particles in thermally grown SiO_2 films. Their structure is identified as that
of cristobalite, with some epitaxial relationships with the substrate having
been observed.

REFERENCES

1. W. D. Kingery, H. K. Bowen, and D. R. Uhlmann, "Introduction to Ceramics,"
    2nd edition (John Wiley Sons, New York, 1976), pp. 73, 274-5.
2. N. G. Ainslie, C. R. Morelock, and D. Turnbull, in "Proceedings of the
    Symposium on Nucleation and Crystallization in Glasses and Melts," edited by
    M. K. Reser (The American Ceramic Society, Columbus, Ohio, 1962), pp 97-107.
3. N. Nagasima and E. Enari, Japanese J. of Appl. Phys. 10, 441 (1971).
4. L. N. Aleksandrov and F. L. Edelman, Surf. Sci. 86, 222 (1979).
5. L. N. Aleksandrov and F. L. Edelman, Thin Solid Films 66, 85 (1980).
6. T. Sugano, Surf. Sci. 98, 145 (1980).
7. T. Sugano, J. J. Chen and T. Hamano, Surf. Sci. 98, 154 (1980).
8. F. A. Ponce, Appl. Phys. Lett. 41, 371 (1982).
9. R. W. G. Wyckoff, "Crystal Structures," 2nd edition (Interscience, New York,
    1963), Vol. 1, pp. 316-9.

III
DEFECTS IN IIII-V COMPOUNDS

POINT DEFECT THERMODYNAMICS OF COMPOUND SEMICONDUCTORS AND
THEIR ALLOYS

F. A. KRÖGER
Department of Materials Science, University of Southern California,
University Park-MC 0241, Los Angeles, California 90089-0241, USA

ABSTRACT

   The physical properties of crystalline solids depend on
the presence of point defects. The concentrations of these
defects in turn depend on the conditions of preparation and
the presence of dopants. Quantitative relations between
these conditions (partial pressures of components, concentrations of dopants, temperature) and the defect concentrations
is arrived at on the basis of defect chemistry. Examples of
pure and doped binary compounds, alloys of binary compounds,
and ternary compounds, are given. Whereas binary compounds
have one composition variable, the alloy systems and the
ternary compounds have two. The role of phase diagrams in
preparing systems of required composition and properties
is stressed.

INTRODUCTION

   The physical properties of crystalline solids depend on the nature of the
material and the presence of point defects. Many semiconducting devices can be
made from elemental semiconductors, primarily silicon. Some devices, however,
can only be made from compounds.
   We distinguish simple compounds such as SiC (a IV-IV compound), GaAs (a III-
V compound) and CdS (a II-VI compound), alloys between these compounds, and
ternary compounds.
   Three types of alloys of binary compounds have technical importance and are
widely studied: 1) Alloys of III-V compounds such as (Ga,Al)As, Ga(As,P), and
(Ga,Al)(As,P); 2) alloys (Pb,Sn)Te, Pb(Se,Te), and (Pb,Sn)(Se,Te); 3) alloys
(Hg,Cd)Te. Alloys 1) are important as solar cells or light emitting diodes;
alloys 2) and 3) have band gaps varying with composition from those of PbTe and
CdTe to zero and are used as infrared detectors with variable wavelength range
of sensitivity. Four-component systems are used to equalize lattice constants
at different band gap widths.
   Examples of ternary compounds are $CuInSe_2$ (used to make solar cells) and
$CdCr_2Se_4$, a magnetic semiconductor. The preparation of materials with well-
defined properties has two aspects: the growth of the material as a single
crystal in bulk or as a film and the regulation of its defect structure, often
by a post-growth anneal. The former requires knowledge of the phase diagram,
the latter involves defect chemistry.

PHASE DIAGRAMS OF BINARY COMPOUNDS [1]

   Complete representation of the phase diagram of binary compounds with variables P, T and composition requires a three dimensional space. If no foreign
species such as air or inert gases are present and the material does not
completely fill the container, the pressure is equal to the sum of the partial

Mat. Res. Soc. Symp. Proc. Vol. 14 (1983) © Elsevier Science Publishing Co., Inc.

208

pressures of the components.

Phase diagrams normally used are sections through the space figure or projections of three-phase surfaces. In P-T projections of systems with a pressure minimum, some areas represent two different situations. Such an overlap can be prevented by plotting one of the partial pressures rather than the total pressure P. E.g. for PbS, p_{Pb} or p_{S_2} ∴ T. Fig. 1 shows such a plot. All possible compositions of the solid are represented by points inside the loop formed by the three-phase lines solid + liquid + vapor and solid + solid + vapor. Compounds with certain compositions such as the exactly stoichiometric one ($\delta=0$) or certain deviations from stoichiometry ($\delta>$ or <0) are represented by lines. Crystals represented by points on the loop can be grown from the melt; those represented by points inside the loop can only be made by equilibration with a vapor.

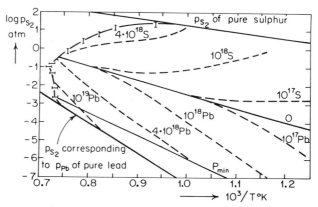

Fig. 1 log p_{S_2} vs. T^{-1} projection for PbS [2]

Although exactly stoichiometric compounds can be grown from the melt at the point where the line $\delta=0$ intersects the three-phase line, the temperature at this point is relatively high and the concentrations of native defects are large; as a result small variations in the experimental conditions lead to marked variations in the composition and the material tends to be inhomogeneous. It is preferable to prepare stoichiometric crystals by equilibration with the vapor at lower temperatures where the extent of disorder is less and therefore inhomogeneities are less severe. P-T or partial pressure-T projections have been determined for PbS [2], PbSe and PbTe [3-5], CdSe [6], CdTe [7-10], HgSe [11], HgTe [12-14], SnS [15], SnTe [16], MnTe [17], Bi_2Te_3 [18-20], AlAs [21], GaAs [22-25], and GaP [22-24].

PHASE DIAGRAMS OF ALLOYS OF TERNARY COMPOUNDS

Alloys of binary compounds have three or four components and thus require a four or five dimensional space to completely represent the P, T phase relations. In many cases the pressure is neglected and temperature-composition diagrams are presented: Ga-In-As [27,28], Ga-Bi-P [29], Ga-In-Sb [30,31], Ga-In-P [32,33], Ga-As-Sb [34], Ga-Al-Sb [35,36], Ga-As-P [37,38], Al-Ga-P [39], Al-Ga-As-P [40].

Pseudo-binary phase diagrams have been given for PbTe-GeTe [41,42] and HgSe-MnSe [43]. Fig. 2 shows a schematical T-x space diagram for $(Pb_{1-x}Sn_x)_{1-x}Te$ showing uninterrupted solid solubility in the solid (Pb,Sn)Te phase. Such diagrams are characterized by the shape of the liquidus surfaces and the correlation between the liquidus and solidus surfaces: so called tie lines, indicating which solid phase is in equilibrium with a given liquid phase: Laugier et al. [44] and Harris et al. [45] have computed both characteristics for the Pb-Sn-Te and Pb-Sn-Se systems and represented their results by two sets of curves, one representing liquidus isotherms (i.e. intersections of the liquidus plane with planes of T = constant), the other giving compositions of the liquid phase from which at different temperatures a solid phase of required composition can be grown (iso-solidus composition lines). Fig. 3 shows results of Harris et al. for Pb-Sn-Te. Comparison of these curves with experimental results by Groves [46], Tamari and Shtrikman [47], Astles et al. [48] and Hatto et al. [49] shows good agreement at larger SnTe contents but give tin contents of the solid that are too large at lower tin contents of the liquid. Agreement with Laugier's results is poor. The curves presented in Fig. 3 refer to the cation/cation ratio but neglect variation in the cation/anion ratio. Experimental solidus lines for different tin concentrations, showing Pb+Sn/Te non-stoichiometry in the solid phase as determined from the concentrations of electrons or holes, are shown in Fig. 4 [50]. Similar information was obtained for (Pb,Sn)Se [51]. By growing from melts with different concentrations of tin and selenium, the ratio Pb/Sn in the solid and the conductivity type could be varied, p-type being obtained at high, n-type at low growth temperature (Fig. 5).

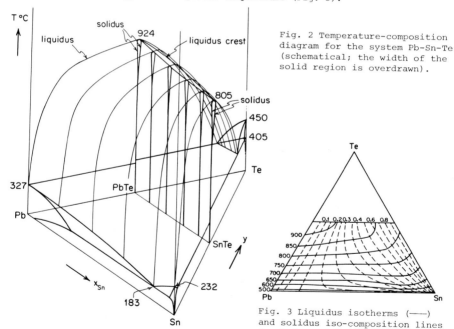

Fig. 2 Temperature-composition diagram for the system Pb-Sn-Te (schematical; the width of the solid region is overdrawn).

Fig. 3 Liquidus isotherms (———) and solidus iso-composition lines (---) for metal rich Pb-Sn-Te calculated by Harris et al. [45].

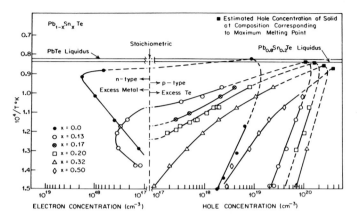

Fig. 4 Solidus lines for $(Pb_{1-x}Sn_x)_{1-y}Te_y$ with different SnTe concentration as determined from carrier density. [50]

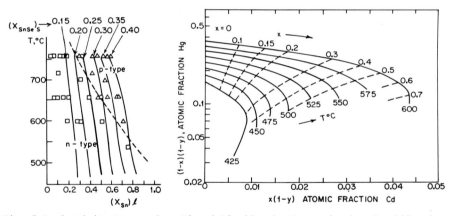

Fig. 5 Conductivity type and equicomposition lines for solid (Pb,Sn)Se grown from liquids with different $(x_{Sn})_\ell$ at temperatures T varied by changing x_{Se}. [51]

Fig. 6 Liquidus isotherms (——) and solidus iso-composition lines (---) for the Te-rich corner of the system Hg-Cd-Te according to Harman. [52]

Liquidus isotherms and solidus isocomposition lines for $(Hg_{1-x}Cd_x)_{1-y}Te_y$ with y>0.5 were determined by Harman [52] (Fig. 6). Note that the way of plotting differs from that in Fig. 3

In order to minimize the pressure of the most volatile component, Te in (Pb, Sn)Te, Hg in (Hg,Cd)Te, crystals of (Pb,Sn)Te are grown from metal-rich melts, crystals of (Hg,Cd)Te from Te-rich melts. P-T projections similar to those for binary compounds can also be given for the alloys. Fig. 7 shows such curves for (Hg,Cd)Te [53]. The maximum Hg pressure increases with decreasing [Hg] in the solid, indicating that the activity coefficient of HgTe increases with increasing cadmium concentration. The Hg pressure at the stoichiometric point where [Hg]+[Cd]=[Te] is almost independent of Cd content and given by [54-56] $\log(p_{Hg}/atm) = -3,099/T + 4920$. Partial pressures of mercury and selenium over (Hg,Cd)Se were determined by Iwanowski and Bak [57]. The required semiconductor properties of (Hg,Cd)Te crystals usually are established by a post-growth anneal in Hg vapor [58]. However, as for (Pb,Sn)Se and (Pb,Sn)Te, with careful choice of the growth conditions, crystals with the required properties can be obtained without such an anneal [59,60]. Precipitation of Te may occur when the samples contain an excess of Te [61]. In the system PbTe-CdTe there is limited inter-solubility in the solid phase (Fig. 8) [62]. Fig. 9 shows the fields of sta-bility of the PbTe and CdTe phases at 700°C in terms of the partial pressures P_{Pb} and P_{Cd} or the activity of lead, a_{Pb}, in liquid alloys containing x_{Pb} lead and x_{Cd} cadmium.

DEFECT CHEMISTRY

For the formulation of reactions involving defects, the atomic symbols intro-duced by Kröger and Vink are often used [63-65]. Superscripts: +, -, and 0 indicate actual positive, negative and zero charges and · (dot), ' (dash), and x (multiplication sign) indicate positive, negative, and zero effective charges. Point defects can be formed by disorder, by non-stoichiometry, or by doping with aliovalent elements. Vacancies of electronegative constituents, interstitial electropositive elements, and foreign or native atoms replacing native atoms with a smaller number of valence electrons act as donors; vacancies of electro-positive elements, interstitial electronegative elements and foreign or native atoms replacing native atoms with a larger number of valence electrons act as acceptors.

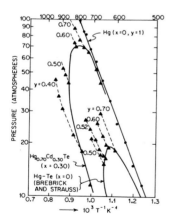

Fig. 7 Mercury pressures as f(T) along the three-phase lines in the system $(Hg_{(1-x)}Cd_x)_yTe_{1-y}$ for x=0 and x=0.3 [53].

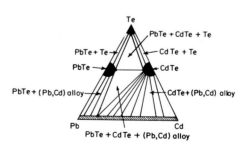

Fig. 8 Schematic phase diagram of the system Pb-Cd-Te at 700°C [62].

212

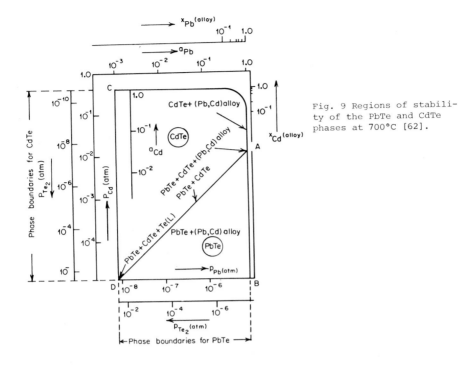

Fig. 9 Regions of stability of the PbTe and CdTe
phases at 700°C [62].

Ionization is governed by the position of energy levels in the forbidden gap.
Dopants substituting isoelectronic native constituents in general give rise to
deep donor and/or acceptor levels and are relatively ineffective as donors or
acceptors in conductivity, but play a role as traps or recombination centers as
found in ZnS:O [66] and GaP:N [67]. In only one case, Y in Al_2O_3 [68], has an
isoelectronic substitutional dopant been found to be an efficient donor.

An isoelectronic dopant can be an effective donor, however, when it occupies
interstitial sites [69,70]. Labeling of levels is often confusing because a
given level is related to two different centers, dependent on whether the level
is occupied or empty. The uncertainty can be removed either by marking the level with both charge states or by labeling the level with the symbol of the center present when the level is occupied by an electron. The latter convention is
favored by the author.

If multiple ionization occurs, the level of the most negative species is
usually above that of less negative (more positive) species: i.e. the A" level
is above the A' level, and the D^X level is above the D· level. This, however,
is not always the case: ionized defects polarize the surroundings, causing a
decrease in energy, and it may occur that the difference in energy between subsequent ionization steps is smaller than the difference between the polarization
energies in the two states [71]. Then the level positions are inverted, the A"
level being below the A' level, or the D· level above the D^X level. In such
situations the center with the intermediate charge is unstable, disproportioning

spontaneously into centers with lower and higher charges. This situation has been found for V^X, $V^{\cdot}$ and $V^{\cdot\cdot}$ in p-type Si [72,73] and for dangling bonds [74] or valence alternating pairs [75,77] in amorphous glasses. Other possible examples are In in In_2Te_3, and Na in Na_xWO_3 [78].

Disorder Processes

For binary compounds there are nine types of atomic disorder: three involving defects of the same kinds, e.g. vacancies (Schottky disorder), interstitials (interstitial disorder), and anti site defects, (anti-structure disorder). In addition there are six mixed types involving interstitials and vacancies (Frenkel disorder), interstitials and misplaced atoms, and vacancies and misplaced atoms. Schottky and Frenkel disorder have been most commonly considered, but recently antistructure defects have found defenders [79,80], and disorder processes involving such defects may well gain further in importance in the future.

Table I shows values of the enthalpies of formation of neutral vacancies and anti site defects and the corresponding values of the enthalpies of Schottky and antistructure disorder for III-V and II-VI compounds AB [81,82]. It is seen that for the II-VI compounds and for AlAs, GaN, and GaP, $H_S^X \ll H_{AS}^X$, indicating predominance of Schottky disorder with reasonable agreement between theoretical and experimental values in several cases. For InAs and InSb, $H_{AS}^X \ll H_S^X$ and therefore antistructure disorder should dominate. Values differing less than 1 eV are found for AlSb, GaSb, GaAs, and InP. A preference for antistructure disorder is indicated for GaSb and experimental results support this prediction

Table I

Theoretical values in eV of the enthalpies of formation of single neutral vacancies and of Schottky disorder for neutral defects (H_S^X), and of antistructure disorder for neutral and charged defects (H_{AS}^X and H_{AS}^Z) according to van Vechten [79,80] and experimental values of H_S^X [81].

	$H(V_A^X)$	$H(V_B^X)$	H_S^X calc.	obs. [81]	$H(B_A^X)$	$H(A_B^X)$	H_{AS}^X	H_{AS}^Z
SiC(c)	4.94	2.92	7.86		-0.06	0.96	0.89	--
AlN					14.18	14.48	28.66	16.86
AlP					5.50	5.60	11.10	6.10
AlAs	2.76	2.75	5.51		5.05	4.90	9.95	5.35
AlSb	2.30	2.81	5.11		3.46	3.12	6.58	3.48
BN					10.6	9.8	20.40	11.20
BP					4.49	3.75	8.24	4.24
BAs					3.37	2.34	5.71	2.91
GaN	5.10	2.73	7.83		9.07	9.71	18.78	11.78
GaP	2.98	2.64	5.62	3.84;5.06	5.08	5.38	10.46	5.76
GaAs	2.59	2.59	5.18	4	3.05	3.05	6.10	3.40
GaSb	2.03	2.56	4.59		1.68	1.92	3.60	2.00
InN					7.33	7.80	15.13	10.33
InP	3.04	2.17	5.21		3.02	3.49	6.51	3.91
InAs	2.61	2.07	4.68		1.03	1.27	2.30	1.60
InSb	2.12	2.12	4.24		0.61	0.61	1.22	0.88
CdS	3.56	2.69	6.25	5.60	6.48	6.86	13.34	8.34
CdSe	3.18	2.65	5.83		4.93	5.10	10.03	6.35
CdTe	2.75	2.75	5.50	4.80	4.56	4.56	9.12	5.90
ZnO	5.41	3.00	8.41	6.29	12.86	12.31	25.17	18.29
ZnS	3.47	3.13	6.00		9.25	9.54	18.79	11.0
ZnSe	3.09	3.09	6.18		6.91	6.91	13.82	8.16
ZnTe	2.54	3.60	6.14		5.74	5.56	11.30	6.52

[82-84]. For the other compounds Schottky disorder energies are the smaller ones but antistructure disorder may still turn out to be dominant. There is evidence in favor of antistructure disorder for GaAs [85]. A hybrid disorder process, with antistructure defects for an excess of one component and vacancies for an excess of the other component remains also possible. For CdS, antistructure defects S_{Cd} have been proposed to explain an increase of the Fermi level with increasing p_{S_2} at high sulphur pressure [86]. Since most disorder processes create both donors and acceptors, disorder generally forms ionized rather than neutral defects, with a correspondingly smaller enthalpy of formation. This correction applies to most types of disorder in the same way and thus does not alter the conclusions arrived at above; only for hybrids involving antistructure defects are only donors or acceptors formed and ionization will not necessarily occur.

DEFECT CHEMISTRY OF BINARY COMPOUNDS

Relations between defect concentrations and the conditions of preparation (partial pressures of constituents, temperature) are obtained by writing formation reactions for all defects expected to play a significant role, apply the law of mass action to these reactions, and solve the equations obtained under the restrictions of electroneutrality. The law of mass action is based on ideal behavior of all reaction partners and random distribution of the defects. If equilibrium is not maintained, e.g. when dopants are present at a fixed concentration, an atom balance equation has to be satisfied for each equilibrium that is frozen in. Although complete solutions can be obtained, it is often advisable to first derive approximate solutions on the basis of neutrality and balance equations approximated by their dominant members (Brouwer approximation [87]). The approximate solutions for the concentrations of defects j are of the form $[j] = \prod_i K_i^\nu p_A^n p_F^m$ or $[j] = \prod_i K_i^\nu p_A^n [F]^m$ where K_i are equilibrium constants, p_A is the partial pressure of A, p_F and $[F]$ are respectively the partial pressure and the concentration of the dopant F, and ν, n and m are small integers or simple fractions. Therefore, isothermal plots of $\ln[j]$ against $\ln p_A$, $\ln p_F$, or $\ln[F]$ are straight lines with slope n or m. Since usually $K_i = K_i^0 \exp(-H_i/RT)$, plots of $\ln[j]$ versus T^{-1} at constant p_A, p_F, or $[F]$, have slopes $\sum_i \nu_i H_i/R$.

Once defect models have been determined, exact solutions are obtained by eliminating from the complete neutrality conditions all concentrations except one (usually the electron concentration). When the value of the electron concentration satisfying the neutrality condition has been found, each term of the neutrality condition gives the concentration of the defect it represents. Defect models are most easily checked by comparing theoretical isotherms with corresponding experimental results, using Hall effect or thermoelectric power to determine concentrations of electrons or holes, and self diffusion or diffusion-dependent properties such as sintering rate or creep to provide information regarding atomic defects. Additional information can be obtained from magnetic susceptibility or electron spin resonance, optical absorption, and solubility studies (for dopants). Luminescence lacks the proportionality between intensity and concentrations of involved defects which characterizes the other effects, but is a powerful tool for the determination of level positions.

Densities can be used when the deviations from stoichiometry are sufficiently large. Thus density measurements show Ti_i to be the dominant defect in $TiS_{2-\delta}$, V_{Co} and V_{Ni} that in $CoS_{1+\delta}$ and $NiS_{1+\delta}$ [27]. Detailed defect models have been determined for CdS [89,90], CdTe [91,92], CdSe [93-95], PbS [94-102], PbSe [103-107], PbTe [103,108-114], SnS [115], SnSe [116], SnTe [116-119], HgTe [120], GaP [80,81,121-124] and TiO_2 [125,126] on the basis of combined studies on pure and doped materials. For CdS and CdTe the structure involves interstitials and vacancies of both types in various states of ionization and, for CdS, S_{Cd}^x [86].

Two types of uncertainties often remain: (1) uncertainty of the nature of the defect mainly responsible for the compensation of charged dopants; (2) uncertainty in the purity of "pure" material: in several cases effects have been attributed to native defects which later proved to have been caused by small concentrations of impurities. For instance in ZnTe effects attributed to V_{Zn}'' [127,128] later proved to be due to Cu_{Zn}' [129]. Similar errors have been made in ZnSe. This does not mean that native defects are not present or are electrically inactive: it only shows that often their effect is overshadowed by that of impurities. The presence of V_{Zn}'' can in fact be inferred from the fact that doping with a donor D creates centers $(DV_{Zn})'$ and $(DV_{Zn})^x$ acting as luminescence and recombination centers in ZnS [130], ZnSe [131-133], and ZnTe [134].

DEFECT CHEMISTRY OF ALLOYS OF COMPOUNDS

In view of the uncertainty in the defect structure of the III-V compounds individually, little can be said about the defect structure of their alloys. For the II-VI compounds the situation is more favorable.

References to papers dealing with the defect structure of various alloys of II-VI compounds are given in Table II. Work on $(Hg_{1-x}Cd_x)Te$ with x=0.18 [158], 0.16-0.25 [58], 0.20 [159] or 0.4 [160] shows that equilibration at T $\geq$ 300°C in

TABLE II

Alloy	Subject and References
(Pb,Cd)Te	electrical properties [135]; Schottky constant as f[Cd] [136]; defect structure [62]; D_{Cd} [137]
(Pb,Zn)Te	defect structure [138]
(Pb,Sn)Te	m_h^* as $f(E_{gap})$ [139]; band structure [140]; K_i [141,142]; electrical properties $\therefore$ T_{anneal}, T_{growth} [143]; [h'] at Te-rich boundary [144]; D_{Sn}, D_{Te} $\therefore$ p_{Te_2} [145]
(Pb,Sn)Te:Cd	electrical properties [146-148]; defect structure [149]; D_{Cd} [150,153]
(Pb,Sn)Te:Zn	electrical properties [148]
(Pb,Sn)Te:In	electrical properties [152]
(Pb,Sn)Se	K_i [141,142]; laser review [145]
(Pb,Sn)Se:Sn+Na	[h'] [153]
Pb(Se,Te):In	electrical properties [154]
(Pb,Sn)(Se,Te)	constant unit cell dimensions at different band gap [155,156]
(Hg,Cd)Te	electrical properties [58,157-159]; defect structures [159,160] Hg $\therefore$ p_{Hg}, T, defect model [161]
(Hg,Cd)Te:In	electrical properties, defect structure [162]; D_{Hg} $\therefore$ p_{Hg},T; defect model [161]
(Hg,Cd)Te:Cu	electrical properties, defect structure [159]
(Hg,Cd)Se	electrical properties [57,163]
(Hg,Cd,Mn)Te	band gap [164]

atmospheres rich in either Hg or Te lead to material showing p-type conductivity after cooling with [h'] decreasing with increasing p_{Hg} (Fig. 10). The results are explained by a defect model governed at the equilibration temperatures by electrons and holes in $(Hg_{0.8}Cd_{0.2})Te$ [159], but by $2[V_{Hg}^H] \simeq [h^\cdot]$ in $(Hg_{0.6}Cd_{0.4})$ Te [160]. Computed values of the concentrations of holes in $Hg_{0.8}Cd_{0.2}Te$, cooled after annealing under mercury pressures inside the stability loop are shown in Fig. 11. N-type behavior observed after equilibration at temperatures <300°C (Fig. 10) is attributed to the presence of donor impurities [58,157-160].

Investigations of the electrical properties of PbTe:Cd show that cadmium is present largely as electrically inactive Cd_{Pb}^x but to a small extent as Cd_i (which is a donor).

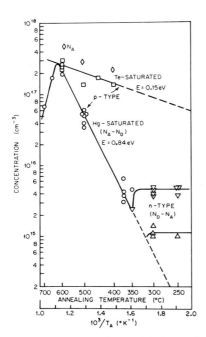

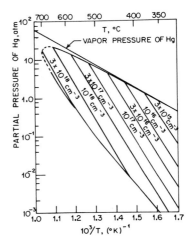

Fig. 11 Computed values of hole concentrations in $Hg_{0.8}Cd_{0.2}Te$ in crystals cooled after equilibration under mercury pressures inside the stability loop [159]; reprinted with permission of the publisher, The Electrochemical Society, Inc.

Fig. 10 Concentrations of holes and electrons in $Hg_{0.6}Cd_{0.4}Te$ cooled after annealing at the highest allowable Hg pressure (holes, o, electrons Δ and ∇) or Te pressure (holes, □). [157]

Ranges governed by various approximations to the neutrality condition and Cd balance are shown in Fig. 12 (p_{Pb} and a_{Cd} as variables) [62]. A similar figure can be drawn with p_{Pb} and [Cd] as variables. The two differ because $Pb_{Pb}^x + Cd$ (alloy) → $Cd_{Pb}^x + Pb(g)$ leads to $[Cd_{Pb}^x] \propto p_{Pb}^{-1} a_{Cd}$. Figures 13 and 14 show 700°C isotherms as a function of p_{Te_2} for PbTe: $3 \times 10^{20}cm^{-3}$ Cd, and for PbTe:Cd as $f(a_{Cd})$ with $p_{Pb} = 8 \times 10^{-6}atm$ [62]. Similar range diagrams have been constructed for $(Pb_{0.8}Sn_{0.2})Te:Cd$ [149].

DEFECT CHEMISTRY OF TERNARY COMPOUNDS

The properties of ternary semiconducting compounds differ from the alloys of binary compounds in that a variation in the ratio between the cations not only varies the width of the forbidden gap and the values of disorder constants, but

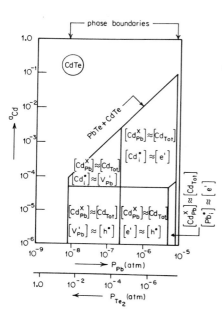

Fig. 12 Range diagram for validity of different approximations of the neutrality condition and Cd-balance for PbTe at 700°C with a_{Cd} and P_{Pb} the variable [62].

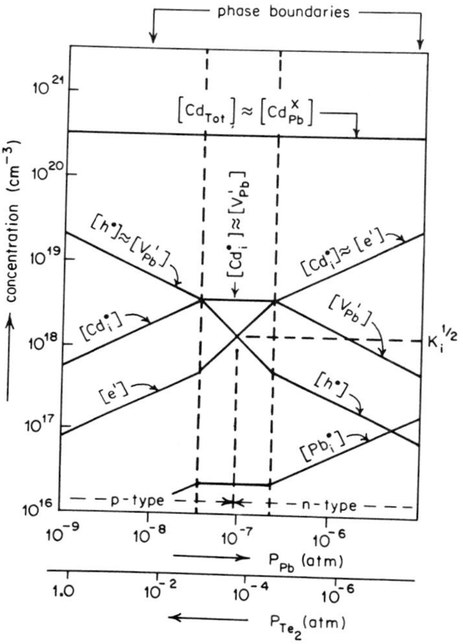

Fig. 13 Defect concentration isotherms as $f(p_{Te_2})$ for PbTe 3×10^{20} cm^{-3} Cd at 700°C [62].

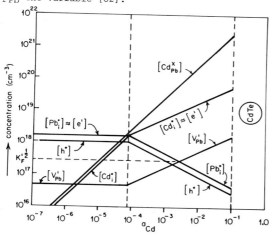

Fig. 14 Defect concentration isotherms as $f(a_{Cd})$ for PbTe:Cd under $p_{Pb}=8 \times 10^{-6}$ atm at 700°C [62].

also has a marked effect on the concentrations of atomic defects due to the fact that the binary subcompounds of the ternary compound have different compositions. [165,166]. In addition to such cation/cation non-stoichiometry (sometimes called "molecularity") [168] which leaves all chemical bonds saturated and does not produce donors or acceptors, there may be normal cation/anion non-stoichiometry leading to donor or acceptor formation.

As in binary compounds, dopants affect the properties in a manner determined by the point defects present in the absence of the dopants.

Table III lists various ternary compounds in which the properties have been changed by changing the cation/anion ratio, the cation/cation ratio, or both. The list is far from complete even as far as chalcogenides are concerned. Oxides such as $LiNbO_3$, $BaTiO_3$, tungstates, molybdates and ferrites have not been included. For a more extensive list see [167], chapters 20 and 21.

The existence range of a compound is determined by the nature of competing phases. Fig. 15 demonstrates this for $CuInSe_2$. Fig. 16 shows the existence range as a function of the activities of copper and indium. A possible set of isotherms as $f(a_{In_2Se_3})$ for p_{Se_2} = constant corresponding to tracks mn in Fig. 16 is shown in Fig. 17. For several of the compounds listed in Table III the cation/anion ratio was established at the two phase boundaries, but complete isotherms were not determined. Such isotherms have been determined for various ternary oxide semiconductors not included in the present survey, e.g. for the titanates of Ba [191], Sr [192], and Ca [193]. For a survey see [194].

TABLE III

Papers describing variation of the properties of ternary semiconductor compounds by variation of the cation/anion and/or the cation/cation ratio.

Compound	Variation of		
	cation/anion	cation/cation	both cation/cation and anion/anion
$AgGaS_2$	168		
$AgInSe_2$		169	
$AgInTe_2$	170		
$CdCr_2Se_4$	171	188	188
$CdCr_2Se_4$:In	172,173		
$CdSiP_2$	174		
$CuCr_2Se_4$		189	
$CuGaS_2$	175,176	177	
$CuGaS_2$:Fe	178		
$CuGaSe_2$	179		
$CuInS_2$	169,180-182	185	180,183,184,195,196
$CuInSe_2$	186		186
$ZnGeP_2$	174		
$ZnIn_2S_4$		187	

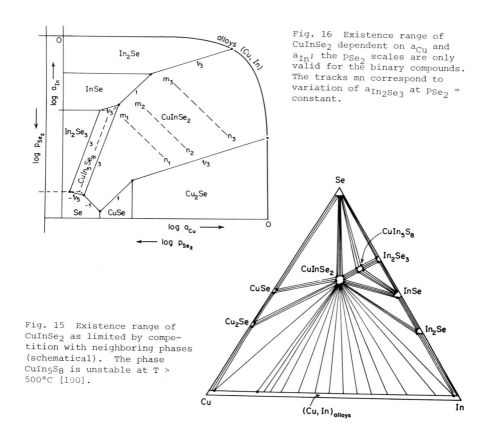

Fig. 16 Existence range of $CuInSe_2$ dependent on a_{Cu} and a_{In}; the p_{Se_2} scales are only valid for the binary compounds. The tracks mn correspond to variation of $a_{In_2Se_3}$ at p_{Se_2} = constant.

Fig. 15 Existence range of $CuInSe_2$ as limited by competition with neighboring phases (schematical). The phase $CuIn_5S_8$ is unstable at T > 500°C [190].

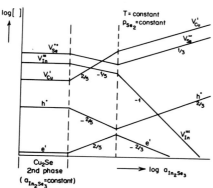

Fig. 17 A possible set of isotherms for $CuInSe_2$ as a function of $a_{In_2Se_3}$ at p_{Se_2} = constant.

220

REFERENCES

1. F. A. Kröger, The Chemistry of Imperfect Crystals (North-Holland Publ. Co.) vol. 1, 175 a.f. (1973).
2. J. Bloem and F. A. Kröger, Z. Phys. Chem. (N.F.) 7, 1 (1956).
3. V. P. Zlomanov, E. V. Masyakii and A. V. Novoselova, J. Crystalgr. 26, 261 (1975).
4. B. J. Sealy and A. J. Crocker, J. Mat. Sci. 8, 1737 (1973).
5. R. F. Brebrick and A. J. Strauss, J. Chem. Phys. 40, 3230 (1964).
6. R. A. Burmeister, Jr. and D. A. Stevenson, J. Electrochem. Soc. 114, 394 (1967).
7. D. de Nobel, Philips Res. Rept. 14, 361, 430 (1959).
8. M. R. Lorenz, J. Phys. Chem. Solids 23, 939 (1962).
9. I. V. Korneeva, A. V. Belyaev and A. V. Novoselova, Neorg, Khim (russ.) 5, 1 (1960).
10. R. F. Brebrick, J. Electrochem. Soc. 118, 2014 (1971).
11. Z. A. Munir, D. J. Meschi, and G. M. Pound, J. Crystalgr. 15, 263 (1972).
12. H. Rodot, J. Phys. Chem. Solids 25, 85 (1964).
13. R. F. Brebrick, and A. J. Strauss, J. Phys. Chem. Solids 26, 989 (1965).
14. A. J. Strauss and R. F. Brebrick, J. Phys. Chem. Solids 31, 2293 (1970).
15. H. Rau, Ber. Bunsenges. Phys. Chem. 69, 731 (1965).
16. R. F. Brebrick and A. J. Strauss, J. Chem. Phys. 41, 197 (1964).
17. J. van den Boomgaard, Philips Res. Rept. 24, 284 (1969).
18. C. B. Satterthwaite and H. B. Ure, Phys. 108, 1164 (1957). (Corrected by F. A. Kröger, J. Phys. Chem. Solids 7, 277 (1958)).
19. R. F. Brebrick, J. Phys. Chem. Solids 30, 719 (1969).
20. R. F. Brebrick and F.T.J. Smith, J. Electrochem. Soc. 118, 991 (1971).
21. W. Kischio, Z. Anorg. Allgem. Chem. 328, 187 (1964).
22. K. Weiser in "Preparation of 3-5 Compounds" (R. K. Willardson and H. L. Goering editors), Rheinhold Pub. Co., New York 1962, p. 471.
23. D. Richman, J. Phys. Chem. Solids 24, 1131 (1963).
24. C. D. Thurmond, J. Phys. Chem. Solids 26, 785 (1965).
25. J. R. Arthur, J. Phys. Chem. Solids 28, 2257 (1967).
26. R. F. Brebrick and R. J. Panlener, J. Electrochem. Soc. 121, 932 (1974).
27. M. A. Pollack, R. E. Nahory, L. V. Deas, and D. R. Wonsidler, J. Electrochem. Soc. 122, 1550 (1975).
28. M. B. Panish, J. Electrochem. Soc. 117, 1202 (1970).
29. A. S. Jordan, Metallurg. Trans 7B, 191 (1977).
30. M. F. Gratton and J. C. Woolley, J. Electrochem. Soc. 125, 657 (1978).
31. S. Szapiro, J. Phys. Chem. Solids 41, 279 (1980).
32. T. Sugiura, H. Sugiura, A. Tanaka, and T. Sukegawa, J. Crystalgr. 49, 559 (1980).
33. K. Kajiyama, Jap. J. Appl. Phys. 10, 561 (1971).
34. M. F. Gratton and J. C. Woolley, J. Electrochem. Soc. 127, 55 (1980).
35. K. Osamura, K. Nakajima and Y. Murakami, J. Electrochem. Soc. 126, 1992 (1979).
36. A. Jouillé, P. Gautier and E. Monteil, J. Crystalgr. 47, 100 (1979).
37. K. Osamura, J. Inoue and Y. Murakami, J. Electrochem. Soc. 119, 103 (1972).
38. G. A. Antypas, J. Electrochem. Soc. 117, 700 (1970).
39. M. Ilegems and M. B. Panish, J. Crystalgr. 20, 77 (1973).
40. M. Ilegems and M. B. Panish, J. Phys. Chem. Solids 35, 409 (1974).
41. S. G. Parker, J. E. Pinnell and L. N. Swink, J. Mat. Sci. 9, 1829 (1974).
42. D. H. Hohnke, H. Holloway and S. Kaiser, J. Phys. Chem. Solids 33, 2053 (1972).
43. A. Pajaczkowska and A. Rabenau, Mat. Res. Bull. 12, 183 (1977).
44. A. Laugier, J. Cadoz, M. Faure and M. Moulin, J. Crystalgr. 21, 235 (1974).
45. S. Harris, J. T. Longo, E. R. Gertner and J. E. Clarke, J. Crystalgr. 28,

334 (1975).

46. S. H. Groves, J. Electron. Mat. 6, 195 (1966).
47. N. Tamari and H. Shtrikman, J. Crystalgr. 43, 653 (1978).
48. M. G. Astles, P. Hatto and A. J. Crocker, J. Crystalgr. 47, 379 (1979).
49. P. Hatto, A. J. Crocker and J. Winn, ibid. 57, 507 (1982).
50. T. C. Harman, J. Nonmetals and Semicond. 1, 183 (1973).
51. L. P. Bychkova, G. G. Gegiadze, O. I. Davarashvili, V. P. Zlomanov, R. I. Chikovane and A. P. Shotov, Sov. Phys. Dokl. 26, 688 (1981).
52. T. C. Harman, J. Electron. Mat. 9, 945 (1980).
53. J. Steininger, ibid. 5, 299 (1976).
54. J. P. Schwartz, T. Tung and R. F. Brebrick, J. Electrochem. Soc. 128, 438 (1981).
55. T. Tung, L. Golinka and R. F. Brebrick, ibid. 451.
56. ibid. 1601
57. R. J. Iwanowski, Acta Phys. Polon. 47A, 583 (1975).
58. B. E. Bartlett, P. Capper, J. E. Harris and M.J.T. Quelch, J. Crystalgr. 49, 600 (1980).
59. P. Becla, J. Lagowski, H. C. Gatos, and H. Ruda, J. Electrochem. Soc. 128, 1171 (1981).
60. P. Becla, J. Lagowski and H. C. Gatos, ibid. 129, 1103 (1982).
61. C. J. Gilham and R. A. Farrar, J. Mat. Sci. 12, 1994 (1977).
62. H. R. Vydyanath, J. Appl. Phys. 47, 4993 (1976).
63. F. A. Kröger and H. J. Vink in Solid State Phys. (F. Seitz and D. Turnbull editors) 3, 307 (1956).
64. F. A. Kröger, The Chemistry of Imperfect Crystals, North-Holland Publ. Company, vol. 2, 2 (1974).
65. Semiconductors (N. B. Hannay editor) Reinhold Pub. Co., New York, (1959).
66. F. A. Kröger and J.A.M. Dikhoff, J. Electrochem. Soc. 99, 144 (1952).
67. J. J. Hopfield, P. J. Dean, and D. G. Thomas, Phys. Rev. 158, 748 (1967).
68. M. M. El-Aiat and F. A. Kröger, J. Am. Ceram. Soc. 65, 280 (1982).
69. A. J. Rosenberg and F. Wald, J. Phys. Chem. Solids 26, 1079 (1965).
70. H. R. Vydyanath, J. Appl. Phys. 47, 4993 and 5003 (1976).
71. P. W. Anderson, Phys. Rev. Letters 34, 953 (1975).
72. G. A. Baraff, E. O. Kane and M. Schlüter, Phys. Rev. Letters 43, 956 (1979).
73. G. D. Watkins and J. R. Troxell, Phys. Rev. Letters 44, 593 (1980).
74. R. A. Street and N. F. Mott, Phys. Rev. Letters 35, 1293 (1975).
75. M. Kastner, D. Adler and H. Fritsche, Phys. Rev. Letters 37, 1504 (1978).
76. M. Kastner and H. Fritsche, Phil. Mag. 37 B, 199 (1978).
77. H. Fritsche, P. J. Gaczi and M. Kastner, Phil. Mag. 37B, 593 (1978).
78. I. A. Drabkin and B. Ya. Moizhes, Sov. Phys. Phys. Semicond. 15, 357 (1981).
79. J. A. van Vechten, J. Electrochem. Soc. 122, 419 (1975).
80. J. A. van Vechten, J. Electrochem. Soc. 122, 423 (1975).
81. F. A. Kröger, Ann. Rev. Mat. Sci. 7, 449 (1977).
82. F. J. Reid, R. D. Baxter and S. E. Miller, J. Electrochem. Soc. 113, 713 (1966).
83. R. D. Baxter, F. J. Reid and A. C. Beer, Phys. Rev. 162, 718 (1967).
84. M. H. van Maaren, J. Phys. Chem. Solids 27, 472 (1966).
85. V. Swaminathan, private communication.
86. H. R. Vydyanath and F. A. Kröger, J. Phys. Chem. Solids 36, 509 (1975).
87. G. Brouwer, Philips Res. Rept. 9, 366 (1954).
88. J. Benard and M. Huber, Proc. Brit. Ceram. Soc. 1, 1 (1964).
89. G. H. Hershman, V. P. Zlomanov and F. A. Kröger, J. Solid State Chem. 3, 401 (1971).
90. V. Kumar and F. A. Kröger, J. Solid State Chem. 3, 387 (1971).
91. S. S. Chern, H. R. Vydyanath and F. A. Kröger, J. Solid State Chem. 14, 33 (1975); 15, 369 (1975).
92. S. S. Chern, F. A. Kröger, J. Solid State Chem. 14, 44 (1975).

222

93. T. H. Rau, J. Phys. Chem. Solids 39, 879 (1978).
94. R. Baubinas, Z. Januskevicius, B. Petretis, A. Sakalas, and J. Viscakas, Phys. Stat. Solidi 15a, 591 (1973).
95. R. Baubinas, A. Martinaitis and A. Sakalas, Phys. Stat. Solidi 30a, K181 (1975).
96. J. Bloem, Philips Res. Rept. 11, 273 (1956).
97. W. W. Scanlon and G. Lieberman,   Proc. IEE 47, 910 (1959).
98. G. Simkovitch and J. B. Wagner, Jr., J. Chem. Phys. 38, 1368 (1963).
99. M. S. Seltzer and J. B. Wagner, Jr., J. Phys. Chem. Solids 24, 1525 (1963).
100. ibid. 26, 233 (1965).
101. K. R. Zanio and J. B. Wagner, Jr., J. Appl. Phys. 39, 5686 (1968).
102. A. Lichanot and S. Gromb, J. Chim. Phys. 68, 891 (1971).
103. M. S. Seltzer and J. B. Wagner, Jr., J. Chem. Phys. 36, 139 (1962).
104. V. P. Zlomanov, O. V. Matveev and A. V. Novoselova, Vesn. Mosk. Univ. (khim) 5, 81 (1967).
105. M. Abrams and R. N. Tauber, J. Appl. Phys. 40, 3868 (1969).
106. Y. Ban and J. B. Wagner, Jr., J. Appl. Phys. 41, 2818 (1970).
107. R. S. Guldi and J. N. Walpole, J. Appl. Phys. 44, 4896 (1973).
108. E. L. Brady, J. Electrochem. Soc. 101, 466 (1954).
109. V. S. Gavrilov and V. A. Shutilov, Sov. Phys. Solid State 8, 501 (1966).
110. A. M. Gaskov, O. V. Matveev, V. P. Zlomanov, and A. V. Novoselova, Vesn. Mosk. Univ. (Khim) 2, 114 (1969).
111. ibid. Neorg. Mater. (Russ.) 5, 1889 (1969).
112. ibid. Izv. Akad. Nauk SSSR, Neorg. Mat. 11, 1889 (1969).
113. M. P. Gomez, D. A. Stevenson and R. A. Huggins, J. Phys. Chem. Solids 32, 335 (1971).
114. H. Heinrich, et al. in "Lattice Defects in Semiconductors", Inst. of Phys. 1974, p. 264.
115. H. Rau, J. Phys. Chem. Solids 27, 761 (1966).
116. H. Schmidtke and V. Leute, Z. Phys. Chem. (N.F.) 103, 101 (1976).
117. H. Scherrer, G. Pineau, and S. Scherrer, Physics Letters 75A, 118 (1979).
118. H. Scherrer, S. Weber and S. Scherrer, Physics Letters 77A, 189 (1980).
119. D. Shaw, J. Phys. (C) 14, L 869 (1981).
120. F. A. Zaitov, A. V. Gorshkov and G. M. Shalyapina, Sov. Phys. Solid State 20, 927 (1978).
121. A. S. Jordan, H. R. von Neida, R. Caruso and M. D. Domenico, Jr., Appl. Phys. Letters 19, 394 (1971).
122. A. S. Jordan, A. R. von Neida, R. Caruso and C. K. Kim, J. Electrochem. Soc. 121, 153 (1974).
123. H. G. Grimmeiss and H. Koelmans, Philips Res. Rept. 15, 290 (1960).
124. U. Kaufmann, J. Schneider, and A. Räuber, Appl. Phys. Letters 29, 312 (1976).
125. J. F. Baumard and E. Tani, Phys. Stat. Solidi 39a, 373 (1977).
126. N. G. Eror, J. Solid State Chem. 38, 281 (1981).
127. F. J. Bryant and H.T.J. Baker, Phys. Stat. Solidi 11a, 623 (1972).
128. N. Hammond, A. Kohn, J. D. Debrun and H. Rodot, J. Phys. Chem. Solids 34, 1069 (1973).
129. P. J. Dean, H. Venghaus, J. C. Pfister, B. Schaub, and J. Marine, J. Luminescence 16, 363 (1978).
130. K. Urabe and S. Shionoya, J. Phys. Soc. Japan 24, 543 (1968) (with further references given there).
131. D. J. Dunstan, J. E. Nicholls, B. C. Cavenett and J. J. Davies, J. Phys. (C) 13, 6409 (1980).
132. S. S. Ostapenko, M. A. Tanatar and M. K. Sheinkman, Sov. Phys. Solid State 23, 720 (1981).
133. H. Bjerkeland and I. Holwech, Phys. Nerwegica 6, 139 (1972).
134. T. Taguchi, Phys. Stat. Solidi 96b, K33 (1979).

135. A. L. Dawar, O. P. Taneja, A. D. Sen and P. C. Mathur, J. Appl. Phys. 52, 4095 (1981).

136. A. J. Crocker, J. Mat. Sci. 3, 534 (1968).

137. E. Silberg and A. Zemel, Appl. Phys. Letters 31, 807 (1977).

138. H. R. Vydyanath, J. Appl. Phys. 47, 5010 (1976).

139. D. B. Kushev, N. N. Zheleva, and S. P. Yordanov, Phys. Status Solidi 100b, 731 (1980).

140. W. Hoerstel and K. H. Herrmann, ibid. 61a, 425 (1980).

141. W. Kaszuba and A. Rogalski, Acta Phys. Polon. 59A, 397 (1981).

142. M. Grudzien and A. Rogalski, ibid. 58A, 765 (1980).

143. C. Pickering, J. Phys. (D) 10, L73 (1977).

144. K. Sugiyama, Japan J. Appl. Phys. 7, 961 (1968).

145. H. G. Tang, B. Lunn and D. Shaw, J. Mater. Sci. 16, 3508 (1981).

146. E. Silberg and A. Zemel, J. Phys. (D), 15, 275 (1982).

147. S. P. Chashchin, I. P. Gushova, N. S. Baryshev and Yu. S. Kharionovskii, Sov. Phys. Semicond. 15, 324 (1981).

148. K. J. Linden, J. Electrochem. Soc. 120, 1131 (1973).

149. H. R. Vydyanath, J. Appl. Phys. 47, 5003 (1976).

150. R. Behrendt and R. Wendlandt, Phys. Stat. Solidi 61a, 373 (1980).

151. E. Silberg and A. Zemel, J. Electron. Mat. 8, 99 (1979).

152. A. Zemel, D. Eger, H. Shtrikman, and N. Tamari, ibid. 10, 301 (1981).

153. L. V. Prokof'eva, M. N. Vinogradova, and S. V. Zarubo, Sov. Phys. Semicond. 14, 1304 (1980).

154. B. F. Gruzinov, I. A. Drabkin and E. A. Zakomornaya, ibid. 15, 190 (1981).

155. L. A. Bovina, V. P. Ponomarenko, and V. I. Stafeev, ibid. 12, 1313 (1978).

156. D. M. Zayachuk and P. M. Starik, ibid. 14, 310 (1980).

157. J. L. Schmit and E. L. Stelzer, J. Electron. Mat. 7, 65 (1978).

158. J. Nishizawa, K. Suto, M. Kitamura, M. Sato, Y. Takase and A. Ito, J.Phys. Chem. Solids 37, 33 (1976).

159. J. L. Schmit and E. L. Stelzer, J. Electron. Mat. 7, 65 (1978).

160. H. R. Vydyanath, J. C. Donovan, and D. A. Nelson, J. Electrochem. Soc. 128 2625 (1981).

161. F. A. Zaitov, A. V. Gorshkov, and G. M. Shalyapina, Sov. Phys. Solid State 21, 112 (1979).

162. H. R. Vydyanath, J. Electrochem. Soc. 128, 2619 (1981).

163. R. J. Iwanowski and J. Bak, Acta Phys. Polon. 59A, 323 (1981).

164. U. Debska, M. Dietl, G. Grabecki and E. Janik, Phys. Stat. Solidi 64a, 707 (1981).

165. H. Schmalzried and C. Wagner, Z. Phys. Chem. (N.F.) 31, 198 (1962).

166. H. Schmalzried, in Progress in Solid State Chemistry, H. Reiss editor, Pergamon Press 2, 265 (1965).

167. F. A. Kröger, The Chemistry of Imperfect Crystals, North-Holland Publ. Co. vol. 2 (1974) chapter 20.

168. P. W. Yu, J. Manthurutil and Y. S. Park, J. Appl. Phys. 45, 3694 (1974).

169. B. Tell and H. M. Kasper, ibid. 5367

170. B. Tell, J. L. Shay and H. M. Kasper, Phys. Rev. B 9, 5203 (1974).

171. L. L. Golik, L. N. Novikov, T. G. Aminov, and E. A. Zhegalina, Sov. Phys. Solid State 19, 1655 (1977).

172. L. Treitinger, H. Pink and H. Göbel, J. Phys. Chem. Solids 39, 149 (1978).

173. A. I. Merkulov, S. I. Radautsan, and V. E. Tezlevan, Phys. Stat. Solidi 53a, K129 (1979).

174. U. Kaufmann, J. Schneider and A. Räuber, Appl. Phys. Lett. 29, 312 (1976).

175. B. Tell and H. M. Kasper, J. Appl. Phys. 44, 4988 (1973).

176. H. J. von Bardeleben, A. Goltzene and C. Schwab, Phys. Stat. Solidi 76b 363 (1976).

177. J. L. Shay, P. M. Bridenbaugh and H. M. Kasper, J. Appl. Phys. 45, 4491 (1974).

178. H. J. von Bardeleben, A. Goltzene, B. Meyer and C. Schwab, Phys. Stat.

Solidi 48a, K145 (1978).

179. L. Mandel, R. D. Tomlinson, and M. J. Hampshire, J. Crystalgr. 36, 152 (1976).

180. D. C. Look and J. C. Manthurutil, J. Phys. Chem. Solids 37, 173 (1976).

181. B. Tell, J. L. Shay and H. M. Kasper, J. Appl. Phys. 43, 2469 (1972).

182. B. Tell, J. L. Shay and H. M. Kasper, Phys. Rev. B4, 2463 (1971).

183. P. M. Bridenbaugh, and P. Migliorato, Appl. Phys. Lett. 26, 459 (1975).

184. B. Tell and F. A. Thiel, J. Appl. Phys. 50, 5045 (1949).

185. A. W. Verheijen, L. J. Giling, and J. Bloem, Mat. Res. Bull. 14, 237 (1979).

186. T. Irie, S. Endo and S. Kimura, Japan. J. Appl. Phys. 18, 1303 (1979).

187. V. F. Zhitar and N. A. Moldovyan, Sov. Phys. Semicond. 11, 1175 (1971).

188. A. I. Merkulov, S. I. Radautsan, and V. E. Tezlevan, Sov. Phys. Solid State 22, 523 (1980).

189. N. A. Tsvetkova, L. I. Koroleva, A. A. Babitsyna and T. I. Konesheva, Sov. Phys. Solid State 22, 214 (1980).

190. L. S. Palatnik and E. I. Rogacheva, Sov. Phys. Dokl. 12, 503 (1967).

191. S. M. Smyth, J. Solid State Chem. 20, 359 (1977).

192. N. H. Chan, R. K. Sharma and D. M. Smyth, J. Electrochem. Soc. 128, 1762 (1981).

193. U. Balachandran, B. Odekirk, and N. G. Eror, J. Solid State Chem. 42, 185 (1982).

194. F. A. Kröger, paper presented at the Chapman Conference on Point Defect in Minerals, Fallen Leaf Lake, Sept. 1982.

195. G. Massé, N. Lahlou, and C. Butti, J. Phys. Chem. Solids 42, 449 (1981).

196. J. J. M. Binsma, L. J. Giling, and J. Bloem, J. Luminescence 27, 35 (1982).

ESR OF DEFECTS IN III-V COMPOUNDS

JÜRGEN SCHNEIDER
Fraunhofer-Institut für Angewandte Festkörperphysik, D-7800 Freiburg, Germany

ABSTRACT

A survey is given on deep defects in GaP, GaAs and InP identified by electron spin resonance (ESR). Defect structures to be discussed are (i) 3d transition metals, (ii) antisite defects and (iii) radiation induced centers. The relevance of such defects for the III-V materials technology will be illustrated by representative examples.

INTRODUCTION

The analysis of point defects in semiconductors by the technique of electron spin resonance (ESR) has its long tradition. The early work which started in the late fifties emphasized ESR studies on shallow donors and acceptors as well as on transition metal impurities in the elemental semiconductors. Here, silicon became the favourite subject of most extensive investigations. Radiation induced paramagnetic defects in silicon, as studied by ESR, soon developed into an additional topic, still remaining of high current interest. For a review of the early ESR work on elemental semiconductors as well as on, yet marginal, studies on compound semiconductors we can refer to the classical monography by Ludwig and Woodbury [1].

ESR assessment of defects in III-V semiconductors has proceeded much more slowly. In these compounds the main experimental obstacles were and are (i) large ESR linewidths resulting from unresolved ligand hyperfine interaction, and (ii) low solubility of certain classes of deep impurities, as transition metals. The latter restriction is also valid for Si and Ge, but not for the more ionic II_B - VI semiconductors. Additional experimental difficulties may arise for n-type GaAs and InP, where the ionisation energy (5 - 7 meV) of shallow donors is so low, that incomplete carrier freeze-out at low temperatures may render an ESR measurement impossible.

Complementary to conventional ESR, the more recent and sophisticated technique of optically detected magnetic resonance (ODMR) now finds increasing application for defect assessment in semiconductors. By this method, also diamagnetic defects can be detected and investigated, via their excited paramagnetic triplet states. An additional advantage of the ODMR technique is the possibility that very small sample volumes, limited by the focal area and penetration depth of the exciting laser beam, can be investigated. Thus, ODMR is ideally suited for defect characterisation in epitaxial layers.

This review is organized as follows: In its first part a summary of present knowledge on $3d^n$ transition metal impurities in GaP, GaAs and InP, as obtained by ESR, will be given. The second part discusses in detail current ESR and ODMR investigations on the anion antisite defects in GaP and GaAs. These basic and native point defects are very characteristic for III-V compounds; they are also suspected to play a key role in devices, in a deleterious as well as in a beneficial manner. Finally, the present state of ESR studies on radiation induced defects in GaP, GaAs and InP will be summarized.

ESR studies on shallow donors and acceptors have been omitted from the discussion. Their magnetic behaviour is dominantly determined by that of the highly delocalized electron or hole, which is essentially an intrinsic property of the solid.

Mat. Res. Soc. Symp. Proc. Vol. 14 (1983) © Elsevier Science Publishing Co., Inc.

226

TRANSITION MATERIALS

$3d^n$ transition metals in III-V semiconductors have been the subject of extensive ESR investigations [2]. As a result, a fairly clear and complete understanding about their electronic structure in III-V compounds has been reached by now. The salient features can be summarized as follows:

(1) In all cases where ESR could be detected the $3d^n$-ion was found to occupy the substitutional cation site. This parallels the behaviour of $3d^n$-ions in the tetrahedral II_B-VI compounds. In contrast $3d^n$-ions in silicon are known to show preference for the interstitial site.

(2) The electronic ground state of an individual $3d^n$ configuration is obtained after crystal field splitting and spin-orbit coupling have been taken into account, as being the case for $3d^n$-ions in II_B-VI hosts. The weak crystal field regime was found to be valid. A notable exception is observed for $V^{2+}(3d^3)$ in GaP and GaAs, where the strong field, i.e. low spin, limit is reached.

(3) All 3d-impurities observed by ESR in GaP, GaAs and InP act as deep acceptors and can consequently occur in two charge states A°/A^-, e.g. $Fe^{3+}(3d^5)/Fe^{2+}(3d^6)$, depending on the position of the Fermi level. For Ni, Fe and Cr in GaP also the double acceptor states $Ni^+(3d^9)$, $Fe^+(3d^7)$ and $Cr^+(3d^5)$ were detected by ESR.

(4) ESR of tetravalent chromium, $Cr^{4+}(3d^2)$ has also been detected in GaP, GaAs and InP. Thus this 3d-element is electrically amphoteric in the above semiconductors, being active as deep acceptor (Cr^{3+}/Cr^{2+}) as well as a deep donor (Cr^{3+}/Cr^{4+}).

(5) Association of the deep $3d^n$ acceptors with shallow donors, often present as background impurities, is likely to occur. By ESR this complexing could sofar only be detected for the manganese acceptor, in its $A^-(3d^5)$ state. Nearest neighbour A^--D^+ pairs of type Mn-S (C_{3v}) were observed in GaP, and of type Mn-Li (D_{2d}) in GaAs and GaP, the lithium donor being at an interstitial site.

It should be added that for the nickel acceptor in GaP and GaAs complexing with nearest group-VI (S, Se, Te) and group-IV (Si, Ge, Sn) shallow donor impurities could be observed by infrared spectroscopy.

The electrical activity of shallow donor - $3d^n$ acceptor pairs is not yet known.

(6) Complementary to the ESR technique, optical spectroscopy has been extensively applied for analysing the electronic structure of $3d^n$-elements in GaP, GaAs and InP. It is found that the A^--states exhibit intra-3d shell (crystal field) transitions of rather narrow linewidth, which had often enabled further Zeeman- and piezo-spectroscopic investigations. For nickel in GaP and GaAs crystal field transitions were also observed in its A^{--} state $Ni^+(3d^9)$.

(7) ESR studies on the system GaAs:Cr had their strong technological motivation because the deep midgap A°/A^- acceptor level $Cr^{3+}(3d^3)/Cr^{2+}(3d^4)$, introduced in GaAs by chromium doping during melt growth, is of paramount importance for the fabrication of semi-insulating material. Its task is to compensate unavoidable silicon shallow donor impurities, which enter the crystal from the quartz crucible walls. In contrast, semi-insulating InP is preferentially grown by iron-doping, since here the Fe^{3+}/Fe^{2+} acceptor level lies closer to midgap than the level Cr^{3+}/Cr^{2+}.

Rare earth ions

The first example of ESR detection of a 4f rare earth ion in a III-V semiconductor was recently reported by Kasatkin et al. [3]. A rather narrow line, $\Delta H \leq 9G$, at g = 3.313 was observed in ytterbium doped InP and assigned to $Yb^{3+}(4f^{13})$ on indium site. Hyperfine interaction arising from the isotopes ^{171}Yb and ^{173}Yb was also observed.

These findings are quite remarkable since ESR in III-V compounds, and particularly in InP, is generally impeded by much larger linewidths, often exceeding 100 G, which result from overlap of the paramagnetic orbitals with the nuclei of the surrounding anions and cations. This source of ESR line broadening is drastically reduced for a rare earth ion, since its 4f-orbitals are screened from the surroundings by the outer closed $5s^2$ and $5p^6$ shells.

ANTISITE DEFECTS

In a binary compound antisite defects are formed when anions occupy cation sites, or vice versa. Their occurrence is favoured for crystals in which the electronegativity difference between the constituent atoms is relatively small, as it is the case in III-V compounds. Beside vacancies and interstitials, antisite defects represent a third basic possibility for deviations from stoichiometry in a binary crystal which so far has been mostly ignored.

Antisite defects are the most prominent and characteristic representatives of native point defects in III-V compounds. Their concentration depends critically on the stoichiometry conditions during crystal growth. Thus, in GaAs the As_{Ga} anion antisite defects are preferentially created if crystal growth proceeds under conditions of arsenic excess. On the other hand gallium excess will enhance formation of the complementary cation antisite defects, Ga_{As}. Typical As_{Ga}:GaAs and P_{Ga}:GaP antisite concentrations in LEC-grown bulk material, as determined by ESR, are in the $10^{15} - 10^{16}$ cm^{-3} range.

A group-V atom on a group-III site in a III-V compound, i.e. an anion antisite defect, should act as a double donor. This parallels the behaviour of substitutional group-VI elements (S, Se, Te) in the group-IV elemental semiconductors (Si, Ge). A double donor can exist in three charge states, D^0, D^+, D^{++}, and may consequently introduce two energy levels in the gap: D^0/D^+ and D^+/D^{++}. Only the one-electron state D^+ is paramagnetic and accessible by ESR. This, however, requires proper positioning of the Fermi level by doping with acceptors and/or by optical excitation. However, the two-electron neutral donor may be detected by ODMR in its excited paramagnetic S = 1 triplet state.

For the following it should be kept in mind that the charge states D^0, D^+, D^{++} have to be identified with the ionic charge states P_{Ga}^{3+}, P_{Ga}^{4+}, P_{Ga}^{5+} or As_{Ga}^{3+}, As_{Ga}^{4+}, As_{Ga}^{5+}. The paramagnetic tetravalent state has also been observed by ESR in a number of ionic hosts; the radicals PO_4^{4-} and AsO_4^{4-} being typical examples.

The complementary cation antisite defects, e.g. Ga_P:GaP or Ga_{As}:GaAs, should behave as double acceptors. Again, only the singly ionized defect, A^-, is paramagnetic in its ground state. ESR of cation antisite defects in III-V compounds has not yet been identified.

Association of antisite defects with other stoichiometric defects, as vacancies or interstitials, or with impurity atoms, should also be considered. The most simple example is that of the antistructure pair, e.g. $Ga_P P_{Ga}$:GaP or $Ga_{As} As_{Ga}$:GaAs; this may be viewed as an isoelectronic molecule, similar to the nearest D-A pair O-Zn and O-Cd in GaP.

The P_{Ga} antisite defect in GaP

Its ESR spectrum, first observed by Kaufmann et al. [4] in as-grown bulk material is shown in Fig. 1. Rather similar ESR-spectra were also observed [4] for the chalcopyrite compounds $CdSiP_2$ and $ZnGeP_2$, which form a tetragonal superstructure of the cubic zincblende lattice. In these hosts, formation of anion antisite defects, e.g. P_{Si} or P_{Ge}, seems even more likely than in III-V compounds, since the electronegativity difference between group-V and group-IV atoms is smaller than that between group-V and group-III atoms.

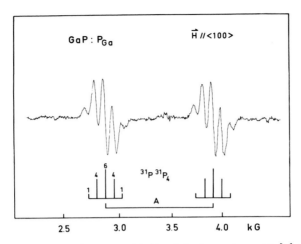

Fig. 1. ESR spectrum of the P_{Ga} antisite defect as observed [4] in as-grown liquid encapsulation Czochralski (LEC) grown GaP. A decomposition of the spectrum into the 2 x 5 hyperfine components of the $^{31}P^{31}P_4$ cluster is shown in the lower part.

The very characteristic 2 x 5 ("fingerprint") hyperfine (hf) structure pattern of the P_{Ga}:GaP antisite ESR-spectrum shown in Fig. 1 arises from the inter-action of the donor electron with the central P_{Ga}^{4+} ion and the four P ligands. From the hf data information about the spatial distribution of the unpaired electron's wave function can be obtained. Using a simple molecular orbital model it was estimated [5] that 26% and 66% of the donor electron's density are localized at the central P_{Ga}^{4+} ion and at the four P ligands, respectively. This strong localisation in space reflects the deep level character of the P_{Ga} donor in its paramagnetic D^+ state.

In p-type GaP, the P_{Ga} double donor is deprived of both electrons and, con-sequently, no ESR is detected in the dark. However, the $P_{Ga}^{4+}(D^+)$ ESR-spectrum (Fig. 1) appears after in situ optical excitation of the crystal at sufficient-ly low temperature ("photo-ESR"). The spectral dependence of this effect, which arises from the photo-ionisation (see Fig. 2)

$$D^{++} + h\nu \rightarrow D^+ + e^+ ,$$

was investigated by Kaufmann et al. [6], and a threshold energy of $h\nu \gtrsim 1.25$ eV was determined for the photo-excitation of the P_{Ga}^{4+} antisite ESR spectrum. In this way, the second donor level, D^{++}/D^+ (i.e. P_{Ga}^{5+}/P_{Ga}^{4+}) is determined to be located at $E_v + 1.25$ eV. The holes created become subsequently trapped at the ionized shallow zinc acceptors, $e^+ + A^- \rightarrow A^\circ$, see Fig. 2.

The reverse process, $D^+ + A^\circ \rightarrow D^{++} + A^-$, may proceed either by back diffusion of the hole to P_{Ga}^{4+} [6], or by direct donor-acceptor recombination. The latter process also results in infrared emission at $h\nu \leq 1.25$ eV. The associated photo-luminescence band was recently monitored by ODMR and its donor-acceptor origin was clearly demonstrated (Killoran et al. [7], O'Donnell et al [8]).

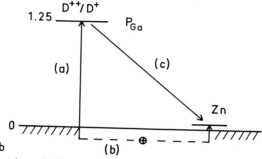

Fig. 2. Photokinetics of the P_{Ga} antisite defect in p-GaP:Zn. (a): ESR excitation [6], $D^{++} + h\nu \rightarrow D^+ + e^+$, followed by hole diffusion to ionized Zn acceptors [6], $e^+ + A^- \rightarrow A^\circ$. (c): $D^+ - A^\circ$ radiative recombination detected by ODMR [8].

Furthermore, the presence of zinc as the shallow acceptor species involved was proven by detection of the Zn-acceptor ODMR under uniaxial stress [9]. These ODMR experiments also confirmed the suspicion that the P_{Ga} antisite defect acts as an electron trap in direct competition with band edge recombination processes, thus lowering the quantum efficiency of light emitting diodes.

We finally note that the first donor level D^+/D° (i.e. P_{Ga}^{4+}/P_{Ga}^{3+}) of P_{Ga}:GaP is estimated to lie in the upper third of the energy gap [6]. Self-consistent Green's function calculations locate the level at $E_C - 0.6$ eV [10]. In their ODMR study of P_{Ga}^{4+}, Killoran et al [7] and O'Donnell et al. [8] also observed a very characteristic $S = 1$ triplet spectrum, see Fig. 3. It arises from a trigonally distorted defect; its unpaired spins exhibit hf interaction with one central ^{31}P nucleus and with three equivalent ^{31}P ligands. It is not yet ascertained whether the trigonal distortion is caused by a static Jahn-Teller effect of the $P_{Ga}^{3+}P_4$ defect in its excited state, or by an associated defect, $P_{Ga}^{3+}P_3X$, whose chemical nature remains to be identified. Apart from a P-sited acceptor, as C_p or Si_p, the antistructure pair, $P_{Ga}^{3+}P_3Ga_p$, could also be considered as a likely candidate.

The As_{Ga} antisite defect in GaAs

ESR identification of the As_{Ga} antisite as native point defect in GaAs was first reported by Wagner et al [11]. Concentrations up to 10^{16} cm^{-3} were observed in semi-insulating as-grown bulk material. There is now an increasing number of experimental observations indicating that the As_{Ga} defect is responsible for a deep midgap electron trap at $E_C - 0.75$ eV, now commonly labelled EL2. This level plays a key role in the compensation processes occurring in semi-insulating GaAs crystals which are grown by the liquid encapsulation Czochralski (LEC) technique from pyrolytic boron nitride (pBN) crucibles -

without the usual chromium dopant. It is assumed that a critical concentration of As_{Ga} defects, i.e. EL2 traps, is required to compensate shallow carbon acceptors which are typical background impurities in pBN-grown GaAs.

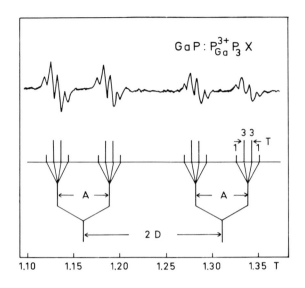

Fig. 3. Triplet ODMR of the $P_{Ga}^{3+}P_3X$ defect in p-GaP:Zn at 35 GHz [8]. The spectrum arises from C_{3v}-distorted centers oriented parallel to H. The build-up of the spectrum due to (i) fine structure splitting 2D, (ii) central ^{31}P hyperfine splitting A and (iii) ligand ^{31}P hyperfine splitting T is illustrated.

The concentration of As_{Ga} defects in GaAs can be enhanced by at least two orders of magnitude, into the 10^{18} cm^{-3} range, either by fast-electron or fast-neutron irradiation (Goswami et al. [12], Wörner et al. [13]).

Remarkably, fast-neutron irradiation of GaP did not result in preferential creation of P_{Ga} defects. One possible explanation for this failure may be due to the fact that replacement collisions leading to antisite formation are less likely to occur in GaP than in GaAs, because of dissimilar anion and cation masses. Weber et al [14] have observed that As_{Ga} antisite defects are also formed under plastic deformation of GaAs at 400° C. Here the As_{Ga} defect production proceeds during the climb of dislocations [14]. Such processes may also be involved in the degradation phenomena of light emitting devices. It should be added that a critical concentration of As_{Ga}:GaAs, as mentioned above for P_{Ga}:GaP, can strongly decrease the near-gap photo-emission yield of GaAs [14]. The ESR-spectrum of the As_{Ga}^{4+}:GaAs antisite defect is shown in Fig. 4. Instead of the 2 x 5 hyperfine (hf) pattern for P_{Ga}^{4+}:GaP, arising from the nuclear spin I = 1/2 of ^{31}P (100%), we now expect a 4 x 13 hf pattern because the nuclear spin of ^{75}As (100%) is I = 3/2. However, the 13 line ligand hf structure is not resolved in the ESR spectrum, see Fig. 4. The non-equidistance of the four hf components of the central ^{75}As nucleus arises from second order effects [13].

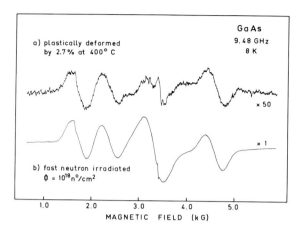

Fig. 4. ESR spectrum of the As_{Ga} antiside defect in (a) plastically deformed and (b) fast neutron irradiated GaAs. Spectrum (a) corresponds to 5×10^{16} As_{Ga}^{4+}/cm^3. The broad line superimposed on one of the As_{Ga} hyperfine lines in the irradiated sample does not appear for the deformed sample. The narrow signal at 3.4 kG is a background line from the cavity [10].

Fig. 4 reveals that almost identical ESR spectra are observed in plastically deformed (a) and fast neutron irradiated GaAs (b). However, in the fast neutron irradiated sample an additional broad ESR signal at g = 2.06 is apparent. Its origin has not yet been elucidated, but it is suspected to arise from another, radiation-induced, basic point defect in GaAs.

Photo-ESR experiments on As_{Ga}:GaAs, similar to those discussed above for P_{Ga}:GaP [6] were performed by Weber et al. [14]. Results are shown in Fig. 5 for two plastically deformed GaAs samples, one semi-insulating (a), the other originally p-type (b). In the latter enhancement of the As_{Ga}^{4+} ESR intensity is seen to start at photon energies $h\nu \geq 0.52$ eV, resulting from the hole ionisation process $D^{++} + h\nu \rightarrow D^+ + e^+$. No ESR photo-enhancement is observed for the semi-insulating sample (a). This indicates that its Fermi level is sufficiently high to render the D^{++}/D^+ level of the As_{Ga} double donor fully, and its D^+/D° level partially occupied. Both samples exhibit rather sharp ESR quenching onsets at 0.75 eV and 1.00 eV, see Fig. 5. Thus, the two ionisation energies of the As_{Ga} double donor which can be inferred from these photo-ESR thresholds are [14]:

$$D^+/D^\circ \text{ (i.e. } As_{Ga}^{4+}/As_{Ga}^{3+}) \quad : E_c - 0.77 \text{ eV}$$

$$D^{++}/D^+ \text{(i.e. } As_{Ga}^{5+}/As_{Ga}^{4+}) \quad : E_v + 0.52 \text{ eV,}$$

$$\text{with } E_g = 1.52 \text{ eV (LHeT).}$$

The insert in Fig. 5 illustrates the position of the two As_{Ga} donor levels in the gap, and their optical filling and depleting.

232

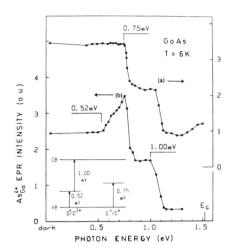

Fig. 5. Intensity of the As_{Ga}^{4+} ESR signal as a function of monochromatic in situ illumination. Trace (a): semi-insulating pBN-GaAs sample, trace (b): p-GaAs:Cd sample with a room-temperature resistivity of 150 Ωcm before deformation. The transitions shown in the insert correspond to the thresholds of photoquenching and photoenhancement of the As_{Ga}^{4+} ESR [14].

It is noted that the position of the first donor level, D^+/D^o, of the As_{Ga} anti-site defect agrees with that of the midgap electron trap EL2, which was extensively studied by transient spectroscopy. Further support for the assignment of EL2 to As_{Ga} has been given by Weber et al. [14]. These authors have also pointed-ed out that the two donor levels of As_{Ga} correspond to the Fermi level pinning energies at Schottky barrier contacts on n-GaAs (E_C - 0.75 eV) and p-GaAs (E_V + 0.55 eV), see e.g. [15].

RADIATION INDUCED DEFECTS

    The bombardment of silicon with high energy particles is known to create a microcosmos of radiation induced lattice defects. ESR has been extremely successful in the past decades in unravelling their electronic structure. However, one should be aware that this goal could only be reached because nature has decided that only a comparatively rare isotope of silicon, ^{29}Si (4.7%) carries nuclear spin and moment. Consequently, ESR lines in silicon are not strongly broadened by unresolved ^{29}Si ligand hyperfine interaction; typical ESR linewidths being in the range 1 - 10 G, at sufficiently low temperatures. Unfortunately, no zero-spin isotopes of group-III nor group-V elements exist. Therefore, ESR in III-V compounds is seriously impeded by relatively large line-widths, which may be far in excess of 100 G. As a result, only very few examples

- to be discussed below - for ESR assessment of radiation defects in GaP, GaAs and InP could sofar be reported.

On the other hand, extensive studies by transient spectroscopy (DLTS) on radiation-induced defects in these compounds have been undertaken, particularly in GaAs. However, DLTS data can not readily be assigned to specific microscopic defect structures. An instructive example for this difficulty is the recent re-assignment [16] of the electron trap family E1-E3, observed by DLTS in fast-electron irradiated GaAs, to defects formed by atomic displacements in the As-sublattice - and not in the Ga-sublattice, as assumed before.

Gallium Phosphide

The first example for ESR detection of a radiation-induced defect in a III-V semiconductor was reported in 1978 by Kennedy and Wilsey [17]. These authors observed a characteristic ESR-spectrum in fast electron irradiated GaP, which they assigned to the isolated gallium vacancy, V_{Ga}. By electron irradiation the V_{Ga} center is preferentially introduced into p-type GaP. In situ optical excitation, $h\nu \gtrsim 1.7$ eV, is required to convert the defect into its paramagnetic charge state V_{Ga}^{2-}. The occupancy level related to V_{Ga}^{2-} lies in the upper half of the band gap [18]. Although the ESR spectrum of the V_{Ga}^{2-} defect exhibits pronounced angular dependence, arising from anisotropic ^{31}P ligand hyperfine interaction, the overall symmetry of the V_{Ga} center was found to remain strictly cubic (T_d), even at 1.4 K [18]. This is a rather surprising observation since, in contrast, isolated vacancy centers in silicon, as well as isolated V_{Zn} centers in ZnS and ZnSe, are known to undergo a strong static Jahn-Teller distortion.

In a very recent study by Kennedy and Wilsey [19] on the behaviour of V_{Ga}^{2-} under uniaxial stress it was reported that the defect is in fact tetragonally distorted, but in a rapidly reorienting state. Furthermore, mass spectroscopic and local vibrational mode spectroscopic data were taken as evidence against a suggestion of Scheffler et al. [20], that the ESR spectrum assigned to V_{Ga} by Kennedy and Wilsey, arises from a deep carbon donor occupying a gallium site, C_{Ga}, and not from V_{Ga}. In contrast to the native antisite defects discussed above, the V_{Ga} center in GaP was only observed after fast electron irradiation. Kaufmann and Kennedy [21] failed to detect the V_{Ga} defect in as-grown LEC GaP crystals down to the 10^{15}/cc level. It was further found that the V_{Ga} center is not an important defect in electron irradiated n-type GaP [18]. Instead, in such material a new P_{Ga} antisite related defect of type $P_{Ga}^{4+}P_3X$ is detected by ESR [22]. The nature of the associated defect X has not yet been identified. However, the $P_{Ga}^{4+}P_3X$ center seems not to be the S = 1/2 state of the center $P_{Ga}^{3+}P_3X$, which was detected by ODMR in its S = 1 triplet state [7, 8].

Gallium Arsenide

In spite of serious efforts, no irradiation-specific defects could sofar be identified by ESR in GaAs. However, fast-electron [12] or fast-neutron [13] irradiation of GaAs can enhance the concentration of native ESR-detectable As_{Ga} antisite defects - and presumably that of the complementary Ga_{As} defects - by orders of magnitude above their level typical in as-grown material.

Schneider and Kaufmann [23] have observed that As_{Ga} defects are also formed under slow neutron irradiation of GaAs. In this case, antisite defects are created by the γ- and β^--emissions which follow thermal neutron capture. The nuclear reactions responsible are

$$^{69}Ga(n\gamma) \rightarrow {}^{70}Ga \xrightarrow{\beta^-} {}^{70}Ge$$

$$^{71}Ga(n\gamma) \rightarrow {}^{72}Ga \xrightarrow{\beta^-} {}^{72}Ge$$

$$^{75}As(n\gamma) \rightarrow {}^{76}Se \xrightarrow{\beta^-} {}^{76}Se;$$

234

they can be exploited for n-type neutron-transmutation doping of GaAs. Electrical activation of the shallow donor species Ge and Se occurs after thermal anneal at T ≧ 500° C, accompanied by thermal destruction of the radiation induced antisite defects [13].

Indium Phosphide
    After exposure of iron-doped InP to fast electron irradiation Kennedy and Wilsey [24] observed ESR of a new trigonal paramagnetic defect exhibiting a highly anisotropic g-factor. Simultaneously the ESR intensity of substitutional $Fe^{3+}(3d^5)$ was found to be strongly reduced.
    These findings indicate that certain primary radiation defects in InP are mobile at room temperature and that these may subsequently become trapped at iron impurity sites. Similar ESR observations were made by Igelmund [25] on fast electron irradiated GaP:Fe and GaAs:Cr.

REFERENCES

1. G. W. Ludwig and H. H. Woodbury in: Solid State Physics, Vol. 13, F. Seitz and D. Turnbull, ed., (Academic Press 1962).

2. For a review covering the numerous literature up to 1980 see: U. Kaufmann and J. Schneider in: Advances in Electronics and Electron Physics, Vol. 58, C. Marton, ed., (Academic Press 1982); see also J. Schneider and U. Kaufmann, Inst. Phys. Conf. Ser. 59, 55 (1981) and Ref. [5].

3. V. A. Kasatkin, V. F. Masterov, V. V. Romanov, B. E. Samorukov and K. F. Stelmakh, Sov. Phys. Semicond. 16, 106 (1982).

4. U. Kaufmann, J. Schneider and A. Räuber, Appl. Phys. Lett. 29, 312 (1976).

5. U. Kaufmann and J. Schneider, Festkoerperprobleme 20, 87 (1980).

6. U. Kaufmann, J. Schneider, R. Wörner, T. A. Kennedy and N. D. Wilsey, J. Phys. C 14, L 951 (1981).

7. R. Killoran, B.C. Cavenett, M. Godlewski, T. A. Kennedy and N. D. Wilsey, J. Phys. C 15, L 723 (1982).

8. K. P. O'Donnell, K. M. Lee and G. D. Watkins, Solid State Commun., in press.

9. K. P. O'Donnell, K. M. Lee and G. D. Watkins, to be published.

10. M. Scheffler, Festkoerperprobleme 22, 115 (1982).

11. R. J. Wagner, J. J. Krebs, G. H. Stauss and A. M. White, Solid State Commun. 36, 15 (1980).

12. N. K. Goswami, R. C. Newman and J. E. Whitehouse, Solid State Commun. 40, 473 (1981)

13. R. Wörner, U. Kaufmann and J. Schneider, Appl. Phys. Lett. 40, 141 (1982).

14. E. R. Weber, H. Ennen, U. Kaufmann, J. Windscheif, J. Schneider and T. Wosinski, J. Appl. Phys. 53, 6140 (1982).

15. W. E. Spicer, I. Lindau, P. Skeath, C. Y. Yu and P. Chye, Phys. Rev. Lett. 44, 420 (1980).

16. D. Pons and J. Bourgoin, Phys. Rev. Lett. 47, 1293 (1981).

17. T. A. Kennedy and N. D. Wilsey, Phys. Rev. Lett. 41, 977 (1978).

18. T. A. Kennedy and N. D. Wilsey, Phys. Rev. B 23, 6585 (1981).

19. T. A. Kennedy and N. D. Wilsey, Phys. Rev. B, in press.

20. M. Scheffler, S. T. Pantelides, N. O. Lipari and J. Bernholc, Phys. Rev. Lett. 47, 413 (1981).

21. U. Kaufmann and T. A. Kennedy, J. Electron. Mater. 10, 347 (1981).

22. T. A. Kennedy and N. D. Wilsey, Inst. Phys. Conf. Ser. 46, 375 (1979).

23. J. Schneider and U. Kaufmann, Solid State Commun. 44, 285 (1982).

24. T. A. Kennedy and N. D. Wilsey, Inst. Phys. Conf. Ser. 59, 257 (1981).

25. A. Igelmund, Thesis (1979), TH Aachen (unpublished).

SELF-DIFFUSION IN COMPOUND SEMICONDUCTORS

ARTHUR F.W. WILLOUGHBY
Engineering Materials Laboratories, The University, Southampton, SO9 5NH, U.K.

ABSTRACT

Self-diffusion studies are vital in the elucidation of atomic mechanisms of diffusion, as well as in the better understanding of device fabrication processes, such as the annealing of ion-implanted layers. This review outlines first the major reasons for interest in self-diffusion in III-V and II-VI compounds. It discusses the main differences with elemental semiconductors, including the wide variety of possible defects in the compounds, the role of departures from stoichiometry, and the value of tracer and inter-diffusion studies. Self-diffusion studies in III-V compounds are next reviewed, including recent measurements in GaAs, where more information on diffusion mechanisms is becoming available. Interdiffusion between different III-V compounds is also discussed in the light of self-diffusion studies. Next, recent progress on self-diffusion in certain II-VI compounds is discussed, where interdiffusion studies have also provided a significant contribution. The review concludes by suggesting areas where research is urgently needed to clarify diffusion mechanisms.

INTRODUCTION

Self-diffusion studies are vital in the elucidation of atomic mechanisms of diffusion, as well as in the better understanding of device fabrication processes. In the III-V compounds such as GaAs, although p-n junctions are prepared by means of impurity donors and acceptors, the migration of the constituent atoms Ga and As is of considerable practical importance. For example, annealing of the damage produced by ion implantation often employs encapsulating surface layers to avoid the evaporation of the volatile arsenic from the surface, but the migration of the constituent atoms through these encapsulating surface layers can have a striking influence on properties such as luminescence and p-n junction reverse leakage current (Streetman [1]). In the II-VI compounds, whose electronic properties are often dominated by point defects, self-diffusion can be the rate-controlling processes in changes in the degree of non-stoichiometry used to fabricate p-n junctions, for example in the fabrication of $Cd_x Hg_{1-x} Te$ infra-red detectors.

On a more fundamental level, the widespread use of impurity diffusion and heat treatments of these compounds requires a much better understanding of these processes than exists at present. Self-diffusion studies are vital to this understanding, and the following review will attempt to summarise the major progress in this area since the last major reviews by Casey and Pearson [2] in 1975, and Casey [3] in 1973.

COMPONENT SELF-DIFFUSION AND INTERDIFFUSION IN III-V and II-VI COMPOUNDS

(a) Self-diffusion

In a binary compound such as GaAs the self-diffusion coefficients D_{As} and

Mat. Res. Soc. Symp. Proc. Vol. 14 (1983) ⓒ Elsevier Science Publishing Co., Inc.

D_{Ga} can be measured by radiotracer techniques. Compared with elemental
semiconductors such as Si and Ge, however, the number of possible defect types
are increased, together with possible charge states, and thus there are
similarly wider possibilities for the mechanisms by which self-diffusion may
take place.

Possible among such mechanisms are direct exchange or ring mechanisms,
which, of course, do not directly involve any lattice defects at all. Turning
to defect mechanisms, vacancy mechanisms of diffusion in the sphalerite
structure, in which many III-V and II-VI compounds crystallise, are not as
straightforward as in the group IV semiconductors. For example, considering
the sphalerite structure of GaAs shown in Fig.1, it is clear that diffusion
of an arsenic vacancy V_{As} via nearest neighbours, will create antistructure
defects by its motion since its first jump will involve the reverse movement
of a gallium atom on to an arsenic site, and so on. Alternatively, it has
been suggested that a vacancy, such as V_{As}, might migrate on its own sublattice,
and thus not create antistructure defects. In the same way, several interstitial mechanisms can also be suggested for self-diffusion. Corbett and
Bourgoin [4] have discussed the migration of interstitial atoms, which do not
bond with the lattice, via the tetrahedral (T_R), a hexagonal (H), and T_x
interstitial positions shown in Figs. 1 and 2. In the event that the interstitial undergoes bonding with the lattice, Corbett and Bourgoin [4] speculated
that a bonded or split-interstitial might migrate via a mixed split-interstitial
as shown in Fig. 3. As pointed out by Hu [5] , however, if self-diffusion is
effected by a direct interstitial mechanism, then the lattice site-interstitial
exchange must be the rate controlling mechanism. Another possibility is the
interstitialcy mechanism, where the interstitial atom moves into a normal
lattice site, and the atom which originally occupied this site is pushed into
an interstitial site. Finally, if antistructure defects are present in
significant numbers, we note that they need another defect such as a vacancy
to help them diffuse, if direct exchange is not possible.

Each of the defect mechanisms of diffusion could involve charged and neutral
defects, and so, in principle, we have to consider an extremely large number of
components of the diffusive flux before deciding which is dominant. For
example, if we consider only two mechanisms, diffusion via neutral and acceptor
arsenic vacancies on the arsenic sublattice, we can write:-

$$D_{As} = f \, D(V^X_{As}) \, [V^X_{As}] + f \, D(V'_{As}) \, [V'_{As}] \tag{1}$$

where f is the correlation factor, $[V^X_{As}]$ is the mole fraction of neutral
arsenic vacancies, etc., and $D(V^X_{As})$ is the diffusion coefficient of neutral
arsenic vacancies, etc., using the Kröger notation. For the interstitial-
substitutional mechanism, Casey and Pearson [2] showed that the radiotracer
self-diffusion coefficient via neutral arsenic interstitials could be written

$$D_{As} = D(As_i^X) \, [As_i^X] \tag{2}$$

and similar expressions can be derived for diffusion via charged interstitials.
One of the ways of assessing the charge state of the defect involved in self-
diffusion, therefore, is to measure the effect of altering the background
doping level (and thus the Fermi level) on the self-diffusion coefficient, but
this of course can have an effect on many different defect types in the lattice.

A factor of great importance in the compounds is the fact that often they
have volatile species which are readily lost when the samples are heated, and
the stoichiometry and defect equilibria are directly related to the component

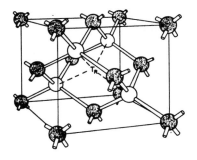 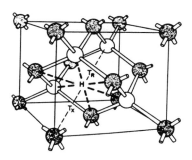

Fig. 1   The T_R tetrahedral interstitial   Fig. 2.   The H hexagonal inter-
position in the sphalerite   stitial site, shown joining
lattice (Corbett and Bourgoin [4])   a T_R and T_x site (Corbett
                                                         and Bourgoin [4])

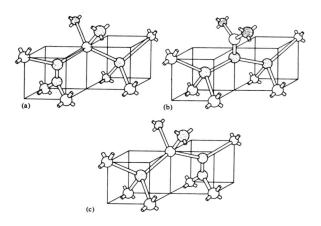

Fig.3.   Possible migration mechanism of a split-interstitial
(a) in the sphalerite lattice, via a mixed split-
interstitial shown in (b).
(Corbett and Bourgoin [4])

pressure. It is therefore important to control and monitor component pressure during diffusion and, preferably, measure diffusion coefficients as a function of component pressure. Such experiments should strictly employ a pre-anneal at the same component pressure to ensure that the defect equilibrium is fully established before the diffusion of the radiotracer species. Examples will be given below where such studies of the effect of component pressure have given valuable information on diffusion mechanisms, but this has only been carried out in a limited number of compounds.

Throughout this discussion of self-diffusion, we have referred to radio-tracer methods of determining diffusion coefficients and profiles. It must be noted, at this stage, however, that it is in principle possible to study self diffusion following the diffusion of an isotope which is not necessarily radioactive, by means of secondary ion mass spectrometry (SIMS) which is increasingly being used to determine dopant profiles. When considering this technique, however, it is important to note that the concentration of the isotope being diffused must, of course, exceed the natural concentration of that isotope in the material originally, so that one can obtain an adequate concentration range of the diffusing isotope.

(b) Interdiffusion

Although radiotracer studies are fundamental to any characterisation of self-diffusion mechanisms, they are rarely sufficient to provide definitive evidence in ruling out all the competing possibilities. It is usually necessary, therefore, to examine other kinds of evidence in such a considera-tion. Interdiffusion studies, which involve the exchange of dissimilar elements across an interface between, say, GaAs and AlAs, have received little attention until recently, in contrast to the face centred cubic metal systems, where the discovery of the Kirkendall effect led to such important conclusions about chemical diffusion (e.g. LeClaire [6]).

In a radiotracer self-diffusion experiment, the tracer isotope is diffusing in the absence of a chemical concentration gradient, and its diffusion coefficient is thus representative of random jumps of that element in the host lattice, apart from the relatively small effect of the isotopic mass. Interdiffusion experiments, on the other hand, involve a chemical concentration gradient of both diffusing species. Such chemical diffusion data, however, is of great value in the elucidation of diffusion mechanisms, and, of course, is of great technological interest in the preparation and heat treatment of heterostructure device structures such as GaAs/GaAl As heterostructure lasers, etc.

Under a chemical concentration gradient the diffusion coefficient of component 1 (D_1) is related to the radiotracer self-diffusion coefficient (D_1^x) by the relation

$$D_1 = D_1^x (1 + \frac{d \ln \gamma_1}{d \ln N_1})$$

(3)

where γ_1 is the chemical activity of component 1, and N_1 is the mole fraction of component 1. The interdiffusion coefficient $(\tilde{D})$, obtained by a Boltzmann-Matano analysis of the concentration profile across an interdiffusion couple, is then expressed by the Darken equation [7]

$$\tilde{D} = (D_1^x N_2 + D_2^x N_1) \{ 1 + \frac{d \ln \gamma_1}{d \ln N_1} \}$$

(4)

where N_2 is the mole fraction of component 2, etc.

Knowledge of the thermodynamic factor $\dfrac{d \ln \gamma_{\pm}}{d \ln N_1}$, from independent

measurements can thus give valuable information about the self-diffusion coefficients D_1^x and D_2^x , from measurement of the interdiffusion coefficient $\tilde{D}$ as a function of composition. In conjunction with Kirkendall experiments, as will be explained below, these studies can provide invaluable evidence about the mechanism of self- and chemical-diffusion.

SELF-DIFFUSION IN III-V COMPOUNDS

Renewed interest in self-diffusion in III-V compounds has been generated by the recognition of its importance in such processes as the growth and heat-treatment of superlattice structures, etc. Available data on the binary III-V's is summarised in Table I and Figs. 4 and 5 showing that there is still a very limited range of measurements available. No data at all is apparently available on the important compounds GaP and AlSb. Details of the techniques used to obtain the data quoted in ref. 8 for InAs were not given. In the case of InSb, there is a considerable discrepancy between the residual activity measurements of Eisen and Birchenall [8] and those of Kendall and Huggins [9] as shown in Fig. 4. This is clearly an area where further measurements are necessary to establish the diffusion constants, preferably as a function of degree of non-stoichiometry, although we note that Kendall and Huggins reported that the diffusion constant of neither component was dependent on the ambient Sb vapour pressure. As noted by Kendall [10] , Millea , Tomizuka and Slifkin [11] were not able to reproduce Eisen and Birchenall [8] measurements, and this casts considerable doubt on whether the latter results are typical of volume diffusion. Detailed discussion of these experimental factors has been made by Kendall [10].

Much of the recent discussion of self-diffusion mechanisms has rested on a comparison of the experimental activation energies for diffusion (Q), with values calculated theoretically for the various possible mechanisms. For example, Van Vechten [12] and others have calculated the enthalpies of formation of the two types of monovacancy in each of the III-V compounds, and Bublik [13] noted that such values in InSb were higher than the activation energy for self-diffusion of the corresponding components obtained by Boltaks and co-workers [14] . However, such an approach relies heavily on the experimental values of activation energy, and this author contends that the range of temperatures over which these are determined render these values subject to considerable uncertainty, particularly in the case of InSb and GaSb, where the temperature range is only of order $40^{\circ}C$.

In this author's view, the comparison of the diffusion constants and activation energies of the two components, and the effect of component pressure, is likely to yield more information on the diffusion mechanisms. Kendall and Huggins [9] noted that the diffusion activation energy via III-V mixed divacancies should be the same for both components, since, as noted by Vorob'ev et al. [15] a divacancy must temporarily drift away from the labelled atom in the sphalerite structure, after the first jump, and this part of the diffusion process is limited by the less mobile component. This argument rules out the mixed divacancy mechanism for both components almost certainly in InP, and probably also for GaAs and GaSb. As noted by Vorob'ev et al [15] , however, this argument does not rule out this mechanism for one of the two components in these compounds.

TABLE I
Reported self-diffusion measurements in III-V compounds

Compound	D_o $(cm^2 sec^{-1})$	Q (ev)	Temp. Range	Ref.	Technique
GaSb					Residual activity
Ga	3.2×10^3	3.15	680-700°C	a	Radiotracer
Sb	3.4×10^4	3.45	660-700°C	a	Radiotracer
[Sb	8.7×10^2	1.13		b	Radiotracer]
InP					
In	1×10^5	3.85	838-980°C	c	Radiotracer
P	7×10^{10}	5.65	903-1010°C	c	Radiotracer
InSb					
[In	1.8×10^{-9}	0.28		d	Radiotracer]
In	5×10^{-2}	1.82	480°-520°C	a	Radiotracer (Residual acty.)
In	1.8×10^{13}	4.3	475°-517°C	e,f	Radiotracer
[Sb	1.4×10^{-6}	0.75		d	Radiotracer]
Sb	5×10^{-2}	1.94	480-520°C	a	Radiotracer (Residual activity)
Sb	3.1×10^{13}	4.3	475-517°C	e,f	Radiotracer
InAs					
In	6×10^5	4		g	Not known
As	3×10^7	4.45		g	Not known

(N.B. Data enclosed in square brackets has been considered to be influenced by
short-circuit diffusion [10].)

References: Table I and Fig. 4.

a. F.H. Eisen and C.E. Birchenall, Acta Met 5, 265 (1957)
b. B.I. Boltaks and Yu.A. Gutorov, Sov. Phys. - Solid State 1, 930 (1960)
c. B. Goldstein, Phys. Rev., 121, 1305 (1961).
d. B.I. Boltaks and G.S. Kulikov, Sov. Phys. - Tech. Phys. 2, 67 (1957).
e. D.L. Kendall, results quoted in "Semiconductors and Semimetals", ed.
    R.K. Willardson and A.C. Beer, Vol. 4 (1968) p.163.
f. D.L. Kendall and R.A. Huggins, J. Appl. Phys. 40 (1969), 2750.
g. Measurement quoted by V.M. Vorob'ev, V.A. Murov'ev and V.A. Panteleev,
    Sov. Phys. - Solid State 23, 653, (1981).

Turning now to the dependence of self-diffusion on component pressure, we
note that this is only available in recent results for arsenic in GaAs,
measured by Palfrey et al. [16] . Diffusion profiles after diffusion under
two different arsenic pressures are shown in Fig. 6, and show that increasing
the arsenic pressure (p As$_2$) from 0.75 to 3.0 atm. reduces the diffusion
constant from 1.5×10^{-15} cm^{2}s^{-1} to 5.5×10^{-16} cm^{2}s^{-1}. This immediately rules
out the interstitial mechanisms discussed above, since equations such as (2)
predict that increase in p As$_2$ should increase the interstitial concentration
and hence the diffusion coefficient; in fact $D_{As} \propto p\ As_2^{\frac{1}{2}}$ by these mechanisms.
Considering vacancy mechanisms, the dependence could be consistent with a
monovacancy mechanism on the arsenic sublattice; for this mechanism, considering
neutral arsenic vacancies $D_{As} \propto p\ As_2^{-\frac{1}{2}}$.

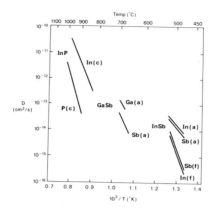

Fig. 4.   Collected self-diffusion coefficients in III-V binary
compounds plotted as a function of reciprocal temperature
(for references see Table I)

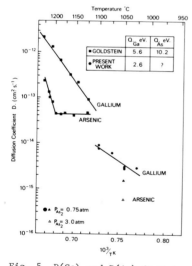

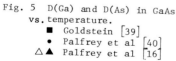

Fig. 5   D(Ga) and D(As) in GaAs
vs. temperature.
   ■ Goldstein [39]
   • Palfrey et al [40]
   △▲ Palfrey et al [16]

   (Palfrey et al [16])

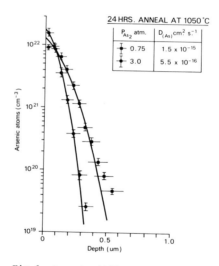

Fig.6.   Arsenic diffusion profiles in GaAs
at 1050°C, under   different
arsenic pressures (Palfrey et al [16])

If we consider diffusion involving antisite defects, there are various possibilities to consider. Firstly, a process involving the migration of an $As_{Ga} - V_{Ga}$ pair can be considered, as envisaged by Vorob'ev et al. [17] in their mechanism 1(b). This process involves migration on the gallium sublattice only, and the defect pair has to split in order to migrate. The dependence of the self-diffusion coefficient on arsenic pressure via this mechanism, however, is $D \propto p As_2^{3/2}$ (considering the dependence of the concentration of each defect on arsenic pressure) and hence is not consistent with the measurements of Palfrey et al. [16] . Secondly, we note that the mechanism 1(c) discussed by Vorob'ev et al [17] is, in fact, the migration of an $As_{Ga} - V_{As}$ pair which occurs by a five stage process involving both nearest neighbour and next-nearest neighbour jumps. The V_{Ga}, which is envisaged by Vorob'ev as the starting defect, does not in fact migrate during the process. The dependence of the self-diffusion coefficient on arsenic pressure via this mechanism is $D \propto p As_2^{\frac{1}{2}}$ and thus is not consistent with the measurements either. We must also consider diffusion via the opposite defect pairs $Ga_{As} - V_{As}$ and $Ga_{As} - V_{Ga}$, but in both cases the migration of arsenic is only via arsenic vacancies, and not via antistructure defects – hence the process is equivalent to diffusion via arsenic vacancies.

Finally, we note that Van Vechten [18] has pointed out that, by nearest neighbour jumps only, the diffusion of a single arsenic vacancy requires an 11 step process to move V_{As} one place on its sublattice without leaving a trail of antistructure disorder. This has been criticised [17] on the grounds of too great a departure from short range order.

In view of this complex local disorder during diffusion, the precise effect of arsenic pressure on diffusion must depend on the nature of the defects created. However, one might still expect the diffusion coefficient to relate primarily to the arsenic vacancy concentration and hence it is clear that the present evidence favours strongly the self-diffusion of arsenic via arsenic monovacancies, although the precise migration mechanism is still not clear.

Thus it is clear that much further work is needed in this area. Dependence of component diffusion on component pressure could give valuable information on diffusion mechanisms, while changing the Fermi level by doping could indicate how charged defects can be involved. These latter experiments must employ a slow-diffusing dopant, by comparison with the component, and would be the first of their kind in these compounds. A further objective must be to reconcile the data with information on defect migration, of the kind reported by Chiang and Pearson [19].

INTERDIFFUSION IN III-V COMPOUNDS

One of the systems which has attracted great interest, because of its role in heterostructure lasers, is the GaAlAs/GaAs system. Since such layers are often subjected to heat treatment during fabrication and growth, it is of considerable importance to obtain a better understanding of interdiffusion across such an interface. Superlattice structures grown by MBE provide an added reason for interest in this phenomenon. As will be seen later in our review of the CdTe/HgTe system, such experiments could also give considerable information on mechanisms of diffusion.

The most comprehensive experimental study in this area was by Chang and Koma [20] . Epitaxial structures composed of thin layers of GaAs and AlAs ($\sim$ 2000 Å thick) were grown by molecular beam epitaxy. The interdiffusion coefficient was measured by Auger-electron profiling after annealing the structures under As-rich conditions over the temperature range 850–1100°C. The measurements are shown in Fig. 7 showing that the interdiffusion coefficient was found to be extremely small, in the region of 10^{-18} cm^2/s at the lower temperatures. The dependence of $\tilde{D}$ on x in Ga$_{1-x}$Al$_x$As was obtained from a Boltzmann – Matano analysis of the profiles, and shows that $\tilde{D}$ decreases with an increase in Al as shown in Fig. 7.

The dependence on x becomes stronger at higher temperatures, giving rise to activation energies of 4.3 and 3.6 eV in the limits of GaAs and AlAs respectively.

These earlier results by Chang and Koma have recently been confirmed by thermal annealing experiments (875–925°C, excess As in the diffusion ampoule) performed on a number of Ga$_{1-x}$Al$_x$As $\neq$ GaAs quantum-well heterostructures grown by metalorganic chemical vapour deposition [21] . Studies of (GaAs)$_n$ (AlAs)$_m$ multilayer structures grown by molecular beam epitaxy, where the layers were only a few monolayers thick, have been made by various techniques and confirmed a very slow interdiffusion rate from 800°C to 930°C (e.g. Fleming et al. [22] who used an x-ray diffraction method). Fleming et al. [22], however, reported a larger temperature dependence of the diffusion coefficient than Chang and Koma, but these differences might have been associated with differences in overpressure of arsenic, during the diffusion anneal.

Although these studies are still at an early stage, it is now possible to make some general comments and suggestions for further work. In Fig. 8 we plot the composition dependence of the interdiffusion coefficient at 1088°C reported by Chang and Koma [20]. In this figure we have noted that the self-diffusion coefficient of Ga in GaAs at 1088°C reported by Palfrey et al.[40] lies below the interdiffusion coefficient at x = 0 (which is D_{Al} in GaAs) and hence we can sketch a suggested curve of D_{Ga} versus x, since D_{Ga}^{Al} in AlAs is the value of $\tilde{D}$ at x = 1. Although the self-diffusion coefficient of Al in AlAs is not available, we believe that $D_{Al} > D_{Ga}$ over the whole composition range, as indicated. This is an important conclusion and confirms the suggestion of Fleming et al.[22] made from the interpretation of their data.

A more complete analysis requires measurements, by radiotracer techniques, of D_{Ga} and D_{Al} into the ternary Ga$_{1-x}$Al$_x$As, together with the unknown self-diffusion coefficient of Al in AlAs. These studies, in conjunction with Kirkendall experiments such as those discussed below, could provide a determination of the thermodynamic factor in equation 4 for comparison with the theoretical value. Ideally, these measurements should be carried out as a function of the arsenic pressure.

Finally, we note that Laidig et al.[21] have shown that the diffusion of Zn into an AlAs-GaAs superlattice structure radically enhances the inter-diffusion of Al and Ga, even at low temperatures (< 600°C). Laidig et al.[21] suggested an explanation of this involving (Zn$_i$-vacancy) pair diffusion. More recently, Van Vechten [23] has proposed another explanation in terms of the interaction of Zn interstitials, with the antisite defect complexes resulting from his own nearest neighbour hopping mechanism of As vacancy diffusion. On this model of As vacancy diffusion, with or without zinc, an 11 step process is required to move an arsenic vacancy one place on its sublattice without leaving a trail of antistructure disorder (Fig. 9). The zinc inter-stitial, in the Zn$_i^{+2}$ charge state, is envisaged as reducing the energy of the 6 member ring of the zinc-blende lattice on which the vacancy diffusion is occurring. In this As-vacancy diffusion 3 Al or Ga and 2 As atoms are intermixed on their respective sublattices. Van Vechten also discussed the effect of zinc diffusion on divacancy diffusion via nearest neighbour jumps, and showed that this should favour the cation (Al or Ga) sublattice.

The above models assume a direct effect of zinc interstitials on defect diffusion, but the present author wishes to propose an alternative mechanism. Ball, Hutchinson and Dobson [24] have shown, by TEM studies of zinc diffusion in GaAs, that gallium interstitials are generated at the diffusion front. This occurs even in undoped material and they proposed that it results from zinc interstitials using up all available gallium vacancies, so that Frenkel disorder leaves excess gallium interstitials. Such a mechanism could therefore enhance the diffusion of either gallium or aluminium if it occurs by an interstitial-substitutional or interstitialcy mechanism. Such an effect, in which native defects are generated, would be similar to the effects of some dopant diffusions in silicon which subsequently enhance other dopant diffusions [25].

SELF-DIFFUSION IN II-VI COMPOUNDS

In view of the large volume of literature on self-diffusion in the II-VI compounds, this review will be restricted to comments on specific systems which are illustrative of an approach which is felt to be of value in other systems. For a more detailed review, readers are referred to the comprehensive review by Stevenson [26].

Self-diffusion in CdTe

Self-diffusion studies in this system are discussed, following a more extensive review of the $Cd_xHg_{1-x}Te$ system by Brown and Willoughby [27], since measurements of the partial pressure dependence of diffusion coefficients are available.

Te self-diffusion

Fig. 10 shows measurements of Te self-diffusion obtained by radiotracer techniques, by two different groups [28][29], and the diffusion coefficients obtained are in good agreement. Fig. 10 shows the results plotted against 1000/T which give two lines for high and low tellurium pressure, the diffusion coefficient increasing with tellurium pressure at all temperatures between 500 and 900°C. The results of a more extensive determination of the dependence on tellurium pressure [29] at 800°C using the isotope ^{123}Te are shown in Fig.11, together with the data of Borsenberger and Stevenson [28] for comparison.

As Borsenberger and Stevenson [28] pointed out, this dependence on tellurium partial pressure strongly suggests an interstitial type of mechanism for the self-diffusion of tellurium in CdTe, since our equation 2 would predict such an increase. Borsenberger and Stevenson represented the dependence by the following equation for diffusion via a neutral Te-interstitial:-

$$D_{Te} = D(Te_i^x) Ki(T) p^{\frac{1}{2}}_{Te_2}$$

where $D(Te_i^x)$ is the diffusion coefficient of neutral Te interstitials, Ki is the mass action constant for the formation of neutral Te interstitials, and p_{Te_2} is the partial pressure of the Te_2 species.

These conclusions are further reinforced by the activation energy for self-diffusion, shown on Fig. 10 as ranging from 1.38-1.42 eV, and obtained over a wide range of temperature by comparison with the above values for III-V compounds. Van Vechten [30] has summarised the calculated and empirically

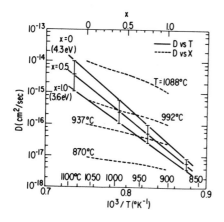

Fig.7. GaAs/AlAs interdiffusion coefficient
vs. temperature and composition
(Chang and Koma [20])

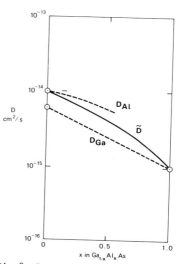

Fig.8. Suggested $D_{Al} + D_{Ga}$ vs. x
in $Ga_{1-x}Al_xAs$.

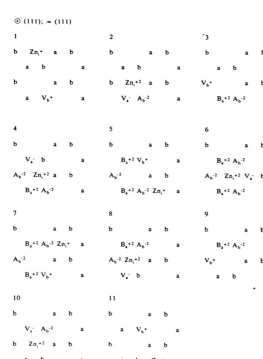

Fig.9. Nearest neighbour
hopping mechanism of
As vacancy (V_b)
diffusion in the
presence of Zn
interstitials.
(Van Vechten [23])

determined estimates for the enthalpy of formation of V_{Te} in CdTe, and these range from 2.3-2.4eV, i.e. greater than the activation energy for self-diffusion. Since the activation energy for self-diffusion (under constant composition conditions), by a monovacancy mechanism, is the sum of the formation and migration enthalpies of the vacancy, this would appear to rule out this mechanism. Diffusion via antisite defect - vacancy pairs might, however, be a possible mechanism although Van Vechten's calculations suggest that diffusion via antisite defects is likely to be less important in III-V's than in II-VI compounds. Near Cd saturation it can be seen that the data depart from the theoretical line predicted for diffusion via neutral Te interstitials, and it has been speculated that a Te-vacancy mechanism might be operative here.

These results, therefore, show a much more complete picture than the III-V compounds. It is still, however, of considerable interest to study the effect of Fermi level changes on self-diffusion so that the role of charged defects could be evaluated.

Cd Self-diffusion

In this case, the agreement between the results of two reports [28][31] is less good than for the case of tellurium, although both reports found Cd self-diffusion to be independent of Cd pressure. Whelan and Shaw |31| found an activation energy of 2.07 + 0.08eV over the temperature range 700-1000°C, while Borsenberger and Stevenson [28] reported activation energies of 2.44 and 2.67eV respectively for Te saturated and Cd saturated conditions over the temperature range 510-920°C. At 775°C they found no change in the diffusion coefficient for Cd pressures from 10^{-4} to 1 atm.

This independence of Cd self-diffusion with Cd vapour pressure (p_{Cd}), is a valuable piece of evidence on the diffusion mechanism. Whelan and Shaw [31] pointed out that this rules out diffusion via Cd vacancies or interstitials as well as any charged defect, on the basis that the concentration of these defects would have to vary with p_{Cd}. Whelan and Shaw [31] thus concluded that either a ring or exchange mechanism (which involve no defects) or a neutral defect complex such as $(Cd_i V_{Cd})^x$ was possible. On the other hand, Borsenberger and Stevenson [28] proposed that Cd self-diffusion was due to a pressure-independent native defect which arises from crystal charge neutrality conditions. However, their model of ionised Frenkel disorder on the Cd sub-lattice in which $Cd_i^{\cdot} = V_{Cd}'$ is not consistent with the dependence of electron concentration on cadmium vapour pressure measured by Chern and Krüger [32]. Charged defects may be involved at lower temperatures, however, where donor (Al) and acceptor (Au) dopants were found to enhance D_{Cd}.

Thus we see that rotation or exchange mechanisms remain a real possibility for the diffusion mechanism of Cd in CdTe. It is clear, however, that we need further kinds of evidence to isolate the mechanism more completely, and we will now discuss how interdiffusion studies can assist.

Interdiffusion in the CdTe/HgTe system

Increasing interest in the growth and processing of $Cd_x Hg_{1-x} Te$ layers on CdTe substrates, for infra-red detector applications, has highlighted the phenomenon of interdiffusion in this system. Interdiffusion, involving the metal atoms Hg and Cd, is an important factor in controlling the composition of layers grown at elevated temperatures, and has an important role in the choice of growth method, and so it is now of greater importance to obtain a better understanding of this effect. A review of data in this system was recently undertaken by the present author [33], but this system is chosen as an example of how interdiffusion can assist to clarify diffusion mechanisms.

The earliest report of interdiffusion studies in this quasi-binary semi-conductor alloy (Cd,Hg)Te appears to have been by Bailly et al [34]. Parallelepipeds of HgTe and CdTe were placed in contact in a sealed quartz ampoule and concentration profiles obtained by electron microprobe analysis scans across the diffusion couple were analysed by Bailly [35].

In 1974, Leute and Stratmann [36] reported the results of some most interesting experiments designed to measure the interdiffusion coefficient, and to establish whether the Kirkendall effect could take place in this system. A diffusion couple was prepared between Te-rich CdTe and HgTe crystals, and tungsten wire markers (10 μm diameter) were used for Kirkendall effect experiments. The marker movement at the CdTe/HgTe interface was established by using comparison markers at CdTe/CdTe interfaces which acted as static reference points. The couple was enclosed in a poly-crystalline HgTe matrix at 590°C for about 15 hours, again under Te excess. After diffusion the markers were observed to have shifted towards the HgTe side of the couple by ∿35 μm, corresponding to a migration velocity ∿3.4 x10^{-8} cm/s. Concentration profiles across the boundary were also obtained by electron microprobe analysis scans, and the concentration dependent interdiffusion coefficient was determined by the Boltzmann-Matano method. The interdiffusion coefficient measurements obtained as a function of composition by Leute and Stratmann are shown in Fig. 12 and the individual diffusion coefficients at a mole fraction x_{CdTe} = 0.23 were D_{Hg} = 1.0 x 10^{-9} cm^2/s and D_{Cd} = 1.7 x 10^{-10} cm^2/s. The measurements of interdiffusion coefficient $\tilde{D}$ of Bailly [35] at 580°C are compared with those of Leute and Stratmann [36] at 590°C in Fig. 12 showing that there is reasonable agreement over most of the composition range.

The Kirkendall experiments of Leute and Stratmann [36] are of considerable interest in showing how such experiments can clarify the mechanism of inter-diffusion. Although it is not clear how the contact interfaces used in these experiments will compare with epitaxial interfaces, or whether vapour transport might have been significant, we can make some general comments.

(a) The fact that the Kirkendall effect has clearly been measured in this system indicates that we can rule out exchange or ring mechanisms of diffusion for Cd and Hg under these chemical diffusion conditions, since marker movement requires annihilation of lattice planes which could not occur by these mecha-nisms. This is particularly significant in this system because of the indications, discussed above, that ring or exchange mechanisms are possible for Cd in CdTe. As pointed out by Leute and Stratmann [36] , the Kirkendall effect has always been explained in metallic systems, following the Darken interpretation, as an indication of the vacancy mechanism, but this conclusion also relied on other evidence (LeClaire [6]). LeClaire pointed out [6] , that an interstitial mechanism might also be invoked to give a satisfactory explanation of Kirkendall phenomena in metals, if taken on their own. However, it is more important to note that, provided the tungsten markers can be assumed to have had no influence on the interdiffusion process, we can rule out exchange or ring mechanisms.

(b) Leute and Stratmann [36] pointed out that the direction of marker movement in their Kirkendall experiment showed that Hg atoms are more mobile than the Cd atoms under the conditions of their experiment. This is also a very significant conclusion, since radiotracer studies of Hg and Cd diffusion in $Cd_x Hg_{1-x}Te$ do not provide a very clear picture at present [27].

250

Following these conclusions, it is of vital importance that measurements of interdiffusion between $Cd_xHg_{1-x}Te$ layers and CdTe substrates are made, where more is known about the nature of the interfaces, as has been done for $Ga_xAl_{1-x}As$ layers on GaAs [20]. Ideally these experiments should study the effects of metal/tellurium ratio on interdiffusion, and the effects of doping, both of which have been studied recently for contact diffusion couples by Leute et al. [37]. It is also of great importance to check that grain boundary or dislocation short-circuit diffusion is not significant, as has been assumed.

CONCLUSIONS

We have seen the cogent need in this field for further careful experimental work to clarify self-diffusion mechanisms in the compound semiconductors. In the III-V compounds, in particular, the basic data is very sparse and studies of the dependence of self-diffusion coefficient on component pressure and Fermi level could give much needed information, as has been shown in the recent work on GaAs. Interdiffusion between different III-V compounds is of increasing importance, and it seems possible here that Kirkendall experiments, using epitaxial interfaces, might be attempted, which would give a much greater understanding of the mechanism of interdiffusion.

Turning to the II-VI compounds we see that, in particular cases such as CdTe, the dependence of self-diffusion coefficients on component pressure is well studied. However, in many II-VI compounds there is a need to carry out experiments on well characterised material to rule out short-circuit diffusion down grain boundaries and/or dislocations. Similarly, Kirkendall experiments on interdiffusion couples, such as those carried out on the CdTe/HgTe system, must be checked using good epitaxial interfaces and material of good crystalline perfection.

Finally we note that there has apparently been no attempt to study the isotope effect on self-diffusion in any of these compounds, as was carried out for germanium [38] despite the fact that this can give such direct information on diffusion mechanisms. While these experiments present considerable experimental difficulties, over and above the problems of volatile components which have already placed severe limitations on the information for III-V compounds, it is hoped that such studies perhaps in a compound like InSb where these problems are minimised, will be undertaken to provide further important information on the mechanism of self-diffusion.

ACKNOWLEDGMENTS

The author wishes to acknowledge discussions with present and former colleagues at Southampton University and particularly with Dr. Helen Palfrey for discussions on self-diffusion mechanisms in GaAs. Comments by Dr.Derek Shaw on the dependence of self-diffusion coefficients on component pressure are gratefully acknowledged.

Thanks are also expressed to Dr. J.A. Van Vechten and Dr. Helen Palfrey for permission to quote unpublished work.

REFERENCES

1. B.G. Streetman, IEEE Trans. Nucl. Sci. NS-28, 1742 (1981)

2. H.C. Casey and G.L. Pearson, in Point Defects in Solids, Vol.2, J.H.Crawford and L.M. Slifkin eds. (Plenum Press 1975), pp.163-255.

3. H.C. Casey in Atomic Diffusion in Semiconductors, D. Shaw ed. (Plenum Press, London and N.Y. 1973), pp.351-430

4. J.W. Corbett and J.C. Bourgoin, in Point Defects in Solids Vol. 2, J.H. Crawford and L.M. Slifkin eds. (Plenum Press 1975), pp.1-161

5. S.M. Hu in Atomic Diffusion in Semiconductors, D. Shaw ed.(Plenum Press, London and N.Y. 1973), pp.217-350.

6. A.D. LeClaire, Progress in Metal Physics 4, 265 (1953)

7. L.Darken, Trans.A.I.M.E., 174, 184 (1948)

8. F.H. Eisen and C.E. Birchenall, Acta Met., 5, 265 (1957)

9. D.L. Kendall and R.A. Huggins, J. Appl. Phys, 40, 2750 (1969)

10. D.L. Kendall, in Semiconductors and Semimetals, R.K. Willardson and A.C.Beer eds. (Academic Press N.York, 1968), pp.163-259

11. M.F. Millea, et al., unpublished data discussed by G.P. Williams and L.M. Slifkin, Acta Met. 11, 319 (1963)

12. J.A. Van Vechten, J.Electrochem.Soc. 122, 419 (1975)

13. V.T. Bublik, Phys. Stat. Sol. (a) 45, 543 (1978)

14. B.I.Boltaks and G.S. Kulikov, Sov.Phys.-Tech.Phys. 2, 67 (1957)

15. V.M.Vorob'ev et al., Sov.Phys.-Solid State 23, 653 (1981)

16. H.D. Palfrey, et al., Paper presented at Electronic Materials Conference, Ft. Collins, June 1982 (to be published in J. Electronic Materials)

17. V.M. Vorob'ev, et al., Sov. Phys.-Solid State 23, 2055 (1981)

18. J.A. VanVechten, Czechoslovak Journal of Physics B30, 388 (1980)

19. S.Y. Chiang and G.L. Pearson, J.Appl. Phys. 46, 2986 (1975)

20. L.L. Chang and A.Koma, Appl.Phys.Lett. 29, 138 (1976)

21. W.D. Laidig, et al., Appl. Phys. Lett. 38, 776 (1981)

22. R.M. Fleming, et al., J.Appl.Phys., 51, 357 (1980)

23. J.A. Van Vechten, J.Appl. Phys. (to be published)

24. R.K.Ball, et al., Phil Mag. 43A, 1299 (1981)

25. A.F.W. Willoughby, J.Phys. D.; Appl. Phys., 10, 455 (1977)

26. D.A. Stevenson, in Atomic Diffusion in Semiconductors, D. Shaw ed. (Plenum 1973), pp.431-541

27. M.Brown and A.F.W. Willoughby, J.Cryst. Growth 59, 27 (1982)

28. P.M.Borsenberger and D.A. Stevenson, J.Phys. Chem. Solids 29, 1277 (1968)

29. H.H. Woodbury and R.B. Hall, Phys. Rev. 157, 641 (1967)

30. J.A. Van Vechten in Handbook on Semiconductors Vol.3, S.P. Keller ed. (North-Holland, Amsterdam 1980) pp.1-111

31. R.C.Whelan and D. Shaw, in II-VI Semiconducting Compounds, D.G.Thomas ed., (Benjamin 1967), p.451

32. S.S. Chern and F.A. Krüger, J.Sol.St.Chem.14, 44 (1975)

33. A.F.W. Willoughby, Materials Letters 1, 58 (1982)

34. F. Bailly, et al., Comptes rendues (Acad.Sci.) 257, 103 (1963)

35. F. Bailly, Comptes rendues (Acad.Sci.), 262, 635 (1966)

252

36. V. Leute and W. Stratmann, Zeit. Phys. Chemie 90, 172 (1974)

37. V. Leute, et al., Phys. Stat. Sol. (a) 67, 183 (1981)

38. D. Shaw, Phys. Status Solidi (b) 72, 11 (1975)

39. B. Goldstein, Phys. Rev. 121, 1305 (1961)

40. H.D. Palfrey, et al.,J. Electrochem. Soc. 128,2224 (1981)

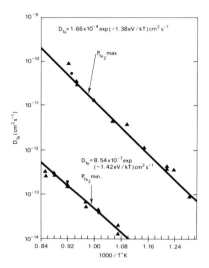

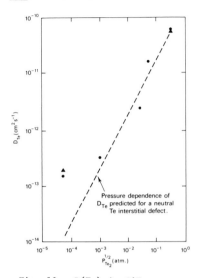

Fig. 10. D(Te) in CdTe vs. temperature and component pressure
▲ Borsenberger and Stevenson [28]
● Woodbury and Hall [29]
(Borsenberger and Stevenson [28])

Fig. 11. D(Te) in CdTe vs. component pressure at 800°C (key as Fig.10) (Borsenberger and Stevenson [28])

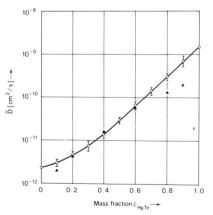

Fig. 12. CdTe/HgTe interdiffusion coefficient vs. composition
O Leute and Stratmann [36] at 590°C
▲ Bailly [35] at 580°C
(Willoughby [33])

DEEP RADIATIVE TRANSITIONS IN InP

H. TEMKIN AND B. V. DUTT

Bell Laboratories, Murray Hill, N. J. 07974

ABSTRACT

Results of a detailed photoluminescence study of deep radiative transitions in InP crystals prepared by bulk and epitaxial techniques with a variety of dopants are reported. In order to understand the origin of the photoluminescence spectra, bulk samples were subjected to isothermal anneals at different partial pressures of phosphorus. Similarly, the LPE wafers were grown with and without phosphorus in the gas stream. The electrical nature of some of the species responsible for the PL emission was inferred by a study of Cd-diffused bulk samples.

Based on these experiments the following tentative assignments are proposed. The photoluminescence band at 0.99 eV, common to all samples, is due to emission from a donor-like level related to the P-vacancy-impurity complex. Bands at 1.21 eV and 1.15 eV appear to be due to emission to native acceptor levels associated with the In-vacancy. The 1.08 eV band is attributed to emission to a level involving the complex of the donor (0.99 eV) and acceptor (1.21 eV) species. The relationship between these bands and residual impurities is discussed. A comparison with the work of other investigators tends to support these assignments.

INTRODUCTION

In recent years InP and InP based alloys have become important for applications in optical and microwave devices. The technological interest, and the subsequent availability of high quality InP, have stimulated a number of papers on the subject of deep radiative centers in InP.[1-6] Photoluminescence studies have revealed a number of spectral features in the region of 1.3 to 0.6 eV. Particular attention has been given to the $\sim$1.15 eV band, variously reported between 1.0 and 1.2eV, prominent in bulk and epitaxial InP. This band has been initially associated with the phosphorus vacancy (V_P)[7] and more recently with its complexes with Fe[3] or to deep acceptor states of Fe and Mn.[4] Other deep levels found in melt grown InP have been attributed to native defect-impurity complexes or to specific chemical impurities.[1,2] Since the device performance is affected by the presence of deep levels in the material, it is necessary to ascertain the nature and origin of these levels. It is the goal of this review to critically

Mat. Res. Soc. Symp. Proc. Vol. 14 (1983) © Elsevier Science Publishing Co., Inc.

examine the photoluminescence results reported by various workers and to resolve some of the differences in the proposed models. Particular attention is given to the residual impurities such as Mn, Fe and Cu and their complexes with the native defects.

EXPERIMENTAL

Crystals of InP investigated in this work were prepared by the techniques of melt growth (MG), liquid encapsulated Czochralski (LEC) and liquid phase epitaxy (LPE). Melt growth is a direct synthesis technique in which elemental In and P are combined to yield polycrystalline InP. The resulting ingots are used as a starting charge in the growth of single crystal InP boules pulled by the LEC technique. The LEC substrates were used in our liquid phase epitaxial growth, in which both LEC and MG material was used as the source material in the melt charge. It is therefore of interest to study any similarities, or differences, between PL spectra obtained on InP produced by these different techniques. In addition to the residual contamination, the widely different growth conditions may result in samples of varying degrees of non-stoichiometry. Thus the MG material used in our work was reacted at 1070°C under a phosphorus pressure of 28-30 atm and synthesised over a period of 36-72 hours, depending on the cooling rate employed.[8] The LEC grown InP was prepared in a chamber kept at the melting point of InP and pressurized with dry N_2 to ~38 atm.[9] In contrast to the high temperature, high pressure bulk techniques, epitaxial layers of InP are typically deposited at 650°C in an ambient of purified H_2 or N_2.[10] Some growth experiments were performed with small amounts of phosphorous added to the gas stream.

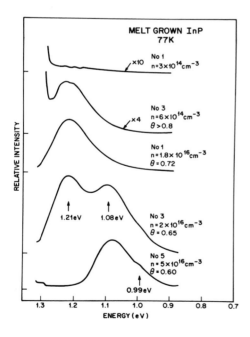

Fig. 1

Evolution of the low temperature PL spectrum of melt grown InP as a function of impurity concentration and position within the ingot (nos. 1, 3, and 5). The traces are vertically displaced for clarity.

Figure 1 presents a series of typical photoluminescence traces obtained on single crystal grains of melt grown InP. The samples were selected from the first, middle, and last to freeze sections, labeled as Nos. 1, 3, and 5, respectively, of ~12 cm long ingots. Their room temperature carrier density increases smoothly from $n=1\times10^{16}cm^{-3}$ at the first frozen end to ~$5\times10^{16}cm^3$ at the last frozen section. The degree of compensation in these unintentionally doped samples, calculated from Hall mobilities, varies systematically from $\theta=0.72$ to 0.60. In addition to this typical material a number of samples were obtained from a highly purified ingot, in which the free carrier density was lowered to $(3-5)\times10^{14}cm^{-3}$ level by careful purification of the starting material. As depicted in Figure 1 the PL spectra evolve as the net carrier concentration increases. No PL features could be observed at the very lowest impurity level shown in the top trace. With increasing carrier concentration a weak band at 1.21 eV is first observed. Its intensity increases by a factor of ~4 and reaches a maximum in samples with $n\approx2\times10^{16}cm^{-3}$ background level. With a further increase, a new PL feature, a broad band at 1.08 eV, appears in samples taken from the center section of MG ingots. In addition an unresolved shoulder at 0.99 eV can be noticed. Finally, in the last to freeze end the 1.21 eV band is usually unobservable and the spectrum is dominated by the 1.08 eV band always accompanied by a prominent 0.99 eV shoulder. It is apparent that such systematic spectral changes could be associated with the residual impurity distribution and may reflect distribution of nonstoichiometry or a combination of the two across the melt grown InP ingot.[11] For instance, small amounts of unreacted In is often found at the bottom of the last to freeze portion, indicative of the lower phosphorus pressure at the time of solidification of that section. This point will be addressed later in more detail.

The PL results described above are in good agreement with the previous work of Negreskul et al. [1] Their 1.10 eV band appears to be similar to our 1.08 eV band in that the bands increase in intensity at the lower phosphorus vapor pressure. The very broad band observed at 1.13 eV in 10^{19} cm^{-3} Si doped InP[1] may be due to superposition of the relatively weaker 1.08 eV present in a spectrum with a stronger 1.21 eV band.

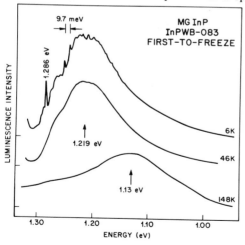

Temperature dependence of the 1.21 eV band of melt grown InP.

Another possibility is suggested in Figure 2, in which the temperature dependence of the 1.21 eV band is examined in detail. At temperatures below 20 K very complicated fine structure is often observed. This consists of a well resolved sharp line at 1.286 eV, followed by a number of what appears to be phonon sidebands with a spacing of 9.7 to 12 meV. This energy is much lower than that of optical phonons of InP. The fine features become much weaker after a few temperature cycles between 6K and 300K and are not seen above 30K. Above that temperature much weaker phonon structure is observed with sidebands separated by $\sim$38 meV, the energy of the transverse optical phonon. Above 90K a new band at 1.13 eV is seen and it becomes the dominant PL feature at higher temperatures, as shown in the lowest trace of Fig. 2. This peak could be equivalent to the 1.13 eV band observed by Negreskul et al.[1] Measurements of the temperature dependence of the 1.08 eV peak intensity, which decreases above 150K with an activation energy of 0.17 eV, do not reveal any new features. In the temperature range of 6 to 250K, above which deep level emission cannot be observed, the shift in peak positions of all of our bands is smaller than the experimental resolution of $\sim$5 meV. Thus the phosphorus vapor pressure dependence appear to suggest at the 1.08 band is related to P-deficiency. On the other hand, the PL spectra of the cleaner samples, top trace Figure 1, and the spectra of Si-doped InP[1] suggest that it may· be due to Si. However, the impurity incorporation influences the nonstoichiometry as well.[2] We cannot therefore preclude the possibility that the bands observed in the 1.0-1.25 eV[1] are related to Si$-V_p$ complexes, and their distribution across the ingot.[2,6]

The PL spectrum of the LEC grown InP is characterized by a single deep transition variously reported between 1.10 eV and 1.16 eV.[3,4,5] This band is characterized by a well resolved phonon structure, with the phonon energy of $\sim$38 meV. This energy corresponds to the flat part of the dispersion curve of a transverse optical phonon and therefore to the highest phonon density of states. The variations in the spectral positions of the LEC band have been attributed either to changes in the electron-phonon coupling strength[3,6] or different chemical impurities involved, such as Mn or Fe.[4] Figure 3 shows the experimental results of Eaves et al[4] obtained on Fe doped, Mn doped and non-intentionally doped LEC material. Below the bandgap region, the PL spectrum of the Fe doped InP is dominated by a band centered at $\sim$1.10 eV, as shown in trace (c) of Figure 3. The spectrum of undoped LEC sample, curve (b), shows a band with a similar phonon structure, the center of which is shifted to $\sim$1.15 eV. In these two cases, the intensity of bandgap emission is $\sim$20 times stronger than that of the deep level. In a spectrum of Mn doped InP however, the 1.15 eV band is the strongest PL feature observable, as shown in trace (a), considerably more intense than the bandgap emission. These results have suggested that the 1.15 eV peak observed in undoped InP might be also due to Mn contamination.

We have examined the low temperature PL spectrum of undoped, Mn, Fe, and Sn doped LEC InP. The results are summarized in Figure 4. Curves (a) and (b) depict spectra obtained on samples from two different boules of LEC material. The PL spectra show a band at $\sim$1.15 eV, with similar phonon structure, even though the absolute intensities of these bands differ by $\sim$2 and differences are evident in the relative strength of the phonon sidebands. In agreement with Yu[3] and Eaves et al[4] this band is 20 to 40 times weaker than the band-edge luminescence. Mass spectroscopy analysis of these samples indicates a reversal in their Mn to Fe ratio, with both impurities present at less than a ppm level. However, curves (a) and (b) do not show any related

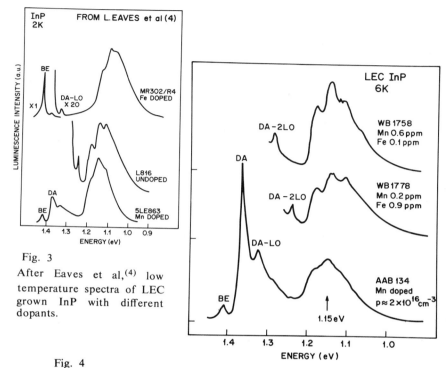

Fig. 3

After Eaves et al,[4] low temperature spectra of LEC grown InP with different dopants.

Fig. 4

Low temperature spectra obtained on our undoped and Mn doped LEC grown wafers. Mass spectroscopic analysis results are indicated.

changes in the PL spectra. The same spectra are obtained on samples doped with Sn (up to $n \approx 1 \times 10^{18} cm^{-3}$) and Fe. The incorporation of iron is confirmed by high electrical resistivity of $\sim 10^8 \, \Omega \, cm$. Typically, intensity of the 1.15 eV peak is low in our Fe doped samples. Curve (c) shows the PL emission of a sample doped with Mn to $p \approx 2 \times 10^{16} cm^{-3}$ level. Again, the deep level emission is dominated by the same ~ 1.15 eV band. In this respect Mn doped InP is different than the p-type material obtained by doping with shallow acceptors such as Zn or Cd, in which the 1.15 eV band is unobservable.[6] However, the absolute intensity of this band in Mn doped InP is the same as in our n-type (undoped or Sn doped) layers. The relative increase in the band intensity is obtained by a sharp decrease in the bandgap luminescence, presumably due to the low distribution coefficient of Mn and the subsequent luminescence quenching upon Mn doping. Thus while the fine structure details of the 1.15 eV band vary from sample to sample, its position or intensity does not appear to change with specific dopants. The 1.10 eV peak cannot be found in any of our LEC samples. The integrated intensity of the 1.15 eV band decreases above 100K with an activation energy of 0.07 eV, identical to that of the 1.21 eV peak found in melt grown InP.[6]

258

Finally, the 6K PL spectra of layers grown by liquid phase epitaxy are shown in Figure 5. The upper curve (a) shows a spectrum typical of undoped ($n \approx 5 \times 10^{16} cm^{-3}$) layers grown in either nitrogen or hydrogen ambient. The most intense feature is a band at 1.15 eV, apparently identical to the same energy peak of LEC InP. In addition, two other bands are seen at 0.99 eV and 0.75 eV. A partially resolved shoulder at 0.99 eV has been also observed in MG InP. Under Nomarski contrast microscopy these epitaxial layers show very small, less than $5 \mu m$ long, thermal decomposition pits known to arise through localized phosphorus loss[10] Curve (b) is typical of LPE layers grown under the same temperature cycle but with phosphorus added to the gas stream. The partial pressure of phosphorus is estimated at $\sim 3 \times 10^{-3}$ atm. The main effect of the increased phosphorus pressure is a decrease in the intensity of the 0.99 eV band. Morphologically, epitaxial layers grown under phosphorus overpressure are completely free of decomposition pits.

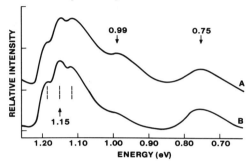

Fig. 5   PL spectra characteristic of liquid phase epitaxial layers grown (a) under nitrogen gas and (b) under phosphorus rich ambient.

THERMAL ANNEALING AND DIFFUSION

The identification of the chemical nature and the charge state of the species responsible for the individual PL features is often attempted through thermal annealing, changes in the growth procedures and diffusion of impurities. For instance, it is assumed that a thermal treatment under a low partial pressure of phosphorus would result in an increased concentration of centers related to the phosphorus vacancy. By contrast, anneals at a high P pressure would result in an increased concentration of defects related to either In-vacancies or P-interstitials. Furthermore, introduction of shallow acceptors by doping or diffusion is expected, on the basis of intersolubility considerations[12], to reduce the native acceptor concentration or increase the native donor concentration or both, depending on the doping level, charge states of the dopants and the thermal history of the sample. Thus diffusion of Cd should result in reduced V_{In} concentration since Cd is a shallow acceptor substituting on an In site. On the other hand, n-type doping, with Sn for instance, should reduce the V_p concentration. Since the $V_p V_{In}$ product is a constant at any given temperature, n-type doping would tend to increase the V_{In} concentration.

This reasoning neglects the presence of residual impurities and their interactions. Thus a high temperature anneal may result in an increased concentration of anomalously diffusing species near the semiconductor surface. Diffusion of an elemental impurity may also result in formation of complexes with a residual impurity already present. Such complexing has been recently found between Mn and Ge in III-V alloys, for example, in InGaAs layers.[13]

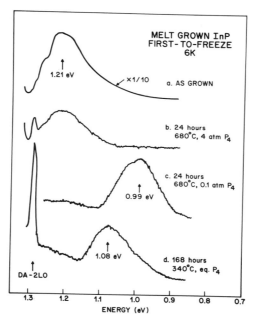

Fig. 6   Anneals carried out on the first to grown section of MG InP initially dominated by the 1.21 eV band.

With these caveats in mind we will discuss the results of various heat treatment experiments. It has been observed by Negreskul et al[1] that the relative intensities of the PL bands changed with crystal growth conditions. Thus a reduction in the phosphorus vapor pressure above the melt below the equilibrium value resulted in increased intensity of the 1.10 eV band, observed in our samples at 1.08 eV. On the other hand, spectra of samples grown under high phosphorus pressure were dominated by the 1.21 eV band. These observations are consistent with the results shown in Figure 1 and similar to our post-growth anneals presented in Figure 6. The 24 hour long anneals at 680°C and a phosphorus pressure of 4 atm were performed on samples sealed in closed ampoules. In the case of a first to freeze sample, initially characterized by the 1.21 eV band, this anneal resulted only in a lower intensity of this band, as shown in trace (b) of Figure 6. The decreased intensity is attributed to a surface deterioration due to high annealing temperature. By contrast, the 1.08 eV band of the last to freeze sample was completely eliminated and replaced by a 1.21 eV band, normally absent in such samples. The two samples were then subjected to a second anneal at 680°C but at a low phosphorus pressure of 0.1 atm. In both samples this low pressure anneal resulted in a spectrum showing only the 0.99 eV, band previously an unresolved shoulder. The spectrum is shown in trace (c) of Figure 6. Thus these high temperature anneals resulted in the PL spectra dominated by the 1.21 eV band at high P-pressure and by the 0.99 eV band at a low P-pressure, irrespective of the initial distribution of deep level intensities.

The remaining PL feature, at 1.08 eV, appears different than the two other bands. It was eliminated in the high phosphorus pressure anneal and it could not be re-introduced by a second, low pressure, anneal. Suspecting a complex involving both 1.21 eV and 0.99 eV species, a prolonged anneal was carried out at a low temperature of 340°C on a sample characterized initially by the 1.21 eV peak. Such a treatment was expected to promote pairing by enabling the participating species to migrate. However, unlike the high temperature treatment, the thermal energy would not be large enough to dissociate the pairs. This low temperature treatment indeed succeeded in introducing the 1.08 eV band, as shown in trace (d) of Figure 6.

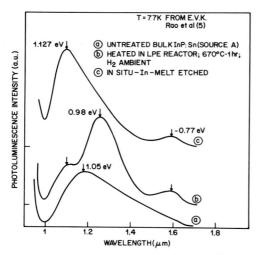

Fig. 7   Heat treatment induced spectral changes in an LEC grown wafer, after Rao et al.[5]

Similar high temperature treatment of LEC grown wafers was reported by Rao et al.[5] Their results are shown in Figure 7. Trace (a) shows the PL spectrum of as-grown, commercially obtained wafer with a single broad band peaking at 1.05 eV. This seems to be an unresolved spectrum of the $\sim$1.15 eV band of LEC InP and a stronger than usual 0.99 eV peak. A 1 hour heat treatment at 670°C, designed to simulate the conditions of liquid phase epitaxy, resulted in a spectrum with a strongly enhanced 0.98 eV band, a weaker $\sim$1.13 eV peak and a weak band at 0.77 eV shown in trace (b).

Removal of a ~15μm thick surface layer, by In etching, resulted in suppression of the 0.98 eV band, as shown in trace (c) of Figure 7. Thus a strong 0.98 - 0.99 eV band was again observed after a low p-pressure treatment, this time in the LEC wafer. This finding is in excellent agreement with our results on melt grown and epitaxial InP. It should be stressed that unlike in the annealing experiments, the results of Figure 5, where P-over-pressure was introduced during epitaxial growth, cannot be ascribed to diffusion of residual impurities at the surface.

Finally, results of two diffusion experiments are discussed. First, we have performed diffusions of Cd, a known shallow acceptor in InP, in order to clarify the electrical nature of species responsible for the 1.21 eV and 1.15 eV bands. Second, Cu diffusion experiments of Rao et al[5] are compared with our high temperature annealing experiments. Since copper interacts strongly with native defects, its diffusion often reveals detailed information about their nature.

Cd was diffused into two samples of melt grown InP, from both ends of an ingot, and an LEC wafer sealed in separate ampoules containing Cd and phosphorus. Diffusion was carried out at 680°C for 4 hours under a phosphorus pressure of 4 atm. Without the presence of Cd these temperature and pressure conditions result in a PL spectrum dominated by a 1.21 eV band, as shown in curve (b) of Fig. 6. Cd diffusion however results in a complete suppression of this band. Similarly, Cd diffusion into the LEC grown sample results in an elimination of the main 1.15 eV peak. Only a weak, and fully resolved peak at 0.97 eV is observed. Based on the rationale briefly discussed at the beginning of this section, this result is interpreted as consistent with an acceptor like nature of the 1.15 and 1.21 eV bands.

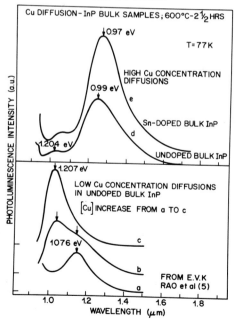

Fig. 8

Results of Cu diffusion into LEC wafer, after Rao et al.[5]

The results of Cu diffusion into LEC wafers, performed by Rao et al,[5] are shown in Figure 8. The PL emission of an undoped sample at the lowest, unspecified, Cu concentration is characterized by a band at 1.076 eV shown in trace (a). This also appears to be the spectrum of an untreated sample. Increase in the Cu concentration results in a buildup of a new PL band at 1.20 eV, as shown in traces (b) and (c). Further increases in the Cu content cause a gradual increase in the PL intensity of the 0.99 eV peak and a relative decrease in the 1.20 eV band. While the exact conditions of Cu diffusions were not given, these results seem consistent with the previous results of Negreskul et al[1] on Cu diffusion into melt grown InP. The trend exhibited by the PL spectra upon introduction of Cu is very similar to that shown by melt grown samples subjected to a succession of high temperature anneals. During those anneals the 1.08 eV band of the as-grown material gave way to the 1.21 eV peak, at high P-pressure, and then to the 0.99 eV peak after a low P-pressure treatment. The behavior of the bands at 1.20 eV and 0.99 eV in Cu-diffused samples is also similar to that in Cd-diffused samples. The 1.20 eV band is acceptor-like and the 0.99 eV band is donor-like as proposed earlier.[6] The chemical identity of the species remains uncertain. Furthermore none of these bands are present in our LEC material pulled from melt-grown ingots.

DISCUSSION

In addition to the well known band-edge luminescence, the PL spectra of InP show a number of deep level transitions. In the as-grown crystals their intensity depends on the preparation technique employed and the background impurity level. By varying the growth conditions and performing high temperature anneals at well defined phosphorus pressures the relative intensities of the PL bands can be changed, with some bands created and others eliminated. Very similar results can also be obtained by diffusion of impurities such as cadmium or copper.

The 1.21 eV band present in melt grown InP is eliminated by low phosphorus pressure heat treatment and preserved or enhanced by a similar anneal carried out at a high partial pressure of phosphorus. The intensity of this band has been observed to increase upon introduction of impurities as chemically diverse as Si[1] and Cu[5] and its relative intensity depends on the background impurity level as shown in Figure 1. The fine structure superimposed on the broad peak, shown in Figure 2, may be the signature of a specific impurity. In addition, Cd or Cu diffusion results[2,5,6] are consistent with an acceptor-like nature of the defect responsible. While the P-interstitial cannot be excluded, it appears likely that this PL band is related to a native acceptor such as an In-vacancy (V_{In}) complexed with impurities.[2] Such complexes are well known in ZnS[14] and considerable evidence for their existence in GaAs has been presented.[15] Since the Zn-vacancy is an acceptor in ZnS, it can form complexes, or pairs, with a substitutional donor on the second neighboring lattice site due to Coulombic interaction. Transitions from the excited state to the ground state of such a complex result in broad, thermally activated luminescence. A number of different impurities could give rise to such complexes, with only small energy shifts, depending on their charge state rather than chemical identity. Our assignment of the 1.21 eV band to V_{In} related complex is in surprisingly good, and perhaps not entirely fortuitous, agreement with studies of InP metal-oxide-semiconductor structures in which states attributed to In vacancies were observed at 1.2 eV.[16]

In contrast to the 1.21 eV band, the 0.99 eV peak is enhanced by heat treatment carried out under low partial pressure of phosphorus. It is observed in the last-to-freeze sections of melt grown ingots where morphological evidence suggests solidification under lower phosphorus pressure. Its presence in the LPE layers can be virtually eliminated by the addition of very small amounts of phosphorus to the gas stream, as shown in Figure 5. This experiment does no involve any changes in the thermal cycle of LPE growth. On the other hand, the intensity of the 0.99 eV band depends on the impurity level, Figure 1, and can be greatly enhanced by diffusion of Cu or Cd as shown in Figure 8. This band seems most probably associated with complexes involving phosphorus vacancies (V_P)[2,6] and impurities such as, for instance $Cu_{In} - V_P$.[5] However, the presence of this band in undoped InP could be due to other impurities, such as Si complexing with V_P. Interstingly, the P-vacancy is thought to be connected with an 0.9 eV level observed in studies of metal-InP interfaces.[16]

The 1.08 eV band which we have initially associated with V_P[17] is different than the above two peaks. As discussed previously, this band is readily eliminated by high temperature treatment, irrespective of the phosphorus pressure used and can be introduced only by prolonged anneals at low temperatures, Figure 6. It has been attributed to a complex involving species responsible for the 1.21 and 0.99 eV bands such as $V_{In}-V_P$.[2,6] Similar conclusion was reached by Rao et al.[5] who proposed a complex defect of the $V_{In}^- - D^+$ type. The donor could be residual Si.[1] It is interesting to note that both Cu and Si exhibit amphoteric behavior in III-V alloys.[12] More work is needed for better understanding of this interesting defect.

Finally, the role of transition metal acceptors has been discussed recently in connection with the 1.15 eV band. In undoped crystals, Eaves et al.[4] associated it with low levels contamination with Mn and attributed to the recombination of holes bound to the Mn acceptor on an In site with electrons loosely bound to shallow donors. This model was supported by annealing experiments on Mn doped InP which have resulted in dramatic increases, up to a factor of 4000, in the 1.15 eV band intensity relative to the near bandgap PL. Similar results were also obtained by Rao et al.[5] on a 1.14 eV band attributed to Mn. Spectral position of this band is consistant with the binding energy of the Mn acceptor,[18,19] at least under doping conditions typical of LEC,[13] as deduced from Hall measurements. Furthermore, the low temperature photoconductivity can be modeled consistently by assuming a connection between the 1.15 eV band and Mn.[20] Similarly, the 1.10 eV band has been attributed to recombination at a complex involving Fe on an In site and a phosphorus vacancy[3] and later to either Fe^{3+} ion or an (Fe,X) complex.[4]

While our results on Cd diffusion infer an acceptor-like nature of the 1.15 eV peak, a specific chemical species appears elusive. In contrast to Eaves et al,[4] our PL data of Figure 4, do not show either absolute intensity changes nor spectral shifts with specific impurities. This holds for less than a ppm level of Fe or Mn of undoped crystals as well as for intentionally doped layers. High temperature treatment of undoped LEC samples does not show any enhancement of the 1.15 eV band attributable to outdiffusion of residual Mn. On the contrary, intensity of the peak decreases relative to that of the 0.99 eV band. The 1.10 eV peak is not observed in our samples, even upon intentional Fe doping. In case of either metal, a transition from a localized metal impurity cannot be sustained in view of a strong phonon coupling and large linewdith of the luminescence band. On the other hand an excellent fit to the lineshape and width can be readily calculated assuming a molecular-like center coupled to the lattice.[6] The differences in peak position follow naturally from variations in the electron-phonon coupling strength parameter S.[3,6] Evidence for a range of S-values is seen in sample-to-sample variations of the distribution of phonon-sidebands intensities shown in Figure 3 and 4. Certain similarities between 1.15 eV and 1.21 eV peaks should also be noted. The 1.21 eV band shows a ~38 meV phonon structure although much less prominent, Figures 2 and 6, than that of the 1.15 eV band. The temperature dependence of intensities of these two bands is characterized by a single activation energy of 0.07 eV. These results suggest that the center responsible for the 1.15 eV emission is also a complex involving impurities coupled to an In-vacancy. Transition metal acceptors may be efficiently coupled into such a center.[3,4]

ACKNOWLEDGEMENTS

We would like to thank W. A. Bonner, V. G. Keramidas and A. A. Ballman for generously supplying us with InP crystals and many stimulating discussions.

References

[1] V. V. Negreskul, E. V. Russu, S. I. Radanstan, A. G. Cheban, Sov. Phys. Semicond., 9, 587 (1975).

[2] H. Temkin, B. V. Dutt, W. A. Bonner, Appl. Phys. Lett., 38, 431 (1981).

[3] P. W. Yu, Solid State Comm., 34, 183 (1980).

[4] L. Eaves, A. W. Smith, M. S. Skolnick, B. Cockayne, J. Appl. Phys., 53, 4955 (1982).

[5] E. V. K. Rao, A. Sibille, N. Duhamel, Int. Conf. on Defects in Semiconductors, August 1982, Amsterdam and private communication.

[6] H. Temkin, B. V. Dutt, W. A. Bonner, V. G. Keramidas, J. Appl. Phys. November 1982, to be published.

[7] J. B. Mullin, A. Royle, B. W. Straughan, P. J. Tufton, E. W. Williams, 1972 Symposium on GaAs, Inst. Phys. Conf. Ser. No. 10, p. 119.

[8] W. A. Bonner, J. Cryst. Growth, 54, 21 (1981).

[9] W. A. Bonner, Mat. Res. Bull, 15, 63 (1980).

[10] V. G. Keramidas, H. Temkin, W. A. Bonner, Appl. Phys. Lett., 40, 731 (1982).

[11] J. van den Boomgard, Phillips Res. Report, 11, 91, (1956).

[12] F. A. Kröger, The Chemistry of Imperfect Crystals (North-Holland, Amsterdam, 1964).

[13] E. Silberg, T. Y. Chang, E. A. Caridi, GaAs and Related Compounds Conf., Alburquerque, NM, 1982.

[14]   D. Curie and J. S. Prener, p. 445 in Physics and Chemistry of II-VI Compounds, ed. M. Aven and J. S. Prener (North-Holland, Amsterdam, 1967).

[15]   H. Barry Bebb, E. W. Williams, p. 321 in Semiconductors and Semimetals, ed. Willardson and A. C. Beer, Vol. 8 (Academic Press, NY 1972).

[16]   W. E. Spicer, I. Lindau, P. Sheath, C. Y. Su, P. Chye, Phys. Rev. Lett., *44*, 420 (1980) and references contained therein.

[17]   H. Temkin, W. A. Bonner, J. Appl. Phys., *52*, 397 (1981).

[18]   S. P. Staroselteva, V. S. Kulov, S. G. Metreveli, Sov. Phys. Semicond., *5*, 1603 (1972).

[19]   H. Asahi, Y. Kawamura, M. Ikeda, H. Okamoto, Japn. J. Appl. Phys. *20*, L187 (1981).

[20]   L. Eaves, A. W. Smith, P. J. Williams, B. Cockayne, W. R. Mac Evan, J. Phys. C, *14*, 5063 (1981).

COMPLEX DEFECTS IN GaAs AND GaP

P. J. LIN-CHUNG
Naval Research Laboratory, Washington, D.C. 20375 USA

ABSTRACT

A study of the electronic states associated with divacancy defects and with the defect complexes involving an anion antisite with a group IV atom (A_c-IV) in GaAs and GaP is reported. The local densities of states have been determined using the large cluster recursion approach. The properties as well as the position of the gap states of the divacancy defect in GaAs are found to be consistent with the experimental results for the EL2 level. The change of the position of the defect levels of (A_c-IV) as a result of the change of bonding is analyzed. The effect of GaAs-AℓAs interface on the (A_c-IV) defect level is also examined.

1. INTRODUCTION

Defects in III-V semiconductors are known to be responsible for producing recombination centers and for degrading optoelectronic devices. Frequently, simple native point defects are trapped near an impurity atom or near another defect to form defect complexes. These complexes are likely to be related to previously unidentified deep levels observed in experiments [1]. Evidence suggesting the existence of complexes involving an anion antisite with a group IV atom (A_c-IV) has been obtained from EPR data on GaP [2]. The divacancy defects (DV) are also known to be stable at room temperature in many semiconductors. Thus there is a great need for a theoretical investigation of these complexes.

During the past decade, many theoretical techniques have been developed to treat isolated point defects. Most of the techniques require rather heavy computation even for a point defect. The present calculations have been made using the large cluster recursion method in a tight-binding framework [3] which has been applied with ease to problems involving a defect near an interface [4] and is suitable for the treatment of problems with substantial degrees of non-periodicity such as defect complexes near interfaces or surfaces. The defect levels in the gap and the changes of the local density of states at different sites are determined for the (DV) and (A_c-IV) defects in GaAs and GaP. The (DV) defect in GaAs is of particular interest because the present study suggests that (DV) defect is responsible for the EL2 center observed in deep level transient spectroscopy (DLTS). Investigation of several (A_c-IV) complexes in GaAs and GaP also provides a framework for understanding the major chemical trends in the deep level energies when the chemical bondings alter.

2. METHOD

A detailed discussion of the large cluster recursion method can be found in Ref. 3. The tight-binding Hamiltonians used here contain all first and second neighbor interactions. The Slater-Koster parameters (SKP) between interacting atoms in the host crystals are obtained from Daw, Osbourn and Smith [5], and the SKP between impurity atoms and host atoms are derived from recently developed semi-empirical expressions for the SKP which provide the explicit dependence of SKP on the atomic characteristics, on the dielectric constants of the

Mat. Res. Soc. Symp. Proc. Vol. 14 (1983) Published by Elsevier Science Publishing Co., Inc.

solids and on the interatomic separations [6]. A cluster containing 512 atoms is used to find the energy levels as well as the local density of states at different sites of the cluster. The deviation of the local density of states from the density of states in a pure crystal is then obtained and denoted by ΔLDOS.

3. DIVACANCY DEFECTS

A (DV) defect in a III-V semiconductor is formed when both a cation atom and one of its near neighbor anion atoms are removed. It is suggested here that the (DV) defect in GaAs gives the 0.82 eV electron trap (EL2 level) because its properties are found to be consistent with the experimental results [7] for EL2 level. These properties include: (a) The (DV) can be annihilated by either a group IV impurity (Si) or a group VI impurity (S, Se, Te) (the former are preferentially incorporated on Ga sites and the latter on As sites. Thus the DV is consistent with the observation of the annihilation of the deep trap by doping with Si, Se, S or Te. (b) It was observed that the concentration of the EL2 trap was greater in crystals doped with Se+Zn than in crystals doped with the same concentration (N) of Se. If Zn and Se prefer to occupy adjacent sites, their presence can eliminate only ~N/2 (DV) defects instead of N defects. (c) It was observed that oxygen can induce indirectly this trap or enhance its concentration. Since oxygen suppresses the Si contamination from the quartz container, it may in turn suppress the annihilation of (DV) defects. On the basis of the above arguments the (DV) assignment seems to be consistent with the observations. Thus this makes the study of the electronic states of (DV) defects most desirable. An alternative assignment of EL2 was given by Lagowski et al. [7] as the isolated As_{Ga} defects. However, this assignment is inconsistent with previous theoretical gap state results for As_{Ga} defects [4.8].

The results of the present calculations for (DV) defect in GaAs give two e states at 0.13 eV and 0.82 eV and a a_1 states at 0.48 eV and 1.17 eV. (Throughout this work, the valence band edges are set at zero energy.) The lowest e state is filled for the neutral defect. For the charged DV defect, which is a deep donor, the upper e state becomes an electron trap at 0.82 eV (or 0.7 eV below the conduction band edge). This level is very close to the observed EL2 level. The (DV) defect in GaP produces two e states at 0.26 eV and 1.17 eV, and two a_1 states at 0.49 eV and 1.52 eV. They all originate from the T_2 gap states of the isolated anion vacancy and cation vacancy. Other features of the defect states away from the energy gap can be found in Fig. 1 which displays ΔLDOS at different sites near the defect.

4. (A_c-IV) DEFECT

The deep levels in the gap obtained in this work are 1.4 eV for As_{Ga}-Ge in GaAs and 1.35 eV, 1.87 eV for P_{Ga}-C and P_{Ga}-Si, respectively, in GaP. All of them are non-degenerate a_1 states and are localized mainly on the antisite. They are related closly to the a_1 gap states at 1.7 eV and 1.25 eV for isolated P_{Ga} in GaP and As_{Ga} in GaAs, respectively. From these results an apparent chemical trend exists among the a_1 gap state energies when the chemical bonding between A_c and its neighbor impurity changes. When the binding between the group IV atom and the A_c is weaker, the a_1 state is located higher in energy.

In Fig. 2 and Fig. 3 the ΔLDOS are shown at different sites for the A_c-IV complexes in GaAs and GaP. A comparison of these figures with the corresponding ΔLDOS for an isolated A_c defect [4] reveals several interesting features: (a) New structures appear below the valence band edges at the group IV impurity site (Fig. 2b, Fig. 3b and Fig. 3d) and extend slightly onto the A_c sites (Fig. 2a, Fig. 3a and Fig. 3c). (b) The As_{Ga}-Ge complex at the GaAs-AℓAs (100)

interface is also examined (Fig. 2a, dashed curve). The presence of an interface with another semiconductor is found to have only a modest effect on the
ΔLDOS. It simply moves the gap state to slightly higher energy. (c) Fig. 3b
and Fig. 3d display slightly greater delocalization of the s-like electron as
compared with isolated P_{Ga}, which is consistent with the hyperfine constant
measurement in Ref. 2. In addition, the calculated delocalization is greater
in the case of P_{Ga}-Si than in the case of P_{Ga}-C.

Because C and Si are both common contaminants in GaP, and because $(P_{Ga}$-IV)
is a neutral single donor, $(P_{Ga}$-IV) complexes are considered to be the most
likely defects responsible for the EPR spectrum observed in electron-irradiated
n-type GaP samples [2]. By using a realistic model interaction, one can now
determine and analyze the $(A_C$-IV) defect induced electronic states. The information obtained from this calculation may help in future identification of the
defect complexes observed experimentally.

REFERENCES

1. T. A. Kennedy and N. D. Wilsey, Phys. Rev. B23, 6585 (1981).

2. T. A. Kennedy and N. D. Wilsey, Inst. Phys. Conf. Ser. 46, 375 (1979).

3. R. Haydock, V. Heine and M. J. Kelly, Solid State Physics 35, ed.
   H. Ehrenreich, F. Seitz and D. Turnbull (Academic, New York, 1980).

4. P. J. Lin-Chung and T. L. Reinecke, Phys. Rev. (in press) and Solid State
   Commun. 40, 285 (1981).

5. M. S. Daw and D. L. Smith, Phys. Rev. B20, 5150 (1979); and private communication; G. C. Osbourn and D. L. Smith, Phys. Rev. B19, 2124 (1979).

6. Yuan Li and P. J. Lin-Chung, Bull. Am. Phys. Soc. 27, 494 (1982), and to
   be published.

7. J. Lagowski, H. C. Gatos, J. M. Parsey, K. Wada, M. Kaminska and W. Walu-
   kiewicz, Appl. Phys. Lett. 40, 342 (1982).

8. M. Jaros, J. Phys. C 11, L213 (1978); J. Bernholc, N. O. Lipari,
   S. T. Pantelides and M. Scheffler, Inst. Phys. Conf. Ser. 59, 1 (1980).

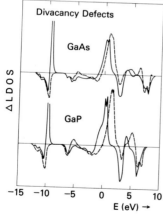

Fig. 1. The change of local density of states (ΔLDOS) due to the
presence of divacancy defects in
GaAs and GaP. The solid curves
and the dashed curves are for
ΔLDOS at the anion site and
cation site next to the defect,
respectively.

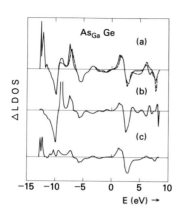

Fig. 2. The ΔLDOS due to the presence of an As_{Ga}-Ge pair defect in GaAs (a) at the As_{Ga} site of the complex in bulk solid (solid curve) and in GaAs-AℓAs interface (dashed curve), (b) at Ge site, (c) at As site next to the As_{Ga} defect.

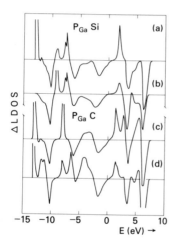

Fig. 3. The ΔLDOS due to the presence of P_{Ga}-IV defect complexes in GaP (a) at P_{Ga} site of P_{Ga}-Si defect, (b) at Si site of P_{Ga}-Si defect, (c) at P_{Ga} site of P_{Ga}-C defect, (d) at C site of P_{Ga}-C defect.

This work was supported in part by an ONR contract.

ANALYSIS OF DEFECTS IN HEAVILY–DOPED MBE–GaAs

C.B. CARTER, D.M. DESIMONE, H.T. GRIEM AND C.E.C. WOOD
Department of Materials Science and Engineering, Bard Hall and Department
of Electrical Engineering and NRRFSS, Phillips Hall, Cornell University,
Ithaca, New York 14853

ABSTRACT

GaAs has been grown by molecular-beam epitaxy (MBE) with
large concentrations ($\sim 10^{18} cm^{-2}$) of Sn, Si, Ge, and Mn as
dopants. The heavily-doped n-type material has been found
to contain regions of a very high dislocation density. An
analysis of the less complex defect areas shows that the
dislocations originate in the MBE-grown layer. These observations and others on more complex defect clusters are compared
with recent studies of defects in material grown by liquid
phase epitaxy (LPE). The more heavily doped p-type material
contains discs of Mn-rich material at the surface of the MBE-
grown epilayer. Both the structure and composition of these
regions have been examined.

INTRODUCTION

One of the most attractive capabilities of MBE for the growth of III–V compounds is that it allows layers with very well-defined thickness to be grown
and, in principle, allows the composition to be accurately varied [1]. Thus
layers of different composition and doping levels can be alternately grown in
a continuous process i.e. simply by varying the flux of individual elements to
the growth surface. A difficulty encountered using not only this technique,
but also other III–V compound growth processes is that it is often difficult
to obtain very heavily-doped n-type or p-type material. Si, Ge, and Sn can be
used successfully in the $10^{18} cm^{-3}$ range to give n-type material, but doping
levels in excess of this leads, for example, in the case of Sn to the formation
of inhomogeneous layers; specifically Sn forms Sn-rich phases on the surface
[2]. Possible candidates for p-type dopants are Mn, Mg, and Be. Mn has long
been known to have an apparent usuable limit of $\sim 10^{18} cm^{-3}$; Mg has a somewhat
higher value, but both show clear signs of surface degradation at such levels.
Be has thus become the most common p-type dopant, but it would be most desirable
to find an alternative in view of the serious toxicity problem associated with
Be.

The present study is, therefore, concerned with understanding the factors
which limit the incorporation of Mn, Si, Ge, and Sn with the object of extending
the range of dopant levels which may be used. Since the techniques often used
to assess the physical characteristics of a grown layer are optical microscopy
and Reflection High Energy Electron Diffraction (RHEED) (the latter being used
during growth), attention will be focused on the final surface region. Growth
defects can arise at the initial substrate surface and propagate through the
MBE-grown layer, they may develop at the growing surface and then propagate or
induce further defects, or they may be only observable at the final surface as
found in the build-up of Sn during MBE growth. In order to correctly interpret
such observations, careful attention must be given to the handling of the final
specimen surface.

Mat. Res. Soc. Symp. Proc. Vol. 14 (1983) ©Elsevier Science Publishing Co., Inc.

EXPERIMENTAL

The materials examined in this study were grown using a Varian MBE 360 system. The molecular beam sources were Knudson effusion cells in resistance wound Ta shielded furnaces and the sources were held in graphite or pyrolytic boron nitride (PBN) crucibles. The PBN crucible was necessary for the more reactive Mn in order to reduce incorporation of oxygen and carbon in the layers [3].

Samples were prepared for examination by transmission electron microscopy (TEM) by etching with a bromine-methanol solution in the usual way. Jet-polishing was carried out from one side only so that the final growth surface could be examined. The structure of defects was studied using a Siemens 102TEM operating at 125kV while the composition was investigated using energy dispersive spectroscopy (EDS) in a JEM200CX Jemscan operating at 200kV and equipped with a Tracor Northern 70° take-off angle detector.

OBSERVATIONS AND DISCUSSION

The Mn-doped layer had a thickness of 1μm and was grown on a 0.5μm thick, nominally undoped, buffer layer at a substrate temperature of 580°C with a growth rate of 1.5μm/hr. At Mn-dopant levels of $\sim5\times10^{18}$cm^{-3}, the surface is observed to be seriously degraded [3]; the reason for this appearance can be understood from fig. 1: particles formed on the surface are predominantly disc-shaped and similar in appearance to the second-phase particles reported by Harris et al. for Sn-doped GaAs [2]. The particles in the Mn-doped sample ranged in diameter from ~0.1μm to 1μm. The important questions which arise concern the composition and structure of the discs and their location relative

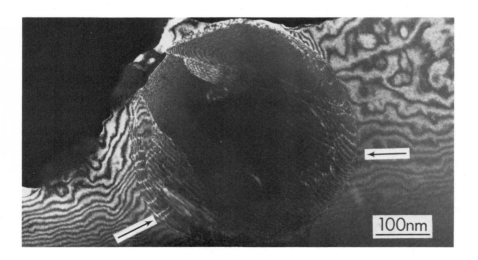

Fig. 1. Particle of Ga$_x$Mn in the surface of Mn-doped GaAs; note the dislocations (arrowed) in the inclined interface.

to the GaAs surface. Part of the disc has been removed by the chemical polish together with the GaAs matrix. Dislocations are present along an inclined interface between the GaAs and the disc. This observation together with several other similar observations demonstrates that the disc is actually situated in the predominantly GaAs material unlike the model proposed for Sn-rich phases on Sn-doped GaAs [2].

Electron diffraction patterns taken from two discs with the GaAs matrix oriented so that the electron beam is nearly parallel to the 001 foil normal are shown in fig. 2. Each pattern shows a square superlattice of reflections, but a close inspection reveals that the two superlattices do not correspond to the same d-spacings. The superlattice in fig. 2a indicates a topotactic relationship between the GaAs matrix and the enclosed disc while that in fig. 2b does not. Such observations are of considerable interest since they suggest that the particles are not the same material.

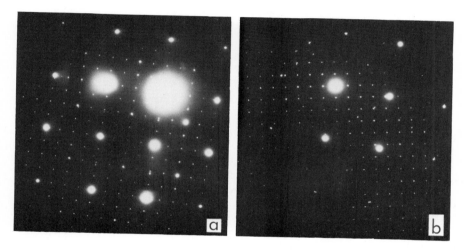

Fig. 2. Diffraction patterns close to the GaAs [001] pole showing two different square arrays of superlattice reflections.

EDS observations on regions of particles which did not overlap the GaAs matrix confirm the interpretation of the diffraction patterns i.e. the several compositions can occur and spectra from two such particles are illustrated in fig. 3. In the two examples shown, there is no detectable As present. In all such cases the particle is a Ga-rich Mn-Ga compound. It is possible that light element, namely C and O, are also present; this possibility will be examined in further work using electron energy loss spectroscopy (EELS). Thus while at first sight the surface particles appear similar to those found in Sn-doped GaAs [2], the particles are actually very different since Harris et al. found either 'pure' Sn or Sn-As phases.

The defect structure in the heavily-doped n-type material examined in this study was very different from that discussed above for Mn-doped GaAs. The most common type of defect is shown in fig. 4 and consists of very heavily strained region with a localized high density of dislocations. In other regions the density of dislocations was lower, but still highly localized. These types of

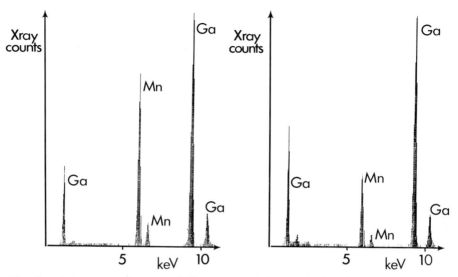

Fig. 3. EDS spectra from two different particles on Mn-doped GaAs.

defects are very similar to those observed in Si- and Ge-doped $Ga_{1-x}Al_xAs$
(x=0.1) epitaxial layers grown by liquid phase epitaxy (LPE) [4]. Kotani et al.
interpreted such defects (see their fig. 2) as arising from mechanical damage
due to mishandling the specimen. Wagner observed similar defect clusters (see
in particular his fig. 3b) in Ge-doped $Ga_xAl_{1-x}As$ (x=0.6) which had also been
grown by LPE, but subsequently annealed for 15 hr. at 820°C [5]. In that study,
the defect clusters were proposed to originate at substantial stress centers

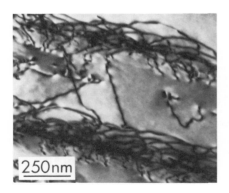

Fig. 4. Heavily-strained regions in
Si-doped GaAs.

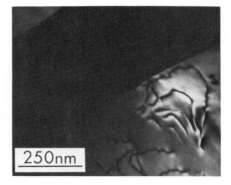

Fig. 5. Bowed dislocations and
associated surface ridges at the
surface of Sn-doped GaAs.

which it was suggested might be associated with GeAs or GeAs$_2$. The occurrence of such defects is not limited to GaAs-based materials having also been observed in S-doped InGaAsP [6] and Sn-doped InP [7]; in both of these studies the defect clusters were attributed to localized stress centers associated with inclusions.

Conclusive evidence that these defect clusters are not necessarily a result of surface damage is given in fig. 5. The final surface of this Sn-doped GaAs layer is not uniform, but consists of troughs separated by ridges. The origin of these features has not yet been ascertained although there are clear similarities to effects known to be due to thermal etching [8]. The feature of particular relevance to this study is the dislocation cluster present in the trough; such positioning of the defect cluster was common and clearly not explicable by the mishandling argument.

CONCLUSIONS

Examples of defects in GaAs layers grown with a high level of Mn, Si, Ge, and Sn dopant have been presented to illustrate the techniques being used to understand the crystallography, chemistry and growth mechanism of these defects. The Mn-doped material shows surface degradation at high dopant levels due to the surface segregation of Mn. The Mn is present at the surface in a number of different Ga-Mn compounds. This result is in sharp contrast to that previously reported for Sn-doped GaAs where either Sn or an Sn-As compound is formed at the layer surface.

It has been found that defect clusters due to a highly localized stress concentration occur in the heavily-doped n-type MBE-grown material and are very similar in appearance to the defect clusters observed in material grown by LPE which were attributed to a second phase.

ACKNOWLEDGMENTS

This research has been supported in part by the NSF through the Materials Science Center at Cornell University (C.B.C. and H.T.G.), by the U.S. Army Research Office (C.B.C.), and by the Air Force Office of Scientific Research under Cornell's Joint Services Electronics Program (D.M.D. and C.E.C.W.). The authors also acknowledge the planning committee of NRRFSS for the use of the MBE facility.

REFERENCES

1. C.E.C. Wood, Phys. Thin Films 11, 35 (1980).

2. J. J. Harris, B. A. Joyce, J. P. Gowers, and J. H. Neave, Appl. Phys. A28, 63 (1982).

3. D. M. DeSimone, Ph.D. Thesis, Cornell University (1981).

4. T. Kotani, O. Ueda, K. Akita, Y. Nishitani, T. Kusunoki, and O. Ryuzan, J. Crystal Growth 38, 85 (1977).

5. W. R. Wagner, J. Electrochem. Soc. 128, 2641 (1981).

6. S. Mahajan and A. K. Chin, J. Crystal Growth 54, 138 (1981).

7. P. D. Augustus and D. J. Stirland, J. Electrochem. Soc. 129, 614 (1982).

8. W. Y. Lum and A. R. Clawson, J. Appl. Phys. 50, 5296 (1979).

IMPROVED MELTBACK PROCEDURES FOR LIQUID-PHASE-EPITAXIAL GROWTH OF PLANAR AND BURIED HETEROSTRUCTURES

B. H. CHIN, A. K. CHIN, M. A. DIGIUSEPPE, J. A. LOURENCO, AND I. CAMLIBEL

Bell Laboratories, Murray Hill, New Jersey

ABSTRACT

In the liquid phase epitaxy on indium phosphide, the substrate, just prior to the growth of the first epitaxial layer, is commonly etched back with an indium melt to remove any thermally-degraded surface and to ensure uniform and consistent wetting. This procedure, however, often produces defects which degrade both planar and buried heterostructure devices. For planar edge-emitters and lasers, the resulting rippled surface morphology degrades device performance by scattering light. For buried heterostructures, the meltback in the regrowth step leads to indium-rich inclusions. The effects of meltback on material quality are presented, and a new multiple meltback procedure which maintains flat surface morphology and eliminates inclusions is described.

INTRODUCTION

InP/InGaAsP double heterostructure devices emitting at a wavelength of 1.3 μm have become important components in lightwave communication systems [1]; these heterostructures are principally grown by liquid phase epitaxy (LPE) [2]. Just prior to epitaxial growth, the substrate is commonly etched with an indium melt [3] to remove any thermally-degraded surface and to ensure uniform and consistent wetting. In this paper, we discuss the defects introduced by conventional meltback techniques during the growth of both planar and buried heterostructures. Novel multiple meltback procedures which minimize or eliminate these defects will be presented.

EFFECT OF CONVENTIONAL MELTBACK ON PLANAR HETEROSTRUCTURES

In LPE practice, a clean, undegraded surface with uniform and consistent wetting is commonly produced by melting back the substrate with pure indium just prior to the first epitaxial growth [3]. A pure indium melt, however, attacks the substrate rapidly and unevenly; the resulting surface of the substrate is not flat but rippled. We have found that surface rippling for a $<100>$ - InP substrate increases with the depth of meltback and also increases with the size of the substrate. In addition to making subsequent device processing easier, a flat surface is also critical for the performance of many devices. As an example, we consider double heterostructure edge emitters and lasers; the typical structure consists of an InGaAsP active layer ($\sim$0.2 μm thick)

Mat. Res. Soc. Symp. Proc. Vol. 14 (1983) © Elsevier Science Publishing Co., Inc.

sandwiched between an n-InP buffer and a p-InP confining layer. In the ideal (flat surface) case, the emitted light is transmitted along a straight waveguide. If the substrate initially is not flat, however, the subsequent epitaxial layers are generally not flat. In this case, the waveguide is not straight; and the surface ripples strongly scatter the light.

Addition of phosphorus to indium has been used by a number of workers [4] for a more controlled meltback. In our experience, we have found that a nearly saturated In/P melt produces flat surfaces; however, non-uniform wetting often occurs. We will now present a new, *double* meltback procedure which maintains flat surface morphology and yields uniform and consistent wetting.

DOUBLE MELTBACK PROCEDURE

Ideally, meltback should accomplish *two* things: (1) remove all gross contaminants (dust, etching residue, and oxides) from the substrate surface (2) planar-etch away the thermally-degraded surface region. Once we have reduced the problem in this manner, an improved meltback results if we use a two-step procedure with each melt optimized for each objective. The new double meltback procedure is carried out in a horizontal slider boat as follows. Initially the substrate sits under an InP cover piece; the first two wells contain a pure indium and a slightly undersaturated In/P melt, respectively, for the double meltback. The substrate is first passed rapidly through the pure indium melt to remove the gross contaminants; it is important for the residence time in the first melt to be kept brief ($\sim$1s) to minimize attack on the substrate. The substrate is then brought under a nearly saturated In/P solution to controllably and uniformly planar-etch away the degraded surface. The final result is a clean, flat, undegraded surface with uniform and consistent wetting for the first epi-growth. In our experience, without the first indium melt, the nearly saturated solution does not etch the surface uniformly and consistent wetting does not occur; presumably, the nearly saturated solution does not effectively remove gross contaminants.

The improvement in surface flatness produced by the double meltback procedure is quite pronounced on both a macroscopic and a microscopic scale. Fig. 1 compares two $<100>$ - InP substrates after a shallow ($\sim$5 μm) layer has been melted away from each. In Fig. 1a, we note the prominent ripples characteristic of a surface after a pure indium meltback. In contrast, the ripples are largely absent from the surface after double meltback (Fig. 1b). On a microscopic scale of more direct consequence for device performance, the improvement in surface morphology is equally dramatic. In Fig. 2, we show stylus profiles of three surfaces. On the vertical axis here, 1 div = 0.2 μm, a typical active layer thickness; the total lateral span is 160 μm, a typical device length. Plot (a) shows the smooth, as-polished surface for reference. For a wafer which has had $\sim$5 μm removed by single meltback (Plot (b)), the surface roughness is on the same order as an active layer thickness. On the other hand, a wafer which has had $\sim$5 μm removed by double meltback (Plot (c)) is considerably smoother, with peak-to-peak undulations $< \sim$.02 μm.

APPLICATIONS OF MULTIPLE MELTBACKS TO BURIED HETEROSTRUCTURES

An extension of the double meltback procedure also improves the LPE growth of buried heterostructures. Buried heterostructures are commonly fabricated by a two-step LPE process; however, the regrowth step has been problematical [5,6]. As an example, in Fig. 3, we schematically show the growth of a buried heterostructure surface emitter. First (Fig. 3a), a planar three-layer structure, consisting of a n-InP buffer ($\sim$4 μm thick), an InGaAsP active ($\sim$1 μm thick), and a p-InP confining ($\sim$6 μm thick) layer, is grown on a <100> n-InP substrate. Next (Fig. 3b), mesas are formed in the p-InP confining layer by photolithography and chemical etching. With different etchants, we have formed either circular mesas or mesas with <100> faceted side walls (lateral dimension $\sim$40–250 μm). The wafer with the etched mesas is then subjected to a second LPE step. After a meltback to etch the surface and to remove the portions of the active layer not covered by mesas (Fig. 3c), a p-InP burying layer and a p-InGaAsP contact layer are grown (Fig. 3d).

When pure indium is used for meltback (Fig. 3c), subsequent analyses by scanning electron microscopy (SEM) and X-ray microanalysis reveal the presence of indium-rich inclusions clinging to the mesa walls. When a burying layer is grown over such inclusions, the portion of the layer around the periphery of a mesa has an irregular morphology and a high dislocation density.

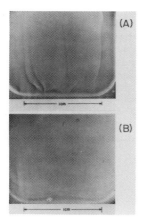

Fig. 1 Substrates after removal of $\sim$5μm by meltback. (a) After single meltback. Note characteristic ripples. (b) After double meltback. Surface is flat.

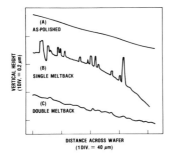

Fig. 2 Stylus profiles of substrate surfaces. (a) As-polished wafer. (b) Wafer after single meltback. (c) Wafer after double meltback.

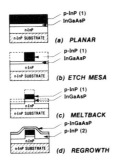

Fig. 3   Growth of buried heterostructure light-emitting diode by two-step LPE.

Elimination of the inclusions requires complete wipeoff after meltback. We should note that achieving clean wipeoff is more difficult on a wafer with etched mesas than on a planar, polished substrate: (1) the additional processing steps needed to form the mesas probably lead to greater surface contamination and (2) the side walls of the mesas act as traps for contaminants. Any contaminants which are not fully dissolved during meltback may settle back onto the mesas and prevent complete wipeoff. In principle, then, multiple meltbacks which more thoroughly remove contaminants from the wafer should lead to better wipeoff. We have, in fact, found that the double meltback procedure described above reduces, but does not eliminate, the inclusions.

Complete elimination of the inclusions has been achieved by the following *triple* meltback: the wafer is first rapidly passed (without stopping) under two successive pure indium melts and is then brought under a slightly undersaturated In/P melt. Subsequent analyses by optical microscopy, SEM, and cathodoluminescence have verified that buried heterostructures grown after triple meltback are free of inclusions and associated regions of irregular growth.

CONCLUSIONS

A multiple meltback procedure consisting of a brief etch in one or two pure indium melts followed by a controlled etch in a slightly undersaturated In/P melt has been shown to be superior to conventional meltback. For planar heterostructures, a double meltback yields uniform and consistent wetting and maintains flat surface morphology. In the growth of buried heterostructures, use of a triple meltback in the regrowth step eliminates inclusions.

ACKNOWLEDGEMENTS

We would like to thank W. A. Bonner for the InP substrates.

REFERENCES

[1] For a recent review, see, for example, H. Kressel, Radio Science **16**, 445 (1981).

[2] For a recent review, see, for example, P. A. Houston, J. Materials Science **16**, 2935 (1981).

[3] V. Wrick, G. J. Scilla, L. F. Eastman, R. L. Henry, and E. M. Swiggard, Electron Lett. **12**, 394 (1976).

[4] See, for example,

 (a) K. E. Brown, J. Crystal Growth **20**, 161 (1973).

 (b) A. G. Dentai, T. P. Lee, C. A. Burrus, and E. Buehler, Electron. Lett. **13**, 484 (1977).

 (c) E. Kuphal, J. Crystal Growth **54**, 117 (1981).

[5] O. Mikami, H. Nakagome, Y. Yamauchi, and H. Kanbe, Electron. Lett. **18**, 237 (1982).

[6] D. B. Darby and M. E. Harding, "Surface Preparation Techniques for LPE Growth with Minimal Etchback in Buried Heterojunction Laser Structures," talk presented at 10th International Symposium on GaAs and Related Compounds (Albuquerque, NM, Sept. 19-22, 1982).

PROCESS-INDUCED DEFECTS IN HIGH-PURITY GaAs

P. M. Campbell, O. Aina* , B. J. Baliga & R. Ehle
General Electric Research & Development Center
Schenectady, New York

ABSTRACT

High temperature annealing of Si_3N_4 and SiO_2 capped high purity LPE GaAs is shown to result in a reduction in the surface carrier concentration by about an order of magnitude. Au Schottky contacts made on the annealed samples were found to have severely degraded breakdown characteristics. Using deep level transient spectroscopy, deep levels at $E_c-.58eV$, $E_c-.785eV$ were detected in the SiO_2 capped samples and $E_c-.62eV$, $E_c-.728eV$ in the Si_3N_4 capped samples while none was detected in the unannealed samples. The electrical degradations are explained in terms of compensation mechanisms and depletion layer recombination-generation currents due to the deep levels.

INTRODUCTION

High temperature diffusion and ion implantation annealing pose special problems in GaAs device fabrication because GaAs decomposes at these high temperatures. GaAs decomposition results in the formation of high concentrations of Ga or As vacancies and thus can enhance impurity redistribution or result in the formation of physical defects. Impurity redistribution in Cr-doped GaAs has been shown[1] to result in an increase in surface carrier concentration while the high temperature annealing of undoped semi-insulating GaAs has been shown to also lead to surface resistivity reduction and type conversion[2-5]. This has been attributed to the outdiffusion of deep levels associated with the semi-insulating properties of the material. Most of the above results were obtained, however, using undoped semi-insulating GaAs with relatively high concentrations of defects already in the material before annealing. They are, therefore, indicative only of the redistribution properties of these defects. A study of the high temperature annealing of high purity GaAs, on the other hand, should allow an understanding of the introduction and formation of process-induced defects.

We present here a study of process-induced defects in high purity GaAs layers grown by liquid phase expitaxy (LPE). Our results show surface carrier concentration reduction by as much as a factor of five after annealing Si_3N_4 and SiO_2 capped LPE GaAs layers at 800^oC. In addition, the breakdown voltage of Schottky contacts made on annealed samples was less than 100 volts compared with 150 volts for unannealed material. Deep level transient spectroscopy measurements (DLTS) showed no deep levels for the unannealed samples while several deep levels were found in the

*Now at Bendix Advanced Technology Center
   Columbia, Maryland 21045

Mat. Res. Soc. Symp. Proc. Vol. 14 (1983) Published by Elsevier Science Publishing Co., Inc.

284

annealed samples. Carrier concentration profiles, breakdown
voltage distributions and DLTS data will be presented to show
these electrical degradations and their possible origin.

EXPERIMENTAL PROCEDURE

The high purity n-type GaAs layer was grown by liquid phase
epitaxy on n+ Si-doped substrates (n ∼ 10^{18} cm^{-3}). The carrier
concentration of the LPE layer was about 3 x 10^{15} cm^{-3} as Figure 1a.
shows. Different samples were covered with 900Å of Si$_3$N$_4$ or 4000Å
of SiO$_2$ and phosphosilicate glass (PSG), and were annealed in a
furnace in N$_2$ ambient at 800°C and 900°C for 30 minutes. The
Si$_3$N$_4$ was obtained by RF plasma deposition at 400°C using a TEGAL
Si$_3$N$_4$ reactor while the SiO$_2$ and PSG were deposited by the
reaction of silane and O$_2$. After annealing, the caps were
removed with buffered HF and carrier concentration profiles were
obtained using a Polaron Instruments electrochemical profiler.
After profiling, Au dots were deposited on the same samples to
form Schottky barrier diodes on areas not etched by the Polaron
profiling. Breakdown voltage and DLTS measurements were made using
these Au Schottky diodes. The DLTS set-up is of the same type
used by D. V. Lang[6] which included a 50MHz capacitance bridge for
transient capacitance measurements, and a PAR boxcar averager for
the signal averaging.

RESULTS

The samples generally looked shiny and undamaged after
annealing for 30 minutes at 800°C and 900°C for all capped
samples, except for the PSG-capped samples which sometimes showed
some physical damage after annealing at 900°C.
Figure 1a. shows that the carrier concentration for an
unannealed sample is relatively constant within the LPE layer.
After annealing with Si$_3$N$_4$ cap at 800°C for 30 minutes
(Figure 1b.), the near-surface carrier concentration was reduced
by almost an order of magnitude to below 4 x 10^{14} cm^{-3}. A similar
but less severe type of surface carrier concentration reduction
was found for the SiO$_2$ covered samples as Figure 1c. shows.
Annealing with PSG caps at 800°C, however, resulted in a higher
carrier concentration at the surface than in the bulk
(Figure 1d.). Reliable carrier concentration measurements could
not be made on any samples annealed at 900°C because the Schottky
contacts formed by either metal deposition or electrolytic contact
(using the Polaron Plotter) resulted in leakage currents that were
too high to make meaningful measurements. It should be noted that
samples annealed with Si$_3$N$_4$ deposited under conditions that may
have resulted in different compositions, showed little or no
decrease in surface carrier concentration evident in Figure 1b.
Data on these results will be presented in a future publication.

Figure 2a. shows the breakdown voltage distribution for Au
Schottky dots on unannealed samples with a high population of high
voltage Schottky diodes. Schottky diodes made on samples annealed
at 800°C and 900°C were generally severely degraded, showing soft
and leaky breakdown characteristics. Figure 2b. shows the breakdown
distribution for SiO$_2$ capped samples annealed at 800°C. All of the
diodes now have breakdown voltages less than 100 volts.

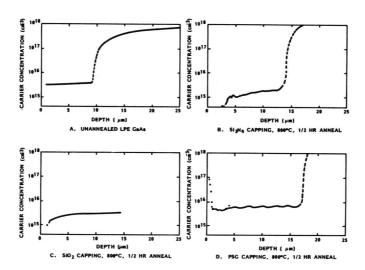

Figure 1. Carrier concentration profiles of annealed and unannealed samples.

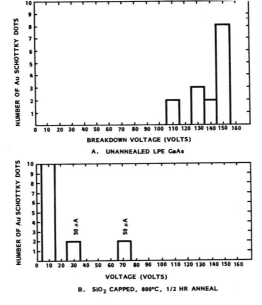

Figure 2. Breakdown voltage distribution of annealed and unannealed samples.

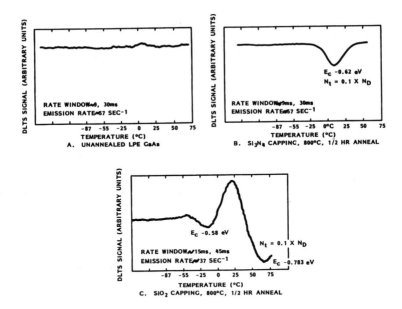

Figure 3. DLTS spectra of annealed and unannealed samples.

The DLTS spectrum of an unannealed high purity LPE sample with no observable deep level peak is shown in Figure 3a. This indicates that there are no deep levels (i.e. defects and impurities) with concentration greater than one-thousandth of the sample carrier concentration (i.e. $N_t < 10^{12}$ cm^{-3}). Figure 3b. shows the DLTS spectrum of a Si_3N_4 capped sample annealed at 800°C. A deep level at $E_c-.62eV$ was detected in the bulk when the DLTS bias was 6 volts and the pulse height was 2 volts. When the pulse height was 5.5 volts, another level at $E_c-.728eV$ was measured (not shown in Figure 3b.). This indicates that there is a deep level ($E_c-.728eV$) close to the surface whose concentration in the bulk is low. Two deep levels at $E_c-.58eV$ and $E_c-.785eV$ were found in the 800°C annealed SiO_2 capped samples both in the bulk and at the surface (Figure 3c.). The concentration of the level at $E_c-.58eV$ was estimated to be about one hundredth of the carrier concentration (i.e., $N_t \sim 0.014N_D$). For the other levels however, the concentration, N_t was $0.1N_D$. The leakage current of the PSG-capped samples and all 900°C annealed samples were too high for DLTS measurements to be made.

DISCUSSION

The surface carrier concentration reduction after high temperature annealing cannot be due solely to compensation by shallow acceptors since one would expect a compensated material with low carrier concentration to also have a higher breakdown voltage than an uncompensated material. The breakdown voltage degradation and high leakage currents are however more typical of the effect of deep-lying recombination centers (such as Au) in silicon p-n junctions[7]. Such recombination centers not only reduce the carrier concentration (at a sufficiently high deep level concentration), but also give rise to considerable charge generation within the depletion layer. This leads to high leakage currents and thus lower breakdown voltages.

The deep levels detected in the annealed samples may be due to defects or defect-impurity complexes which are acceptor recombination centers. Such levels could be responsible for both the carrier concentration reduction and the breakdown voltage degradation. On the other hand, it is quite possible that some of these levels are acceptors (in addition to possible shallow acceptors) and some are deep donors. For example, the level at E_C-.785eV may be due to oxygen which has been associated with a deep donor[5] variously measured at E_C-.75eV,[8] E_C-.79eV and E_C-.83eV[5] in GaAs. Oxygen contamination in our samples could come for example, from the in-diffusion of free oxygen from the SiO_2 during the annealing. In this case, the carrier concentration reduction could be caused by the deep acceptors (or possible shallow acceptors) while the breakdown voltage degradation may be due to deep donor recombination centers. It is not possible from our data to determine which of the two explanations is appropriate.

SUMMARY

It has been shown that the high temperature annealing of Si_3N_4 and SiO_2 capped high purity GaAs at 800°C resulted in severe degradation in the surface carrier concentration and in the breakdown voltage. Deep levels at E_C-.58eV, E_C-.785eV were detected in the SiO_2 capped samples and E_C-.62eV, E_C-.728eV in the Si_3N_4 capped samples while none was found in the unannealed samples. These deep levels may be responsible for the reductions in surface carrier concentration and in the breakdown voltage.

ACKNOWLEDGEMENT

The authors would like to thank Bea Hatch for her help in sample preparation , Patty Mowers for SiO_2 and PSG deposition and Lorraine Walker for manuscript preparation.

REFERENCES

1. A.M. Huber et al. Appl. Phys. vol. 34, p. 858 (1979).
2. B. Hughes & C. Li, International Symposium on GaAs and Related Compounds, Albuquerque, New Mexico, Sept. 19 - 22, 1982
3. S. J. Pearton et al. Electron. Lett. Vol. 18, p. 715 (1982).
4. L. B. Ta, A. M. Hobgood, A. Rohatgi, R. N. Thomas, J. Appl. Phys. Vol. 53, p. 5771 (1982).
5. S. Makram-Ebeid, D. Gautard, P. Devilland, G. M. Martin, Appl. Phys. Lett. Vol. 40, p. 161 (1982).
6. D. V. Lang, J. Appl. Phys. Vol. 45, p. 3023 (1974).
7. S. K. Ghandhi, "The Theory and Practice of Microelectronics". John Wiley & Sons, Inc. New York, 1968.
8. D. V. Lang et al., J. of Electronic Materials, Vol. 4, #5 (1975).
9. M. G. Emerson et al. Electron. Lett. Vol. 15, p. 553 (1979).

ENHANCED INP SUBSTRATE PROTECTION FOR LPE GROWTH OF INGAASP DH LASERS

P. Besomi, R. B. Wilson, W. R. Wagner and R. J. Nelson
Bell Laboratories, Murray Hill, New Jersey 07974

ABSTRACT:

Thermal degradation of InP single crystal substrates prior to LPE growth has been virtually eliminated by using an improved protection technique. The phosphorus partial pressure provided by a Sn-In-P solution localized inside an external chamber surrounding the InP substrate prior to growth prevents thermal damage to the surface. Nomarski contrast photomicrographs, photoluminescence and X-ray diffractometric measurements indicate that InP substrates protected by this method suffer   negligible deterioration, in contrast to the results of the more commonly used cover wafer method.

INTRODUCTION:

Due to the low absorption coefficient of optical fibers in the 1.3-1.6 μm wavelength region, $In_{1-x}Ga_xAs_yP_{1-y}$ alloys have become one the most promising system for lightwave communications. To date, liquid phase epitaxy (LPE) has been the predominant method used to fabricate such devices, and this growth technique has been often described in the literature (1), (2). Among the possible causes of irreproducibility in LPE growth, thermal decomposition of InP substrates is of prime importance. The loss of phosphorus from the substrates at high temperatures results in an indium rich region which creates a perturbed epitaxial layer.

The first method proposed to remove the damaged substrate surface was an in situ etching step, or melt-back technique (3). This method, however, leads to nonplanar surface morphology and cannot be used when micron size features, such as etched mesas or channels, are present on the wafer. Other techniques have been proposed to prevent thermal decomposition. The use of an InP cover wafer in close proximity with the InP substrate has been shown to significantly reduce the density of etch pits resulting from thermal etching (4). Other methods were recently described to provide excess phosphorus to the InP substrate by using elemental phosphorus powder (5) or phosphine (PH_3) gas (6). Both methods, however, introduce excess phosphorus in the entire growth system. Consequently, the development of a method providing excess phosphorus localized to the immediate vicinity of the InP substrate is of interest.

Recently, Antypas (7) reported on the use of a graphite basket containing a solution of Sn-In-P for totally eliminating visible decomposition of InP substrates before and after LPE growth. In this study, we report on the protection of InP substrates by using a quartz saturation chamber containing a Sn-In-P solution. Nomarski contrast photographs, photoluminescence (PL) and X-ray diffraction measurements were performed on S doped as well as Sn doped substrates to evaluate the efficiency of this method.

Mat. Res. Soc. Symp. Proc. Vol. 14 (1983) © Elsevier Science Publishing Co., Inc.

EXPERIMENTAL:

In Figure 1, the InP substrate is seen inside a cylindrical quartz chamber, downstream from a typical graphite LPE boat. A solution of Sn-In-P was prepared by combining InP and Sn and loaded into the quartz chamber, under the slider. The ratio of InP to Sn was always in excess of the solubility of InP in pure Sn at the temperatures of interest so that the solution was saturated. The temperature cycle used was 40 min. at 720°C, followed by 1 hr 20 min. at 645°C. Newly polished InP substrates, InP substrates heat treated while protected using the quartz saturation chamber, and substrates protected using the cover wafer method, were examined with single crystal and double crystal diffractometry. The photoluminescence of the substrates after various treatments was recorded using an Argon laser for excitation.

RESULTS AND DISCUSSION:

Thermal decomposition of InP substrates begins as soon as a substrate is exposed to the high temperatures (650°C to 720°C) needed for the various melts to be fully homogenized. Since the thermal decomposition of InP is due to a loss of phosphorus, a method providing an excess phosphorus vapor pressure contained inside a small volume in the immediate vicinity of the substrate should eliminate the phosphorus loss and reduce the substrate thermal degradation during the thermal cycle. Moreover, the excess phosphorus vapor is contained inside chamber, in the immediate vicinity of the InP substrate to be protected, therefore maximizing the protection of the substrate while minimizing the contamination of the melts by excess phosphorus. This last point is of particular importance when InGaAsP or InGaAs layers are grown from slightly supersaturated indium melts (8).

It was found that S doped substrates protected by a cover wafer (see Fig. 2) exhibit a photoluminescence intensity 20% of the value for untreated substrates. The S doped substrate protected by the quartz saturation chamber, however, retained almost all its photoluminescence efficiency. Table 1 gives a summary of the integrated photoluminescence peak intensities for these various substrates. It was found that S doped and Sn doped substrates were well protected by the phosphorus saturation chamber when compared to the results of the cover wafer (5). Furthermore, the use of hydrogen instead of an inert gas ambient does not increase the thermal degradation when the saturation chamber is used, in contrast to previous result using a cover wafer (9), (10). This, again, exemplifies the substrate protection enhancement brought by a localized excess phosphorus partial pressure.

In order to determine the depth of the thermal damage suffered by the substrates protected by the quartz saturation chamber, the PL intensity was measured after progressive removal of the damaged surface by chemical etching in 10:1:1 $H_2SO_4/H_2O_2/H_2O$. It was found that a surface layer of 300 to 500Å has significantly greater damage than the bulk of the substrates in the case of Sn doped substrates. This photoluminescence reduction is not present in S doped substrates however, suggesting that different surface properties influence the photoluminescence intensities measured for S doped or Sn doped substrates. However, secondary ion mass spectroscopy (SIMS) characterization indicated that the photoluminescence losses could not be accounted for by large dopant concentration gradients near the surface.

A possible mechanism which could lower the PL intensity is related to the presence of a surface layer with high recombination velocity. For a given doping density, a drastic reduction of the PL intensity occurs when the surface recombination velocity is larger than 10^3-10^4cm/sec, the widely reported bulk value for N type InP (11). The cause of this high surface recombination velocity may possibly be due to a high concentration of nonradiative recombination centers, such as phosphorus vacancies and phosphorus vacancy complexes, as well as microprecipitates surrounding dislocations. In any event, improving the substrate protection before LPE growth is a necessary step to eliminate the degraded surface layer where nonradiative recombination centers are present.

The double crystal diffractometric measurements of the heat-treated InP substrate protected with the quartz saturation chamber or without any protection resulted in diffraction peaks of 9 and 10 arc sec half-peak width, respectively, a negligible variation. The X-ray diffractometric measurements therefore indicate that thermally treated InP substrates do not suffer any substantial lattice changes during pre-epitaxial heat treatment.

Using this new protection method $In_xGa_{1-x}As_yP_{1-y}/InP$ double heterostructure wafers have been prepared. High quality material was grown, with active layer mismatch reproducibility in the 10^{-4} to 10^{-5} range, and cw threshold currents from gain guided lasers as low as 105 mA were measured.

ACKNOWLEDGEMENTS:

The authors wish to thank S. Maynard, S. G. Napholtz and R. G. Sobers for technical assistance with the LPE growth. Many helpful discussions with J. Degani and P. D. Wright are also acknowledged.

REFERENCES

[1]   H. Kressel and H. Nelson, Physics of Thin Films, 7 (1973), 115.

[2]   G. A. Antypas and R. L. Moon, J. Electrochem. Soc., 120, (1973), 1574.

[3]   V. Wrick, G. J. Scilla, L. F. Eastman, R. L. Henry and E. M. Swiggard, Electron. Lett., 12 (1976), 394.

[4]   A. Doi, N. Chinone, K. Aiki and R. Ito, Appl. Phys. Letters, 34, (1979), 393.

[5]   M. A. DiGuiseppe, H. Temkin and W. A. Bonner, J. of Cryst. Growth, 58, 279, (1982).

[6]   A. R. Clawson, W. Y. Lum and G. W. McWilliams, J. of Cryst. Growth, 46, (1979), 300.

[7]   G. A. Antypas, Appl. Phys. Lett., 37, (1980), 64.

[8]   R. J. Nelson, Appl. Phys. Letters, 35, (1979), 654.

[9]   K. Pak and T. Nishinaga, Jap. J. Appl. Phys., 18, (1979), 1859.

[10]   J. A. Lourenco, unpublished.

[11]   H. C. Casey, Jr. and E. Buehler, Appl. Phys. Letters, 30, (1977), 247.

Table I: Average PL intensity ratio for thermally treated InP substrates.

Type of substrate	Type of protection	Temperature (°C)	PL intensity ratio (a)
Sn:InP	saturation chamber	720	0.35
Sn:InP	saturation chamber	600	0.53
S:InP	saturation chamber	720	1.00
S:InP	InP cover wafer	720	0.23
S:InP	none	720	<0.02

(a) non-thermally cycled substrate is taken as a reference.

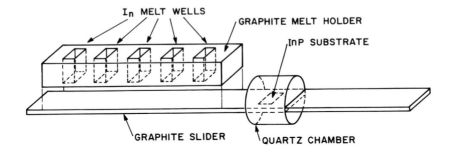

Fig. 1:   Schematic of LPE boat with external chamber containing Sn-In-P solution.

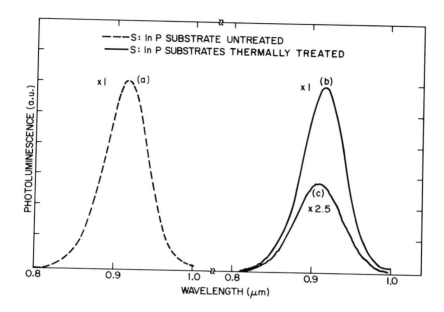

Fig. 2:   Photoluminescence spectra of InP substrates a) untreated, b) protected by Sn-InP and
c) protected by a cover wafer.

SUMMARY:

The use of a Sn-In-P solution inside a protection chamber provides an excess phosphorus partial pressure localized in the vicinity of the substrate to be protected, thus reducing thermal degradation prior to growth. Photoluminescence spectra as well as X-ray diffraction spectra of S doped and Sn doped substrates protected by the protection chamber were shown to be essentially identical to those of untreated substrates. It was found that S doped substrates remained undamaged, while Sn doped substrates still exhibited a small loss of photoluminescence efficiency.

DYNAMIC OBSERVATION OF ATOMIC-LEVEL EVENTS IN CADMIUM TELLURIDE BY HIGH RESOLUTION TEM

T. Yamashita and R. Sinclair
Department of Materials Science and Engineering, Stanford University
Stanford, California 94305   USA

ABSTRACT

A conventional 120 keV high resolution TEM equipped with a TV camera has been used to make lattice resolution video recordings of cadmium telluride in the <110> orientation. The motion of dislocations and the migration of atomic species at the edge of the crystal have been studied. These processes are thought to be induced in the crystal by localized heating of the specimen by the electron beam. The observations are made on a TV monitor at a total magnification of 15 million times.

INTRODUCTION

The recent trend towards the use of TV recording systems and bright electron sources, such as the LaB_6 filament, in a high resolution TEM (HRTEM) has added a new dimension to the application of the lattice imaging technique to the analysis of materials. For instance it makes possible the TV-rate recording of dynamic processes occuring in the bulk and on the surface of the crystal at near atomic resolution. Two major advantages of TV recording capability are the fast recording times (~30 frames per second) and the added magnification which can be obtained on the TV monitor (up to 10-20 million times). The first use of such recordings was reported by Hashimoto et al. on dislocation motion and twinning in gold [1]. We have also reported recently the recording of dislocation motion in CdTe at lattice resolution using the Cambridge 500 keV HRTEM [2].

By coincidence, CdTe is a crystal uniquely suited for studies of this kind. It has the zinc blende structure, and its lattice parameter of 6.47 A is the largest in the II-VI semiconductor group. This combined with the large atomic numbers of its constituents and relatively clean surface provides the necessary conditions for obtaining excellent high contrast lattice images in the <110> projection. CdTe crystals grown by the present technology contains a very high density of defects (> 10^{14} cm^{-3}), and numerous stacking faults of both intrinsic and extrinsic character can be observed in end-on projection in <110> orientation foils. Similar types of defects are also present in industrially more important semiconductors, but with far lower densities. The fact that the defects can be induced to move in CdTe offers a unique opportunity to study various defect reactions.

CdTe also exhibits some interesting surface phenomena under electron beam irradiation. We have observed the motion of image spots corresponding to Cd-Te atomic column pairs at the edge of the foil [3]. This was displayed only indirectly however, by means of sequential TEM micrographs taken several minutes apart. Viewed under lower magnification, the surface of CdTe appears to ablate after some period of electron beam irradiation, leading to redeposition and growth of new CdTe crystallites elsewhere on the surface. This effect was attributed to electron beam induced heating of the specimen, since CdTe has a

Mat. Res. Soc. Symp. Proc. Vol. 14 (1983) ©Elsevier Science Publishing Co., Inc.

low thermal conductivity (0.06 W/cm^{o}K at 297° K). The localized heating and the resultant thermal stresses are probably responsible for the dislocation motion as well. Other II-VI compounds such as CdS have been reported by Cockayne et al.[4] to be unstable to electron beams. In the same work, they describe the observation of dislocation motion by the weak-beam imaging method.

In this work, we describe the recent results on the observation of dislocation motion and migration of atomic species at the edge of the crystal in CdTe, obtained in a conventional 120 keV HRTEM equipped with a TV camera.

EXPERIMENTAL METHODS

The CdTe crystal used in this work was grown by a modified Bridgman method at the Center for Materials Research, Stanford University. The crystal was doped with phosphorus to approximately 10^{17} cm^{-3} carrier density for p-type conductivity. TEM samples with <110> foil orientation were prepared by a chemical jet-thinning technique with a 2% bromine-in-methanol solution. The specimens were briefly ion-milled at 4 keV to obtain a thin profile at the edge of the foil, and immediately placed into the microscope for observation. The microscope used in this work was a Philips EM400ST equipped with a conventional side-entry stage and LaB$_6$ filament, and operating at 120 keV. The point-to-point resolution of the microscope is approximately 3.5 Å. Nineteen beams were allowed through the objective aperture to form the image, but image calculations indicate that only the transmitted and the four (111) beams contribute to the image, giving rise to bright image spots which correspond to either the Cd-Te atomic column pairs or to channels through the structure [5]. The vacuum at the specimen chamber is approximately 2 x 10^{-7} torr, and the typical current density on the specimen was between 5-15 A/cm^2. The image pick-up system, manufactured by Philips Electronic Instruments, attaches directly under the microscope column. The microscope is usually operated at its maximum magnification of 900,000 times, and the TV imaging system magnifies this by another 17 times, so that final magnification on the TV monitor is approximately 15 million times. A standard 3/4-inch video tape recorder was used to record the images. The original video recordings were digitally enhanced with a Quantex model DF-80 digital video filter and a DS-50 digital video processor, and still-frame photographs were taken directly from the TV monitor during playback of the processed recordings.

OBSERVATION OF ATOMIC MIGRATION AT THE EDGE OF THE CRYSTAL

Most semiconductor materials prepared for HRTEM imaging usually have a thin film of amorphous contamination or an oxide layer which obscures the lattice image at the edge of the foil. CdTe exhibits similar behavior, but with careful specimen preparation and good microscope vacuum, some areas can be obtained which are relatively free of surface films. In these areas, a sharp lattice image can be obtained which extends to the very edge of the foil. It is also possible to remove the surface film by an intense irradiation of the sample. Apparently, the CdTe surface oxidizes very slowly [6,7], and has a low sticking coefficient (10^{-12}) for oxygen [7]. For an already oxidized surface, an XPS study indicates that oxygen can be removed from the surface by heating the crystal at 400°C in a vacuum of 10^{-6} or better [7]. Another source of contamination is carbon, but its behavior on a CdTe surface has not been studied in any detail. When clean foil edges are irradiated by an intense beam, a rapid rearrangement of the edge profile is observed, which eventually leads to formation of facets of various configurations. The (111) facet formation was found to be the most common, followed by (110) facets. During the rearrangement, rapid appearance and disappearance of individual image spots at the foil edge can be observed, with each event taking place within one TV frame (0.03 sec.). Generally, the individual appearance and disappearance of image spots seems coordinated with another event occuring nearby, giving the illusion of spot motion. Once the facets form, the process slows considerably, but local changes still continue to occur. One phenomenon which was observed frequently was the formation or removal of an entire layer of atoms on the (111) facet. This process appears to be nucleated at the corner of two facets, at a kink site or at ledges. A set of micrographs in figure 1 shows some of the changes taking place on a (111) facet over time. The presence of a surface film at the edge of the crystal always stabilizes the edge, and facet formation is then never observed, even when such a film is extremely thin (< 5 Å).

OBSERVATION OF DISLOCATION MOTION

Many of the intrinsic and extrinsic faults in the sample extend out to the edge of the foil. After a few minutes of electron beam irradiation, the motion of dislocations bounding stacking faults was sometimes observed. The slip of a Shockley partial dislocation bounding intrinsic stacking fault was the most common, and an example is shown in figure 2. This particular dislocation passed through the field of view (~200 Å) within 3 TV frames (0.1 second), which

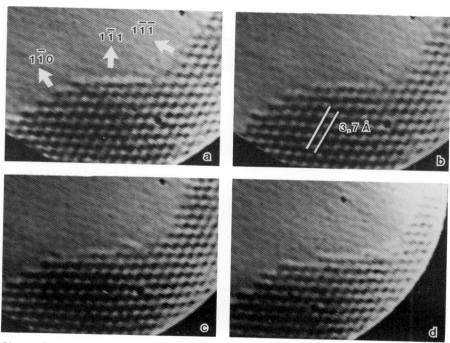

Figure 1. a through d shows the change in the profile at the edge of the foil taking place within a span of approximately 2 minutes. Crystallographic directions of the facets are indicated in 1a.

298

converts to a speed of 2 X 10⁻⁵ cm/sec. Viewed in real time, the motion appears as a blurred white spot passing from left to right as the stacking fault is annihilated.

Another class of dislocation motion is the climb of 1/3<111> Frank dislocations out of the crystal, thereby eliminating an extrinsic stacking fault in the process. The example shown in figure 3 took approximately two minutes to climb 50 Å. Dislocation motion in the interior of the crystal was recorded only on one occasion. Several defect reactions were observed such as elimination of a Cottrell lock followed by the motion of released partials away from each other. The local temperature was probably very high, and the image contrast was very poor for this observation. Generally, dislocation motion was observed in only a small fraction (< 5%) of the total defect population, and those which moved were mainly near the edge of the foil.

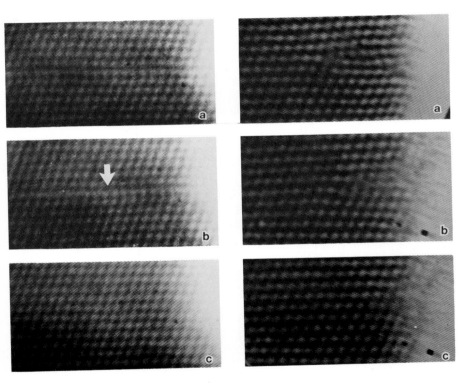

Figure 2. Sequence showing a Shockley partial dislocation (arrowed in b) passing out of the crystal, from left to right. The sequence takes place in 0.1 sec. (unprocessed).

Figure 3. a-c shows 1/3<111> Frank dislocation passing out of the foil. The process takes 2 minutes. The images have been digitally processed.

DISCUSSION

This work indicates that variety of phenomena occuring on the surface and in the bulk of the crystal can be recorded at lattice resolution in a conventional HRTEM equipped with a TV recording system. Most of the observations made in this work still lack proper explanation, and more analysis and experimentation are required. The rapid appearance and disappearance of image spots at the edge of the crystal for example, is not clearly understood. Since each image spot represents a number of atoms in projection, their appearance or disappearance implies addition or removal of entire column of atoms. The number of atoms in the column needed in order for them to be seen as an image spot is not known accurately because of the difficulty in assigning an absolute contrast scale in any image simulations. Other observations are what one might expect, such as the preferential formation of (111) facets (since it is the surface with highest atomic density and hence the most stable configuration).

The current density used in order to induce these phenomena is roughly 10 times higher than those used in typical HRTEM imaging. A simple heat-flow calculation [8] using the experimental parameters indicates however that the expected temperature increase is still considerably below the $110^{\circ}C$ required for the sublimation process to become noticeable [3]. Experimentally, the temperature rise which can be achieved is very high. At one point, the actual melting of the specimen was observed (MP $= 1092^{\circ}C$). Therefore, much of the temperature increase can be attributed to the specimen geometry and the low thermal conductivity of CdTe. To date, similar observation have been made only in CdS. Whether this technique is applicable to many other materials remains to be tested. To induce defect motion in a gold film apparently requires a very high current density of 100 A/cm-2 [1].

ACKNOWLEDGEMENT

We gratefully acknowledge support from the Basic Energy Sciences Division of the Department of Energy and from the Department of Materials Science at Stanford University. The image pick-up system was loaned by the Philips Electronic Instruments Inc. and digital image processing was performed by P. Bliven of Quantex Corporation. We thank F.A. Ponce of Hewlett-Packard Laboratories for thoughtful comments.

REFERENCES

1. H. Hashimoto et al. Jap. J. Appl. Phys. 19, L1 (1980)

2. R. Sinclair et al., Nature, 298, 127 (1982)

3. R. Sinclair, T. Yamashita and F.A. Ponce, Nature, 290, 386 (1981)

4. D.J.H. Cockayne, A. Hons and J.C.H. Spence, Phil. Mag., 42, 773 (1980)

5. T. Yamashita, F.A. Ponce, P. Pirouz and R. Sinclair, Phil. Mag., 45, 693, (1982)

6. J.G. Werthen et al., Submitted to J. Vacuum Science

7. M.H. Patterson and R.H. Williams, J. Phys D, 11, L83 (1978)

8. L.W. Hobbs, Introduction to Analytical Electron Microscopy, J.J. Hren, J.I. Goldstein and D.C. Joy ed. (Plenum Press New York 1979) pp. 437-480

ELECTRICALLY ACTIVE DEFECTS IN CID IMAGING ARRAYS FABRICATED ON $Hg_{0.7}Cd_{0.3}Te$

H.F. Schaake and A.J. Lewis
Central Research Laboratories, Texas Instruments Incorporated
P.O. Box 225936, Dallas, TX 75265

ABSTRACT

CID imaging arrays were fabricated on $Hg_{0.7}Cd_{0.3}Te$ produced by the solid state recrystallization technique. It was found that the most serious source of dark current was sub-grain boundaries. SEM studies of the microstructure revealed by etching showed that boundaries with a high denisty of dislocations were detectable sources of dark current, while those boundaries with a low density of dislocations, as well as individual dislocations were not. TEM showed that all dislocations were free of precipitates, and most were not dissociated. The sub-grain boundaries were found to arise from misorientation between dendrites which form during the solidification from the melt.

INTRODUCTION

Semiconducting alloys in the system $Hg_{1-x}Cd_xTe$ have particular usefulness in high performance infrared detector applications. In the past, the primary use of these materials has been in photoconducting devices. More recently, there has been considerable interest in their use in the fabrication of charge-coupled, photocapacitive, and photovoltaic devices, particularly one- or two-dimensional arrays. In photocapacitive devices, photo-generated minority carriers are collected in the depletion region of an array of MIS capacitors, and detected by an appropriate addressing mechanism. These devices place stringent requirements on the material used in their fabrication, as they are sensitive to dark currents generated in either the depletion region of the device, or in neighboring regions within a minority carrier diffusion length. From experience with other semiconductors, it is well known that point, line, and two-dimensional lattice defects can seriously degrade the minority carrier lifetime, leading to a potentially harmful increase in the dark currents. In previous work on photo-capacitive devices on $Hg_{0.7}Cd_{0.3}Te$ anomalous dark currents were noted, and it was suggested that they might arise from crystallographic defects in the material [1]. This suggestion has been pursued to establish an understanding of the relationship between lattice defects and dark current, so that minimum requirements on the types and densities of these defects may be specified for a particular device application.

In this paper, we report on some of our investigations of the sources of dark current in the HgCdTe alloys. Our approach has been to use a CID imaging array as a probe for the detection of dark current sources at particular locations on the material. Subsequently, these sources are examined using a chemical etch to reveal crystallographic defects. Transmission Electron Microscopy (TEM) is used as a supplemental technique to identify the defects from their etch patterns, and to examine defects for evidence of impurity precipitates.

EXPERIMENTAL

For this work $Hg_{1-x}Cd_xTe$ alloys with $x = 0.3$ were prepared by the standard solid state recrystallization technique [2,3]. The material was n-type, with a carrier concentration at 77°K of 4-6 x 10^{14} cm^{-3}. CID arrays, 32 columns by 32 rows, were fabricated on this material [1,4]. After DC probe at room temperature, a good device was mounted in a header package and bonded to adjacent silicon signal processing circuitry. The entire package was then mounted in a dewar, and operated 77°K.

In the operation of this device, 32 capacitors in a column are biased into depletion, and the dark currents in these capacitors integrated for a period of time. Using circuitry on the row read lines, the integrated leakage charge in each capacitor in this column is sensed, and processed through the peripheral circuitry to form a serial video signal. The remaining columns are sequentially operated in the same manner. The resulting video signal is then displayed on a CRT. Dark current is displayed as black on a white background.

After operation, the device structure fabricated on top of the semiconductor was removed by a combination of ion-milling and etching using concentrated HCl. Crystallographic defects were then revealed using chemical defect etching. In this work, a proprietary defect etch was used [5]. Many of the same results could be obtained using the Polisar 2 etch discussed by Brown and Willoughby [6]. The defects revealed by etching were also studied on similar material using the TEM. The microstructure revealed by etching was then compared with the dark current video display, pixel by pixel, and the cause of the dark current determined.

RESULTS AND DISCUSSION

The video display of the dark current patterns of one device is shown in Fig. 1. The predominant feature in this figure is a line of dark current running diagonally from the top to the bottom of the image. In addition, a second line of dark current is partially visible in the upper right corner, and there are also a few isolated pixels with high dark current (the 2 columns and 3 rows which appear black in this figure result from electrical problems within the device, and not from dark current).

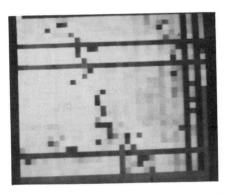

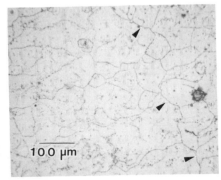

100 µm

Fig. 1. Video display of dark current

Fig. 2. Nomarski micrograph of etched surface of portion of device shown in Fig. 1.

An optical micrograph showing a portion of the upper left hand corner of this device after defect etching is shown in Fig. 2. The predominant crystallographic defects were found to be sub-grain boundaries. Some isolated dislocations were also noted. In this imager, the majority of sub-grain boundaries were found to run from right to left across the imager. There were only two sub-grain boundaries which ran from top to bottom. (One of these is indicated by arrows in Fig. 2). Comparison of the optical micrographs of the etched surface with the dark current video pattern showed that the lines of dark current correlated exactly with the pixels through which these top-bottom sub-grain boundaries ran. Further examination of the etch features was made in the scanning electron microscope. It was found that the top-bottom sub-grain boundaries were etched as a V-shaped groove, while all of the left-right sub-grain boundaries etched as individual dislocation etch pits (figure 3). The dislocation density in the right-left sub-boundaries may be approximately determined from the spacing of the etch pits in this figure as 10^4 cm^{-1} (10^4 cm of dislocation line per cm^2 of sub-boundary). Averaged over an entire pixel (50 μm by 50 μm), such a sub-grain boundary (which accomodates a misorientation of 0.02 -0.03°) would be equivalent to a dislocation density of ~ 2 x 10^6 cm^{-2}. From experience with the defect etch, it has been determined that dislocations separated by 10^{-5} cm or more are resolvable in the SEM, even if the etch pits overlap. Thus the dislocation density of the top-bottom sub-boundaries is of the order of 10^5 cm^{-1} or greater. This density will accommodate a misorientation of 0.2 - 0.3° and be equivalent to a dislocation density of ~ 2 x 10^7 cm^{-2} averaged over a pixel. The origin of the dark current, confirmed in many other observations [7], is thus the large number of dislocations present in the pixels through which the large misorientation sub-grain boundaries run.

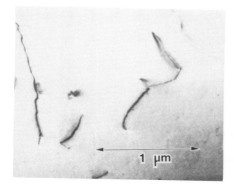

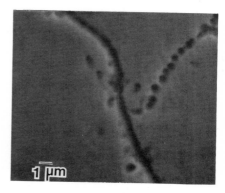

Fig. 3. SEM micrograph of portion of Fig. 2

Fig. 4. TEM micrograph of typical dislocation structure.

Transmission Electron Microscope study of similar materials revealed a complex structure for both isolated dislocations and dislocations on sub-grain boundaries (Fig. 4). All dislocations were found to be free of precipitates. Weak beam studies of the dislocations showed that very few were dissociated into Shockley partial dislocations. Thus, it would appear that the dislocations themselves are the source of the dark current, although electrically active impurities attached to the dislocation core, or trapped in the strain field surrounding the dislocation cannot be completely eliminated.

The finding that sub-grain boundaries separating two regions of sufficiently large misorientation are significant sources of leakage is in agreement with the recent results of Takagawa et al. [8], who found that in their structures, only boundaries between sub-grains with a misorientation of 0.1° or more were significant sources of dark current.

The origin of the sub-grain boundaries in the solid state recrystallization process has also been examined. In this process, the appropriate amounts of the elements are compounded and thoroughly mixed at a temperature above the liquidus. The material is then quenched to room temperature. Defect etching of the material after the quench reveals the existence of sub-grain structure (Fig. 5). It is well established that solidification of this material begins initially by dendritic growth of cadmium rich spines [9]. As freezing progresses, the remaining mercury rich material crystallizes in the interdendritic regions. As material which freezes last possesses the same crystal structure as the cadmium rich spine, a compositionally non-uniform single phase material is produced during the quench. The defect etch which we

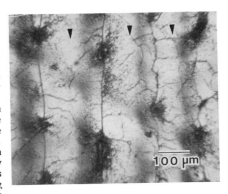

Fig. 5. Nomarksi micrograph of material after solidification. Arrows indicate cadmium-rich dendritic spines. Dark "blobs" are tellurium inclusions.

have used etches the cadmium rich regions faster than the mercury rich regions, giving rise to an etch trough along the cadmium rich spines. These troughs are revealed in Nomarski micrographs as shading within the major sub-grains. This shading parallels the sub-grain structure (Fig. 5), showing that at the center of each sub-grain there is a cadmium rich spine. The major sub-grain boundaries thus result from the impingement of slightly misoriented dendrites during the freezing of the alloy. During the subsequent recrystallization anneal, the major change which occurs is the homogenization of the material, with little change in the sub-grain structure. Microstructural changes which occur during the quench from the recrystallization anneal and during the subsequent post-anneal to convert the material to n-type are discussed in another paper [10].

The elimination of the harmful sub-grain boundaries therefore requires that suitable steps be taken during the solidification to eliminate the impingement of misoriented dendrites.

SUMMARY AND CONCLUSIONS

1. A major source of dark current in CID imagers fabricated on HgCdTe alloys is the extremely high density of dislocations along some sub-grain boundaries.

2. All dislocations studied have been free from impurity precipitates.

3. In the solid state recrystallization process, sub-grain boundaries originate from the impingement of misoriented dendrites during the solidification of the material from the melt.

ACKNOWLEDGMENTS

The authors would like to thank M. Brau and M. Williams for providing the material used in this investigation, J. Schaefer for operation of the devices and providing the video image shown in Fig. 1, and J.H. Tregilgas and R.F. Pinizzotto for many helpful discussions and for their review of the manuscript.

REFERENCES

1. A.J. Lewis, R.A. Chapman, E. Schallenberg, A. Simmons, and C.G. Roberts, Proc. IEDM Conf., pp. 571-573, (1979).

2. S.G. Parker and H. Kraus, U.S. Patent 3 486 363 (1969).

3. W.F.H. Michlethwaite, in Semiconductors and Semimetals, Vol. 18, R.K. Willardson and A.C. Beer, eds., (Academic Press, New York, 1981), p. 65.

4. R.A. Chapman, S.R. Borrello, A. Simmons, J.D. Beck, A.J. Lewis, M.A. Kinch, J. Hynecek, and C.G. Roberts, IEEE Trans. Electron Devices, ED-27, 134 (1980).

5. H.F. Schaake, unplublished.

6. M. Brown and A.F.W. Willoughby, J. Phys. Colloque C6, 40, 151 (1979).

7. P. Everett, H.F. Schaake, and A.J. Lewis, to be published.

8. H. Takagawa, T. Akamatsu, T. Kanno, and R. Tsunoda, Proc. IEDM Conf., 172 (1981).

9. W.F.H. Michlethwaite, ibid., p. 72.

10. H.F. Schaake and J.H. Tregilgas, to be published.

IV
DISLOCATIONS AND INTERFACES

INTERACTION OF DISLOCATIONS WITH IMPURITIES AND ITS INFLUENCE ON THE MECHANICAL PROPERTIES OF SILICON CRYSTALS

KOJI SUMINO
The Research Institute for Iron, Steel and Other Metals, Tohoku University,
2-1-1 Katahira, Sendai 980, Japan

ABSTRACT

A review is presented of the work on the influence of impurities on the dynamic behavior of dislocations in silicon
crystals and also on the resulting effects in the mechanical
strength performed by the author's group. Special emphasis
is laid on the effects of light element impurities such as
oxygen, nitrogen and carbon. Although all of these impurities
do not affect the velocities of dislocations moving under
relatively high stresses, oxygen and nitrogen atoms are found
to lock slowly moving dislocations and dislocations at rest
very effectively. Such locking of dislocations results in the
decreases in the activities of generation and multiplication
centers for dislocations, leading to the strengthening of the
crystals. The high strength of silicon crystals brought about
by the impurities is lost when the crystals are subjected to
the heat treatments which allow the precipitation of the impurities on a macroscopic scale. This softening is shown to
be caused by dislocations punched out from precipitates and by
the exhaustion of the impurities dissolved in the matrix crystal.

INTRODUCTION

Semiconductor crystals are the typical material having high Peierls barrier
for dislocation motion. Mobility of dislocations in semiconductor crystals is
very sensitive to the temperature and rather insensitive to the stress compared
with other type of materials. Such characteristics in the mobility of individual dislocations are reflected directly in the mechanical properties of these
crystals that are measured to be sensitive to the temperature and the strain
rate. In this respect, semiconductor crystals are the most suitable material
for the study of quantitative relationship between macroscopically observed
mechanical properties and microscopically occurring dislocation processes in
crystals. Such situation has stimulated dislocation-dynamical studies on plastic deformation of semiconductor crystals [1-7]. Developments in in situ X-ray
topography [8] and high voltage electron microscopy with a high temperature
stressing stage [9-11] have made it possible to follow directly dislocation
processes in semiconductor crystals occurring under stress at elevated temperature. Many important characteristics in the mechanical behavior of pure semiconductor crystals have now been interpreted quantitatively in terms of the
dislocation theory.

Meanwhile, the mechanical strength of Si crystals has recently attracted a
great amount of attention from the practical point of view concerning the
occurrence of slip and warping of wafers during device production process.
It has been recognized that the mechanical stability of Si wafers depends
sensitively on the concentration and state of impurities involved in them.
Wafers of CZ Si are known to show a higher resistance against the deformation
caused by thermal stress than those of FZ Si [12]. This may be ascribed to the

Mat. Res. Soc. Symp. Proc. Vol. 14 (1983) ©Elsevier Science Publishing Co., Inc.

interaction of dislocations with O atoms in CZ Si crystals [13,14]. Very recently, the technique to control the concentration of O atoms in Si crystals in the range of 2∿20 at.ppm has been established by growing crystals by the Czochralski method in a transverse magnetic field [15]. FZ Si crystals doped with N atoms at concentrations lower than about 0.1 at.ppm have also been grown successfully and have been reported to show a high resistance to the occurrence of thermal slip [16]. These developments in the technique of impurity controlling have facilitated the study on the interaction between dislocations and light element impurities in Si crystals to a great extent.

The author's group has been conducting the investigation on the characteristics in the dislocation-impurity interaction in Si crystals in detail with the technique of in situ X-ray topography. The mechanical behavior of impurity-doped Si crystals has also been studied simultaneously. The essentials in the interrelation between the dislocation-impurity interaction and the solution hardening in Si crystals seem now to have been clarified on the basis of such works. This paper gives a brief review on these works.

INFLUENCE OF IMPURITIES ON THE DISLOCATION MOBILITY

In situ X-ray topography has been applied to investigate the influence of various types of impurities on the dynamic behavior of dislocations in Si crystals [17]. The experimental system consists of a high-power X-ray generator, a high-temperature tensile stage, a large topographic goniometer and a T.V. system including a highly sensitive PbO vidicon X-ray camera [8].

The velocity v of dislocations in high purity FZ Si crystals is found to be described as a function of the stress τ and the temperature T by the following equation in the stress range $1.2 \sim 40.0$ MN/m^2 and the temperature range 600 ∿ 800°C:

$$v = v_0 \tau \exp(-E / k_B T) \quad . \tag{1}$$

The magnitudes of v_0 are 1.0×10^4 and 3.5×10^4 m^3/MN·s and those of E 2.20 and 2.35 eV for 60° and screw dislocations, respectively, and k_B the Boltzmann constant. It is noticeable that v is linear against τ down to a stress as low as 1.2 MN/m^2 which is the minimum stress applicable with the tensile stage used in the experiment.

Effects of light element impurities, C, N, and O, as well as those of electrically active ones, P, Sb, As, and B, on the dislocation mobility have also been investigated. The influence of light element impurities on the velocity of 60° dislocations is shown in Fig. 1 in which the data for high purity FZ Si crystals are shown by open circles. C atoms at a concentration of 2 at.ppm have no influence on the dislocation mobility. Dislocations in FZ Si crystals doped with N atoms at a concentration of 0.11 at.ppm move at velocities which are equal to those in the high purity crystals in the stress range higher than about 4 MN/m^2. However, dislocations which have been moving under high stresses cease to move and become immobile when they are brought under stresses lower than the above value. The critical stress for immobilization of about 3 MN/m^2 exists also in crystals doped with O atoms at a concentration of 15 at.ppm (determined by infrared absorption with the Kaiser-Keck calibration curve). In this case, though the dislocation velocities in the high stress range are almost equal to those in high purity crystals, they decrease with the decrease in the stress more rapidly than in the high purity crystals when the stress approaches the critical stress for immobilization. Figure 2 shows the velocity-stress relations for 60° dislocations at 647°C in crystals doped with O atoms of various concentrations. Broken lines drawn vertically show the critical stresses below which dislocations originally moving under high stresses become immobile. Both the deviation of the velocity from that in the high purity

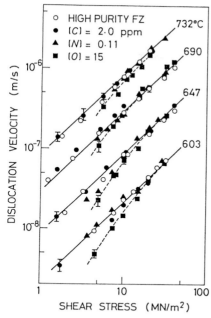

Fig. 1. Velocities of 60° dislocations at various temperatures plotted against the shear stress for high purity FZ Si crystals and also for Si crystals doped with C, N and O atoms of which concentrations are shown in the figure (Imai and Sumino, Ref. 17).

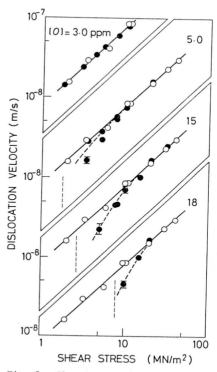

Fig. 2. The stress-dependence of the velocity of 60° dislocations at 647°C for intrinsic Si crystals containing O atoms of various concentrations shown in the figure. Open marks are for high purity FZ Si crystals (Imai and Sumino, Ref. 17).

crystals in the low stress range and the critical stress for the immobilization of dislocations increase with the increase in the concentration of dissolved O atoms.

The shape of moving dislocations in the high purity FZ crystals is observed always to be a regular hexagon or half-hexagon of which segments are straight along <110> over the wide stress range investigated. This is true also for the impurity-doped crystals when dislocations are moving at the velocities equal to those in the high purity crystals in the high stress range. However, whenever the velocities of dislocations deviate from those in the high purity crystals under relatively low stresses, the shape of the moving segments is observed to be perturbed from the <110> straight lines and to become irregular. Such perturbation in the shape of moving dislocations is found to be reversible. When the crystal in which moving segments are perturbed from the <110> straight lines under a low stress is brought under a high stress, the segments restore the straightness along <110>. Upon reducing the stress, the shape is perturbed again. On the contrary, once dislocations cease to move under a low or zero

310

stress in the impure crystals, the stress needed to start the dislocations increases with increases in the period and the temperature where the dislocations have been kept at rest, depending on the species and the concentration of the impurities involved. This effect will be mentioned in the next section.

The shape perturbation starts to take place at higher stresses in the impure crystals in which the cease of dislocation motion occurs at higher stresses. Since the shape perturbation and the immobilization of dislocations are characteristic of the impure crystals, the phenomena are thought to be related to the interaction between dislocations and solute atoms. Probably, the perturbation in the shape of moving dislocations is caused by

TABLE I

Magnitudes of the critical stress τ_c for immobilization of dislocations in various types of Si crystals (Imai and Sumino, Ref. 17)

Materials	Impurity (at.ppm)		$\tau_c (MN/m^2)$
FZ Si	B :	0.00004	< 1.2
	C :	2.0	< 1.2
	N :	0.11	4.2
	P :	240	8.5
MCZ Si	O :	3.0	< 1.2
(Magnetic Field CZ)	O :	5.0	1.8
CZ Si	O :	15	3.0
	O :	18	8.0
	P : 120	(O : ~12)	4.5
	P : 300	(O : ~12)	11.0
	B : 280	(O : ~14)	5.0

the local locking of dislocations by solute atoms and the cease of motion by the locking closely along the dislocation line. Table I shows the magnitudes of the mean critical stress for the immobilization of dislocations in the temperature range 600~800°C for crystals doped with various types of impurities. It may be concluded that O and N atoms in Si crystals have strong interactions with dislocations.

Essentially the same behavior as that of 60° dislocations is observed for screw dislocations in impure Si crystals.

Doping of donor impurities leads to the increase in the velocities of both 60° and screw dislocations in accordance with previously published data [18-21]. This is achieved through decreases in the magnitudes of v_o and E in Eq. (1). Doping of P atoms results in the immobilization of dislocations under low stresses, showing that substitutional P atoms have a strong locking effect on dislocations. B atoms of a concentration as high as 280 at.ppm have almost no influence on the dislocation velocity. The locking action of B atoms for dislocations is found to be weaker than that of P atoms.

LOCKING OF DISLOCATIONS BY IMPURITIES

The in situ X-ray topographic observations have been applied to investigate quantitatively how the locking of dislocations is related with the development of impurity atmosphere around the dislocations [22]. Isolated loops of fresh dislocations in impurity-doped Si crystals are first moved under a high stress and, then, subjected to aging at various temperatures in absence of applied stress. The stress necessary to start such aged dislocations is measured as dependent on the period and temperature of aging and also on the species and concentration of impurity atoms. This stress is defined as the locking stress. The locking stress is sensitive to the distribution of impurity atoms around the dislocations and to the temperature but insensitive to the releasing rate of the dislocations from the impurity atmosphere.

Figure 3 shows the growth of the locking stress at 647°C with the period of aging at the same temperature for crystals doped with various impurities. Solid lines are for intrinsic crystals with various concentrations of O atoms

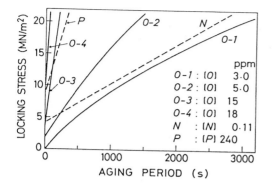

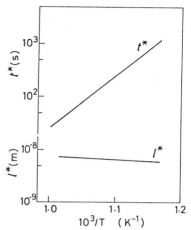

Fig. 3 Growth of the locking stress at
647°C with the period of aging at 647°C
for crystals doped with various types of
impurities, of which species and concentrations are shown in the figure (Sumino
and Imai, Ref. 22).

Fig. 4 The aging period t* and the
mean diffusion distance l* of O
atoms resulting in the locking
stress of 20 MN/m² at 647°C in the
intrinsic Si crystal with an O concentration of 15 at.ppm plotted
against the reciprocal of the aging
temperature T (Sumino and Imai, Ref.
22).

shown in the figure and broken lines for
FZ Si crystals doped with N atoms at a
concentration of 0.11 at.ppm and also
those doped with P atoms at a concentration of 240 at.ppm. In all types of the
crystals, the locking stress is seen to
increase with the increase of aging
period. For O-doped crystals, the in-
creasing rate is higher for higher concentrations of O atoms. It is surprising
to note that the locking behavior in the crystal with a N concentration as low
as 0.11 at.ppm is roughly equivalent to that in the crystal with an O concentration of 3.0 at.ppm. This seems to show an extraordinarily strong locking
effect due to N atoms compared with O atoms in Si crystals. The effect of P
atoms at a concentration of 240 at.ppm is roughly equivalent to that of O atoms
at a concentration of 15 at.ppm. In this case, the growth rate of the locking
stress due to the P atoms is smaller than that due to the O atoms. This is
thought to be related to a much smaller value in the diffusion rate of P atoms
compared with that of O atoms in Si crystals. No locking is found to take place
for dislocations in high purity FZ Si crystals by aging. C atoms in FZ Si crystals at a concentration of 2.0 at.ppm show almost negligible effect for the dislocation locking.
    Figure 4 shows the aging period t* to obtain the locking stress of 20 MN/m²
at 647°C plotted against the inverse of aging temperature T for intrinsic CZ Si
crystals with an O concentration of 15 at.ppm. It is evident that the period
of aging giving rise to a certain value of locking stress decreases as the aging
temperature increases. The mean diffusion distance l* of O atoms during such
aging can be evaluated from t* and the diffusion constant reported by Haas [23]
and is shown in the bottom of the figure. It is known that the diffusion distance of O atoms related to the same value of locking stress is approximately
constant with respect to the aging temperature. This result may be interpreted
to show that the locking stress is determined by the number of O atoms gathered

on a unit length of dislocations if the concentration of O atoms in the crystal is specified. The same is found also for N atoms in FZ Si crystals.

The number of impurity atoms accumulated in a unit length of dislocations is determined by both the concentration and the diffusion distance of the impurity atoms during aging. The diffusion distance l* of impurity O atoms related to the locking stress of 20 MN/m^2 at 647°C is found to decrease approximately linearly with the increase in the concentration of O atoms in the crystals. The values of l* for N and P atoms in FZ Si crystals turn out to be much smaller than that of O atoms if the l* value of the latter is extrapolated to that at the concentrations of the former impurities. This seems to show that the interactions of a dislocation with individual N and P atoms are much stronger than that with individual O atoms. The interaction of C atoms with a dislocation is found to be smaller than that of O atoms.

It is interesting to see how the locking stress is related to the number of impurity atoms accumulated in a unit length of dislocations during aging. The latter quantity can be evaluated by solving the diffusion equation in the stress field of a dislocation assuming that the core region of the dislocation acts as a perfect sink for impurities arrived there. This calculation has been done for O atoms in Si crystals by means of numerical method using a computer [22]. It is found that the concentration N* of O atoms at the dislocation core is expressed by

$$N*/c_o = k \ (\ D \ t \ / \ d^2 \ )^{0.787} \qquad (2)$$

where c_o is the concentration of O atoms in the crystal, D the diffusion constant of O atoms, t the aging period, and d the spacing of interstitial sites. k is a constant determined by the energy of interaction between the dislocation and an O atom and the temperature, being about 10 at 647°C.

The locking stress τ_ℓ in Fig. 3 is thus related to N*. Figure 5 shows the obtained τ_ℓ - N* relations for crystals with various concentrations of O atoms, in which N* is expressed in a unit of the number of O atoms per unit length of dislocations. The locking stress in all the types of crystals is seen to increase linearly with N*. The τ_ℓ - N* relations are almost identical for crystals with relatively low concentrations of O atoms. However, in crystals with high concentrations of O atoms, the same value of the locking stress is achieved with smaller numbers of accumulated O atoms for crystals with higher O concentrations. This effect may be attributed to inhomogeneous distribution of O atoms in highly concentrated crystals.

The interaction energy for a dislocation and an O atom can be estimated from the results in Fig. 5 on the basis of the discussion on thermally activated

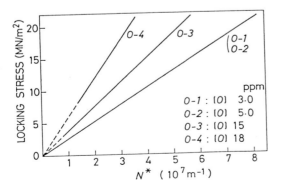

Fig. 5 The relation between the locking stress τ_ℓ at 647°C and the number N* of O atoms accumulated in a unit length of the dislocations for intrinsic crystals with various concentrations of O atoms of which values are shown in the figure (Sumino and Imai, Ref. 22).

release of the dislocation from its impurity atmosphere. Suppose that O atoms are distributed uniformly along the dislocation line at the interval equal to the reciprocal of N* calculated above. The locking stress τ_ℓ is then given by

$$\tau_\ell = N^* [E_o - k_B T \ln(L N^* \nu / \Gamma)] / b^2 \qquad (3)$$

where E_o is the maximum of the interaction energy for the dislocation and an O atom, L the length of the dislocation, ν the frequency of the dislocation vibration, Γ the release rate of the dislocation from the atmosphere, and b the magnitude of the Burgers vector of the dislocation. Since the term $\ln(LN^*\nu/\Gamma)$ is a weak function of N*, τ_ℓ may be regarded to be linear with respect to N* in accordance to the results in Fig. 5.

Rather uniform distribution of O atoms along dislocations may be realized in crystals with low concentrations of O atoms. Thus, Eq. (3) is fitted to the $\tau_\ell - N^*$ relation for the crystals with low O concentrations in Fig. 5. Substitution of suitable values to L, ν, Γ and b leads the magnitude of E_o of 3.0 eV. This magnitude of the interaction energy is much larger than that of the elastic interaction due to the size misfit of an O atom and an interstitial site. The origin of such a high interaction energy may be attributed to a strong binding of an O atom with Si atoms at the dislocation core. The interaction energy of 3.0 eV in this case may be interpreted to be the difference between the energy of an O atom locating an ordinary interstitial site and that at the dislocation core.

Alternatively, the following model for the interaction may be conceivable. Clusters of O atoms developed in the early stage of precipitation are known to act as donors. Dislocations in Si crystals are reported to act as acceptors [24-26]. If O atoms accumulated on the dislocations are gathered together rapidly by means of diffusion along the dislocation core resulting in the formation of clusters having the donor function, then both the elastic and electrostatic interactions may be available between such clusters and the dislocations. If the clusters each consisting of, for instance, four O atoms are formed along the dislocations at the interval of 4/N*, the above type of discussion leads to the interaction energy of 3.5 eV for each cluster. The electrostatic interaction between a singly ionized donor and a charged acceptor site on the dislocation in Si crystals amounts to the interaction energy of about 2 eV. Then, the energy of the elastic interaction between the cluster and the dislocation turns out to be about 1.5 eV, which seems to be a reasonable magnitude. At present, it is not known which of the above two pictures is more probable. Works are needed to clarify the state of O atoms locating at the dislocation core.

The origins for the strong interactions of N and P atoms with dislocations are not well understood. In the case of P atoms, the electostatic interaction with acceptor sites on the dislocations is thought to play an important role. It may also be probable that P atoms captured at the dislocation core gather fast diffusing impurities in the crystal and form complexes having a high interaction energy. N atoms in FZ Si crystals used in the present work are electrically inactive and are believed to occupy the interstitial sites. The extraordinarily strong interaction of N atoms with dislocations seems to indicate that there is a large difference in the energies between the state of a N atom locating at the ordinary interstitial site and that at the dislocation core. The understanding of the reaction which may take place at the dislocation core is of great interest.

GENERATION OF DISLOCATIONS AS INFLUENCED BY DISPERSED IMPURITY ATOMS

It has been found that dislocations are generated and undergo multiplication in a dislocation-free crystal with a macroscopically smooth surface under stresses much lower than the ideal strength of the crystal. The former phenomenon is attributed to the presence of minute irregularities on the crystal surface which facilitate the generation of dislocations. Such irregularities are thought to cause stress concentration or to be accompanied by heavily disturbed atomic arrangements favorable for the dislocation generation. The in situ X-ray topographic technique has been applied to follow the generation process of dislocations from surface flaws such as Knoop indentations and scratches drawn by a diamond stylus on the (111) surface of Si crystals at room temperature [8].

Faulted regions with strong contrasts in X-ray topographs are introduced around such flaws. The crystal with such flaws is first stressed at room temperature and is, then, heated to high temperatures with the stress applied on it. Rather small stress concentration is found around individual flaws. Upon heating the crystal, the contrast and the size of the faulted regions around the flaws first diminish significantly. However, the contrast and the size of the faulted regions soon recover by continued heating and start to increase with the passage of time. Such enlarged contrasts are found to consisit of images of many dislocations generated from the flaws.

Detailed examinations of the relationship between the orientation of the flaws and the character of dislocations generated from them lead to the following picture for the generation process of dislocations in Si crystals from surface flaws formed at room temperature. Thin layers with highly disturbed structure are introduced around the flaws at room temperature of which exact nature is not known at present. Such regions start to recover upon heating the crystals to temperatures above about 350°C. At this time, various kinds of dislocations favorable to release the strains induced by the flaws are introduced in such regions. Among these dislocations, having various Burgers vectors, only those subjected to the maximum force due to the applied stress are able to expand outwards preferentially and penetrate into the matrix crystal, thus, leading to the generation of dislocations from the flaws.

An interesting difference in the generation processes of dislocations is observed between high purity FZ Si crystals and impure crystals. When crystals are heated up to various temperatures at a given rate, dislocations are observed to be generated from flaws in high purity FZ Si crystals under any values of applied stress. On the other hand, in impure crystals, there are certain critical stresses below which no generation of dislocations takes place. Figure 6 shows the stress-temperature region where the dislocation generation takes place from Knoop indentations made with a load of 0.25 N at room temperature in intrinsic CZ Si crystals with an O concentration of 8.0 at.ppm heated at a rate of about 300°C/min. When crystals with the surface flaws are heated without applying stress and, then, stressed at elevated temperatures, disloca-

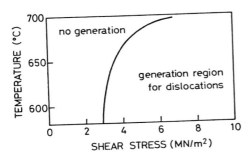

Fig. 6 The stress-temperature region for dislocation generation from the Knoop indentations made with a load of 0.25 N in CZ Si crystals with an O concentration of 8.0 at.ppm (Sumino and Harada, Ref. 8).

tions are generated from the flaws in high purity FZ Si crystals even under very low stresses. On the contrary, no dislocations are generated in impure crystals even when a high stress is applied to the crystals.

The absence of the dislocation generation in impure crystals under low applied stresses is interpreted in terms of the locking of dislocations by impurity atoms mentioned in the preceding section. During the expansion of dislocations in the regions around the surface flaws at the time of the dislocation generation, there may be some stage where the velocity of dislocations is fairly low owing to the interaction with the high density of dislocations introduced. At this stage, dislocations in impure crystals would be locked by impurity atoms and cease to move further on. This gives an explanation for the existence of the critical stress for the dislocation generation from the surface flaws and also for the non-activity of aged flaws for the generation of dislocations in the impure Si crystals.

MECHANICAL STRENGTHS OF DISLOCATION-FREE CRYSTALS

It is well known that wafers of CZ Si crystals show a higher mechanical resistance against thermal stress during high temperature processing compared with those of FZ Si crystals, Such difference between CZ and FZ Si crystals is naturally thought to be caused by the interaction of dislocations with O atoms which are contained in CZ Si crystals. However, a surprising fact is met when one measures the stress-strain curves of dislocation-free crystals of the two types of Si crystals with chemically polished surfaces by ordinary tensile or compressive tests. Little difference is found in the mechanical strengths of the two types of Si crystals [14]. Figure 7 shows the stress-strain curves of dislocation-free crystals of CZ Si with an O concentration of 18 at.ppm and of FZ Si. The stress-strain behaviors of the both crystals are characterized by

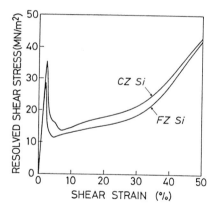

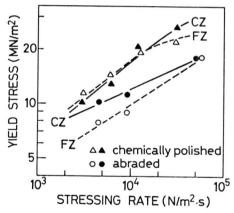

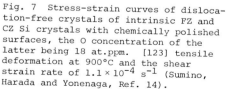

Fig. 7 Stress-strain curves of dislocation-free crystals of intrinsic FZ and CZ Si crystals with chemically polished surfaces, the O concentration of the latter being 18 at.ppm. [123] tensile deformation at 900°C and the shear strain rate of 1.1×10^{-4} s^{-1} (Sumino, Harada and Yonenaga, Ref. 14).

Fig. 8 The yield stress at 900°C plotted against the stressing rate for intrinsic FZ and CZ Si crystals with abraded and chemically polished surfaces. The O concentration of the CZ Si is 12 at.ppm (Sumino and Yonenaga, Ref. 29).

an extremely high upper yield stress, a sharp stress drop after the yielding and a heterogeneous deformation from the upper to the lower yield point by means of the propagation of Lüders bands. Though the yield stress in the CZ Si crystal is somewhat higher than that in the FZ Si crystal, the difference is rather small. The stress-strain curves of dislocation-free crystals with O concentrations of 8.0 at.ppm and 5.0 at.ppm and also that of a FZ crystal doped with N atoms at a concentration of 0.11 at.ppm all coincide approximately with that of the FZ Si crystal in the figure [27,28]. Thus, it may be concluded that both O and N atoms in the dissolved state have little influence on the dislocation processes occurring during the deformation of originally dislocation-free Si crystals.

The difference in the strengths in ordinary mechanical tests is found between dislocation-free crystals of CZ Si and FZ Si when they have rough surfaces and are deformed at low deformation rates [29]. Figure 8 shows the yield stress at 900°C plotted against the stressing rate for dislocation-free crystals of CZ Si and FZ Si having chemically polished surfaces and rough surfaces abraded with #800 carborundum powder. The O concentration of the CZ Si is 12 at.ppm. For the crystals with the abraded surfaces, the yield stress of the CZ Si becomes progressively higher than that of the FZ Si as the stressing rate decreases. For the crystals having chemically polished surfaces, no systematic differene in the yield stresses is observed between the CZ Si and the FZ Si over the range of stressing rates in the figure, except for the highest rate of 3×10^4 N/m$^2 \cdot$s. The crystals with abraded surfaces in the figure all deform homogeneously while those with chemically polished surfaces inhomogeneously by means of the propagation of Lüders bands in the yield region. The deformation of the CZ Si crystal with a chemically polished surface at the stressing rate of 3×10^4 N/m$^2 \cdot$s is extremely inhomogeneous even after the yield region. Thus, the yield stress under this condition may have some physical meaning differing from that under other conditions.

A theoretical analysis of the yield behavior of the crystals with the abraded surfaces in Fig. 8 [7,29] shows that the density of the surface flaws active as the dislocation-generation centers in the CZ Si is only about 2 % of that in the FZ Si under a stressing rate of 4.5×10^3 N/m$^2 \cdot$s. The results in Fig. 8 are interpreted satisfactorily in terms of locking of dislocations by O atoms in the generation process at surface flaws in the CZ Si crystals. As seen in the preceding section, dislocations are locked by O atoms under low stresses in CZ Si crystals. If the increasing rate of the applied force on the dislocations generated at the flaws is lower than the growing rate of the locking force, then most dislocations become locked firmly and do not contribute to deformation, resulting in the mechanical behavior of the crystals having a low density of generation centers with a high yield stress. On the other hand, most dislocations are released and contribute to deformation if the increasing rate of the applied force is higher than that of the locking force. The higher the increasing rate of the applied stress, the less the fraction of dislocations immobilized, resulting in lower yield stresses. In the crystals with chemically polished surfaces, dislocations are nucleated at minute irregularities under an extremely high local stress. The upper yield stress may be determined by the condition under which the propagation of a Lüders band begins. All processes proceed under high stresses and at no stage does the locking of dislocations take place, thus, leading to approximately the same yield stresses for the CZ Si and FZ Si crystals.

MECHANICAL STRENGTHS OF DISLOCATED CRYSTALS

Mechanical behavior of dislocated Si crystals having chemically polished surfaces is found to be very sensitive to the species and concentration of impurities doped in the crystals. Crystals with various densities of randomly

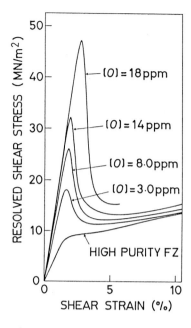

Fig. 9 Stress-strain curves of dislocated
crystals of various O concentrations shown
in the figure. Dislocation density is 1×10^6
cm^{-2}. [123] tensile deformation at 800°C
and the shear strain rate of 1.1×10^{-4} s^{-1}
(Yonenaga, Sumino and Hoshi, Ref. 27).

distributed dislocations are obtained
from those deformed to various strains
upon annealing at 1300°C followed by
rapid cooling. This heat treatment is
effective also to eliminate possible
segregation of impurities which might
occur during the specimen preparation.

The stress-strain behavior of high purity FZ Si crystals is sensitive to the
density of dislocations involved before
deformation tests. With a decrease in
the dislocation density, the magnitudes
of the upper yield stress and the stress
drop from the upper to the lower yield
point increase significantly [6]. The
same type of dependence of the yield behavior on the initial density of dislocations is also observed for impurity-doped
Si crystals [27,30]. However, the magnitudes of the upper yield stress and the
stress drop in the yielding of Si crys-
tals doped with O or N atoms are much larger than those of high purity FZ Si
crystals having the comparable densities of dislocations. These become more
remarkable as the concentration of the impurities increases. Figure 9 shows
stress-strain behavior in the yield region of Si crystals doped with various
concentrations of O atoms having dislocation densities of approximately 1×10^6
cm^{-2} [27]. The stress-strain behavior of the FZ Si crystal doped with N atoms
at a concentration of 0.11 at.ppm having the same dislocation density coincides
almost with that of the crystal doped with O atoms of 3.0 at.ppm [28]. The upper yield stress measured under the same deformation condition is found to increase linearly with the increase in the concentration of dissolved O atoms in
the crystals with the same dislocation densities. C atoms doped at concentrations up to 3.5 at.ppm in FZ Si crystals are found to have no appreciable effect on the yield behavior.

It is to be noted that the stress-strain characteristics of a crystal doped
with O or N atoms are similar to those of a high purity FZ crystal with a dislocation density which is much lower than that in the former. Detailed examinations of the relationship between the upper yield stress and the dislocation
density prior to deformation for crystals with various concentrations of O
atoms show that the yield behavior of the crystals with 3.0 at.ppm of O atoms
is almost the same as that of the high purity FZ crystals whose dislocation
densities are about an order of magnitude lower than those in the former, and
that in the crystals with 18 at.ppm of O atoms as that of the high purity FZ
crystals whose dislocation densities are more than two orders of magnitude
lower. With the results obtained in the preceding section, it may be concluded
that dislocations in the impure Si crystals existing before deformation are
locked firmly by impurities, not being able to act as multiplication centers
for dislocations. The locking is thought to be more effective for crystals

with higher concentrations of O atoms. The similarity of the yield behavior of the crystal doped with 3.0 at.ppm of O atoms and that of the crystal doped with 0.11 at.ppm of N atoms agrees well with the fact that the locking behaviors of the impurities in these crystals are similar as seen in Fig. 3.

Dislocations are found to be activated to a considerable extent in impurity-doped Si crystals showing high upper yield stresses already at the deformation stage before the upper yield point is reached. Thus, the upper yield stress does not mean the stress at which locked dislocations are collectively released. We reach the following picture for the mechanism of solution hardening in Si crystals. The hardening due to O or N atoms in Si crystals is caused by the locking of dislocations by these impurity atoms. The yield strength of the crystal is controlled by the length of dislocations which are free from such locking in the unit volume of the crystal. Thus, the crystals doped with O or N atoms behave like high purity FZ Si crystals with dislocation densities much lower than those of the former. The higher the concentration of the impurity atoms in the crystal, the less the fraction of dislocations which can act as multiplication centers and, consequently, the higher the upper yield stress. If the sites to be occupied by impurity atoms line up along the dislocation core with the interaction energy of E_D, it is readily shown that the probability f of occupation of each site by an impurity atom is given by

$$ f = c_o / [c_o + \exp(- E_D / k_B T_{eff})] \qquad (4) $$

where c_o is the concentration of impurities in the matrix and T_{eff} the temperature at which the occupation of the sites effectively took place.

EFFECT OF IMPURITY PRECIPITATION

It is well known that the mechanical strength of dislocation-free crystals of ordinary CZ Si diminishes drastically upon annealing at temperatures around 1000°C [31]. The stress-strain behavior of the softened CZ Si crystal is characterized by a low upper yield point followed a small stress drop which is characteristic of that in a crystal with a rather high density of dislocations. This softening is shown to be associated with precipitation of SiO_2 particles on a macroscopic scale. Usually, defects such as stacking faults of extrinsic type and loops of punched out dislocations are seen to develop around the precipitate particles in the softened crystals.

Annealing of the precipitation-softened crystals at temperatures above 1200°C leads to the dissolution of the precipitates and also to the diminution of the defects. The high upper yield stress and large stress drop after yielding are restored with the progress of such dissolution treatment. This change in the stress-strain behavior is similar to that caused by the decrease in the density of the dislocation sources. The stress-strain curve of the crystal dissolution-treated at 1320°C for 1 hr coincides with that of the crystal subjected to no precipitation-treatment [32]. It may, thus, be concluded that the precipitation softening in CZ Si crystals is caused by the generation of defects that can act as effective dislocation sources under stress.

Detailed examinations of the relationship between the upper yield stress and the density of stacking faults show that the relationship is different between the course of the precipitation and that of the dissolution. Moreover, the relationship in the course of the dissolution depends on the temperature of the dissolution [32]. This indicates that stacking faults are not the dislocation sources that are responsible for the softening. On the other hand, a fairly unique relation common to both the precipitation and the dissolution courses is found to hold between the upper yield stress and the amount of precipitated O atoms. An example is shown in Fig. 10 for CZ Si crystals with an O concentration of 18 at.ppm. At first sight, this fact seems to suggest that precipitate

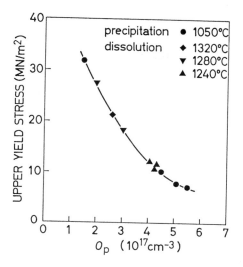

precipitation ● 1050°C
dissolution ◆ 1320°C
▼ 1280°C
▲ 1240°C

Fig. 10 Variation of the upper yield stress with the amount O_p of precipitated O atoms in CZ Si crystals with an O concentration of 18 at.ppm (Yonenaga and Sumino, Ref. 32).

particles act as dislocation sources which lead to the softening due to the precipitation. However, calculations with elasticity theory show that the magnitude of the stress concentration coefficient due to the SiO_2 particles is of order of 2 [33], suggesting that dislocations are not nucleated at the precipitates under stresses such as those shown in Fig. 10.

Analyses in terms of dislocation dynamics have been attempted [32] to describe quantitatively the experimental relation in Fig. 10. With the assumptions that the precipitates themselves act as dislocation generation centers and the softening is caused by the increase in the density of the precipitates, it turns out that the precipitates have to increase in their density by more than 6 orders of magnitude for the change in the amount of precipitated O atoms of a factor $2 \sim 2.5$ to get a good fitting of the calculations with the experiments. This result is unreasonable and, thus, the model is concluded to be inadequate. With the alternative assumption that the density of precipitates is constant in the intermediate stage of precipitation and also during the dissolution of the precipitates, the nucleation rate of dislocations at a precipitate under a given applied stress has to increase by more than 10 orders of magnitude for the change in the size of the precipitate of a factor 2 to have a good fitting. This result is again unreasonable. In any way, it may be concluded that precipitates themselves are not dislocation sources which bring about the precipitation softening.

At present, prismatic loops of dislocations introduced around the precipitates during the growth of the latter by the large misfit strain are thought to be the most probable dislocation sources responsible for the precipitation softening. In the earlier stages of precipitation, each precipitate is smaller in size, and lower densities of dislocations are punched out owing to smaller misfit strains. The punched out dislocations are quickly locked by O atoms and immobilized since higher concentrations of dissolved O atoms remain in the matrix at this stage. As precipitation process proceeds, the precipitates grow and more dislocations are punched out to release the increased misfit strain. The locking for such dislocations is less effective as O atoms in the regions around are now low. At the same time, the O atoms segregated along the dislocations which have been punched out in the earlier stage change to precipitate particles. The O atoms, originally distributed uniformly along the dislocations, are concentrated at discrete positions, leaving O-free portions on the dislocations. Such portions are free from locking and are able to act as dislocation sources under applied stress. The occurrence of these phenomena has indeed been observed by transmission electron microscopy. In the course of dissolution process, the density of punched out dislocations is decreased by annealing and the remaining dislocations are locked effectively by O atoms dissolved into the ma-

320

trix from the precipitates. The above situation brings about the difference in
the density of active dislocation sources of several orders of magnitude between the crystal with a small amount of precipitated O atoms and that with a
large amount, which can well describe the rapid decrease in the upper yield
stress with the increase in the amount of precipitated O atoms seen in Fig. 10.
    The precipitation behavior of O atoms in Si crystals is rather sensitive to
the thermal history of the crystals which affects the density and the distribution of the nucleation sites for precipitates. Generally, thermal treatments
at higher temperatures eliminate the nucleation sites for the precipitates
which are destined to develop at lower temperatures. The precipitation behavior depends, of course, on the concentration of O atoms and also on those of
other kind of impurities. Dislocation-free crystals of Si with various concentrations of O atoms have been subjected to annealing at 1300°C followed by rapid cooling for the purpose of homogenization, and comparisons have been made on
how the precipitation softening proceeds for these crystals [27]. Essentially
no precipitation takes place by annealing at 1050°C in the crystals containing
O atoms at concentrations lower than about 5 at.ppm. In these crystals, no
softening occurs even by prolonged annealing at 1050°C. The same is observed
for FZ Si crystals doped with N atoms at a concentration of 0.11 at.ppm [28].
The precipitation of SiO_2 and, consequently, the softening occur in the Si crystals with O atoms of which concentration is higher than about 8 at.ppm. The
rate of the occurrence of the softening depends very sensitively on the O concentration, being higher in the crystals with higher concentrations of O atoms.
    The relation between the upper yield stress and the amount of precipitated O
atoms as shown in Fig. 10 is found to depend on the concentration of O atoms involved in the crystal. For the same amount of precipitated O atoms, the upper
yield stress is measured to be lower for the crystals with lower concentrations
of O atoms. The cause of this result may be ascribed to the fact that the exhaustion of O atoms dissolved in the matrix is more remarkable in these crystals
: thus, the locking of punched out dislocations takes place less effectively.

ACKNOWLEDGEMENTS

The author is grateful to the members of his research group for their cooperation in the investigations on which this review is based. He also wishes to
express his gratitudes to Hitachi Ltd., Komatsu Electronic Metals Co., Osaka
Titanium Co., Shin-Etsu Handotai Co., and Sony Corp. for affording the Si crystals used. This work was supported partially by Grant-in-Aid for Scientific
Research from the Ministry of Education, Science and Culture in Japan.

REFERENCES

1. H. Alexander and P. Haasen, Solid State Physics 22, 28 (1968).

2. K. Kojima and K. Sumino, Crystal Lattice Defects 2, 147 (1971).

3. K. Sumino and K. Kojima, Crystal Lattice Defects 2, 159 (1971).

4. K. Sumino, S. Kodaka and K. Kojima, Mater. Sci. Engng. 13, 263 (1974).

5. K. Sumino, Mater. Sci. Engng. 13, 269 (1974).

6. I. Yonenaga and K. Sumino, Phys. Stat. Sol. (a) 50, 685 (1978).

7. M. Suezawa, K. Sumino and I. Yonenaga, Phys. Stat. Sol. (a) 51, 217 (1979).

8. K. Sumino and H. Harada, Phil. Mag. A 44, 1319 (1981).

9. M. Sato and K. Sumino, Proc. Fifth Int. Conf. on High Voltage Electron Microscopy, Kyoto, T. Imura and H. Hashimoto eds. (1977) pp. 459.

10. K. Sumino and M. Sato, Kristall u. Technik 14, 1343 (1979).

11. M. Sato and K. Sumino, Phys. Stat. Sol. (a) 55, 297 (1979).

12. S. M. Hu and W. J. Patrick, J. Appl. Phys. 46, 1869 (1975).

13. S. M. Hu, Appl. Phys. Lett. 31, 53 (1977).

14. K. Sumino, H. Harada and I. Yonenaga, Jpn. J. Appl. Phys. 19, L49 (1980).

15. T. Suzuki, N. Isawa, Y. Okubo and K. Hoshi, Semiconductor Silicon 1981, H. R. Huff et al. eds.(1981) pp. 90.

16. T. Abe, K. Kikuchi, S. Shirai and S. Muraoka, Semiconductor Silicon 1981, H. R. Huff et al. eds. (1981) pp. 54.

17. M. Imai and K. Sumino, Phil. Mag. A, in the press.

18. V. N. Erofeev, V. I. Nikitenko and V. B. Osvenskii, Phys. Stat. Sol. 35, 79 (1969).

19. S. B. Kulkarni and W. S. Williams, J. Appl. Phys. 47, 4318 (1976).

20. J. R. Patel, L. R. Testardi and P. E. Freeland, Phys. Rev. B 13, 3548 (1976).

21. A. George and G. Champier, Phys. Stat. Sol. (a) 53, 529,(1979).

22. K. Sumino and M. Imai, to be published.

23. C. Haas, J. Phys. Chem. Solids 15, 108 (1960).

24. H. Weber, W. Schröter and P. Haasen, Helv. Phys. Acta 41, 1255 (1968).

25. V. A. Grazhulis, V. V. Kveder and V. Yu. Mukhina, Phys. Stat. Sol. (a) 43, 407 (1977).

26. H. Ono and K. Sumino, Jpn. J. Appl. Phys. 19, L629 (1980).

27. I. Yonenaga, K. Sumino and K. Hoshi, to be published.

28. K. Sumino, I. Yonenaga and T. Abe, to be published.

29. K. Sumino and I. Yonenaga, Jpn. J. Appl. Phys. 20, L685 (1981).

30. K. Sumino, I. Yonenaga, H. Harada and M. Imai, Proc. Int. Conf. on Dislocation Modelling of Physical Systems, Gainesville, M. F. Ashby et al. eds. (1981) pp. 212.

31. J. R. Patel, Discuss. Faraday Soc. 38, 201 (1964).

32. I. Yonenaga and K. Sumino, Jpn. J. Appl. Phys. 21, 47 (1982).

33. K. Yasutake, M. Umeo and H. Kawabe, Phys. Stat. Sol. (a) 69, 333(1982).

ELECTRICAL PROPERTIES OF DISLOCATIONS AND BOUNDARIES IN SEMICONDUCTORS

HANS J. QUEISSER
Max-Planck-Institut für Festkörperforschung, 7000 Stuttgart 80, FRG

ABSTRACT

Simple models have been suggested to predict electronic properties of lattice defects in semiconductor crystals: dislocations ought to act via the acceptor character of dangling bonds, and small-angle grain boundaries ought to consist of regular arrays of dislocations. The actual situation in most semiconductors is, however, much more complicated. The observed electrical effects of dislocations do not confirm the dangling-bond concept, they are affected by dissociation and reconstruction. There appear to be differences between straight and kinked dislocations. Dislocations owe much of their electronic behavior to clouds and precipitates of impurities; oxygen in silicon plays a significant role. This review summarizes the present status of experimental methods and results, including luminescence and capacitance spectroscopy as well as mapping and imaging techniques using electron-microscopes.

INTRODUCTION

Dislocations are crystalline defects, whose dynamics determine the plastic flow of solid state materials. Additional effects arise in the covalent semiconductors: the density, the mobility, and the lifetime of charge carriers can be strongly modified by dislocations. Such effects are screened to insignificance in typical metals, but in semiconductors the perturbations of the strongly directional tetrahedral bonds and their influence on the properties of the relatively dilute electron gas become severe. Semiconductor devices are thus expected to suffer detrimental consequences.

Early appraisals of the effects of dislocations involved ingeniously simple concepts. Shockley [1] suggested in a talk in 1953 that *edge states* might arise in the diamond crystal structure, with the extra half-plane of atoms terminating *"with broken bonds parallel either to a [111] direction or at tetrahedral directions thereto."* He predicted that bands of states arise and that a one-dimensional degenerate electron gas enables the dislocation to conduct current. Such a prediction was as alarming as it appeared novel and convincing. Semiconductor devices with small dimensions in critical regions were therefore presumed to become highly susceptible to dislocations. A grain boundary in a solid could also be described by the simple model of a regular array of dislocations. Here one expected even stronger disturbances and conduction by a two-dimensional degenerate electron gas.

Dislocations and boundaries were hence considered as pernicious; efforts were devised to eliminate these defects. However, the prophesied disasters never really occured. Dislocations proved almost harmless and sometimes even totally imperceptible, invalidating therefore a general acceptance of the naive models. Damage was observed mostly when impurities interacted with the dislocations. Dislocations were found to affect devices through indirect phenomena, mutilating the diffusion profiles and altering spatial distributions

Mat. Res. Soc. Symp. Proc. Vol. 14 (1983) ©Elsevier Science Publishing Co., Inc.

of dopants and other impurity atoms. These extrinsic effects display a wide variety of phenomena and had to be studied case by case by the materials specialists in device development teams. Macroscopic parametrizations were invoked, while basic research attacked microscopic features of electronic processes at and around dislocations. These two paths toward understanding these rather intricate processes are still being pursued today, they gradually converge.

In this talk, I review the evolution of research on dislocations and boundaries in semiconductors, introduce the refining of the early models, then summarize some of the most recent activities.

EARLY MODELS

Figure 1 shows dangling bonds arising at the core of a dislocation. The missing neighbor atom leaves a lone electron dangling. The tendency to form stable

Fig. 1. Ball-and-spokes model of a dislocation in the diamond structure. Dangling bonds with unpaired electrons visible in the core.

sp^3 covalent bonds by electron pairing suggests that such dangling bonds be acceptors, ready to capture an electron. Yet such acceptors would differ from the usual kind - such as a trivalent impurity like Ga in Si - because the states are no longer randomly distributed throughout the crystal but instead strung up on the onedimensional defect dislocation. These states are spatially close to each other. Wave functions overlap, which generates a band composed of the individual levels. The overlap implies a one-dimensional conductance.

A special treatment of the occupancy statistics for dislocation states becomes necessary. [2] Coulomb repulsion by the accumulated charge density Q makes it increasingly more difficult to place further charge on the dislocation, the fraction f of sites occupied thus influences the energy position relative to the bulk. A potential wall results; the compensating charge is contained in a space charge cylinder around the dislocation, its radius R (in an n-type material) is determined by

$$\Omega = \pi q \, R^2 \, (N_D - N_A) \ , \tag{1}$$

where q is the electron charge and N_D, N_A are the bulk donor and acceptor densities. This local potential increase surrounded by a space charge cylinder, shown in Fig.2, implies significant consequence for electronic effects of dislocations, which are capable to capture charge.

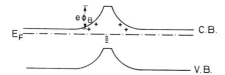

Fig. 2. Potential around a dislocation (center) with captured negative charge, which is compensated by adjacent positive space charge. E_F indicates Fermi energy, CB conduction band, VB valence band.

If the fraction of occupied sites is $f=s(Q,q)$, with s being the distance between adjacent charge-trapping sites, presumably dangling bonds, then the electrostatic energy per dangling bond is

$$E_e = \frac{Q^2 s}{4\pi\varepsilon} \left(\ln \frac{Rf}{s} - \frac{1}{2} \right) \tag{2}$$

with ε denoting the dielectric constant. Equation (2) involves simplifying assumptions, discussed in the literature and reviewed by Labusch and Schröter [3] who also summarize experiments and interpretations of Hall effect data in deliberately deformed Ge and Si with high densities of dislocations.

A more recent counterproposal for the spectrum of dislocation-induced electronic states was presented by Osip'yan and Ryzhkin, [4] who include quantum-mechanical mixing of the localized wave functions at dangling bonds with undisturbed wavefunctions in the vicinity. In contrast to the earlier theories, they find an "*irregular dependence*" of the state density with the energy of the dislocation band and claim better fit to experiments. In particular, they suggest [4] occupancy changes to alter the character of dislocation screening and to lead to gross variations of the electronic effects.

Grain boundaries are two-dimensional defects, which can be reduced to dislocations by a simple model, which assumes that - at least for small misfits a boundary consists of a regularly spaced array of dislocations. Figure 3 depicts this model. Grain boundaries should thus serve conveniently to study dislocation properties. Their effects should be more easily detectable due to the larger number of dangling bonds. A two-dimensional degenerate electron gas is expected, and a plane of trapped charge should be envisioned, with the compensating charge contained in adjacent slabs in the bulk.

Grain boundaries were a concern in the very first transistors of polycrystalline Ge. Soon it was realized that only single crystals provide the reliable materials basis for semiconductor electronics, consequently grain boundaries were avoided and no longer presented problems. Present technology, however, again faces boundaries for two reasons. Polycrystalline Si has become important for silicon-gates in MOS integrated circuits with conductivity controlling device speed. Secondly, poly-Si is less expensive than single crystals, thus potentially advantageous as a viable material for economical large-area solar cells. [5] The electronic conductivity across grain boundaries in Si as well as the recombination processes at boundaries for mino-

Fig. 3. Simple model of a small-angle tilt boundary, consisting of a regularly spaced array of edge dislocations.

rity carriers - such as created by sunlight in a photovoltaic cell - have therefore become elementary processes, where basic research is required.

PAST PREDICTIONS AND PRESENT REALITY

The consequences of the simple models for dislocations and boundaries, all reduced to the notion of dangling bonds, disquieted the semiconductor community. Both the doping level and the properties of the carriers in a device appeared uncontrollably endangered. Especially serious seemed the prospect of a locally enhanced electrical conduction along a dislocation or a boundary. The most sensitive device part was at that time the reverse-biased base-collector junction, which had to show minimal currents despite a large electric field. Any dislocation penetrating such sensitive structures should induce a short circuit, thus irreversibly damage the device.

Initial observations seemed to corroborate this pessimistic prediction. The only clear correlation at that time between device failure and other materials properties relied on the etch pit count, a measure for dislocation density. Crystal regions with slip or "lineage"-boundaries usually yielded no useful devices at all. This general trend of observations [6] stimulated an extensive research to improve the starting material, which culminated in the success of growing perfectly dislocation-free silicon single crystals.[7,8] The tetrahedrally bonded semiconductors Ge and Si can fortunately easily be grown dislocation-free for two reasons: the thermal conductivity is comparatively large, which reduces steep thermal gradients with their detrimental stresses; secondly the dislocation velocity in these materials is relatively slow, the directional bonds offering resistance like a viscous medium, [9] which in turn decelerates motion of dislocations and impedes their multiplication.

Dislocation-free starting material, however, did not provide the ultimate relief, because technological processes of device manufacturing invariably introduce new lattice defects. An emitter of a bipolar transistor, for example, requires heavy doping. The diffusion of large amounts (still only less than 1 at.%) of dopants leads to massive generation of misfit dislocations, which are aligned on slip planes and become easily visible by etching [10] and X-ray topography. [11] Epitaxy at even slightly contaminated interfaces was shown to generate pyramids of stacking faults bounded by partial dislocations. [12] Ion implantation, contact-alloying, mechanical scribing and cleaving, and the mounting of a device all damage the lattice, which presumably cuts device yield or seriously affects device performance, especially under heavy electrical and thermal loads. How miraculous that transistors could be made to operate at all; the dangling bond idea in its simple form could not possibly be correct!

Impurities, in particular heavy metals, were soon identified [13] to promote premature avalanche breakdown, which degrades p-n junctions under reverse bias. Precipitates, rather than individually dissolved atoms, act as sites for local field enhancement and device failure. Such precipitation needs nucleation centers, and those are easily provided by dislocations and boundaries. This hypothesis explains that not all dislocations are electrically active, but only those contaminated. The partials at stacking faults were shown [14] to lead to localized avalanche breakdown, as demonstrated by visible microplasmas, but again not all of the partials proved to be electrically active. Twin boundaries [15] and grain boundaries [16] in silicon behave similarly; excessive recombination at defect levels within the space-charge region of junctions appeared to be correlated with impurities which preferentially precipitated at these defects. Such precipitation was already well-known and was utilized to observe "decorated" dislocations in infrared-transmitting microscopes. [17]

Oxygen in Si is often present in large quantities (up to $10^{18} cm^{-3}$), when the crystal is pulled out of a quartz crucible (rather than float-zone refined). The expected influence of this impurity was also demonstrated, for example by photoresponse measurements at grain boundaries. [18]

Evidence from the enormous experience in the materials and device development laboratories thus favored rejection of the naive dangling-bond model. Reviews of exactly 20 years ago [19,20] emphasize this conclusion: Strong direct electrical effects of the dangling bond hypothesis are missing, they either do not exist or are saturated by impurities; precipitation and other indirect effects govern electrical behavior. A later review by Holt [21] in 1979, based on even more experience and including data on compound semiconductors, reaches identical conclusions: *"the sobering fact is that dislocations in these devices are of little importance in themselves. They cause serious problems only if they nucleate metal precipitates..."*

Saturation of dangling bonds by impurities requires merely modest concentration. A density of 10^4 dislocations/cm^2 demands only about $4 \cdot 10^{11}$ impurities/cm^3. [21] Even extreme densities around $10^8 cm^{-2}$ require less than commonly observed doping levels (which, however assumes a complete migration to the active sites). The reduced dimensionality of the one-dimensional dislocation and the two-dimensional boundary versus the three-dimensional bulk is the reason for this moderate requirement. Despite this evidence from devices, the trend in interpretation of results obtained in strongly deformed bulk semiconductors with "fresh" dislocations adhered to correlating electrical data to electronic intrinsic states at the dislocation core. [3]

The elastic strain around dislocations produces a *Cottrell atmosphere*. [22] An undersized impurity, for example, reduces the total energy of the crystal by occupying an atomic site in the compression region around the dislocation. Impurities thus assemble in clouds around dislocations. Diffusion of impurities is likewise enhanced at dislocations and grain boundaries. This preferential diffusion has been studied in detail for small-angle grain boundaries in Si. [23,24] Figure 4 exemplifies a phosphorous diffusion into a p-type Si crystal with a grain boundary. [24] Twin boundaries lack such enhancement. [15] Dislocations and boundaries thus exert indirectly electrical effects by severely distorting the shape of p-n junctions. Utilization of this preferential diffusion was attempted to fabricate special high-frequency device structures with exceedingly narrow base layers. [20,23] This idea to achieve structuring

328

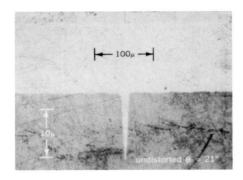

Fig. 4. Bevel-polished and stained junction, produced by phosphorous diffusion (top, white region) into p-type silicon (bottom, dark). Preferential diffusion along a grain boundary causes a spike. Note separate scales, caused by the bevel.

of crystals down to atomic dimensions by using this unique situation at dislocation cores is documented in early papers and patents. [25]

The idea of shaping device structure by utilizing dislocations has not yet brought practical solutions and seems today less promising for submicronstructuring than other, more developed, techniques such as molecular-beam epitaxy. [26] Dislocations, however, are usefully applied today. They "getter" harmful impurities by precipitating them in regions which are noncritical for devices; so-called deliberate "backside damage" is widely used. Today's device designer thus no longer insists on completely dislocation-free materials, he has become more demanding by specifying judicious amounts of defects but restricted to crystal regions where they can be tolerated, yet are sufficiently close to exert beneficial influence by providing sinks for undesirable impurities and provide flexible plastic behavior through dislocation motion.

REFINEMENTS OF THE MODELS

Indirect effects dominate electrical defect properties in macroscopic device structures, yet the question remains how to properly describe the microscopic structure at the dislocations core and how to relate the details of the atomic arrangement to three important parameters: (i) the number and energetic positions of electronic states within the forbidden energy gap, (ii) carrier capture and emission coefficients, to understand the release and trapping dynamics for electrons and holes, and (iii) the degree of localization of the electronic wave functions, to estimate whether individual isolated states arise or overlapping, hence current-carrying delocalized, states form. This ambitious program requires extremely well defined samples and experimental techniques with high spatial resolution as well as advanced solid state theory to handle many atoms in highly irregular, nonsymmetrical arrangements.

Our rather modest advances in this difficult regime rely on refined and more detailed descriptions of the dislocation core, which we shall now review. This review will enable us to summarize next some of the most recent advances. The conspicuous absence of unpaired electrons in dangling bonds must first of all be rationalized.

Reconstruction of the core atoms is one way to avoid dangling bonds. Figure 5 gives one of many examples, which Hornstra [27] published as early as 1958. The lattice disturbance is no longer locally confined to one row of atoms but shared by a grouping of atoms around the core, abounding in distorted, squeezed and stretched bonds. For many dislocations one can devise such configurations.

The cubic face-centered space lattice of the diamond structure has two atoms in the unit cell. Each layer thus consists of two atomic planes, unequally spaced in a ratio 1:3. Thus, two sets of glide planes exist: the *shuffle* set if glide occurs between the widely spaced planes, the *glide* set between the narrow ones. Core structure, stability, and dislocation motions differ for these two types. [28]

Compound semiconductors bring further complication, since different atoms alternate. The important optoelectronic materials GaAs and GaP belong to the A^{III}-B^{V} compounds. The "Hünfeld convention" [29] fixes the terminology for A- or B-dislocations, depending on which atom species lies in the most disordered core position. Thus in these polar compounds, one must distinguish between A(g) and B(g) (in the glide set) or A(s) and B(s) (in the shuffle set). The distinc-

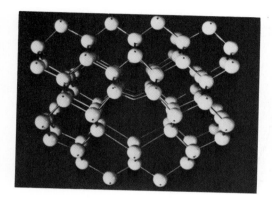

Fig. 5. Model of a dislocation in the diamond structure, dangling bonds avoided by a reconstruction, causing bond stretching, squeezing, and distortions

tions between A- and B-dislocations rise with increasing ionicity of the compound, Haasen first pointed out the differing cores and properties. [30]

The specification of Burgers vector $\vec{b}$ and dislocation line vector $\vec{s}$ again enhances the multitude of possibilities. The *angle* between the two vectors is significant. The symmetry of the diamond lattice gives prominence to $60°$ angles between $\vec{b}$ and $\vec{s}$, each vector pointing into different <110>-directions. The resulting manifold of structural possibilities cannot be covered here. [27,31,32]

Dissociation of a perfect dislocation into two partial dislocations is yet another and particularly important intricacy of dislocation structure. Perfect dislocations have Burgers vectors equal to lattice translational vectors, partials only to fractions thereof. Vectorial conservation laws govern splitting reactions, such as

$$\frac{a}{2} [110] \rightleftharpoons \frac{a}{6} [211] + \frac{a}{6} [12\bar{1}] \tag{3}$$

for the Burgers vectors. The two partials subtend a two-dimensional stacking fault. The energy balance between dislocation line energies and stacking fault energy (about 60 mJ/m^2 for Si) leads to perfect (or "*constricted*") dislocations or to an *extended* dislocation consisting of two partials with stacking fault. Figure 6 gives one example. Again, the lattice distortion is not concentrated onto just a few rows of atoms but rather spread over a wider volume.

330

Fig. 6. Model of a 60-degree dislocation after dissociative split into a 90-degree partial and a 30-degree partial with a stacking fault in between: view onto a (111)-plane. (Courtesy D. Mergel)

Experimental evidence for the elemental [33-36] and compound [37] semiconductors indicates that dissociation is the rule rather than the exception! The necessary refined electron microscopy of high resolution and enhanced sensitivity was provided by the "weak-beam technique". [38]

Any deviation from straightness in a dislocation line adds further complexity. *Jogs* are steps in a dislocation line, for example produced by a few atoms missing in the inserted extra half-plane of an edge dislocation. [39,40] Such jogs play a vital part in dislocation motion and the interaction amongst dislocations and between dislocations and intrinsic point defects. [39,40] Intersection of two dislocations creates jogs. *Kinks* arise if the dislocation direction does not coincide with the direction of minimal energy within the lattice. [41] The dislocation must overcome a Peierls-barrier potential wall to change over to another minimal valley. Jogs and kinks are likely to establish sites along the dislocation, where special electrical activity may be presumed. One electrical effect appears associated with kinks: the mobility of dislocation depends on the semiconductor's carrier concentration. [42] Since mobility is believed to be governed by generation of kinks, a recent interpretation of the influence of doping on mobility involves electronic transitions to kink sites. [43]

Point defect phenomena are difficult to separate from dislocation effects, since each production of dislocations, especially at high densities with many intersections, invariably also generates interstitial atoms and vacancies, whose densities are affected by dislocation climb. [40] Impurities, on the other hand, impede dislocation motion by pinning, especially at precipitates. Recombination of excess electron-hole pairs can release their energy (of the order of the band gap) to defects and thus enhance defect reactions and diffusion. [44]

This short summary was to demonstrate the complexity and difficulty of research on dislocations. Even the most careful preparation cannot succeed in supplying just one kind of dislocation in sufficient density for measurements. Great care, however, is taken to work with well-defined deformed samples and to exchange them. [45,45a] For silicon, deformation along a <123>-axis tends to yield single slip with 60° dislocations, which can dissociate.

One remedy is the combination of local structural determinations with measurements of electronic or optical properties at exactly the same spot; such
techniques require high spatial resolution, such as electron optics. Another
hope lies in using grain boundaries, since they are supposedly composed of arrays of identical dislocations. Indeed, grain boundaries celebrate a renais-
sance presently. [46] earlier work [47] is being continued. Advances in para-
metrizing grain boundaries rely on the concept of the coincidence lattice. [48]
Both grains are imagined to be continued, one then determines how many of the
atoms coincide on sites. A parameter Σ measures the fraction of jointly occupied positions. This notation abbreviates grain boundary description, which
even for plane boundaries needs at least five numbers: a vector triplet for
the relative misorientation and two numbers for the boundary plane.

Theory for electronic effects on dislocations is difficult. Deep levels generated by just one atom in a host are already hard to calculate. [49] How much
more complicated is the situation of many atoms arranged in a highly distorted
low-symmetry environment! Present calculations thus do not enjoy the same acceptance as band structure calculations but must be regarded as preliminary.[50]

RECENT RESULTS

Dislocations and boundaries in semiconductors are currently an active research topic, their literature expands. This chapter reviews selectively -
thus subjectively - some recent work. New experimental techniques are emphasized.

Electron-beam-induced current (EBIC) is measured with high resolution in a
scanning electron microscope. [51] Figure 7 indicates this method. EBIC combines

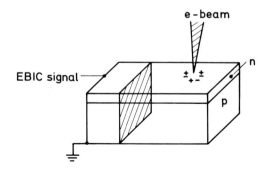

Fig. 7. Principle of imaging
and measuring recombination
of excess carriers by electron-beam-induced current
(EBIC)

direct defect observation with a specific electrical measurement, since the
amplitude displayed on the video screen corresponds directly to the local recombination of excess electron/hole-pairs. Very high resolution is another advantage. The work of Donolato, [52] and Klann,[53] and recently of Marek [54]
and Pasemann [55] strenghtened the theoretical basis to interpret EBIC images:
computer simulations of image strength for dislocations and faults have been
made, [53] in agreement with experiments on stacking faults in Si, [56] which
again showed that the fault is inactive,and the Frank partials vary in activity with impurities.

These quantitative interpretations of EBIC images led to clear distinctions
between screw dislocations (very weak recombination) and two groups of 60° dislocations (stronger effects), although these dislocations were located at dif-

ferent distances from the surface, which represents a competing recombination site. [57] Clean dislocations lack contrast, while contaminated dislocations yield contrasts in agreement with the idea of space charge cylinders surrounding the dislocation; bands of EBIC contrast have been ascribed to point defects, which dominate the temperature-dependence of the minority carrier lifetime in deformed Si. [58] Light as the pair-generating mechanism gives additional information because of the different depth-dependence of the generation rate. A direct comparison of transmission electron microscopy and EBIC [59] indicated markedly different recombination efficiencies in Si between dissociated and undissociated dislocations.

Capacitance Transients inform about energetic positions, cross sections, and concentrations of deep electronic levels. [60] This technique is restricted to space charge layers, thus requires p-n- or Schottky barriers with metallic contacts. A thermally activated process is measured; the temperature of maximal transients yields the energy of the traps; the method is thus more indirect than optical spectroscopy and suffers from rather broad signals. However, much detail on dislocation-related states in Si has recently been obtained, especially useful whenever directly correlated with electron microscopy, such as reported by Kimerling, Patel, Benton, and Freeland. [61] An efficient recombination center ("E(0.68)") with 0.68 eV energy has been correlated with regions of very high dislocation density in slip bands ("Lüders bands"); this acceptor state tends to pin the Fermi level near midgap on the p-type side. This level was observed in high-purity Si and has been related to dislocated regions but not necessarily to the dislocations themselves. [61]

Heavier doping entails more complicated spectra; various defects appear. An important conclusion is that much of the signal, including the E(0.68), originates in point defect clusters which are produced by nonconservative jog motoin. [61] Two states, E(0.38) and possibly H(0.35) are directly relatable to dislocation core structure, but only 2.5 % site occupation is calculated for the Shockley dangling-bond model and only 1% occupation when dissociated 60° dislocations are presumed. The trap activity is thus possibly restricted to special sites, such as kinks. A kink hypothesis is estimated to agree with these results; direct tests are, however, difficult. A variation of EBIC contrast with dislocation line orientation, which had earlier been observed in rather heavily doped samples [62] was not confirmed by the later work. [61]

A direct combination of DLTS with scanning electron microscopy, called SDLTS, has been demonstrated by Petroff and Lang [63] on compound semiconductors.

Luminescence is exceptionally sensitive to detect extrinsic electronic states within the forbidden energy gap. Search for dislocation-related states in GaAs was futile, although enhanced recombination was found near dislocations. [64] This negative result disappoints, because the photoluminescence spectra of this direct-gap semiconductor are rich in features and well understood; any dislocation-related signal would set a promising trace. In the indirect-gap material Si, however, there is luminescence related to the presence of dislocations. Figure 8 shows spectra without (bottom) and with (top) dislocations, here generated by laser-pulse heating. [65] The lines D1 to D4 are typically observed in dislocated Si, they were first reported by Drosdov, Patrin, and Tkachev in mechanically deformed Si. [66] This assignment was recently confirmed in a careful study by J. Weber and R. Sauer (Stuttgart University), by photoluminescence on judiciously deformed samples, checked by transmission electron microscopy, made by E. Weber and H. Alexander (Cologne University). [67] One of their results is shown in Fig.9. High compressive stress at low temperatures, known to produce straight 60°-dislocations and screws, yields strong signals (Fig.9a) at 0.96 eV, 1.01 eV, and a weaker one at 0.90 eV photon energy. Annealing at 390°C relaxes the straight dislocations, the D quartet reappears (Fig.9). The Stuttgart/Cologne-collaboration [67] concludes: (i) the D lines originate from

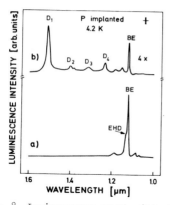

Fig. 8. Luminescence of Si, (b) with laser-induced dislocations, (a) undislocated. BE: bound excitons, EHD: el./hole-drops. (Ref. 65)

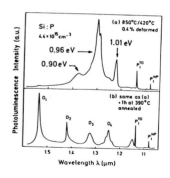

Fig. 9. Luminescence of Si, before (a) and after (b) anneal (Cologne/Stuttgart-collaboration, Ref. 67)

kinks or constrictions, not from straight dislocations (as had also been suspected earlier [68]), (ii) a band between 0.8 and 1.05 eV luminescence is correlated with point defects and is seen for very high dislocation density, (iii) signals at 0.96 and 0.90 eV correlate with straight dislocations, and (iv) the 1:01 eV line is associated with stacking faults. These statements are preliminary, but this detailed spectroscopy, combined with definitive deformation and high-resolution transmission microscopy, is one of the most promising new developments.

Luminescence may be excited by fast electrons in an electron microscope. One gains high spatial resolution and can maintain the good spectral resolution if the sample is at sufficiently low temperature. The high excitation energy, however, is not as favorable as the low level of excitation with photons just barely above the band gap energy. The ease of display of both luminescence output and lifetime has recently been demonstrated by Steckenborn, Münzel and Bimberg, [69] who mapped a local symmetry of the lifetime patterns around dislocations in GaAs. Lifetime and luminescence intensity were oppositely modulated in these samples. Another very useful technique, applied to compound semiconductors and devices, is transmission cathodoluminescence, combining imaging with volume optical analysis. [70]

Luminescent- and laser diodes are the devices most seriously imperiled by disorder. "Dark-line defects" are related to dislocations, their presence and multiplication under load degrades the diodes. This vital subject of optoelectronics is reviewed by Hayashi [71]; a recent proposal concerns self-perpetuating dangling bond traps and anti-site point defects. [72]

Photoplasticity is a phenomenon enjoying renewed interest despite obvious complexity. Dislocation breakaway and influence of mobile and space charges have been invoked to explain the influence of illumination upon the stress-strain curves of CdS, [73] an example of which is shown in Fig.10. CdS crystals become harder to deform when illuminated with appropriate light, strong influence of temperature is seen. [73] In Si, Küsters and Alexander [74] found a strong influence of light on the motion of dislocations at sufficiently low temperatures, where the light-generated carriers are not overwhelmed by intrinsic carriers. Recombination effects are proposed for the effects of different behavior of two types of 60° dislocations; kink migration and impurity effects

334

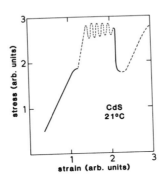

Fig. 10. Example of a stress-strain relation of CdS at
21°C with and without illumination; interpreted as a
photoplasticity effect leading to dislocation-break-
away under illumination; (adapted, after ref. 73)

on kinks are suggested. [74]

Optical excitation, ensuing photoconductivity, and studies of recombination
are other experimental means to determine the energetic position of defect
states, such as in Si performed by Mergel and Labusch. [75]

Electron Paramagnetic Resonance is a technique most powerful to reveal atomic charge distributions and symmetries. Alexander and his collaborators applied
it to deformed Si. [45] Two sharp, so-called "central lines" are found and attributed to a center with S=1/2, symmetrical around the Burgers vector. A defect, possibly with a dangling bond, may be presumed, yet once again its density is only about 1% of the total sites along a dislocation line. [45a] Signals
ascribed to point defects in the immediate and further vicinity are plentiful.
EPR studies thus require extreme care in sample preparation and interpretation. [45,45a] In Ge, a dislocation conduction band has been claimed from ultra-
sensitive observations with self-resonant samples. [76]

Most of the recent results just briefly summarized show firstly the complexities of dislocation effects and secondly invalidate a universal dangling-bond
model. The complexity is increased by indirect effects that defects bring to
bear on semiconductor materials and device structures. *Crystal growth* is strongly promoted by dislocations because they provide nucleation centers for solidification. Bauser and Strunk [77] have recently correlated dislocations observed by
electron microscopy - with step patterns on the surface of GaAs grown by liquid-phase epitaxy. Figure 11 shows a schematic. Presented was the first demon-
stration [78] of edge dislocations as persistent sources for monomolecular steps,
while previously only screw dislocations were suspected. [79] The preferential
growth alters the distribution of dopants and other impurities, as demonstrated
by etching and luminescence. [80] Gettering of impurities by dislocations and
faults is another indirect interaction with point defects, affecting electrical
properties. Absence of dislocations, on the other hand, causes new types of defects to coagulate in the bulk. [81]

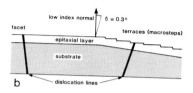

Fig. 11. Growth morphology of a GaAs
layer, epitaxially grown from Ga solution. The facet growth (flat region,
left) results from strictly regular
monomolecular steps, nucleated at dislocations. Terrace growth (right) at
misoriented substrate; (Courtesy
Bauser & Strunk, ref.77.)

Mechanical properties are strongly affected by dislocations, which is noticed
in Si according to its dislocation content, also depend on oxygen in its various forms within the Si matrix. [82] Distortion of wafers by dislocation motion, which may be stimulated or impeded by precipitates, is serious, because
the dimensions of this erratic motion are now approached by those of todays
devices in very-large-scale-integration; lithography is jeopardized! [83]
Study of dislocation motion and impurity influence [9] is necessary, direct
observation by X-ray topography in real-time on a video screen is possible today. [84] Dislocations also scatter phonons; recent results in Ge are explained
with point defects rather than dangling bonds. [85]

Structural observations, especially using the electron microscope, with resolution attaining an imaging of individual atomic planes, are essential to
all progress in understanding the defect details. That topic is summarized
elsewhere in this volume. [86]

Grain boundaries have been discussed extensively [46,54,87], because of their
renewed importance. Two topics are here discussed, carrier recombination and
boundary-surmounting electron transport; both concern solar cells and IC gate
electrodes. EBIC studies by Marek [54] with a truly two-dimensional model for
the boundary plane and a complete image theory exemplify a novel method to
sensitively measure the diffusion lengths of the carriers in the adjoining
grains and to describe the boundary via a macroscopic average recombination
velocity. Figure 12 shows a typical picture, comparing theory and experiment.

Electrical conductance and capacitance of Si-small-angle tilt boundaries
revealed [89] a continuous distribution of states, diminishing from the conduction band edge into the gap - similar to "tail-states". A small density
peak arises near $E_v+0.55$ eV, where vacancy complexes are suspected. [89]
Charge in these states builds up a potential barrier ϕ. Thermionic transgres-
sion of this barrier results in a d.c.-conductivity j as a function of bias U:

$$j = A^*T^2 \exp(-e(\phi+\zeta)/kT)(1-\exp(-eU/kT)) \qquad (4)$$

where A^* is a Richardson constant, ζ the Fermi level.

Investigations of the spatial and energetic distribution of these charge-
capturing states was recently continued by Werner and Strunk. [90]. Electron
microscopy was combined with a quantitative evaluation of the frequency-depen-
dence (up to 10^6Hz) of the complex, capacitance-like conductance. The continu-

336

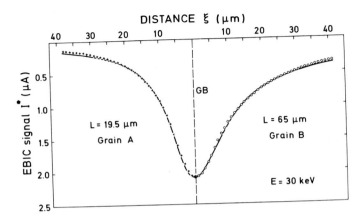

Fig. 12. Measured (solid line) and calculated (points)
EBIC signals for a grain boundary in silicon. Given are
the determined diffusion lengths L in the two grains
(Courtesy J. Marek, ref. 54).

ous distribution in energy was verified, then analyzed with a model of potential fluctuations at the boundary, see Fig. 13. The fluctuations have a minimal spacing of 7 nm, much larger than the regular spacing of grain boundary
dislocations of 1.8 nm! Thus the so-called intrinsic, boundary-building dislocations are inactive; additional extrinsic dislocations, observed to be accu-

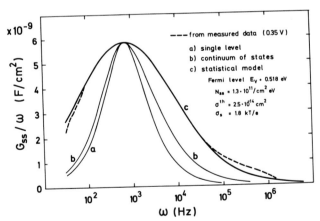

Fig. 13. Admittance versus frequency; data (dashed
line) compared to various model, only a statistical
spatial charge distribution fits; courtesy Werner
& Strunk. (ref. 90)

mulated statistically around the boundary, provide the activity. [90] The naive dangling bond concept is contradicted by this quantitative analysis. For
compound semiconductors, the barrier heights for both grains may differ even
for symmetrical boundaries. [91]

Devices containing grain boundaries or related defects are today usually
described by macroscopic, phenomenological models to account for parameters
such as conductance, majority-carrier characteristics, or photoresponse. These
models have now become sufficiently detailed that comparison can be made to
microscopic features. For example, conclusions on preferential grain boundary
diffusion can be drawn from an analysis of solar cells from poly-Si. [92]
Photo-excitation and recombination of poly-Si was found inconsistent with the
assumption of discrete boundary states in the forbidden gap. [93] A current-
induced resistance decrease in poly-Si has been explained by local melting at
boundaries and impurity distribution. [94] Capacitance data in heavily dislocated Si, taken at two temperatures, inform about microscopic dislocation parameters. [95] These few examples of recent device research indicate an encourag-
ing convergence toward the work on microscopic phenomena in the semiconductor
bulk.

CONCLUSIONS

The initial, simplified models of dislocations as linear arrangements of
dangling bonds and grain boundaries as regular arrays of dislocations are untenable against evidence from studies on semiconductor devices and deliberately deformed bulk material. Reconstruction and dissociation apparently circum-
vent the formation of dangling, unsaturated bonds. Point defects near dislocations play usually a prominent role. Indirect electrical effects, such as precipitation of impurities or junction profiles disfigured by preferential diffusion, exert the most dangerous influence on device electrical performance.
A wide variety of effects between defects and impurities is therefore observed,
but recent tendencies promise more unified descriptions.

ACKNOWLEDGMENTS

Many helped me to collect and appraise current research: H. Alexander,
E. Bauser, P. Haasen, R. Labusch, J. Marek, D. Mergel, R. Sauer, A. Seeger,
W. Shockley, H. Strunk, E. Weber, J. Weber, and J. Werner, their advice and
the patient help of A. Vierhaus are gratefully acknowledged.

REFERENCES

1. W. Shockley, Phys. Rev. 91, 228 (1953) (abstract only).

2. W. T. Read, Jr. Phil. Mag. 45, 775 (1954).

3. R. Labusch and W. Schröter in:"Dislocations in Solids", Vol.5,
    F.R.N. Nabarro, editor (North Holland, Amsterdam 1980) p. 127.

4. Yu. A. Osip'yan and I.A. Ryzhkin, Zh. Eksp. Teor. Fiz. 79, 961 (1980)
    [Sov. Phys. JETP 52, 489 (1980)].

5. B. Authier in: Festkörperprobleme, Vol. XVIII, J. Treusch, ed. (Vieweg,
    Braunschweig 1978) p.1; H. Fischer, ibid. p. 19.

6. W.E. Taylor, W.C. Dash, L.E. Miller and C.W. Mueller (Panel Discussion)
    in: "Properties of Elemental and Compound Semiconductors", H.C.Gatos,
    ed. (Interscience, New York 1960) p.327.

338

7. W.C. Dash, J. Appl. Phys. <u>30</u>, 459 (1959); G. Ziegler, Z. Naturf.<u>16a</u>, 219 (1961).

8. E. Spenke and W. Heywang, "Twenty-Five Years of Semiconductor-grade Silicon", phys. stat. sol. <u>a64</u>, 11 (1981).

9. J.R. Patel and A.R. Chaudhuri, Phys. Rev. <u>143</u>, 601 (1966); P. Haasen, phys. stat. sol. <u>a28</u>, 145 (1975).

10. H.J. Queisser, J. Appl. Phys. <u>32</u>, 1776 (1961); S. Prussin, J. Appl. Phys. <u>32</u>, 1876 (1961); a review on process-induced defects in Si by M. Watanabe, Y. Matsushita, and K. Shibata, Inst. Phys. Conf. Ser. <u>59</u>, Oiso Conf. (1980) R.R. Hasiguti, ed. p.123.

11. G.H. Schwuttke and H. J. Queisser, J. Appl. Phys. <u>33</u>, 1540 (1962).

12. R.H. Finch, H.J. Queisser, G. Thomas, and J. Washburn, J. Appl. Phys. <u>34</u>, 406 and 3153 (1963).

13. A. Goetzberger and W. Shockley, J. Appl. Phys. <u>31</u>, 1821 (1960).

14. H.J. Queisser and A. Goetzberger, Phil.Mag. <u>8</u>, 1063 (1963); recent confirmation on precipitation at faults: H. Strack, K.R. Mayer, and B.O. Kolbesen, Solid-St. Electron. <u>22</u>, 135 (1979) or R. Ogden and J.M. Wilkinson, J. Appl. Phys. <u>48</u>, 412 (1977).

15. H.J. Queisser, J. Electrochem. Soc. <u>110</u>, 52 (1963).

16. H.J. Queisser, Z. Physik <u>176</u>, 313 (1963), concerns the forward characteristics of p-n junctions with defects.

17. W.C. Dash, J. Appl. Phys. <u>27</u>, 1193 (1956).

18. H.J. Queisser, J. Phys. Soc. Jpn. <u>18</u>, Suppl. III, 142 (1963); D. Redfield, Appl. Phys. Lett. <u>40</u>, 163 (1982).

19. H.J. Queisser "Dislocations and Semiconductor Device Failure" in:"<u>Physics of Failure in Electronics</u>" (Proc. Chicago Conf. 1962), M.F. Goldberg and J. Vacarro, editors (Spartan Books, Baltimore 1963).

20. H.J. Queisser, in: "<u>Festkörperprobleme</u>", Vol. II, F. Sauter, editor (Vieweg, Braunschweig 1963), p. 162. (A detailed review, in German, on diffusion and dislocations, junction properties, stacking faults; 95 references).

21. D.B. Holt, J. de Physique <u>40</u>, suppl. 6, C6-189 (1979).

22. A.H. Cottrell "<u>Dislocations and Plastic Flow in Crystals</u>", Clarendon, Oxford (1953), p.56.

23. H.J. Queisser, "<u>Failure Mechanisms in Si Semiconductors</u>", Final Rep.AF 30 (602) 2556 (1963).

24. H.J. Queisser, K. Hubner, and W. Shockley, Phys. Rev. <u>123</u>, 1245 (1961).

25. W.B. Shockley, United States Patent 2,954,307 (filed 1957); H.F. Mataré, Z. Physik <u>145</u>, 206 (1956), for a book on early work, see Ref. 47.

26. K. Ploog, Ann. Rev. Mater. Sci. 11, 171 (1981).

27. J. Hornstra, J. Phys. Chem. Solids 5, 129 (1958).

28. See, for example Ref. 3, p. 183.

29. Proceedings Hünfeld Symposium: J. de Physique 40, suppl. C6, introductory foreword (1979).

30. P. Haasen, Acta Metallogr. 5, 598 (1957).

31. H. Alexander, Ref. 29, p.C6-1.

32. For an introductory summary on structures and statistics of dislocations in compounds, see R. Masut, C.M. Penchina, and J.L. Farvacque, J. Appl. Phys. 53, 4964, 4970 (1982).

33. For a recent summary: P.B. Hirsch, Ref. 29 p. C6-27; also Ref. 3.

34. D.J. H. Cockayne, I.L.F. Ray, and M.J. Whelan, Phil. Mag. 22, 853 (1970).

35. F. Häussermann and H. Schaumburg, Phil. Mag. 27, 745 (1973).

36. R. Meingast and H. Alexander, phys. stat. sol. a17, 229 (1973).

37. P.L. Gai and A. Howie, Phil. Mag. 30, 939 (1974); H. Alexander, Proc. 6th. Eur. Congr. Electr. Micr. Jerusalem, p.208 (1976); H. Gottschalk, G. Patzer, and H. Alexander, phys. stat. sol. a45, 207 (1978).

38. D.J.H. Cockayne, I.L.R. Ray, and M.J. Whelan, Phil. Mag. 20, 1265 (1969).

39. An early textbook on dislocations is: W.T. Read, "Dislocations in Crystals", (Mc.Graw Hill, New York 1953).

40. D. Hull, "Introduction to Dislocations", 2nd ed. (Pergamon, Oxford 1975).

41. A. Seeger, H. Donth, and F. Pfaff, Disc. Faraday Soc. 23, 19 (1957) or Ref. 40, p. 232.

42. P. Haasen, in Ref. 29, p. C6-111.

43. P.B. Hirsch, in Ref. 29, p. C6-117.

44. D.V. Lang and L.C. Kimerling, Appl. Phys. Lett. 28, 248 (1976); reviewed in: L.C. Kimerling, Solid-St. Electron. 21, 1391 (1978).

45. H. Alexander, Abt. f. Metallphysik, II. Phys.Inst.Universität Köln, Fed.Rep.of Germany, organizer, see: H. Alexander, C. Kisielowski-Kemmerich, and E.R. Weber, to be publ. 1982/83 in Physica B.
45a A very recent critique by V.V. Kveder et al., phys. stat. sol. a72, 701 (1982).

46. "Grain Boundaries in Semiconductors", ed. by H.J. Leamy, G.E. Pike, and C.H. Seager (Proc. Mat. Res. Soc. 1981 Meeting), (North Holland, New York 1982).

47. H.F. Mataré, "Defect Electronics in Semiconductors" (Wiley, New York 1971).

48. W. Bollmann, "Crystal Defects and Crystalline Interfaces" (Springer, Berlin 1970); W. Bollmann, Surface Sci. 31, 1 (1972).

49. S.T. Pantelides, Rev. Mod. Phys. 50, 797 (1978); G.D. Watkins in: Proc. XVI th Int. Conf. Phys. Semic., M. Averous, ed. (1982), to be published.

50. M.J. Kirton and M. Jaros, J. Phys. C14, 2099 (1981); J.L. Farvacque, D. Ferre, and P. Lenglart, Int. Phys. Conf. Ser. 59, 389 (1981); R. Jones, S. Öberg, and S. Marklund, Phil. Mag. 43, 839 (1981), also Ref. 32.

51. A comprehensive review: H.J. Leamy, J. Appl. Phys. 53, R51 (June 1982).

52. C. Donolato, Optik 52, 19 (1978), phys. stat. sol. a65, 649 (1981).

53. C. Donolato and H. Klann, J. Appl. Phys. 51, 1624 (1980).

54. J. Marek, J. Appl. Phys. 53, 1454 (1982).

55. L. Pasemann, phys. stat. sol. a69, K199 (1982), also Ref. 57.

56. S. Kawado, Jpn. J. Appl. Phys. 19, 1591 (1980).

57. L. Pasemann, H. Blumtritt, and R. Gleichmann, phys. stat. sol.a70,197(1982).

58. L. Castellani, P. Gondi, C. Patuelli, and R. Berti, phys. stat. sol. a69, 677 (1982).

59. G.R. Booker, A. Ourmazd, and D.B. Darby, Ref. 29, p. C6-19.

60. G.L. Miller, D.V. Lang, and L.C. Kimerling, Ann.Rev.Mater.Sci.7,377 (1977).

61. L.C. Kimerling, J.R. Patel, J.L. Benton, and P.E. Freeland in: "Defects and Radiation Effects in Semiconductors, 1980", R.R. Hasiguti, ed. (Conf.Ser.59, Instit.of Physics, Bristol, 1981), p.401.

62. A. Ourmazd and G.R. Booker, phys. stat. sol. a55, 771 (1979).

63. P.M. Petroff and D.V. Lang, Appl. Phys. Lett. 31, 60 (1977).

64. K. Böhm and B. Fischer, J. Appl. Phys. 50, 5453 (1979); also W. Heinke and H.J. Queisser, Phys. Rev. Lett. 33, 1082 (1974).

65. R.H. Uebbing, P. Wagner, H. Baumgart, and H.J. Queisser, Appl. Phys. Lett. 37, 1078 (1980).

66. N.A. Drosdov, A.A. Patrin, and V.D. Tkachev, Zh. Eksp. Theor. Fiz. Pis'ma Red. 23, 651 (1976) [Sov. Phys. J. Exp. Theor. Lett. 23, 597(1976)].

67. J.Weber, R. Sauer, E.R. Weber, and H. Alexander, to be published.

68. R.H. Uebbing and K.L. Merkle (unpublished).

69. A. Steckenborn, H. Münzel, and D. Bimberg, J. Luminescence 24/25,351(1981).

70. A.K. Chin, H. Temkin, and S. Mahajan, Bell.System.Tech.J.60, 2187 (1981).

71. I. Hayashi, J.Phys. Soc. Jpn. Suppl. A49, 57 (1980).

72. J.D. Dow and R.E. Allen, Appl. Phys. Lett. 41, 672 (1982).

73. V.F. Mayer and J.M. Galligan, Appl. Phys. Lett. 40, 1020 (1982); also see S. Takeuchi and K. Maeda (this volume).

74. K.H. Küsters and H. Alexander, ICDS-12 (Amsterdam 1982) to be published in Physica B.

75. D. Mergel and R. Labusch, phys. stat. sol. a69, 151 (1982).

76. E.J. Pakulis and C.D. Jeffries, Phys. Rev. Lett. 47, 1859 (1981).

77. E. Bauser and H. Strunk, Thin Solid Films 93, 185 (1982).

78. E. Bauser and H. Strunk, J. Crystal Growth 51, 362 (1981).

79. F.C. Frank, J. Crystal Growth 51, 367 (1981).

80. B. Fischer, E. Bauser, P.A. Sullivan, and D.L. Rode, Appl. Phys. Lett. 33, 78 (1978).

81. Dislocation-free Si develops swirl-defects; see e.g. a recent example: M. Futugami, J. Appl. Phys. 52, 5575 (1981), also Ref. 83.

82. K. Sumino in "Semiconductor Silicon 1981", ed. by Electrochem. Soc., Princeton, p. 208; also this volume.

83. Defects in Si VLSI-Technology reviewed: L. Jastrzebski, IEEE Trans.ED-29, 475 (1982).

84. S.L. Chang, H.J. Queisser, H. Baumgart, W. Hagen, and W. Hartmann, Phil. Mag. (in press).

85. S.R. Singh and S. Singh, phys. stat. sol. b112, 51 (1982).

86. J.C.H. Spence (Si interfaces); T. Yamashita and F.A. Ponce (GaAs); this volume (1982 Mat. Res. Soc. Ann. Meeting).

87. Proc. Int. Conf. on Grain Boundaries, Perpignan (1982), to be published.

88. C.H. Seager, this volume (1982 Mat. Res. Soc. Ann. Meeting).

89. J. Werner, W. Jantsch, and H.J. Queisser, Solid State Commun. 42, 415(1982).

90. J. Werner and H. Strunk in Ref.86 and to be published.

91. E. Ziegler, W. Siegel, H. Blumtritt, and O. Breitenstein, phys. stat. sol. a72, 593 (1982).

92. A.Neugroschel and J.A. Mazer, IEEE Trans. ED-29, 225 (1982).

93. E. Poon and W. Hwang, Solid-St. Electronics 25, 699 (1982).

94. K. Kato, T. Ono, and Y. Amemiya, IEEE Trans. ED-29, 1156 (1982).

95. S. Milshtein, phys. stat. sol. a72, K99 (1982).

THE ELECTRICAL BEHAVIOR OF GRAIN BOUNDARIES IN SILICON*

C. H. SEAGER
Sandia National Laboratories, Albuquerque, NM, 87185, USA

ABSTRACT

Despite the fact that lattice imaging studies
have shown that grain boundaries in group IV semiconductors often have structures which are complicated and inhomogeneous on the scale of tens-to-
hundreds of angstroms, simple theories assuming
uniform double depletion layers have recently been
shown to successfully predict many of the majority
carrier transport properties of these defects. On
the other hand our knowledge of the interaction of
grain boundaries with minority carriers is in a
considerably more primitive state. I will describe
recent attempts to understand the effects of illumination on grain boundary potential barrier heights
and the influence of these defects on the optically
generated minority carrier population. Quantifying
this latter interaction is particularly important
in estimating the performance of polycrystalline
solar cells. Simple but elegant scanned excitation
measurements for measuring s, the minority carrier
recombination velocity at grain boundaries, will be
reviewed. I will discuss recent measurements of s
as a function of temperature and illumination intensity and show how these data can be correlated with
zero-bias impedance measurements.

I. Introduction

In the last several years an increased interest in the use of polycrystalline semiconductors in devices such as varistors, IC gates and inter-
connections, and photovoltaic cells has been accompanied by a concomitant
increase in our knowledge of the electronic transport properties of semiconductor grain boundaries[1]. Transmission electron microscope studies of
the detailed microstructure of boundaries have also proliferated with particular emphasis on group IV semiconductors[2-7]. These studies, as well as
experiments involving micro-chemical analysis of fractured (exposed) boundary
interfaces, have raised questions about the origin of grain boundary trapping
states and the relevance of simple, uniform charge models for the band structure near these defects. Some of these issues will be reviewed here.
While several investigators have claimed[9-14] that a simple, uniform
charge, double depletion layer model can successfully account for many of
the majority carrier transport properties of silicon and germanium bicrystals,
essentially no data has been available until recently on the interaction of
minority carriers with these defects. In section III of this paper we shall
discuss recent experimental data taken in our laboratory[15,16] and elsewhere[8]

*This work performed at Sandia National Laboratories supported by the U.S.
Department of Energy under contract #DE-AC04-76DP00789.

on optically illuminated bicrystals. To account for these results we will again use the simplest, zeroth-order model for grain boundary band structure - the uniform double depletion layer picture.

II. Structural Aspects of Semiconductor Grain Boundaries

Prior to the advent of lattice imaging techniques it was postulated that the lattice mismatch at a grain boundary was accommodated by the formation of regular (planar) arrays of dislocations[17]. In this simple picture boundaries with a large twist component would have a high density of screw dislocations, and those having a large tilt component could be expected to contain edge dislocations. Since it has been known for some time that edge dislocations introduced by plastic deformation in silicon and germanium introduce in-gap electrical states[17], the simplest assumption would be to expect that grain boundary trapping states arise in this fashion. There are cautions which must be observed here however. While screw dislocations are expected to contain no dangling bonds, reconstruction or pairing of the dangling bonds at edge dislocations is a strong possibility for reducing their activity[18]; in addition, the dissociation of 60° edge dislocations into two partials enclosing a stacking fault has also been observed[6,19]. It is clear that there is no single dislocation type which can be expected to dominate all grain boundary structures, and furthermore the extent of electrical activity of a given dislocation depends strongly on presently unestablished details such as its degree of bond reconstruction and/or structural dissociation. Nevertheless, observations of well ordered arrays of dislocations has been made at boundaries in silicon[4,6] and germanium[2] and some cases correlated with EBIC (Electron Beam Induced Current) activity[7]. Furthermore the distinct signature of the ESR resonance associated with a non-tetrahedrally coordinated silicon atom (dangling bond) has been recently seen in polycrystalline silicon by Johnson et al[20]. From previous work this resonance has been rather firmly correlated with the presence of donor/acceptor sites in a-Si[21], plastically deformed Si[22], and at Si-SiO$_2$ interfaces[23,24]. It would thus be somewhat surprising if at least some of the electrical activity associated with grain boundaries was not associated with "dangling bonds."

Recent work by Bourret et al[2], Cunningham et al[4], and others[25] has indicated that certain boundaries have a facetted, irregular structure rather than simple planar arrays of dislocations; this gives rise to the nonuniform, sometimes periodic EBIC activity that is frequently seen on both a microscopic scale[7] and sometimes even over macroscopic distances[26]. As we shall discuss in the next section, the scale of these microstructural variations of trap density is crucial in determining the applicability of models[9] assuming uniformly trapped grain boundary charge. What they certainly indicate is that careful measurements of grain boundary microstructure are desirable when applying models designed to explain the electrical transport properties of these defects.

While the data linking grain boundary trap sites with dangling bonds at dislocations is ambiguous and complex, the situation with regard to impurity effects is best described as confused. Kazmerski and co-workers[8,27] have observed evidence of the segregation of oxygen and transition metal impurities at grain boundaries following high temperature anneals. Russell et al[28] have suggested that the increases observed in the grain boundary recombination velocity after such anneals are a direct result of oxygen segregation. Redfield[29] has gone so far as to suggest that clean grain boundaries in silicon are electrically inactive, and that oxygen segregation is responsible for all grain boundary activity.

Many of these assertions are based on observations of changes in grain boundary barrier heights in Wacker Silso polycrystalline silicon. However, the experimental data for this material is mixed. More recent work by Fonash[30] and Seager and Ginley[31] have shown that many grain boundary potential barriers in this material actually <u>decrease</u> with thermal annealing and some of the changes seen take place below 600°C[31]. Fonash[30] has pointed out that properties of the Wacker material vary from lot to lot; this fact may be responsible for these contrasting data. In addition we note that stress relaxation of cut wafers is likely to occur in this material at temperatures above 600°C. The influence of this phenomena on grain boundary activity has yet to be established. Other possible mechanisms include the outgassing of impurities absorbed from the growth atmosphere.

It is clear that careful microstructural observations correlated with electrical activity on grain boundaries in a variety of polycrystalline materials measurements will be needed to resolve some of these issues; it is quite likely that the degree of impurity influence may vary significantly depending on the boundary microstructure. Certainly the degree of grain boundary impurity segregation would be expected to depend strongly on the stress field present at a boundary[17]. Rough calculations of boundary strain energy[32] show that it is usually larger than the electrostatic energy associated with the charged traps. This in turn suggests that mechanical relaxations of boundary structures, particularly in highly stressed thin film or cast polycrystalline material might result in a lowered total energy state of the boundary which could well involve <u>larger</u> electrostatic energies.

It is safe to say that interpretation of most of the grain boundary electrical data which have been reported so far has been a relatively trivial matter when compared to the prospects of sorting out the structural details of these defects.

III. Modeling the Grain Boundary Minority Carrier Interaction

Extensive discussions concerning the formation of the grain boundary potential barrier are available in the literature[1]. We shall only review them briefly here.

If a grain boundary contains electronic states which can accept a majority carrier <u>and</u> which lie below (or above in the case of p-type grains) the equilibrium Fermi level of the bulk, charge transfer to these states will occur until their energies become large enough to stop this process. The simplest assumption to make in modeling the resultant potential barrier is to postulate the existence of trapping states which are uniformly distributed in the plane of the grain boundary. The observations of grain boundary structural inhomogeneities mentioned in the preceeding section suggest that in some cases this hypothesis may have to be modified. However, if these inhomogeneities are on a scale much smaller than the depletion width, or if the lowest trap state densities in the boundary are sufficient to pin the equilibrium Fermi level near the typically observed midgap position, one still expects to find a relatively uniform electrostatic potential at the barrier center. Variations of grain boundary impedances and EBIC response have been seen along boundary planes on a scale of tens to thousands of microns[7], so that applying uniform barrier theories to all macroscopic grain boundary segments is certainly to be avoided. Much of the bicrystal data that we have reported in the past[10-12] has been taken on samples whose barrier uniformity has been checked by techniques such as controlled etching and large area EBIC analyses.

Given the existence of a uniform grain boundary potential barrier we can now address the subject of the interaction of this structure with a population of optically generated minority carriers in the limit where the majority carrier density is relatively unchanged (low injection). Because the minority carriers can alter the height of the grain boundary barrier through changes in the steady state trapped majority carrier population, and conversely recombination at the boundary plane can lower the local concentration of minority carriers, this problem is a coupled one which necessarily involves some tedious mathematical expressions. Various published models[33-35] address these issues in some detail; we shall not reproduce these treatments here - we prefer to concentrate on some special cases of the problem where simple solutions can be found which are illustrative of the physics involved. We shall divide our discussion into two parts, treating first the alterations of the barrier produced by minority carrier excitation and secondly the question of the minority carrier distribution near a boundary. Recent data taken on silicon bicrystals which bear on these issues[8,35] will be discussed.

A. Grain Boundary Barrier Height Alterations During Illumination

We begin this section in what appears to be a rather paradoxical fashion by discussing the flow of minority carriers (holes, for example) into the edge of the grain boundary depletion region. The solution to this simple one dimensional diffusion problem is[35].

$$J_p(\lambda) = \frac{eD_p}{X_L} (p_\infty - p_O) \tag{1}$$

where p_O is the, as yet unspecified, hole concentration at the depletion region edge at $x = \lambda$, D_p the hole diffusion coefficient, X_L the (bulk) minority carrier diffusion length, and p_∞ is the bulk hole concentration. p_∞ is related to the excitation rate, L, and the bulk (grain) lifetime, τ, by:

$$p_\infty = L\tau \tag{2}$$

Using extensive numerical evaluation of the equations describing this problem, Seager[35] has shown that unless ϕ_B is quite small (<0.15eV) p_O can generally be neglected with respect to p_∞. This is equivalent to saying that the grain boundary is effectively a "perfect sink" for minority carriers. This does not mean, however (as we shall show in Section B) that the minority carrier recombination velocity is equivalent for all barriers above 0.15eV. The utility of considering expressions without the p_O term is that it becomes apparent that the hole current crossing the depletion edge scales almost linearly with L (insofar as τ is L independent). As shown by Seager[35] the assumptions of current conservation in the depletion region are well satisfied (negligible generation or recombination) so that Eq. 1 adequately describes the hole current incident on the grain boundary plane.

In equilibrium (L = 0) the net majority carrier flow into the grain boundary trapping states is zero. For wide barriers ($\lambda > 250$ Å) this equilibrium is brought about by a balance of the majority carrier emission from traps and the thermionic capture rate, written as[35]

$$J_{CAP}^n = 2cA \exp[-(\phi_B + \zeta)/kT], \tag{3}$$

where c is the fraction of electrons emitted over the barrier that are captured by the traps, A is a psuedo-Richardson factor, and ζ is the (positive) energy separation of the Fermi level from the majority carrier band edge. With a finite excitation, L, a net capture rate of electrons is required to

balance the minority carrier current (twice Eq. 1 since there are two depletion regions). This requires a decrease in ϕ_B. An important simplification of the problem comes about when it is realized that, for trap states
not too widely spaced in energy, the emission rate of majority carriers is
expected to be a weak function (through the number of filled traps which can
emit) of the total charge in the boundary. The capture rate on the other
hand is exponentially dependent on ϕ_B and grows so rapidly as ϕ_B is
decreased by more than a few kT that, by comparison, electron emission is
essentially negligible. For this reason we have approximated the emission
rate by a constant[35], i.e., its dark (L = 0) value. This approximation is
very good for small alterations in barrier charge Q and gets better for
larger ones (higher L) since the exponentially growing electron capture rate
totally swamps out the weakly decreasing emission term. If we make this
approximation and equate net electron and hole currents into the barrier, we
find:[35]

$$\phi_{BD} - \phi_B = kT \ln [1 + X_L Le^2 (1 - c/2)/kTcG_D]$$
(4)

where the dark, zero-bias conductance is given by:

$$G_D = (1 - c/2) (eA/kT) \exp [-(\phi_{BD} + \zeta)/kT]$$
(5)

where ϕ_{BD} is the equilibrium, dark band bending.

Both Kazmerski[8] and Seager[35] have attempted to verify this prediction
on silicon bicrystals by measuring ϕ_B as a function of L, Kazmerski[8] using
impedance methods and Seager[35] using both impedance and capacitance techniques. Seager's results[35] are shown in Figure 1. These data were obtained
with weakly absorbed light incident along the barrier plane, insuring that
the barrier height alterations were reasonably uniform throughout the sample.
A good functional agreement with equation was seen although X_L was not precisely known for this specimen allowing some adjustment of the theoretical
curve along the abscissa. The limiting (high L) behavior prediction that ϕ_B
should change 2.3kT for every decade increase in L is clearly seen. This is
a simple result of the exponential form for J_{CAP}^n and the linear dependence
of J_p on L. Kazmerski's ϕ_B data[8] show good agreement with this formula
over much of the light intensity range, but flatten off somewhat at higher L
values. This could be due to his use of strongly absorbed radiation, or
perhaps to failure of the model approximations at low barrier heights. It
is clear that, while some data exists which support the simple theory described here, much more experimental verification of this relationship would be
desirable. In fact, measurements of this type are particularly useful in
several ways; not only do they indicate the excitation range where ϕ_B is
constant, which is important in the recombination velocity problem, but they
also could in principle yield information about the dominant mechanism for
majority carrier transport near the barrier. For example, in the case of
more highly doped (>10^{17} cm^{-3}) grains where λ_n is smaller, majority carrier
capture is likely to be influenced by tunnelling processes[36] and the functional form of J_{CAP}^n will be quite different from Eq. 3.

B. The Effect of Grain Boundaries on the Steady State Minority Carrier Dis-
    tribution

In this section we discuss an aspect of the grain boundary recombination
problem which bears directly on the performance of polycrystalline materials

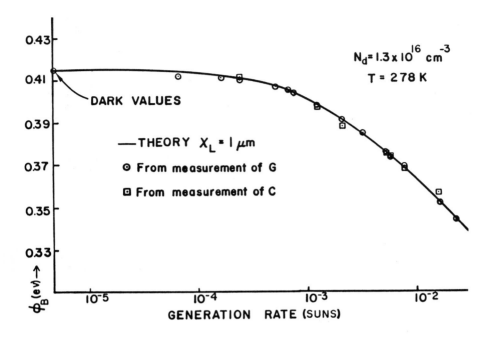

Figure 1. Measured and calculated (Eq. 4) values of ϕ_B as a function of illumination for a silicon bicrystal having $N_d = 1.3 \times 10^{16}$ cm^{-3}. The conductance data were obtained using zero-bias, d.c. conductance measurements and Eq. 5; the capacitance determined points were generated from 1 MHz values and Eq. 43 of reference 35. Dark values are shown at $L = 5 \times 10^{-6}$ sun for convenience of presentation.

in minority carrier devices such as photovoltaic cells. As we have indicated in the preceeding section, it is always possible to solve for the minority carrier population by solving the appropriate diffusion problem (with bulk generation and recombination terms included) outside the grain boundary depletion regions. The effect of the grain boundaries comes in through the boundary value condition at the edge of the space charge regions, i.e., through the p_o term in Eq. 1. We shall make the argument below that p_o can in some simple, but useful limits, be simply related to the hole current flowing into the grain boundary center.

We begin by observing that the most likely centers for hole recombination in the grain boundary are expected to be those that are negatively charged; it is, in fact, commonly observed[37] that the capture cross sections of coulombic traps are from one to two orders of magnitude larger than for neutral sites. Using the Shockley, Read, Hall formalism[38] a reasonable form for J_p at the boundary center can then be written as[35]:

$$J_p(0) = 1/2 \ v^p_{th} \ \sigma_p Q p_b \ , \qquad (6)$$

where v_{th}^p is the hole thermal velocity, σ is the cross section of a trapped electron for capturing a hole, p_b is the valence band hole density at the boundary center, and Q is the trapped majority carrier density (in coulombs/m^2). The factor of 1/2 comes from the fact that we have defined J_p to be the current flowing into the boundary center from one depletion region only.[35] It is clear from the form of Eq. 6 that not only we are specifically ignoring recombination at neutral sites, but also charged sites which are compensated by oppositely charged species in the boundary. In the parlance of the single electron formalism of Seager and Pike[39], these other sites, if they exist, must lie below the socalled "neutral" Fermi position in the grain boundary. As we shall mention below, the only measurements of recombination rate that are presently available are in the regime where Q is not changing (as a function of L), and thus the existence of any recombination current components not proportional to Q is as yet unverified. We can now find an expression for the minority carrier recombination velocity, defined at the depletion edge as:

$$ s \equiv \frac{J_p(\lambda)}{e\, p_0} \tag{7} $$

where $X = \lambda$ is the edge of the grain boundary depletion region. We accomplish this by again ignoring the last term in the parenthesis of Eq. 1 and noting that it has been shown[34,35] that for barrier heights <0.25eV, the hole quasi-Fermi level varies less than kT in the depletion region.

Thus:
$$ p_b \tilde{=} p_0 \exp\left(\phi_B/kT\right) . $$

Using the double depletion layer formula[9] relating Q and ϕ_B,

$$ Q = 2(2\varepsilon\varepsilon_0 N_D)^{1/2}\, \phi_B^{1/2} , \tag{8} $$

we get:

$$ s = e^{-1}\, v_{th}^p\, \sigma_p\, (2\varepsilon\varepsilon\, N_d)^{1/2}\, \phi_B^{1/2}\, \exp\left(\phi_B/kT\right). \tag{9} $$

Note that we have explicitly assumed current conservation[35] in the depletion region by substituting $J_p(0)$ for $J_p(\lambda)$.

Eq. 9 indicates that, as long as the grain boundary potential barrier height remains relatively constant (lower excitation levels), the minority carrier flux at the depletion region edge is proportional to the concentration at that point. This leads to a particularly simple solution to the minority carrier profile from which parameters such as the short circuit current can be calculated in a real polycrystalline solar cell. It also indicates that s is expected to decrease quite rapidly as a function of temperature or at higher illumination levels where ϕ_B is decreased.

Using the data of Fig. 1 where $(\phi_B + \zeta) = .62$eV and Eq. 3. we estimate that barriers displaying an activation energy of ~ 0.30eV or less should have ϕ_B (and hence s) be illumination independent for $X_L = 20\,\mu$m and excitation levels <10^{20} electron-hole pairs per cm^3 per second. This level, defined as 1 sun in reference 35, is roughly the total number of carrier pairs generated per second in AM1 sunlight averaged over a $10\,\mu$m absorption depth and is useful for reference purposes in discussing generation in indirect gap semiconductors like silicon. It is, therefore, clear from this formalism that we expect many of the barriers typically found in p type silicon to be relatively insensitive to AM1 sunlight. At the same time our prior numerical calculations of s (and data that we shall display shortly) indicate that these same barriers can have s values between 10^5 and 10^6 cm/sec, which means that they drastically lower the minority carrier density within a distance roughly X_L from the boundary center. Thus, the common observation that fine-grained (less than 50 μm grain size) polycrystalline silicon solar cells show rather low short circuit currents[40] is to be expected on the basis of this model.

It is possible, using polycrystalline diodes with well separated grain boundaries oriented normal to the collecting junction to accurately measure s by monitoring, I_{sc}, the short circuit current as a monochromatic laser beam is rastered across a particular grain boundary. Zook[41] has given a formula for calculating s from the distance dependence of I_{sc}. Faughnan[42] pioneered the use of this formula in Laser-Beam-Induced-Current (LBIC) studies of s in virgin and passivated grain boundaries in Wacker polysilicon, and Seager has recently shown[15] that this formula accurately describes the wavelength and distance dependence of LBIC scans in Honeywell silicon-on-ceramic n^+/p polycrystalline diodes. Some data extracted from this latter study[15] is shown in Fig. 2. It is worthwhile noting that it was not hard, using higher power levels of the excitation laser, to markedly reduce the apparent values of s; in other words, only at rather low light levels were intensity independent s values observed. Quantitative comparison of the excitation levels where this decrease was seen with theoretical predictions has not yet been made, however.

In an attempt to check the predictions of Eq. 9 we have recently[16] measured the temperature dependence of s values on a variety of grain boundaries and attempted to correlate this data with other measurements. Because $e^{-\zeta/kT}$ is nearly constant, it can be seen from Eq.'s 5 and 9 that the the recombination velocity and the zero-bias dark resistance (= $1/G_D$) should have the same temperature dependence. We have measured G_D and s on several barriers by etching off the n+ layer and back contact on the same laser scanned diodes used to obtain recombination data like that shown in Fig. 2. Using a surface potential probe technique the barrier resistance was then measured at various temperatures. Results for one particular grain boundary are shown in Fig. 3; other data is listed in Table 1. Considering the errors inherent in determining s, the agreement seen between the variations of s and the dark resistance is reasonable. It was also observed[15,16] that the magnitude of g_n predicted from the s values and Eq. 9 was reasonable ($10^{-15} - 10^{-16}$ cm^2) for carrier capture by a charged trap. (Note that since these samples have p-type grains the carrier type indices in Eq. 9 are reversed).

While the recombination data quoted above seems to follow the predictions of Eq. 9, this is by no means a rigorous test of the model. In particular it would be desirable to obtain recombination velocity data in the regime where ϕ_B is not equal to its dark value to ascertain whether s is strictly proportional to Q; if this is not observed, minority carrier capture at neutral or compensated traps might have to be included as a necessary complication to this simple model. In order to avoid resolving the equations treated by Zook[41] with non-linear boundary conditions (excitation dependent s), the LBIC experiments could be done in the high L regime using a small signal technique. That is to say response to a low power, chopped laser beam could be measured with a background source of intense, uniformly absorbed, near band gap light which could be adjusted to decrease the steady state value of ϕ_B. While difficult, such experiments would be quite informative particularly in view of the fact that the barrier heights at some grain boundaries in polycrystalline silicon must surely be functions of illumination near AM1. In addition to direct measurements of this dependence, this effect can be inferred from the frequent observation[40] that I_{sc} is not linearly dependent on light intensity in many polycrystalline silicon photo-voltaic cells, particularly since these non-linearities are in excess of those which would be expected from the normal, small L dependences of τ.

IV. Conclusions

We have discussed the structural aspects of grain boundaries with parti-cular emphasis on group IV semiconductors. In addition to the complexities frequently seen in grain boundary microstructure, there is also considerable controversy considering the nature of the defects which give rise to electri-cal activity. While there is evidence from ESR[20] that the well-known

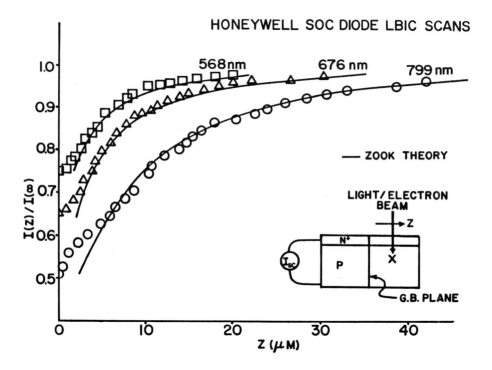

Figure 2. Room temperature, low power laser-beam-induced-current scans of a single grain boundary in a Honeywell silicon-on-ceramic n+p diode at three different excitation wavelengths. I(z) is the diode short circuit current as a function of distance, z, from the grain boundary plane. The theoretical lines were generated using the formula given by Zook (reference 41) using X_L = 40 μm, D = 25 cm²/sec, s = 1.25 x 10⁵ cm/sec and values of the optical absorption coefficient from reference 43. The inset shows the geometry of the scanned excitation experiments.

352

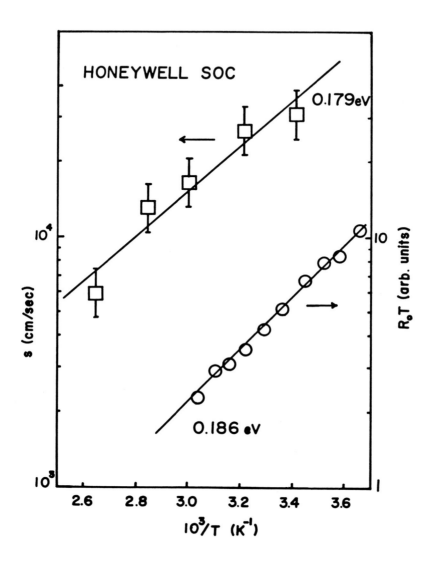

Figure 3. Values of the grain boundary recombination velocity and dark resistance, temperature product plotted versus inverse temperature for grain boundary H211-1-4 (see Table 1). The solid lines are the least squares Arrhenius fits of the experimental data.

TABLE I

Barrier parameters for several grain boundaries in Honeywell polycrystalline silicon. The diffusion length, X_L, is measured in the particular grains adjacent to each boundary. E_S and E_A are the least squares fit activation energies of the grain boundary recombination velocities, s, and the products of the dark zero-bias barrier resistance and absolute temperature, respectively.

Barrier Designation	X_L(300K) (μm)	s (300K) (cm/sec)	E_S (eV)	E_A (eV)
H211-1-4	29	3.1×10^4	0.179	0.186
H211-1-5	13	3.5×10^5	0.215	0.250
H211-1-8	22	1.5×10^5	0.153	0.174

dangling bond defect is present in polycrystalline silicon and is removed by hydrogenation, there is also indirect evidence, from thermal annealing experiments that certain impurities tend to segregate at grain boundaries,[8,27] and that some changes are at the same time observed in electrical properties[28]. We conclude that while it is more likely that the dangling bond defect is the primary source of grain boundary traps, more careful work must be done to establish the possible role of impurities.

Models for the electrical behavior of boundaries were also reviewed with special emphasis on the interaction of these defects with optically generated minority carriers. Recent data on the illumination dependence of the barrier height and the temperature dependence of the minority carrier recombination velocity are in reasonable agreement with a simple model based on the double depletion layer hypothesis. Justification of many of the approximations used to predict this data comes from a previous more detailed study of this problem by the present author[35]. It was concluded that while this model yields some simple formulas that are apparently useful in some regimes, much more data is required to confidently predict minority carrier behavior in polycrystalline semiconductors.

Acknowledgement

The author would like to thank the Solar Energy Research Institute for the support necessary to carry out this work.

References

1. See for instance, Grain Boundaries in Semiconductors, ed. by G. E. Pike, C. H. Seager, and H. T. Leamy (North Holland, New York, 1982).

2. A. Bourret and J. Desseaux, Phil. Mag. A39, 405 (1979).

3. C. B. Carter, J. Rose, and D. G. Ast, Proc. 39th EMSA (1981).

4. B. Cunningham and D. Ast in reference 1, p. 21.

5. Anne-Marie Papon, Maurice Petit, Georges Silvestre, and J. J. Bacmann, reference 1, p. 27.

6. C. B. Carter, reference 1, p. 33.

7. B. Cunningham, H. P. Strunk, and D. G. Ast, in reference 1, p. 51.

8. L. L. Kazmerski, J. Vac. Sci. Technol. 20, 423 (1982).

9. W. E. Taylor, N. H. Odell, and H. Y. Fan, Phys. Rev. 88, 867 (1952).

10. C. H. Seager and G. E. Pike, Appl. Phys. Lett. 37, 747 (1980).

11. C. H. Seager and G. E. Pike, Appl. Phys. Letter. 35, 709 (1979).

12. C. H. Seager, G. E. Pike, and D. S. Ginley, Phys. Rev. Lett. 43, 532 (1979).

13. J. Werner, W. Jantsch, K. H. Froehner, and H. J. Queisser in reference 1, p. 99.

14. R. K. Mueller, J. Appl. Phys. 32, 635 (1961); 32, 640 (1961).

15. C. H. Seager, J. Appl. Phys. 53, 5968 (1982).

16. C. H. Seager, to be published in the Nov. 1 Issue of Applied Physics Letters.

17. See for instance, Defect Electronics in Semiconductors, H. F. Matare, (John Wiley & Sons, New York, 1971).

18. J. Hornstra, J. Phys. Chem. Solids 5, 129 (1958).

19. I. L. F. Ray and D. J. H. Cockayne, Proc. R. Soc. London A325, 532 (1971).

20. N. M. Johnson, D. K. Biegelsen, and M. D. Moyer, Appl. Phys. Lett. 40, 882 (1982).

21. M. H. Brodsky and R. S. Title, Phys. Rev. Lett. 23, 581 (1969).

22. D. Lepine, V. A. Grazhulis and D. Kaplan, Physics of Semiconductors, International Conference, Rome, 1976, p. 1081 (unpublished).

355

23. P. J. Caplan, E. H. Poindexter, B. E. Deal, and R. R. Razouk, J. Appl. Phys. 50, 5847 (1979).

24. P. M. Lenahan and P. V. Dressendorfer, to be published in Applied Physics Letters.

25. C. Fontaine and D. A. Smith, reference 1, p. 39.

26. D. Vaughan, Phil. Mag. 22, 1003 (1970).

27. L. L. Kazmerski, P. J. Ireland and T. F. Ciszek, Appl. Phys. Lett. 36, 323 (1980).

28. P. E. Russell, C. R. Herrington, D. E. Burke, and P. H. Holloway in reference 1, p. 185.

29. D. Redfield, Appl. Phys. Lett. 38, 174 (1981).

30. S. Fonash at the Solar Energy Research Institute Polycrystalline Silicon Subcontractors Meeting, Golden, CO, June 17-18, 1982, SERI CP-211-1648.

31. C. H. Seager and D. S. Ginley, Fundamental Studies of Grain Boundary Passivation with Application to Improved Photovoltaic Devices, A Research Report Covering Work Completed from February 1981 to January 1982, SAND 82-1701.

32. C. H. Seager, unpublished work.

33. H. C. Card and E. S. Yang, IEEE Trans. Elect. Devices ED-24, 397 (1977).

34. J. G. Fossum and F. A. Lindholm, IEEE Trans. Elect. Devices ED-27, 692 (1980).

35. C. H. Seager, J. Appl. Phys. 52, 3960 (1981).

36. C. H. Seager and G. E. Pike, Appl. Phys. Lett. 40, 471 (1982).

37. See for instance, Deep Impurities in Semiconductors, A. G. Milnes (Wiley & Sons, New York, 1973).

38. R. N. Hall, Phys. Rev. 87, 387 (1952); W. Schockly and W. T. Read, Phys Rev. 87, 835 (1952).

39. G. E. Pike and C. H. Seager, J. Appl. Phys. 50, 3414 (1979).

40. See, for example, T. L. Chu, S. S. Chu, K. Y. Duh, and H. I. Yoo, Proc. of the National Workshop on Low Cost Polycrystalline Silicon Solar Cells, May 1976, Dallas, TX, p. 408.

41. J. D. Zook, Appl. Phys. Lett. 37, 223 (1980).

42. B. W. Faughnan at the Solar Energy Research Institute Polycrystalline Silicon Subcontractors Meeting, Colorado Springs, CO, Nov. 17-19, 1980. SERI/CP6141263.

43. W. C. Dash and R. Newman, Phys. Rev. 99, 1152 (1955).

THE INTERFACE STRUCTURE OF GRAIN BOUNDARIES IN POLYSILICON†

S. M. JOHNSON[*], K. C. YOO[**], R. G. ROSEMEIER[***], P. SOLTANI[***], AND H. C. LIN[**]
[*] Solarex Corporation, Rockville, MD 20850
[**] University of Maryland, College Park, MD 20742
[***] Semix, Incorporated, Gaithersburg, MD 20760

ABSTRACT

Normal grain growth in polysilicon material proceeds by a twinning mechanism such that the individual grain orientations are related by single or multiple twinning steps. The twinning relationship between adjacent grains can result in the alignment of their boundary interfaces along certain planar segments which have a very low electrical activity. The varying degree of electrical activity of these boundaries, and boundary positions, is attributed to the smaller dislocation portion of the larger change in orientation across the grains.

INTRODUCTION

In earlier investigations of cast polysilicon material using Laue back-reflection and x-ray topographic techniques, adjacent grain orientations were found to be related by either single or multiple twinning steps which resulted in a varying degree of electrical activity, measured by using a long wavelength scanning photoresponse method, at certain grain boundaries and boundary segments[1,2,3]. Twin-related boundaries have also been studied[4] in polysilicon material fabricated by a variety of other methods: edge defined film-fed growth (EFG)[5], silicon-on-ceramic (SOC), dendritic web, ribbon-to-ribbon (RTR), heat exchanger method (HEM), CVD polysilicon[6], and capillary action shaping technique (CAST)[7]. These studies can be favorably related to a pioneering investigation of twin boundaries in silicon by Kohn[8] who proposed certain special interface structures between grains for both first and second order twinning relationships. This current study presents the analysis of the structural and electrical properties of three segments of a boundary between two grains whose relative orientation is given by a two-step twinning process.

RELATIVE GRAIN ORIENTATIONS

Figure 1 is a optical micrograph of two relatively large grains, designated 1 and 2, located near the top of a cast polysilicon ingot. The relative orientation of grains 1 and 2 was determined by Laue back-reflection measurements and is shown in the superposed stereographic projections of Figure 2. The grains have coincident $[110]_1$ and $[101]_2$ axes with a rotation of $38.9°$ being required to bring the $[1\bar{1}0]_1$ and $[\bar{1}01]_2$ into coincidence by a dislocation-free two-step twinning process[1,8]. The grains also have coincident $[\bar{2}21]_1$ and $[\bar{2}12]_2$ axes with the common $<110>_{1,2}$ being located within the common $\{221\}_{1,2}$. Thus the $\{221\}_{1,2}$ is a symmetrical second order twin plane.

† Work supported by Semix, Incorporated under DOE Cooperative Agreement No. DE-FC01-80 ET 23197.

358

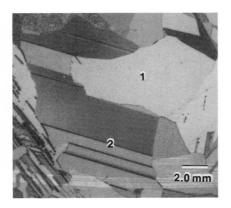

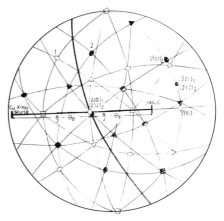

Fig. 1. Optical Micrograph of Grains 1 and 2.

Fig. 2. Superposed Stereographic Projections of Grains 1 and 2 (Orientation is Relative to Fig. 1).

The special orientation relationship between grains 1 and 2 allowed a simultaneous $\{220\}_{1,2}$ topographic image of both grains by surface reflection[9] to be obtained using the x-ray source position shown in Figure 2. The $\{220\}_{1,2}$ surface reflection, shown in Figure 3, was obtained with the Berg-Barrett technique using $CuK\alpha_1$ radiation at 20 KV, 20 mA for an exposure of 1.5 hours on 25 μm nuclear emulsion plate.

BOUNDARY INTERFACE POSITIONS AND THEIR ELECTRICAL ACTIVITY

Figure 4 is an optical micrograph of a portion of the boundary between grains 1 and 2 showing three boundary segments designated A, B, and C. The surface of the grains was chemically polished and Secco-etched to reveal dislocaton etch pits along the boundary. The boundary interface positions for these three segments were determined from an analysis of the trace directions on the crystal surfaces together with the measurement of the boundary positions in adjacent serial sections of the ingot. The boundary planes and their plane normals are identified in the superposed stereographic projections of Figure 5.

The boundary interface of segment A can be described by the near coincidence of the $(\bar{1}15)_1$ with the $(\bar{1}\bar{1}1)_2$ twin plane. The secondary electron image of this boundary segment and its electrical response measured by EBIC are shown in Figures 6A and 6B respectively. This boundary was previously observed in EFG material with TEM methods and its electrical response was measured with EBIC[10]. This boundary is labeled $\Sigma=9$ in the coincidence site lattice (CSL) theory[11] and has been more recently modeled[12]. The relatively weak electrical response at this boundary is attributed to the smaller dislocation portion of the larger change in orientation across the grains. This is confirmed by the observation of an ordered array of dislocation etch pits along the boundary. The localized regions of decreased EBIC response in the grain volumes are not observed in the secondary electron mode or by back-scattered electrons and thus appear not to be associated with localized surface contamination. These regions are currently being investigated.

Fig. 3. Simultaneous $(220)_1$, $(202)_2$ Berg-Barrett X-Ray Surface Reflections on Grains 1 and 2.

The boundary interface of segment B was identified as the symmetric second order twin interface of $(\bar{2}21)_1$ and $(\bar{2}12)_2$ described earlier in Figure 2. The secondary electron image of this boundary segment and its electrical response measured by EBIC are shown in Figures 7A and 7B, respectively. The extremely weak electrical response of this boundary, in comparison with segment A, is in agreement with the observation that this coherent symmetric twin interface ideally does not require misfit dislocations to account for the misorientation between the grains. Thus, relatively few etch pits are observed along this boundary segment. This boundary is also described by $\Sigma=9$ in CSL theory and has been observed more recently with TEM methods[6].

The boundary interface of segment C was identified to be closest to the near coincidence of $(\bar{3}26)_1$ and $(\bar{7}35)_2$. The secondary electron image of this boundary and its electrical response measured by EBIC are shown in Figures 8A and 8B respectively. This higher order asymmetrical boundary has not been previously observed and its structure is presently being investigated.

SUMMARY AND CONCLUSIONS

The twinning mechanism responsible for grain growth in polysilicon material can produce certain interface structures which have a very low electrical activity. Three segments of a grain boundary between two polysilicon grains whose relative orientation is given by a two-step twinning process were investigated. The symmetric coherent {221}/{221} second order twin interface was found to have an extremely weak electrical response due to the relatively few dislocations being required to account for the misorientation between the grains. The asymmetric {115}/{111} and {326}/{735} interface structures were found to have a larger electrical activity which can be accounted for by misfit dislocations in the interface. The varying degree of electrical activity of grain boundaries in polysilion material is attributed to the smaller dislocation portion of the larger change in orientation across the grains.

ACKNOWLEDGEMENTS

The authors are grateful for the advice and guidance of R. W. Armstrong, University of Maryland; for a helpful discussion with D. G. Ast, Cornell University; and to C. Winter for device fabrication.

360

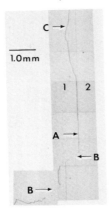

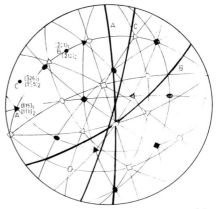

Fig. 4. Secco-Etched Surface of Grains 1 and 2 Showing Boundary Segments A, B, and C.

Fig. 5. Superposed Stereographic Projections of Grains 1 and 2 Showing the Boundary Planes and Plane Normals (Orientation is Relative to Fig. 4).

REFERENCES

1. R. W. Armstrong, M. E. Taylor, G. M. Storti and S. M. Johnson, Proc. 14th IEEE Photovoltaic Specialists Conf., 196, (1980).

2. R. G. Rosemeier, R. W. Armstrong, S. M. Johnson, G. M. Storti and C. Cm. Wu, Proc. 15th IEEE Photovoltaic Specialists Conf., 1331, (1981).

3. S. M. Johnson, R. W. Armstrong, R. G. Rosemier, G. M. Storti, H. C. Lin, and W. F. Regnault in Grain Boundaries in Semiconductors, Eds. H. J. Leamy, G. E. Pike, and C. H. Seager (North Holland, N. Y., 1982) p. 179.

4. B. Cunningham, H. P. Strunk, and D. G. Ast, in ref 3, p. 51.

5. D. G. Ast, B. Cunningham, and H. P. Strunk, in ref 3, p. 167.

6. B. Cunningham and D. G. Ast, in ref 3, p. 21.

7. K. Yang, G. H. Schwuttke, T. F. Ciszek, J. Crystal Growth 50, 301, (1980).

8. J. A. Kohn, American Mineralogist 43, 263, (1958).

9. R. W. Armstrong, in The Characterization of Crystal Growth Defects by X-Ray Methods, Eds. D. K. Bowen and B. K. Tanner (Plenum, London, 1980) p. 349.

10. H. Strunk, B. Cunningham, and D. Ast, in Defects in Semiconductors, Eds. J. Narayan and T. Y. Tan (North Holland, N. Y., 1981) p. 297.

11. W. Bollmann, Crystal Defects and Crystalline Interfaces (Springer-Verlag, Berlin, 1970).

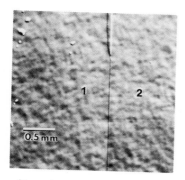

Fig. 6A. Secondary Electron Image
of Segment A.

Fig. 6B. EBIC Response of the Area
in Fig. 6A.

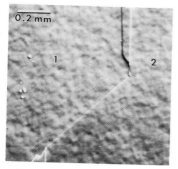

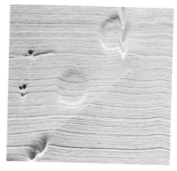

Fig. 7A. Secondary Electron Image
of Segment B.

Fig. 7B. EBIC Response of the Area
in Fig. 7A.

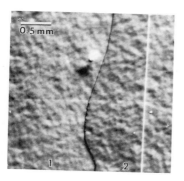

Fig. 8A. Secondary Electron Image
of Segment C.

Fig. 8B. EBIC Responses of the
Area in Fig. 8A.

ELECTRONIC STATES OF GRAIN BOUNDARIES IN BICRYSTAL SILICON[*]

C. M. SHYU[**] AND L. J. CHENG
Jet Propulsion Laboratory, California Institute of Technology
4800 Oak Grove Drive, Pasadena, California, 91109   USA
K. L. WANG
Department of Electrical Engineering, University of California,
Los Angeles, California, 90024   USA

ABSTRACT

Electronic states at a 20° symmetrical (100) tilt boundary
in p-type silicon were studied using deep level transient
spectroscopy (DLTS) and other electrical measurements. The
data can be explained with a model in which the local barrier
height at the grain boundary varies on a scale much smaller
than the boundary plane (~ 1 mm^2) under study. Based on a
relationship between the carrier capture cross section and
energy level deduced from the experimental data, we have been
able to calculate the distribution of the density of states
in the energy bandgap at the boundary, which contains two
groups of continuously distributed states; a major one whose
density of states increases monotonically with the position
of the state from the valance band, and a minor narrow one
whose density of states is centered at E_v + 0.20 eV.

INTRODUCTION

Electrical properties of grain boundaries in silicon, such as conductance
[1,2], capacitance, barrier height, minority carrier lifetime [3], and recombination velocity [4,5] have been investigated. It is known that the electronic
states in the energy bandgap at the grain boundary due to the presence of structural defects actually dictate the behavior of the electrical properties.
Therefore, it is very important to study the electronic states in the energy
bandgap at the grain boundary. One way to study the boundary states is to
examine the transient behavior of the boundary capacitance of bicrystal samples
using DLTS [6]. Our previous results [2,7] have illustrated the applicability
of the technique and revealed the continuous distribution of the density of
states in the bandgap. In addition, based on the observation, we proposed a
model which suggests the existence of a non-uniform spatial distribution of the
boundary states in silicon.
This paper reports new experimental results on a sample with 20° symmetrical
(100) tilt boundaries. A detailed calculation of the density of states at the
boundary based on the experimental results has been performed.

EXPERIMENTAL

The sample discussed in this paper was prepared from a 4.3 Ω-cm CZ-grown
p-type bicrystal silicon wafer with a 20° symmetrical (100) tilt boundary. The

[*]The research described in this paper was carried out for the Flat-Plate Solar
Array Project, Jet Propulsion Laboratory, California Institute of Technology and
was sponsored by the U.S. Department of Energy through an agreement with NASA.
[**]This will be part of Shyu's Ph.D. dissertation in the Department of Electrical Engineering, University of California at Los Angeles.

bar-shape sample has two ohmic contacts on each grain with the boundary running across the middle of the sample. The details of the sample preparation and the experimental techniques were described in previous papers [2,7].

RESULTS AND DISCUSSIONS

(A) ZERO-BIAS RESISTANCE AND CAPACITANCE MEASUREMENTS
    The purpose of the measurement is to determine the activation energy for the carrier transport at the boundary. In order to measure the value at a condition close to the thermal euilibrium, the voltage across the sample was set to be much smaller than q/kT. We applied 5 mV across the sample in the temperature range from 100 °K to 400 °K where q/kT is between 9 and 35 mV. Figure 1 is a typical plot of resistance as a function of temperature measured on the bi-crystal and single crystal silicon samples, respectively. According to the thermionic emission theory, the boundary resistance is an exponential function of the reciprocal temperature. At low temperature, the resistance of the sample with the grain boundary is much larger than the bulk resistance as shown in the diagram. As the temperature increases, the resistance of the sample with the boundary decreases drastically as predicted by the theory and the bulk resistance increases due to the decrease in the mobility. At high temperature, therefore, the bulk resistance and the bicrystal resistance are similar since the bulk resistance predominates. After applying temperature correction, an activation energy of 0.30 ± 0.01 eV was obtained from the slope of the straight portion of a plot of ln(RT) versus 1000/T, where R is the boundary resistance and T is temperature. The slope of the curve decreases gradually at low temperature. This result indicates that some areas of the boundary plane have considerably lower activation energies and that the barrier height at the boundary is not uniform. In addition, it was found that the temperature dependence of boundary resistance is the same for all samples cut from a wafer, indicating that the non-uniformity occurs at the scale much smaller than the sample size.
    Figure 2 shows the boundary capacitance, C (pF cm^{-2}), as a function of temperature. The increase of the boundary capacitance with temperature indicates that the number of carriers, Q_b (# cm^{-2}), trapped at the boundary decreases gradually with increasing temperature. Q_b is calculated from the measured values of capacitance by

$$Q_b = \epsilon_s N_a / C \qquad (1)$$

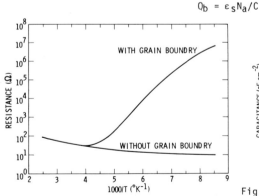

Fig. 1 Temperature dependence of zero-bias resistance of bicrystal and single silicon samples.

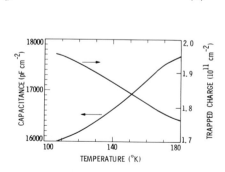

Fig. 2 Measured capacitance and calculated trapped charge density at the grain boundary as a function of temperature.

where ε_s is the permittivity of silicon and N_a is the acceptor concentration in the grain. Q_b versus temperature is also shown in Figure 2. The result indicates that the distribution of the localized states at the boundary is continuous throughout the energy bandgap.

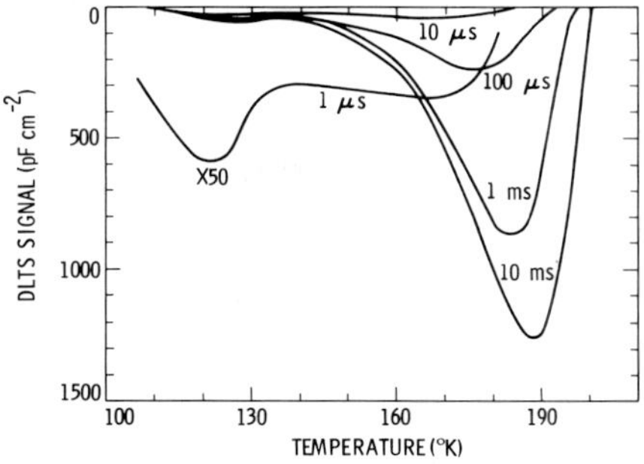

(B) TRANSIENT CAPACITANCE MEASUREMENT

A boundary can be considered as two opposing Schottky diodes in series. The boundary states are assumed to be located between the Schottky barriers. A flow of current through the boundary is required for the DLTS measurement of the density of states at the boundary. The local current density through the boundary can vary strongly if the local barrier height varies, depending on the spatial distribution of the density of states.

Fig. 3 Pulse width dependence of DLTS spectra; measuring condition: 0 Volt bias, 4 Volts pulse height, 20 and 60 ms gate window settings.

In accordance to our non-uniformity model of the grain boundary, which was supported by experimental data [7] the potential barrier can be different from place to place. Therefore, the current flowing through the boundary may not be uniform during the electrical pulsing of the DLTS measurement. The portions of the boundary with lower barrier height have higher current density and more carriers are available to become trapped. As a result of trapping, the local potential barrier becomes higher. Now, more current will flow through the portions of the boundary originally having a higher barrier. As a result of current flow in these portions, the states located at these areas become more observable under the DLTS measurement. In other words, the current will first fill the empty states at the grain boundary where the barrier is lower, and then, fill the states located where the barrier is higher.

Figure 3 shows a typical pulse width dependence of the DLTS signal of bicrystal samples. On increasing the pulse width, the peak amplitudes increase and the peak positions shift towards a higher temperature. When the pulse width is narrow, the current fills only the shallower states located in the low barrier height areas shown in Fig. 3. As the pulse width increases, the states located in the area originally having a high barrier height will have a chance to be filled, and the DLTS signal due to these states becomes observable. These phenomena can be attributed to the observation of additional deep states located where the barrier was originally higher and the local current density was smaller.

The pulse width dependence of the DLTS measurements on bicrystal samples indicates that the shallower states are observed first. This result indicates that there is no direct transition from the shallow level to the deep level, i.e., the boundary states are localized. It reveals that the physical distribution of boundary states is not uniform on a microscopic scale. This result is consistent with the generally known fact from defect studies on as-grown ingots, that the non-uniform distribution of microdefects in silicon always exists [8].

(C) DENSITY OF STATES AT THE GRAIN BOUNDARY

The density of states at the grain boundary can, in principle, be obtained from the DLTS spectra, but the process is very complicated since the boundary states are continuously distributed in the energy bandgap. The relationship between the DLTS signal, $S(T)$, and the density of states, $N_t(E)$, at the grain boundary is approximated by

$$S(T) = \sum_E (C_o^2/\varepsilon_s N_a) \, N_t(E) \, \Delta E \, [\exp(-r_E t_1) - \exp(-r_E t_2)] \qquad (2)$$

where C_o is the boundary capacitance with the observed states being empty, r_E is the emission probability of the carrier at energy level E, t_1 and t_2 are the gate window settings, and the density of states is assumed to be a constant within a ΔE interval at an energy level E above the valance band edge E_v. The peak width of the DLTS spectra shown in Fig. 3 is wider than that calculated for a single defect level, revealing the continuous nature of the density of states at the grain boundary. However, the width is not too much larger than that for a single level. This indicates that the conventional technique can be used to estimate the position of the dominated state for the peak without introducing a considerable error. In order to verfy this statement, we plotted r_E/T^2 vs. 1000/T for several sets of data taken using different pulse widths. We found that the data were always in the form of straight lines in this type of plot. Therefore, we can assume that the following formula is suitable for the grain boundary under the study;

$$r_E = AT^2 \exp(-E/kT), \qquad (3)$$

where A is a constant and depends on energy level. Figure 4 gives the experimental values of A factor as a function of energy level. These A factors were obtained from a group of Arrhenius plots of the emission probability measured under various pulse width conditions. The data seems to indicate that the A factor increases exponentially with energy level. For simplying the calculation of the density of states, we have further assumed that the relationship is appropriate for all the states observed. Then, the density of states can be obtained using curve fitting of the DLTS data with the calculated curve using N(E) as a variable. The result of this curve fitting for the data shown in Fig. 3 is shown in Fig. 5. There are two groups of states clearly shown in the figure. A group of narrowly distributed states centered at $E_v+0.20$ eV can be observed with short electrical filling pulses. The density of states of the other group, which was observed only with longer filling pulses, increases exponentially as the level becomes deeper. Because of the deterioration of the boundary capacitance at higher temperatures, only a part of this group was observed. The observation supports the commonly used assumption that the density of states at the grain boundary increases

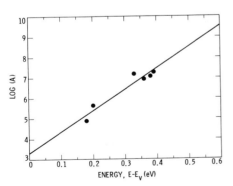

Fig. 4 The pre-exponential factor, A, of emission probability as a function of the energy level extracted from the DLTS data.

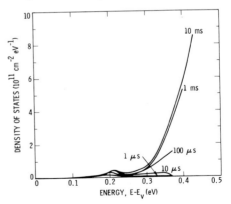

Fig. 5 The density of states at the grain boundary calculated from the DLTS spectra shown in Fig. 3.

exponentially as the state moves away from the band edge [9]. However, the
results from the study do show the existence of noticeable deviations from the
exponential dependence, such as the observation of the narrowly distributed
states centered at E_v+0.20 eV.

As stated in the previous paragraph, Eq. (3) is appropriate for use in the
analysis of data observed in this study. Since the formular is similar to the
one derived from the principle of the detailed balance, one may use this
analogy to think that the A factor is proportional to the capture cross
section. As a consequence, the data shown in Fig. 4 indicates that the deeper
levels have larger cross sections. This argument does not contradict to the
observed fact that the deeper levels were observed only with longer pulses,
because the capture rate of each state is a product of the capture cross
section and the local current density. In addition, the argument indicates
that the local current density through the location of the deep level is small,
i.e., the deep states are located where the barrier height is higher. As a
conclusion for this argument, the spatial distribution of the levels along the
grain boundary may not be uniform.

The maximum density of states observed is about 9×10^{11} cm^{-2}eV^{-1} under the 10
ms pulse width condition (Fig. 5). The maximum density of the states measured
in the group of states centered around E_v+0.20 eV is about 4×10^{10} cm^{-2}eV^{-1},
which is considerably lower. The densities of states of in the present samples
are about 10 times larger than these in the previous samples [2] which could
be attributed to the fact that the two studies were carried out on different
boundaries. However, narrowly distributed states centered around E_v+0.20 eV
were found in both cases, indicating that they could be due to the same kind
of lattice defects at the boundary.

The total number of trapped carriers at the grain boundary is about 2×10^{11}
cm^{-2} at temperature of 105 °K shown in Fig. 2. This could represent the
total density of states located between E_v+0.08 eV, as estimated from the
Fermi level at 105 °K and the middle of the bandgap, where the neutral level
of the grain boundary assumed to be. The integrated density of states under
the curve of 10 ms pulse width in Fig. 5 is about 6×10^{10} cm^{-2} between the
valance band and E_v+0.43 eV. These results illustrate that there are more
states located at deeper levels. It should be noted that the states deeper
than those indicated in Fig. 5 could not be observed because the barrier of
the boundary deteriorates at higher temperatures.

ACKNOWLEDGEMENT

The authors thank Rindge Shima for the bicrystal growth.

REFERENCES

1. C. H. Seager and T. G. Castner, J. Appl. Phys. 49, 3879 (1978)
2. L. J. Cheng and C. M. Shyu, "Semiconductor Silicon/1981", ed. by H. R. Huff
    et al., (Symposium Proceeding of the Electrochemical Society), p. 390
3   H. C. Card and E. S. Yang, IEEE Trans. ED-24, 397 (1977)
4. P. Panayotatos, E. S. Yang and W. Hwang, Solid-State Electronics, 25,
    417 (1982)
5. L. J. Cheng and C. M. Shyu, "Grain Boundaries in Semiconductors", ed. by
    H. J. Leamy et al., (Symposia Proceeding of the Materials Research Society),
    p. 105 (1982)
6. D. V. Lang, J. Appl. Phys. 45, 3014 (1974); 45, 3023 (1974)
7. C. M. Shyu and L. J. Cheng, p. 131 of Ref. 5.
8. P. Rava, H. C. Gatos and J. Lagowski, p. 232 of Ref. 2.
9. P. Panayotatos and H. C. Card, IEEE Electron. Device Lett. 1, 263 (1980)

THE STRUCTURE OF A NEAR COINCIDENCE Σ=5, [001] TWIST BOUNDARY IN SILICON

MARK VAUDIN AND DIETER AST
Department of Materials Science and Engineering, Cornell University,
Ithaca, New York 14853.

ABSTRACT

The dislocation structure of a near coincidence Σ=5 <001>
twist boundary in silicon was studied using transmission
electron microscopy. Secondary dislocations, with localized
cores, were observed in the boundary accommodating a small
deviation (<.5°) from perfect coincidence. The O-lattice
theory for general low angle boundaries was extended to
calculate the expected dislocation content of near
coincidence boundaries. Comparison between predictions and
observations was used to deduce information on the primary
dislocation structure of the boundary.

INTRODUCTION

Grain boundaries in semiconductors influence their electrical properties,
with important consequences in electronic and photovoltaic devices. The
structure of near coincidence boundaries has been the subject of many
experimental and theoretical studies. Secondary dislocations accommodating
deviations from perfect coincidence (low Σ) misorientations have been observed
in a number of materials, e.g. gold [1], stainless steel [2], copper [3] and
ceramic oxides [4]. Secondary dislocations have been observed in a {310} Σ=5,
36.9°/[001] tilt boundary in germanium [5].

This study is concerned with the observation of secondary dislocations in
sintered (001) Σ=5, 36.9°/[001] twist boundaries in silicon. Bicrystals were
fabricated by a hot pressing technique [6] which produced an orientation
relationship across the boundary that deviates from perfect coincidence due to
small errors in crystal alignment. This deviation can be described as a twist
around [001] and a tilt about an axis lying in the (001) plane. If both
components are non-zero, they can combine to a rotation about an axis which is
not a high symmetry direction. The deviation is then accommodated by a network
of secondary dislocation with non-coplanar Burgers vectors. Diffraction
experiments were used to determine the geometric parameters of the deviation
from coincidence, from which the expected dislocation network was calculated
using the O-lattice theory [7]. The secondary dislocation networks were
examined using the weak beam technique of electron microscopy [8], and the
observations were compared with the theoretical predictions.

EXPERIMENTAL DETAILS

A float zone silicon disc, (001) surface orientation, was Syton polished on
both sides. Two sets of wafers, 15mm x 5mm, were cut from the disc, one set
being misoriented 36.9° about [001] from the other. The two sets were
interleaved; the resultant stack of wafers was carefully aligned in a graphite
die and hot pressed for three hours in a H_2:Ar 15:85 mixture at a temperature of
1275°C, and a pressure of about 1 MPa. The stack was cut into slices oriented
approximately 20° to the plane of the wafers and 3mm discs were cut from the
slices. The discs were prepared for TEM by chemical polishing in a HNO_3:HF 9:1
mixture followed by ion beam milling. The specimens were examined in a Siemens

Elmiskop 102 operating at 125 kV. The twist and tilt components of the deviation from perfect coincidence were determined by diffraction experiments. The secondary dislocations accommodating the deviation from coincidence were imaged under a number of weak beam conditions, in particular using diffraction vectors that would be common for a perfect $\Sigma=5$ misorientation.

THEORY

The O-lattice is the lattice of all points of perfect match within two interpenetrating crystal lattices, misoriented by an angle θ about an axis $\underline{t}$. For two crystals of the same structure, the O-lattice is an array of O-lines parallel to the rotation axis. On moving between adjacent O-lines along a vector $\underline{X}_o$ perpendicular to the O-lines, the two lattices rotate apart, or are displaced, by an allowed Burgers vector $\underline{b}$ of the crystal structure. This displacement is concentrated into a dislocation positioned halfway between the O-lines. Bollmann (7) derived the relationship between $\underline{b}$ and $\underline{X}_o$:

$$X_o = (I-A^{-1})^{-1}\,\underline{b} = T^{-1}\,\underline{b} \tag{1}$$

where A is the matrix describing the rotation of crystal 1 anti-clockwise into crystal 2, and I is the identity matrix. Each $\underline{b}$ vector belongs to the b-lattice which, for a low angle boundary, is defined as the translation lattice of crystal 1. The matrix T^{-1} transforms the b-lattice vectors by a clockwise rotation of $90°-\theta/2$ and an expansion of $1/(2\sin(\theta/2))$.

The Burgers vectors of the dislocations that constitute a low angle grain boundary sum to a resultant vector normal to the rotation axis, $\underline{t}$. When $\underline{t}$ is normal to two linearly independent Burgers vectors, the orientation of $\underline{t}$ is termed "special". For example [001] is normal to a/2[110] and a/2[1$\bar{1}$0], allowed Burgers vectors in the diamond cubic structure, and is therefore a special rotation axis for this structure. The plane normal to the rotation axis (the b-plane) is densely occupied by b-lattice points and a planar b-net can be formed. Any orientation of $\underline{t}$ which does not satisfy the above condition is termed "general". The b-plane may contain few b-lattice points and it is expected that the large b-lattice vectors that lie in the b-plane will dissociate into smaller Burgers vectors with components parallel to $\underline{t}$, but in such a way that these parallel components sum to zero over the boundary. In effect, the b-plane is approximated by a stepped b-net [7,9]. For example, figure 1 shows the b-lattice (fcc) for the diamond structure in (010) projection together with stepped approximations to the ($\bar{1}$04), ($\bar{2}$09) and ($\bar{1}$05) b-planes. The b-planes are approximated by facets on (001) and ($\bar{1}$01) planes which contain the Burgers vectors a/2[110], a/2[1$\bar{1}$0] and a/2[101]. Different stepped or faceted b-nets can be devised for the same b-plane. As will be seen, a particular stepped b-net predicts a unique dislocation network, and conversely, the predicted dislocation network that best agrees with the experimental observations in general corresponds with a unique stepped b-net, from which Burgers vectors can be assigned to all the observed dislocations.

The b-lattice points embedded in the stepped b-net are projected parallel to the rotation axis onto the b-plane to produce the planar b-net, which therefore consists of the displacements perpendicular to $\underline{t}$ due to each Burgers vector in the stepped b-net. It is these displacements that cause the rotation, the displacements parallel to $\underline{t}$ automatically summing to zero over one period of the b-net. The planar b-net is transformed by T^{-1} to form the O-network and Wigner Seitz cell walls are inserted halfway between each pair of O-elements. The boundary plane is passed through the two interpenetrating crystals and one crystal lattice is discarded from each side of the boundary to create a bicrystal. Dislocations are predicted to lie where the boundary intersects the Wigner-Seitz cell walls.

To calculate the dislocation network for a given rotation axis $\underline{t}$ (=[hkl]), angle θ and grain boundary normal $\underline{n}$, the stepped b-net is determined and the b-lattice vectors embedded in the stepped b-net are projected normally onto (hkl). It is convenient to express the projected b-vectors $\underline{b}'$ using the basis vectors $\underline{u}_1$ and $\underline{u}_2$ which are unit vectors in the directions $\underline{t} \times \underline{n}$ and $\underline{t} \times (\underline{t} \times \underline{n})$ respectively. The components of $\underline{b}'$ in the $(\underline{u}_1,\underline{u}_2)$ basis are found by premultiplying the $\underline{b}$ vectors by P which is a 2x3 matrix whose rows are the components of $\underline{u}_1$ and $\underline{u}_2$ respectively. Using equation 1, the O-vectors $\underline{X}_o$ are derived from the projected Burgers vectors. Passing the boundary plane through the O-network is equivalent to projecting the O-network parallel to $\underline{t}$ onto the boundary plane. Since $\underline{u}_1$ is perpendicular to $\underline{n}$ and $\underline{t}$, it is a unit vector in both the b-plane and the grain boundary plane. Thus projection of an $\underline{X}_o$ vector onto the boundary plane leaves the $\underline{u}_1$ component unchanged. The component in the $\underline{u}_2$ direction is multiplied by $\sec\phi$ where ϕ is the angle between $\underline{t}$ and $\underline{n}$. We define a second unit vector in the boundary plane, $\underline{u}_3$, parallel to $\underline{n} \times \underline{u}_1$, and also a 3x2 matrix M whose columns are the components of $\underline{u}_1$ and $\underline{u}_3$ respectively. The projected O-vector $\underline{X}_o'$ that corresponds to a Burgers vector $\underline{b}$ can be found in the crystal 1 coordinate system from:

$$X_o = M \begin{pmatrix} 1 & 0 \\ 0 & \sec\phi \end{pmatrix} T^{-1}\, P\, \underline{b} \qquad (2)$$

This equation can be used to calculate the O-network for low angle boundaries, and also small deviations from high coincidence misorientations.

RESULTS

Figure 2 is a weak beam image of a hot pressed $\Sigma = 5$ [001] twist boundary, imaged using the 400_2 diffraction vector. Figures 3(a) and (b) are weak beam images of the same area taken using "common" g vectors $260_1/\overline{2}60_2$ and $620_1/\overline{6}20_2$ respectively. Diffraction experiments showed that the twist component of the deviation from coincidence was a rotation of $.38 \pm .04°$ about [001]. The average tilt component along the boundary was a rotation of $.25 \pm .05°$ about an axis near $[120]_1/[210]_2$.

The boundary structure consisted of three families of dislocations, two orthogonal to each other, and the third, more widely spaced, lying diagonally across the orthogonal sets. The orthogonal dislocations lay along the $[1\overline{3}0]_1/[130]_2$ and $[310]_1/[3\overline{1}0]_2$ directions, and were spaced 20.5 ± 1.0 nm apart. The line direction of the third set of dislocations was approximately $[120]_1/[210]_2$ and their spacing varied between 55nm and 80nm over the boundary. The orthogonal dislocations and the dislocations lying along <120> exhibited contrast features consistent with screw and edge character respectively, when imaged under different diffraction conditions.

To analyze the structure of the observed boundary the b-lattice was assumed to be the DSC lattice for the $\Sigma = 5$ misorientation with unit vectors a/10[310], a/10[$\overline{1}$30] and a/10[125]. The deviation from coincidence was calculated from the twist and tilt components to be a rotation of $.45° \pm .06$ about an axis near [124]. The stepped b-net for a (124) b-plane was determined by inspection and equation 2 was used to calculate the $\underline{X}_o$ vectors that make up the projected O-network lying in the (001) plane. The O-network was plotted out, and the Wigner-Seitz cell walls were inserted to show the predicted positions of the dislocations (figure 4). Burgers vectors were assigned to the dislocations using the relationship between the b-net and the O-network.

DISCUSSION AND CONCLUSIONS

The model predicts a grid of kinked screw dislocations, crossed at regular intervals by near-edge dislocations with Burgers vectors a/10[125]. The screw dislocations consist of segments with a/10[310]$_1$, a/5[120]$_1$ and a/10[$\overline{1}$30]$_1$ Burgers vectors. Comparing the model with the observations shows that there is good qualitative agreement. The O-lattice theory is based on purely geometrical concepts and ignores both the dislocation core energy, and the elastic interactions between dislocations, which will cause the dislocations to relax to a lower energy configuration. Screw dislocations with orthogonal Burgers vectors tend to intersect orthogonally, as is observed experimentally, figure 2. The screw dislocation spacing predicted by the construction is 22 nm which agrees closely with the observations. The angle of tilt varied down the boundary, which causes a change in the orientation of the b-plane and a change in the spacing of the near-edge dislocations as observed.

The electrical activity and energy of grain boundaries in semiconductors are believed to be increased by the presence of broken or distorted bonds in the boundary structure, which create electron-hole recombination sites. Broken bonds have a higher energy and are more efficient recombination sites than distorted bonds. Hornstra [10] showed that it was possible to model screw dislocations in the diamond lattice without broken bonds and so an attempt was made to model the structure of $\Sigma=5$ twist boundaries using a/2<110> screw dislocations. The Burgers vectors and line directions of the secondary dislocations in a near coincidence boundary reflect the primary dislocation structure of the boundary. The secondary dislocations reported in this paper are similar to those found in fcc metals, and one possible boundary structure can be created by taking a computed $\Sigma=5$ twist boundary model (e.g. [11]) and adding an extra atom at 1/4[111] from each fcc atom. Models cannot be formed in this way without broken bonds at each intersection of the two orthogonal sets of primary dislocations. However, it is possible to model twist boundaries in tetrahedrally coordinated materials without broken bonds, using three sets of screw dislocations, on adjoining (004) planes so that the screw dislocations do not directly intersect. The line and Burgers vector directions of the lower, middle and upper dislocation sets are $[110]_1$, $[2\bar{1}0]_1$ and $[110]_2$ respectively. Small twist deviations from perfect coincidence are taken up by systematic variation of the spacing of the three primary dislocation sets, equivalent to introducing secondary dislocations with Burgers vectors a/6$[110]_1$, a/15$[2\bar{1}0]_1$ and a/6$[110]_2$ respectively. These three vectors form a lattice of which the $\Sigma=5$ DSC lattice is a superlattice, and it is therefore plausible that these secondary dislocations could associate to give the DSC secondary dislocations observed. Observations on boundaries with very low deviations from coincidence ($<.1°$) may provide a conclusive test for this model.

ACKNOWLEDGEMENTS

This research was supported by DOE under contract No. 76ER02899 and by DOE/JPL under contract No. 956046. Research facilities were provided by the Materials Science Center at Cornell. Useful discussions with Dr. Brian Cunningham are gratefully acknowledged.

REFERENCES

1. T. Schober and R. W. Balluffi, Phil. Mag. A21, 109 (1970).
2. W. A. T. Clark and D. A. Smith, Phil. Mag. A38, 367 (1978).
3. L. M. Clareborough and C. T. Forewood, Phys. Stat. Sol. (a) 58, 597 (1980).
4. C. P. Sun and R. W. Balluffi, Phil. Mag. A46, 49 (1982).
5. J. J. Bacmann, G. Silvestre, M. Petit and W. Bollmann, Phil. Mag. A43, 89 (1981).
6. H. Foll and D. G. Ast, Phil. Mag. A40, 589 (1979).
7. W. Bollmann, Crystal Defects and Crystalline Interfaces (1970) Springer-Verlag, New York.
8. D. J. H. Cockayne, J. Microscopy 98, 116 (1973).
9. W. Bollmann, Acta Cryst. A33, 730 (1977).
10. J. Hornstra, J. Phys. Chem. Solids 5, 129 (1958).
11. J. Budai, P. D. Bristowe and S. L. Sass, MSC Report #4739 (1982), Cornell University, Ithaca, NY.

373

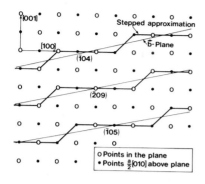

Fig. 1. Stepped approximations to b-planes.

Fig. 2. Weak beam image of dislocation network. Diffraction vector $\underline{g} = 400_2$.

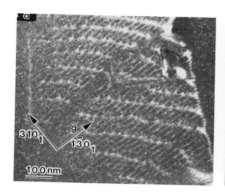

Fig. 3. Weak beam images of dislocation network.
(a) $\underline{g} = 2\bar{6}0_1/\bar{2}\bar{6}0_2$ (b) $\underline{g} = 620_1/6\bar{2}0_2$

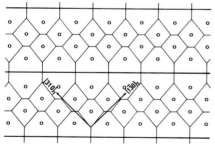

Fig. 4. O-network and Wigner-Seitz cells for $\underline{t} = [124]$, $\underline{n} = [001]$ and $\theta = 0.45°$.

———— = 50 nm.

ELECTRONIC PROPERTIES OF GRAIN BOUNDARIES IN GaAs: A STUDY OF ORIENTED BICRYSTALS PREPARED BY EPITAXIAL LATERAL OVERGROWTH

Jack P. Salerno, R. W. McClelland, J. G. Mavroides, and John C. C. Fan
Lincoln Laboratory, Massachusetts Institute of Technology
Lexington, Massachusetts 02173

A. F. Witt, Department of Materials Science and Engineering,
Massachusetts Institutte of Technology
Cambridge, Massachusetts 02139

ABSTRACT

The electronic properties of tilt boundaries with misorientation angles ranging from 0 to 30° in n-type GaAs bicrystal layers have been investigated. The current-voltage and capacitance-voltage characteristics are consistent with a double-depletion-region model. The height of the grain boundary potential barrier remains constant while the density of grain boundary states varies with misorientation angle. Deep level transient spectroscopy has revealed the presence of two bands of grain boundary states at approximately 0.65 and 0.9 eV below the conduction band. These states are attributed to bond reconstruction at the grain boundary.

INTRODUCTION

Grain boundaries (GBs) have a critical effect on the properties of polycrystalline semiconductors, which are increasingly utilized in electronic device technology. We have previously shown by cathodoluminescence studies that both structure and composition influence the properties of GBs in GaAs [1,2], and we have recently reported the electrical properties of tilt boundaries in oriented GaAs bicrystal layers prepared by vapor-phase epitaxy (VPE) using epitaxial lateral overgrowth [3,4]. This paper reports a further study of the electronic properties of such tilt boundaries and proposes a model for the bonding arrangement at the GB interface.

BICRYSTAL GROWTH AND CHARACTERIZATION

The oriented GaAs bicrystal layers were prepared by VPE on composite substrates formed from two GaAs single crystals cut to the desired orientations and bonded together by a continuous $PbO-SiO_2-Al_2O_3$ glass film about 5 µm thick. The details of the growth technique have been described previously [3]. The configuration of a composite substrate used to prepare a (110) tilt boundary with a $(\bar{1}11)$ boundary plane is shown schematically in the upper diagram of Fig. 1. An SiO_2 stripe masks the bonding glass film and extends 15 µm over each of the GaAs crystals, which are designated as Crystals 1 (left) and 2 (right). Since the lateral growth seeded by Crystal 1 is bounded by a slow-growing $(\bar{1}11)$ facet, the GB occurs close to the edge of the SiO_2 stripe over Crystal 1, as shown in the lower diagram of Fig. 1. As a result of the formation of the facet, the plane of the boundary is a $(\bar{1}11)$ plane, regardless of the misorientation angle θ of Crystal 2.

Mat. Res. Soc. Symp. Proc. Vol. 14 (1983) Published by Elsevier Science Publishing Co., Inc.

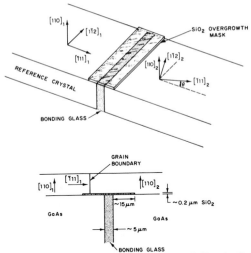

Fig. 1. Schematic diagrams of a composite GaAs substrate (top) and the cross section of the substrate and bicrystal layer grown by vapor-phase epitaxy (bottom).

The electronic properties of GBs in GaAs were determined from measurements on 12 bicrystal layers grown in 4 separate deposition runs. In Run 1 nominally undoped layers were prepared with θ = 10, 24, and 30°. The layers grown in Run 2 were also nominally undoped but with θ = 2.5, 5, and 10°. In this run the substrates were oriented 2° off the (110); in the other runs the substrates were not intentionally misoriented from (110). The last two depositions produced S-doped bicrystal layers containing 10° tilt boundaries. In each run a control layer, designated as a "0° bicrystal," was grown on a composite substrate formed from two crystals both having bonding cross sections cut 10° from the ($\bar{1}$11) plane. In each run a conventional epilayer was grown on a single-crystal (110) substrate. The quoted carrier concentrations, $n_{(110)}$, were measured on these conventional layers.

In order to measure the electrical characteristics of GBs in the bicrystal layers, specimens with 4 ohmic contacts, 2 on each grain, were prepared by the technique described previously [4]. For these specimens electrical conduction between contacts on opposite sides of the GB is restricted to the bicrystal epilayer. The specimens were used both for conventional current-voltage (I-V) measurements and for capacitance-voltage (C-V) measurements, which were made by applying a dc bias voltage with a superimposed 20 mV ac signal at a frequency of 1 MHz. Only low-donor-density samples with θ > 10° displayed sufficient rectification for C-V analysis. The nature of the GB states was investigated by using a modification of the DLTS technique [5]. For this analysis an 0.4-ms, 1-V bias pulse was applied across the GB to fill additional bandgap states. Upon termination of the pulse, the 1 MHz capacitance associated with the GB space charge region was monitored, and the transient accompanying the return to equilibrium was analyzed to obtain the DLTS signal. Unlike conventional DLTS analysis, the traps in the space charge region cannot be emptied to a level below the zero-bias Fermi level position. Thus only bandgap states with energies above the zero-bias Fermi level can be analyzed by this technique. Since the DLTS analysis detects electronic states in the entire space charge region, both GB states and bulk states within the grains appear in the spectra. In order to distinguish the bulk traps from the GB states, a Au-Schottky contact placed on a 0° GB was used for conventional DLTS analysis.

DOUBLE-DEPLETION-REGION MODEL

The room temperature I-V characteristics obtained by measurements across the GBs prepared in Run 1, for which $n_{(110)} = 2.5 \times 10^{15}$ cm^{-3}, have been reported previously [Ref. 4, Fig. 4]. The 0° specimen shows linear I-V characteristics, indicating that no potential barrier is present. As θ is increased to 10° the characteristics become increasingly nonlinear, indicating that a potential barrier is developing. Strong rectification is exhibited by the 10° boundary, consistent with the back-to-back diode characteristic associated with the double-depletion-region model for the electronic band structure.

The I-V characteristics for the 10° GB exhibit an asymmetry which indicates that the carrier concentration is lower in the off-($\bar{1}11$) grain.

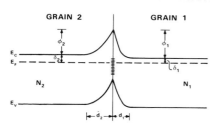

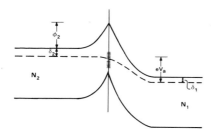

Fig. 2. Band structure model for a grain boundary with $N_1 > N_2$ in an n-type semiconductor for zero bias (top) and for a reverse bias voltage V_a applied to grain 1 (bottom).

Such a difference in concentration is consistent with the orientation dependence of dopant incorporation found in a previous investigation of the epitaxial lateral overgrowth process [6]. Figure 2 shows the electronic band structure of a generalized GB in an n-type semiconductor with effective donor densities N_1 and N_2 in grains 1 and 2, respectively. The respective zero-bias barrier heights are ϕ_1 and ϕ_2, and the positions of the Fermi level below the conduction band are given by δ_1 and δ_2. The upper diagram of Fig. 2 shows the zero-bias band structure for $N_1 > N_2$ and, hence, $\phi_1 > \phi_2$ and $\delta_1 < \delta_2$. The respective depletion regions in each grain have widths of d_1 and d_2.

The lower diagram of Fig. 2 shows the band structure when grain 1 is reverse biased by applying a voltage V_a across the GB. In this case electrons flow from grain 2 to grain 1 in response to the decrease of eV_a in the Fermi level in grain 1. It can be assumed that the drop in applied voltage occurs entirely across the reverse-biased side and that the flow of current occurs due to thermionic emission over the barrier [7]. Under these conditions the barrier height of the forward-biased grain remains constant at the zero-bias equilibrium value while that of the reverse-biased grain changes in response to V_a. The equilibrium Fermi level of the GB itself is

$$\delta_{GB} \equiv \phi_1 + \delta_1 = \phi_2 + \delta_2 , \qquad (1)$$

and the current density moving across the GB is

$$J = A^* \exp(-\delta_{GB}/kT)[1 - \exp(-eV_a/kT)] , \qquad (2)$$

where A^* is a modified Richardson constant. It is seen from Eq. (2) that the current flowing at a given applied voltage is determined by the value of δ_{GB}. In turn, the value of δ_{GB} is primarily determined by the barrier height, since δ_1 and δ_2 are small compared to ϕ_1 and ϕ_2 for highly rectifying GBs.

The electronic charge density trapped by the GB states at zero bias, which is equal in magnitude to the positive space charge density in the depletion regions, is given by [8]

$$Q_0[\text{Coul/cm}^2] = (2\epsilon_0\epsilon_R N_1\phi_1)^{1/2} + (2\epsilon_0\epsilon_R N_2\phi_2)^{1/2} \quad . \tag{3}$$

The excess trapped charge density Q_E due to a reverse-bias voltage V_a applied to grain 1 is given by [7]

$$Q_E \simeq (2\epsilon_0\epsilon_R eN_1V_a)^{1/2} \quad . \tag{4}$$

At a sufficiently high reverse-bias voltage, denoted as the breakdown (or saturation) voltage V_{max}, the available excess GB states become full and the potential barrier collapses [7]. The collapse of the barrier is accompanied by a rapid increase in current with applied voltage. From Eq. (4), the breakdown voltage is given by

$$V_{max} = 2\epsilon_0\epsilon_R eN_1Q_{max}^2 \quad , \tag{5}$$

where Q_{max} is the maximum charge density that can be contained in the GB.

Note that V_{max} is determined by both the density of GB states and the carrier concentration of the reverse-biased side. Thus in analyzing transport across the GBs in the bicrystal layers it is only valid to compare the J-V characteristics for reverse bias of the $(\bar{1}11)$ reference crystal, which has the same carrier concentration for all layers grown in the same run.

Since the GB can be treated as two back-to-back diodes, the GB capacitance C is given as the series sum of the capacitances of the two Schottky diodes represented by the space charge region in each grain [7,9]. Thus C is given by [8]

$$\frac{1}{C} = \left(\frac{2}{e^2A^2\epsilon_0\epsilon_R}\right)^{1/2} \left[\left(\frac{\phi_1 + eV_a}{N_1}\right)^{1/2} + \left(\frac{\phi_2}{N_2}\right)^{1/2}\right] , \tag{6}$$

where A is the GB area and the drop in applied voltage is assumed to occur entirely across grain 1.

RESULTS AND DISCUSSION

Figure 3 shows typical room temperature J-V characteristics for GBs in undoped layers with θ ranging from $0°$ through $30°$. In each case the reverse bias was applied to the $(\bar{1}11)$ oriented grain. The rectification clearly increases with θ, but the characteristics of the 10, 24, and $30°$ GBs indicate that the barrier height is constant. The slight rectification displayed by the 2.5 and $5°$ GBs indicates that the donor density of these layers is sufficiently high to essentially fill the GB states.

The reverse-bias voltage at which J begins to increase rapidly with applied voltage (the breakdown or saturation voltage) is also observed to be a

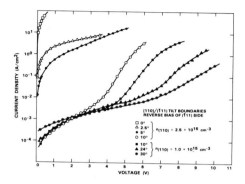

Fig. 3. J-V characteristics for reverse bias of the ($\bar{1}$11) grain as a function of misorientation angle θ.

strong function of θ. This indicates that the density of GB states is directly related to the GB structure. These J-V characteristics clearly indicate that the electronic band structure associated with a GB is influenced by the intrinsic physical structure of the GB.

The J-V characteristics of the 10° tilt boundaries with donor densities of 1.0 and 2.5 x 10^{15} cm^{-3} indicate that ϕ decreases as the carrier concentration in the grains increases. This effect is consistent with the double-depletion-region model based on Fermi level pinning by GB states. The J-V characteristics of four 10° tilt boundaries with $n_{(110)}$ ranging from 1.0 x 10^{15} to 1.0 x 10^{18} cm^{-3} [Ref. 4, Fig. 6] show a decrease in GB rectification as the carrier concentration increases. This is attributed to both a decrease in barrier height and the increased probability of electron tunneling through the barrier.

Although the J-V characteristics show the trends in the barrier height and bandgap state density as a function of θ and carrier concentration, the carrier concentration nonuniformities in the bicrystal layers exclude the use of routine characterization techniques for the quantitative determination of the GB band structure parameters. In particular, conventional C-V analysis is not valid. Therefore a technique of self-consistent analysis [10] was applied to the C-V data for the undoped layers with θ = 10, 24, and 30°. This technique is self-consistent in the sense that the values for N_1, N_2, ϕ_1, and ϕ_2 are assigned to fit the C-V data. These values are then used to obtain δ_{GB} and the number of filled GB states per unit area at zero-bias equilibrium, N_T. The values of δ_{GB} were 0.55 and 0.93 eV for two samples with θ = 10° but ranged from 1.15 to 1.43 eV for a third 10° sample and the 24° and 30° samples. This difference suggests that there are two sets of bandgap states at which the Fermi level can be pinned. The values of N_T were found to be ~ 2.5 x 10^{11} cm^{-2} for the 10° samples and ~ 1.5 x 10^{12} cm^{-2} for the 24° and 30° samples.

The GB capacitance is directly proportional to the charge in the GB states and, hence, the number of filled GB states. The collapse of the GB potential barrier occurs when a sufficient bias voltage is applied to fill all the GB states. Thus the total number of GB traps per unit area, N_{Tmax} should be given by

$$N_{T_{max}} \cong N_T(C_0/C_{V_{max}}) \quad , \qquad (7)$$

where C_{Vmax} is the capacitance measured just before the potential barrier collapse. The average values of C_0/C_{Vmax} were found to be 1.12, 1.54, and 1.59 for the 10, 24, and 30° GBs, respectively. Using these values, the total density of GB states is approximately 2.8×10^{11} cm^{-2} for $\theta = 10°$ and 2.4×10^{12} cm^{-2} for θ in the range of 24 to 30°.

The DLTS spectra, plotted as ΔC versus T, for the 0° and 24° specimens are shown in Fig. 4. The 24° spectrum contains three peaks, the 0° spectrum

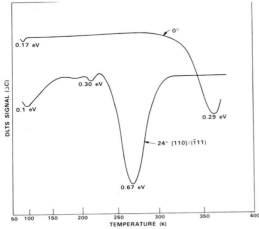

Fig. 4. DLTS spectra for 0° and 24° bicrystal layers. Activation energies of the traps are indicated.

only two. The representation of the peaks as minima indicates that these bandgap states act as electron traps. Thus, by convention, the energies of the trap states are their positions below the conduction band edge. The results of the DLTS analysis are given in Table 1. Bandgap states with

Table 1 DLTS Results for GaAs Tilt Boundary Samples

θ (°)	$n(110)$ (cm^{-3})	$E_C - E_T$ (eV)	Normalized Peak Height ($\Delta C/C$)
10	2.5×10^{15}	0.10	1.2×10^{-4}
		.33	4.0×10^{-4}
		.71	9.5×10^{-4}
10	1.0×10^{15}	.10	1.1×10^{-3}
		.30	2.2×10^{-3}
		.65	3.3×10^{-3}
24	1.0×10^{15}	.10	2.8×10^{-3}
		.30	1.0×10^{-3}
		.67	1.1×10^{-2}
30	1.0×10^{15}	.63	2.5×10^{-2}
		.87	1.9×10^{-2}
0*	2.5×10^{15}	.17	N.A.
		.29	N.A.

* Schottky Contact

energies of approximately 0.1, 0.3, and 0.65 eV were found for the two 10° samples and the 24° sample. States with energies of 0.63 and 0.87 eV were found for the 30° sample (the rate window used for this sample precluded the observation of the 0.1 and 0.3 eV states). The normalized peak height, $\Delta C/C$, is proportional to the density of the trap states. Only the height of the 0.65 eV peak is observed to change significantly with θ. The analysis of the 0° specimen indicates that the 0.1 and 0.3 eV traps are located within the grains; these energies are consistent with reported values for bulk GaAs [11]. The presence of the 0.87 eV trap in the 30° sample is consistent with the C-V results, and its high density relative to the observed bulk states suggests that it is a GB state. It may be that this trap is observed only in the 30° GB because the Fermi level is pinned at a lower energy in this sample.

The DLTS results suggest that there are two discrete bands of states associated with the GB, located about 0.65 and 0.9 eV below the conduction band edge, and that the density of these states depends on the GB orientation. These observations indicate that the states are intrinsic to the GB structure.

The level centered at 0.65 eV has not been observed in bulk GaAs. However, it has been reported for many types of GaAs interfaces, such as surfaces [12], heterojunctions [13], and GaAs-insulator interfaces [14]. A trap at 0.90 eV has also been associated with GaAs surfaces [12]. In addition, GaAs GB states with energies of 0.62 and 0.74 eV have recently been reported [9].

The DLTS analysis of GaAs GBs reported here and by other workers [9] has shown that discrete electronic states are associated with the intrinsic GB structure. This result indicates that a characteristic defect structure is associated with the GB due to the formation of a particular atomic arrangement at the interface. This is interpreted as the result of reconstructed bonding in the formation of the stable GB microstructure. A plausible model for this atomic arrangement can be deduced from these results.

McPherson, et al. [15] have proposed that a symmetric tilt boundary in GaAs is composed of a dislocation array with dangling bonds spaced periodically along the length of the dislocations. Their theoretical analysis predicts a decrease in ϕ with increasing θ for $\theta > 1°$. This is contrary to our J-V results, which show that ϕ is independent of θ. The geometric structure to which their analysis applies differs from that of our tilt boundaries only with respect to the orientation of the GB plane. Thus, the disagreement between theory and experiment suggests that the dislocation array model is not appropriate.

We propose an alternative model for the GB structure based on recent studies of Ge [16,17] and Si [18-20]. According to this model the GB is composed of microfacets along twin orientations, and bond reconstruction results in a periodic arrangement of 5, 6, and 7 member rings, which results in bond length dilation and variations in bond angle. Applied to GaAs, this structure gives rise to like-atom bonds (Ga-Ga and As-As bonds), but there are no associated dangling bonds. The Ga-Ga and As-As bonds are analogous to missing As and Ga atoms, respectively. Bandgap states with energies of 0.65 and 0.90 eV associated with GaAs surfaces have been attributed to missing As and Ga atoms, respectively [12]. Since our data suggest that there are GaAs GB states with similar energies, the proposed model for the GB microstructure is plausible.

382

ACKNOWLEDGEMENTS

We gratefully acknowledge J. P. Donnelly and A. J. Strauss for valuable discussions, and B. DiGiorgio, S. Duda, E. L. Mastromattei, and A. Napoleone for expert technical assistance.

The single-crystal GaAs boules used as the source of substrates were supplied by G. W. Iseler, G. M. Metze, and D. M. Tracy. This work was sponsored by the Solar Energy Research Institute and the Department of the Air Force.

REFERENCES

1. J. P. Salerno, R. P. Gale, J. C. C. Fan, and J. Vaughan, in Defects in Semiconductors, eds. J. Narayan and T. Y. Tan (North-Holland, New York, 1981), p. 509.
2. J. P. Salerno, R. P. Gale, and J. C. C. Fan, in Proceedings of the 15th IEEE Photovoltaic Specialists Conference, 1981 (IEEE, New York, 1981), p. 1174.
3. J. P. Salerno, R. W. McClelland, P. Vohl, J. C. C. Fan, W. Macropoulos, C. O. Bozler, and A. F. Witt, in Grain Boundaries in Semiconductors, eds. H. J. Leamy, G. E. Pike and C. H. Seager (North-Holland, New York, 1982), p. 77.
4. J. P. Salerno, R. W. McClelland, J. C. C. Fan, P. Vohl, and C. O. Bozler, in Proceedings of the 16th IEEE Photovoltaic Specialists Conference, San Diego, 1982 (IEEE, New York, in press).
5. D. V. Lang, J. Appl. Phys. 45, 3023 (1974).
6. C. O. Bozler, R. W. McClelland, J. P. Salerno, and J. C. C. Fan, J. Vac. Sci. Technol. 20, 720 (1982).
7. W. E. Taylor, N. H. Odell, and H. Y. Fan, Phys. Rev. 88, 867 (1952).
8. S. M. Sze, Physics of Semiconductor Devices (Wiley, New York, 1969), Chapter 8.
9. M. G. Spencer, W. J. Schaff, and D. K. Wagner, in Grain Boundaries in Semiconductors, eds. H. J. Leamy, G. E. Pike, and C. H. Seager (North-Holland, New York, 1982), p. 125.
10. J. P. Salerno, Ph.D. Thesis, M.I.T. (1983), unpublished.
11. G. M. Martin, A. Mitonneau, and A. Mircea, Electron. Lett. 13, 191 (1977).
12. W. E. Spicer, P. W. Chye, P. R. Skeath, C. Y. Su, and I. Lindau, J. Vac. Sci. Technol. 16, 1422 (1979).
13. S. R. McAfee, D. V. Lang, and W. T. Tsang, Appl. Phys. Lett. 40, 520 (1982).
14. T. E. Kazior, Ph.D. Thesis, M.I.T. (1982), unpublished.
15. J. W. McPherson, G. Filatous, E. Stefanakos, and W. Collis, J. Phys. Chem. Solids 41, 747 (1980).
16. O. L. Krivanek, S. Isoda, and K. Kobayashi, Phil. Mag. 36, 931 (1977).
17. A-M. Papon, M. Petit, G. Silvestre, and J-J. Bacmann, in Grain Boundaries in Semiconductors, eds. H. J. Leamy, G. E. Pike, and C. H. Seager (North-Holland, New York, 1982), p. 27.
18. D. Valachavas and R. C. Pond, Inst. Phys. Conf. Ser. No. 52, 195 (1980).
19. C. Fontaine and D. A. Smith, Appl. Phys. Lett. 40, 153 (1982).
20. C. Fontaine and D. A. Smith, in Grain Boundaries in Semiconductors, eds. H. J. Leamy, G. E. Pike, and C. H. Seager (North-Holland, New York, 1982), p. 39.

ENHANCED DIFFUSION OF PHOSPHORUS AT GRAIN BOUNDARIES IN SILICON*

L.J. CHENG, C.M. SHYU, and K.M. STIKA
Jet Propulsion Laboratory, California Institute of Technology,
Pasadena, California 91109

ABSTRACT

It is found that the grain boundaries in cast polycrystalline silicon material capable of enhancing diffusion
always have strong recombination activities. Both phenomena
could be related to the existence of dangling bonds at boundaries. Because the enhanced diffusion is an atomic transport
phenomenon and the recombination is an electronic process,
the relationship between the two phenomena is still not clear
at this moment. The present study gives the first evidence
that incoherent second order twins of $\{111/115\}$ type are
phosphorus diffusion-active.

INTRODUCTION

Enhanced diffusion of phosphorus at grain boundaries in polycrystalline silicon
has obtained more attention in recent years, because of the development of low
cost polycrystalline silicon solar cells. It has been suggested [1] that the
enhancement can increase the carrier collection area and reduce the recombination effect. The enhanced diffusion of phosphorus at grain boundaries in silicon was first reported more than twenty three years ago [2]. Since then, a
number of papers have appeared in the literature. Because of its complexity,
there has been only limited knowledge available concerning the detailed mechanism of the phenomena and its relationship to material properties, particularly
those of the grain boundaries. Queisser [3] reported that the enhancement can
not occur at coherent twin boundaries.   It is an established fact that high
angle boundaries are usually diffusion-active. No detailed relationship to the
boundary structure has been reported.

In this paper, we report the results of a study on enhanced diffusion of
phosphorus at grain boundaries in cast polycrystalline wafers produced by Wacker,
one of several polycrystalline silicon materials important for solar cell applications. The results show that the diffusion enhancement is always associated
with the boundaries having strong minority carrier recombination activities. In
addition, we have found that incoherent second order twins of $\{111/115\}$ type are
diffusion-active.

EXPERIMENTAL DETAILS

The samples were wafers of Silso cast polycrystalline p-type silicon with
resistivity of 1-3 ohms-cm, which were purchased from Wacker. The samples were
diffused at 1100°C for 30 minutes using phosphine. A grooving and staining
technique using concentrated hydrofluoric acid as a staining agent was used to

----

The research described in this paper was carried out for the Flat-Plate Solar
Array Project, Jet Propulsion Laboratory, California Institute of Technology and
was sponsored by the U.S. Department of Energy through an agreement with NASA.

reveal the n-type diffused region, including enhanced diffusion at grain boundaries. Hydrofluoric acid stains p-type material, which makes n-type show up as
bright regions; consequently, the enhanced diffusion at grain boundaries mani-
fests itself as white lines in the reflecting light photograph. Detailed features of stained samples were examined under a Zeiss metallurgical microscope.
A detailed correlation study between the images of n-type material from the
staining experiment and those negatively charged phosphorus-silicon complex ions
obtained from an IMS-3F secondary ion mass spectroscope [4] assured that the
white lines appearing beyond the bulk diffused region are truly due to enhanced
phosphorus diffusion at grain boundaries. Images of electron beam induced
current (EBIC) of samples were examined using an ISI-60A scanning electron
microscope to investigate the relationship between recombination activities and
enhanced diffusion at grain boundaries. A Sirtl etch was used to reveal the
structural defects appearing at the surface for examining those boundaries
responsible for the enhancement.

EXPERMENTAL RESULTS

Figure 1 shows a pair of pictures of a sample taken after grooving and staining (Fig. 1a) and Sirtl etching (Fig. 1b). Fig. 1a shows one white curved line
and two white straight segments which are due to the enhanced diffusion of phosphorus. Fig. 1b is the surface appearance of the same area after a Sirtl etching, showing the enhanced diffusion occured at a high angle grain boundary
(curved and heavily etched) and two straight segments of boundaries with heavy
etched marks. In addition, there are apparently a group of coherent twins and/
or microtwins appearing as lightly etched straight lines which cause no enhanced
diffusion and appear to have some angular relationship to the two segment
boundaries having enhanced diffusion. The details of the relationship will be
discussed in the next section. It is noticed in Fig. 1 that the line width
and etched mark of the two segments are significantly thinner than those of
the high angle boundary, indicating that the imperfection of the segment boundaries due to the lattice mismatch is not as severe as those of the high angle
boundary.

Figure 2a shows an EBIC picture of a sample in which there are three dark
curved lines and one thin dark straight line indicating recombination in these
areas. The three curved lines correspond to three high angle boundaries with

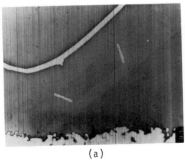
(a)

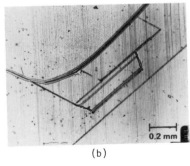
(b)

Figure 1: a)Image of enhanced diffusion at grain boundaries showing as white
lines after grooving and staining and b) the surface appearance of the same
region after Sirtl etching.

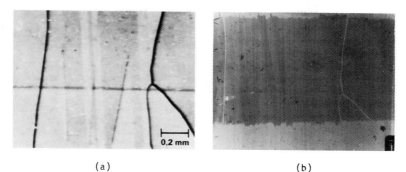

(a)   (b)

Figure 2: a) Image of electron beam induced current of a diffused sample showing heavy dark lines at high angle grain boundaries which represent strong
recombination activities and b) image of enhanced diffusion of phosphorus at
the same boundaries appearing as white lines after grooving and staining. The
straight boundaries with weak recombination activity (the thin dark line in
Fig. 2a) does not have any observable enhanced diffusion.

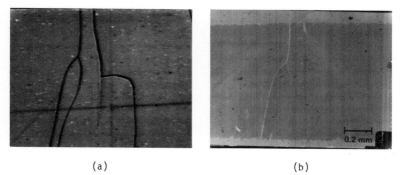

(a)   (b)

Figure 3: Similar pair of images as Figure 2 but of different sample area.
This figure illustrates that the boundaries having similar strong recombination
activities may have different magnitude of diffusion enhancement.

strong recombination activities. The origin of the thin line is not known and
it contributes only a weak recombination activity. Figure 2b is the surface
appearance of the same area after grooving and staining, showing that the diffusion enhancement occurs at three boundaries with strong recombination activities, but not at the one with weak recombination activity. Figure 3 gives a
pair of EBIC and enhanced diffusion images of another area, which contains
three high angle boundaries with strong recombination activities. However,
the depth of the enhanced diffusion varies considerably among the three bound-

386

aries, illustrating that the boundaries having similar strong recombination
activities may have different magnitude of diffusion enhancement. This is
consistent with the reported observation [4] that the depth of the enhanced
diffusion at grain boundaries in several cast polycrystalline materials is
widely distributed.

DISCUSSION

Figures 1,2 and 3 illustrate the fact that high angle boundaries are diffusion-active whereas coherent twin boundaries are not. However, the degree of
the enhancement varies considerably from boundary to boundary, which is consistent with the fact that the high angle boundaries are very complicated and
have an almost infinite number of possible boundary structures. The boundaries
capable of enhancing phosphorus diffusion generally have strong carrier combination activities, but the relationship between the two properties is not clear
at this moment.

The angular relationship of the two diffusion-active straight boundaries and
twins (and/or microtwins) appearing on the surface of the sample after a Sirtl
etching , as shown in Figure 1b, is very interesting. The twins and/or microtwins appearing as lightly etched straight lines in Figure 1b can be separated
into two groups of straight parallel lines intersecting each other with an measured angle of 70 degrees. This observation reveals that the grains containing
the twins share one common surface (110) crystalline plane, say [110], since it
is known that the intersecting angle between two non-parallel coherent twins is
70.53 degrees for this surface orientation. This is a simple method to determine whether or not the common surface of the grains is a (110) plane. After
the determination of the surface orientation, other orientation information
concerning the grains can be known in accordance to the orientation of twin
plane. For example, the coherent twins are along either $[\overline{1}1\overline{2}]$ or $[1\overline{1}\overline{2}]$, as
shown in Figure 4. Fig. 4 retraces the the boundaries for the sake of clear
explanation. An analysis of the picture containing the twins reveals that
there are only two orientations for the twinned crystals, marked T1 and T2.
Both share the common surface plane normal to [110]. The twins between two T1
grains, marked A, can be either coherent first order twins or microtwins containing even numbers of single coherent twins, and the twins between T1 and
T2, marked B, can be either single coherent twins or multiple coherent twins
of odd numbers. The two straight segment boundaries having enhanced diffusion
are likely to be incoherent {111/115} second order twins, as determined

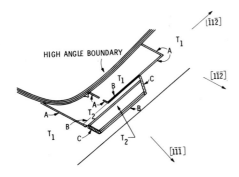

Figure 4: Schematic of boundaries
shown in Figure 1b, where A's are
first order coherent twins or
microtwins with even number of
coherent twins; B's are single
coherent twins or microtwins with
odd number of coherent twins; C's
are incoherent {111/115} second
order twins.

by the angular relationship with the nearby twins. The lower segment is not exactly aligned with the connecting microtwin. It can be explained that the segment consists of a sequence of incoherent {111/115} second order twins displaced gradually toward the T2 crystals. The existence of thin T1 crystals sandwiched between T2 crystals, as shown in Fig. 1b, supports this argument. The experimental result gives the first experimental evidence that incoherent {111/115} secon order twins are diffusion-active.

A recent transmission electron microscopy and EBIC study on EFG silicon ribbon material [5] has shown that incoherent second order twins of {111/115} type act as strong carrier recombination centers. The structural model of the {111/115} interface presented by the authors reveals the existence of high density of dangling bonds at the interface and also an alternating seven member-ed ring structure having relatively open space in the centers of these rings. The authors suggested that the dangling bonds are responsible for the strong recombination activity. Since the enhancement of phosphorus diffusion at grain boundaries is an atomic transport phenomenon, it can not be explained simply by the existence of dangling bonds. It is likely that some combined effect of dangling bonds and open space ring structure, possibly, plus others (such as the chemical nature of phosphorus) are responsible for the enhancement. In addition, it is generally accepted [6] that phosphorus mainly uses vacancies as diffusion vehicles at low temperatures, i.e. less than $1200°$ C. Consequent-ly, the boundaries can act as effective sources for generating vacancies which enhance the diffusion in the material nearby and as open channels in the seven membered rings for phosphorus atoms to move more easily. This argument is consistent with the observed facts that the boundaries capable of enhancing the phosphorus diffusion always have strong recombination activities. The boundaries with weak recombination activities are likely to have fewer dangling bonds and fewer structural defects. As a consequence, they are not able to contribute any significant doping impurity diffusion enhancement.

ACKNOWLEDGEMENT

The authors thanks Prof. Dieter Ast for his valuable assistance on the analy-sis of incoherent {111/115} second order twins.

REFERENCES

1. For example, J. Lindmayer, Proceedings of the 13th IEEE Photovoltaic Special-ists Conference, p.1092, 1978.

2. K. Hubner and W. Shockley in "Structure and Properties of Thin Films", edited by Neugebauer et.al. (John Wiley Sons, Inc., 1959), pp.302.

3. H.J. Queisser, K. Hubner, and W. Shockley, Phys. Rev. 123, 1245 (1961).

4. L.J. Cheng, C.M. Shyu, K.M. Stika, and T. Daud, and G.T. Crotty, Proceedings of the 16th IEEE Photovoltaic Specialists Conference, San Diego, California, 1982 (to be published).

5. B. Cunningham, H. Strunk, and D. Ast, Appl. Phys. Letters 40, 237 (1982).

6. A. Seeger, W. Frank, and U. Gosele, in "Lattice Defects in Semiconductors 1979" (Inst. Phys. Conf. Ser. No. 46), pp. 148.

SHEAR FORCE EFFECTS ON SIDEWALL PENETRATION OF EXTENDED DISLOCATIONS IN PHOSPHORUS IMPLANTED EMITTERS

W. F. TSENG AND G. E. DAVIS
Naval Research Laboratory, Washington, D.C. 20375

ABSTRACT

Phosphorus implanted emitters are known to have a dislocation network formed to relieve the strain caused by the lattice contraction of solute phosphorus in silicon. The extension of the dislocation network into the emitter-base region has also been related to the degradation of leakage current and the generation of excess noise in an NPN bipolar transistor. The penetration of the extended dislocation network is shown in this paper to be related to a shear force. The shear force is the resultant of the compressive force generated from the solute phosphorus and an induced surface force generated from the thick oxide layer over the emitter formed during high temperature heat treatments. This shear force can pull the dislocation network through the junction sidewall.

INTRODUCTION

The device performance of phosphorus implanted NPN transistors is affected by high temperature processing [1,2]. The device characteristics were shown to degrade due to the penetration of the interior emitter dislocations into the emitter-base junction. The penetration varied for the different ambients used in the high temperature processing. This paper will relate the penetration of the dislocations to the shear force present at the emitter-base junction. The shear force is the resultant of the compressive force due to the lattice contraction of the solute phosphorus in silicon (atomic radii: 1.10 $\overset{\circ}{A}$(P) and 1.18 $\overset{\circ}{A}$(Si)) and the surface force due to the presence of the oxide layer formed during a high temperature processing step. The surface force resulting from the silicon dioxide layer can be tensive or compressive depending on the deposition method and the difference in the thermal expansion coefficients. The effect of the shear force is to pull the dislocations through the emitter-base sidewall and terminate the extended dislocation network preferentially in the direction perpendicular to the shear force. A similar shear force effect has also been noted in arsenic implanted emitters [3].

EXPERIMENTAL

The wafers were n-epitaxial layers (8 ohm-cm, 18 micron thick) on n^+(111)-oriented Si substrates. The transistors were patterned on one-half of the wafer. The base was formed by a predeposition and diffusion of BCl_3. The emitter was formed by implantation of 1×10^{16} ^{31}P ions/cm^{-2} at 50 kev. After implantation, the high temperature processing steps were: (i) 900°C for 30 min. in either a steam or dry N_2 ambient to consume or reorder the implant damage, respectively; (ii) 1060°C for 60 min. in a dry N_2 ambient (drive-in step) to establish a base width of one micrometer; and finally (iii) 900°C for 60 min. in steam to permit isolated window Al contacts.

The unpatterned half of the wafer was used for the structural analysis. The samples were investigated by various analytical methods. X-ray rocking curves were used to determine the degree of lattice contractions in the phosphorus diffused regions. Spreading resistance measurements were used to obtain carrier concentration profiles. The Rutherford backscattering spectrometry (RBS) data was used to determine the location of the crystallographic defects. The spatial

distribution of the crystallographic defects was observed by transmission electron microscopy (TEM). Scanning electron microscopy (SEM) channeling patterns
were used to determine the induced surface strains formed by the silicon dioxide
layer during the anneal.

RESULTS AND DISCUSSION

Figures 1a and b show the TEM micrographs across the emitter-base junction.
The distinct hexagonal- and Y-shaped dislocations in the emitter region are
found to be a pure edge type. However, there is apparent difference in the
number of dislocations which penetrate through the sidewall of the emitter-base
junction due to the ambient anneal. The formation of the hexagonal- and
Y-shaped dislocations in the emitter is due to phosphorus substitutionally
replacing the silicon sites causing the lattice to constrict because of the
difference in atomic radii of the solute phosphorus (1.10 Å) and the host silicon (1.18 Å). The x-ray rocking curve of Fig. 2 shows that the (440) lattice
plane of phosphorus diffused regions contracts by 0.17% as compared with the
underlying substrate. The extent of lattice contraction into the sample is
related to carrier (or ionized phosphorus) concentration profile as shown in
Fig. 3. The strain induced by the solute phosphorus is proportional to the
slope of the carrier profiles. The kink (or abrupt change of slope) observed in
the accompanying RBS data for the sample annealed in a steam ambient shown in
this figure is attributed to the difference in diffusivities in the high and low
phosphorus concentration regions. The kink is greatly enhanced in steam oxidation due to the impurity pile-up of phosphorus at the Si/SiO$_2$ interface and the
phosphorus solid solubility limit in silicon. The step-like increase in the
backscattered yield of the RBS signal indicated that the hexagonal-shaped dislocation network is highly localized at a depth of approximately 3500 Å. The
sample annealed in dry N$_2$ did not exhibit the enhanced kink.

An idealized configuration of hexagonal-shaped dislocations based on the RBS
and TEM data is presented in Fig. 4. The strain induced by the phosphorus can
be calculated assuming the average dislocation spacing (D) is 4500Å and the
lattice spacing (d$_o$) below the dislocations is the same as that of the host
undoped silicon (or step-wise carrier profile). Because this model assumes a
highly localized dislocation, the strain calculated in the silicon surface layer
(0.85×10^{-3}) is smaller than the value determined from the rocking curve
(1.71×10^{-3}). The SEM channeling pattern measurements [4] can determine sign and
magnitude of the strain induced at the silicon surface by comparing the electron
channeling patterns of the stressed and unstressed regions. Strain determinations made by this method show that the silicon in the diffused phosphorus
region is under compression (1.62×10^{-3}). The measurements of strain on an
undiffused silicon caused by the presence of an oxide layer indicate that the
silicon under the oxide is clearly under tension (2.3×10^{-3}). In the case of
the thin oxide layer formed in an N$_2$ ambient, the strain in an undiffused silicon substrate near the interface is very large and in compression. The postulated lattice changes for the various test configurations is shown in Fig. 5.
The strain calculated by the various methods for the different configuration is
summarized in Table I.

The combination of the compressive force caused by the phosphorus which
followed the carrier profile and the tensile surface forces caused by the thick
oxide layer is shown in Fig. 6a. The combination of these distributed forces
into the silicon substrate gives rise to a shear force (Fig. 6b) at the emitter-
base junction. The oxide layer grown during the steam ambient anneal becomes
thicker in the heavily doped emitter region than in the less heavily doped
silicon. This is because of the enhanced doping oxidation rate. The magnitude
of the tensile force increases. Thus, in the emitter region the difference
between the tensile force due to the oxide formation and compression force due

to the solute phosphorus becomes larger. In effect, a shear force is formed at the emitter-base junction. This shear force pulls segments of the dislocations off the hexagonal-shaped network in the emitter region and terminates the dislocations in the direction perpendicular to the shear force in the emitter-base region as shown in Fig. 1a. When the annealing is performed in a dry N_2 ambient, the thin oxide induces a compressive force in the silicon. Since these forces are not opposing, there is no shear force generated and the dislocation network remains in the emitter interior.

SUMMARY

NPN bipolar transistors with phosphorus emitters have a dislocation network in the emitter interior. This network is formed to relieve the lattice contraction of the silicon due to the solute phosphorus. High temperature processing in a steam ambient can extend the dislocation network into the emitter-base junction due to an impurity pile-up and the induced surface tension caused by the silicon dioxide layer. The opposing forces in the silicon substrates due to the solute phosphorus (compressive) and the silicon dioxide layer (tensive) create a shear force which tends to pull the interior emitter dislocations through the emitter-base sidewall.

REFERENCES

1.   T. Koji, W. F. Tseng, J. W. Mayer and T. Suganuma, Solid-State Electronics 22, 335 (1979).

2.   T. Koji, W. F. Tseng, J. W. Mayer and T. Suganuma, IEEE Trans. Electron Devices ED-26, 1310 (1979).

3.   S. Mader, J. Elect. Mat. 9, 963 (1980).

4.   G. E. Davis and M. E. Taylor, J. Vac. Sci. Technol. 19, 1024 (1981).

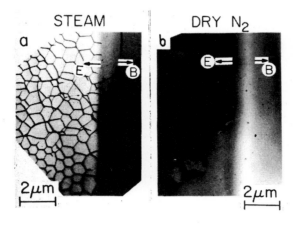

Fig. 1. TEM micrograph of the emitter-base region for a) steam and b) dry N_2 ambient anneal at 900°C.

392

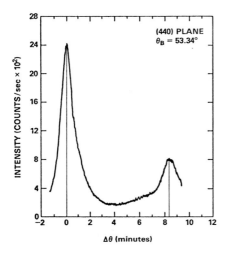

Fig. 2. X-ray rocking curve of solute phosphorus in silicon.

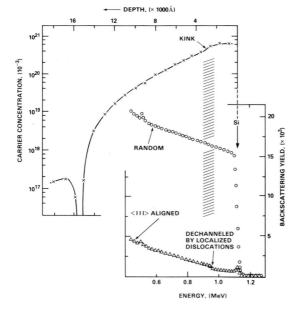

Fig. 3. Carrier concentration profile and RBS data for phosphorus implanted emitter processed with a steam ambient.

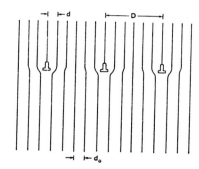

Fig. 4. TEM Network Model for solute phosphorus in silicon.

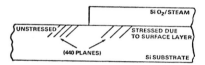

(a)

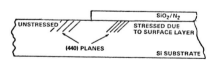

(b)

Fig. 5. Induced strain in silicon substrate for various test configurations; a) solute phosphorus, b) silicon dioxide grown in a steam ambient and c) silicon dioxide layer grown in dry N_2.

(c)

Table I. Strain in the Silicon Substrate

Measurement Method	Strain (+T/-C)
P in Silicon	
Rocking Curve	-1.71×10^{-3}
TEM network Model	-0.85×10^{-3}
SEM Channeling Patterns	-1.62×10^{-3}
SiO_2/Steam on Si(SEM)	$+2.3 \times 10^{-3}$
SiO_2/dry N_2 on Si (SEM)	-2.5×10^{-2}

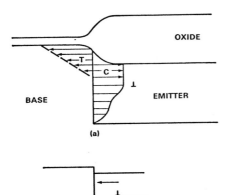

Fig. 6. Proposed strain model: a) distributed compressive and tensive strain due to solute phosphorus and the emitter oxide layer respectively and b) the resulting shear force at the emitter periphery.

STRUCTURAL STUDIES OF METAL-SEMICONDUCTOR INTERFACES WITH HIGH-RESOLUTION ELECTRON MICROSCOPY

J. M. GIBSON, R. T. TUNG AND J. M. POATE
Bell Laboratories, 600 Mountain Avenue, Murray Hill NJ 07974

ABSTRACT

We have studied interface atomic structure in epitaxial cobalt and nickel disilicides on silicon using high-resolution transmission electron microscopy. By employing UHV techniques during deposition and reaction we have grown truly single-crystalline $NiSi_2$ and $CoSi_2$ films on (111) Si and in the former case on (100) Si. These films are shown to be continuous to below 10Å thickness. By close control over preparation conditions, afforded by UHV, we can greatly influence the nucleation and growth of these films to the extent, for example with $NiSi_2$ on (111)Si, of yielding continuous single-crystal films with either of two orientations as desired. Whilst in the (111) $NiSi_2$ on Si system the interfacial structure invariably appears to well-fit a model in which metal atoms nearest to the interface are 7-fold co-ordinated, for (111) $CoSi_2$ on Si agreement is generally better with a model involving 5-fold co-ordination of these atoms. A misfit dislocation core is also imaged. Results are discussed in the light of silicide nucleation and growth. The structure and stability of the (100) $NiSi_2$ on Si interface is also considered.

INTRODUCTION

Recent progress in high-resolution transmission electron microscopy (HRTEM) has made possible useful determinations of interfacial structure for metal silicides on silicon. Not only has instrumental resolving power exceeded some atomic periodicities in Si but also methods for preparing the necessary thin cross-sectional specimens have been perfected [1]. Low-resolution TEM is useful in determining the uniformity and continuity of thin metal layers on Si [2]. High-resolution TEM has been able to identify the sharpness of the "crystallographic" interface [3] and very recently actual atomic structure determinations have been attempted [4,5,6]. This last aim is invaluable in modelling the electronic structure of these interfaces and so important properties such as Schottky Barrier height [7].

Mat. Res. Soc. Symp. Proc. Vol. 14 (1983) Published by Elsevier Science Publishing Co., Inc.

Currently the best available HRTEM instrumental resolution is only sufficient to properly image widely-spaced atomic planes in Si, not the basic interatomic separations. The state-of-the-art 200kV ultra-high resolution JEOL JEM200CX used in this study has a point-to-point resolution (r_{pp}) of 2.5Å, so that the (111) and (200) Si spacings (3.1 and 2.7Å respectively) can be faithfully imaged. Although it would be possible to image with (220) (1.9Å) and even 400 (1.4Å) periodicities these would not be simply interpretable in terms of atomic positions. The <110> orientation is thus invariably used in fcc structures as it provides 2 (111) and one (200) spacings. An objective aperture is generally employed to prevent higher-order Bragg reflections from "confusing" the image. The <110> zone is also very useful for interfacial studies using cross-section specimens since one can be found perpendicular to all commonly-used wafer surface-normals. No very significant improvement will be realised in HRTEM images of semiconductors in this orientation until the resolution limit reaches 1.4Å (the (400) interatomic spacing), which is some way off.

Given the limited available resolution how can we best make use of HRTEM in interface physics? Look for example at figures 5 and 12, HRTEM images taken under the conditions where the resolution approaches the 2.5A limit. In the lower half of each one sees a <110> image of Si. Instead of the detailed atomic structure we see only the tunnels in the structure. Figure 1 demonstrates the relationship between the image and atomic positions under these conditions (specimen thickness <90Å) both for Si and CoSi$_2$. One is limited in studying interfaces to detail on this scale. Nevertheless one must bear in mind that the white dots in figure 1 are accurately located in the centres of the tunnels, under these imaging conditions. Thus for a perfect lattice one can deduce the position of the real (Si) lattice with considerably greater accuracy than the point resolution (or size of the white dots). This corresponds to sensitively measuring the local phase of the scattered electron beams which are certainly defined, by the long-range order, to better than 2π. If one employs the same technique for locating the silicide lattice, for instance in the upper part of fig. 5, one can then measure the relative shift of these two with an accuracy of better than 1Å. (Accuracy is limited by the certainty of imaging conditions and specimen orientation and by projector lens and recording media distortions. It can be

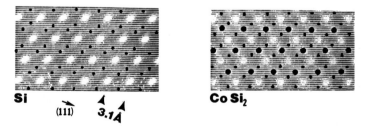

Si ➔ (1̄11) **3.1Å**   **Co Si$_2$**

Fig. 1. The relative location of axial bright-field HRTEM images and the real lattices for Si and CoSi$_2$ in the [1$\bar{1}$0] orientation, calculated using the multi-slice algorithm for the JEOL 200CX, C_s=1.2mm, λ=0.025 Å, δf=-660Å (the Scherzer focus) and sample thicknesses about 60Å. The white dots locate with tunnels in the structures under these conditions.

experimentally determined by several measurements on independent micrographs and appears to be better than 0.3Å typically.) This information can then be used to distinguish possible rigid models of the interface. Such "interferometry" is perhaps the most powerful application of current HRTEM's in interface science and was first demonstrated for the (111)NiSi$_2$ on (111)Si system by Cherns et. al. [5]. Their study was limited to discontinuous, non-single-crystalline (111)NiSi$_2$/Si films. Here we extend this to UHV grown continuous single-crystalline films of both NiSi$_2$ and CoSi$_2$, both on (111) and in the latter case (100) Si.

CRYSTALLOGRAPHY OF MSi$_2$ ON Si (M=Ni or Co)

Both CoSi$_2$ and NiSi$_2$ have the cubic calcium fluoride structure with lattice parameters within 2% of Si. It is therefore not surprising that both can be grown epitaxially on Si [8,9]. The commonly employed method for growing them involves room-temperature vacuum deposition of the metal and subsequent annealing >700°C for about 1 hour. In both cases the resultant films on (111) Si suffer from the phenomenon of "double-positioning" [2,10]. That is, they contain silicide grains with [111]MSi$_2$||[111]Si but of two differing orientations: A in which [1$\bar{1}$0]MSi$_2$||[1$\bar{1}$0]Si and B in which [$\bar{1}$10]MSi$_2$||[1$\bar{1}$0]Si. We have reported that the growth of CoSi$_2$ under UHV conditions (that is deposition on atomically-clean surfaces and subsequent in-situ annealing) avoids the double-positioning problem [10,11,12]. The resulting continuous films are single-crystals which, somewhat unexpectedly, have the B orientation.

On (100) Si previous attempts to grow good continuous films of NiSi$_2$ and CoSi$_2$ have proven unsuccessful due to gross faceting of the interface on (111) planes [2,3,13].

For NiSi$_2$ the simple use of UHV deposition and annealing conditions does not avoid the problem of double-positioning [11,14]. Because NiSi$_2$ has a considerably smaller misfit with Si than does CoSi$_2$ this suggests that misfit stress can be a driving force for epitaxial growth in these systems: i.e. that dislocations may play an important role[12,4]. Another difference between the two disilicides which may have relevance to their growth is that NiSi$_2$ appears to have the greater nucleation barrier[13]. However we have recently found that uniform single-crystal NiSi$_2$ films can be grown, not only on (111) but on (100) Si, by novel UHV techniques [15,16]. It was discovered that very thin layers of Ni (<30Å) deposited near room temperature on atomically-clean Si can react to form single-crystal layers of NiSi$_2$ at relatively low temperatures ($\sim$ 400°C). These layers then act as "templates" if Ni is deposited directly on them at elevated temperatures (>600°C), yielding thick single-crystalline NiSi$_2$ films. Incredibly the orientation of these templates and subsequent thick silicide layers is determined by the initial deposited nickel thickness. For 3Å$<t_{Ni}<$10Å the orientation is B whereas for 12Å$<t_{Ni}<$20Å it is A. At other thicknesses it is A/B (doubly-positioned). On (100) Si, the orientation is fixed (A) but the template can be used to stabilise the interface and produce a uniform thick (100) silicide layer.

The details of template fabrication and their effect on NiSi$_2$ epitaxial growth will be described elsewhere [15,16]. In this paper only high-resolution TEM studies of these layers will be described, mostly but not exclusively concerning interfacial structure.

MODELLING INTERFACIAL STRUCTURE IN MSi$_2$/Si

There are a large number of possible models for the atomic arrangements at these interfaces. In order to restrict these we will invoke the constraint that all Si atoms retain tetrahedral co-ordination. For the case of (111) MSi$_2$/Si, either A or B orientation, there are as pointed out by Cherns et.al. [5] only two models within this constraint. These are shown in figure 2 for the A epitaxy and are referred to by the co-ordination number of the metal atoms nearest to the interface as either 7-fold (a) or 5-fold (b). In figure 3 (a) and (b) the same models are depicted for the B epitaxial orientation. Figure 3(c) and (d) show computed HRTEM images which demonstrate the validity of direct interpretation even at the interface (conditions as figure 1). For the (100) epitaxy there is only one possibility, shown in figure 4, on the assumption that the interfacial plane is (100) - otherwise one of the two models above might apply on (111) facets. This (100) interface has 6-fold co-ordination and shows two variants (depending on angle of view). If this model were correct one might expect antiphase boundaries within the silicide film. All of these models take no account of any reconstruction at the interface but are considered to be reasonable based on the strength and rigidity of the Si-Si bond.

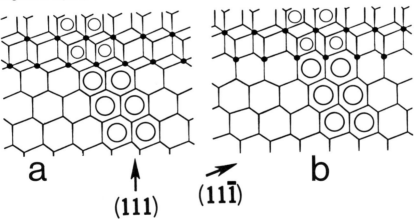

Fig. 2. Models of the (111) MSi$_2$ interface for the A orientation, (a) with 7-fold co-ordination of the metal atoms nearest the interface and (b) with 5-fold co-ordination. The metal atoms are the larger dots. Circles indicate the approximate position of the white dots in an HRTEM image (taken under appropriate conditions). Note the opposite senses of shift across the interface for (111) and (11$\bar{1}$) planes between models.

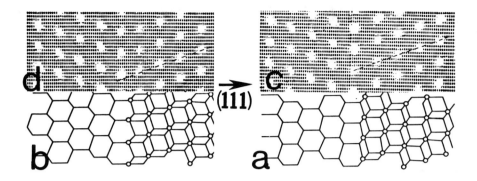

Fig. 3. Models of the (111) MSi$_2$ interface for the B orientation, (a) 7-fold co-ordination and (b) 5-fold. (c) and (d) show the respective calculated HRTEM images (under conditions described in text) which demonstrate the validity of the simple graphical method used to locate tunnels in the structures, even at the interface itself. The calculations use the full multi-slice algorithm and take account of all important effects.

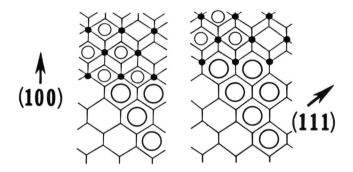

Fig. 4. Models of the (100) MSi$_2$ on Si interface assuming that is on a (100) plane. Both are 6-fold co-ordinated for the metal atoms nearest the interface and are in fact simply related by a 90° rotation about (100) and a shift.

In discussing the use of these models for analysis of HRTEM images some description of the imaging process must be given. In axial bright-field illumination of thin specimens ("phase" objects) the image is a distorted, magnified representation of the object's projected atomic potential. The distortion is most easily represented as a filtering function in fourier space, known as the contrast transfer function (CTF) [17].

Thus certain spatial periodicities are transmitted well and others not at all or reversed in sign of contrast. At the so-called Scherzer focus the first uniform interval of this transfer function extends to its maximum spatial frequency. The inverse of this frequency is the point resolution r_{pp} because all information up to this limit is transmitted faithfully. (The HRTEM amplifies spatially-varying signals like a conventional amplifier does for time-varying signals: the CTF is analogous to its frequency spectrum and the point-resolution to its response-time.) The CTF is sensitively dependent on microscope parameters such as spherical aberration, defocus, coherence, residual astigmatism and specimen drift during exposure.

The best way to visualise this CTF is to look at the image of an amorphous object which essentially has a "white-noise" fourier spectrum (diffraction pattern). For example, the inset to figure 5 shows the optically-obtained diffraction pattern (ODP) of an amorphous region overlapping the crystal of interest. The diffuse background is uniform out to at least the spatial periodicity of the Si lattice (3.1 and 2.7Å) - so this image is sufficiently close to Scherzer focus for direct interpretation of the lattice image. If the defocus was incorrect or the astigmatism was not properly corrected this would reveal itself on the ODP as one or more black rings (or ellipses) between the zero beam and the lattice spots, e.g. the inset to figure 6 at a higher defocus shows a reversal in contrast of the lattice (each ring corresponds to a change in sign of the CTF). To achieve full resolution (r_{pp}) the following conditions must be satisfied: axial illumination, Scherzer focus, sufficient spatial and temporal coherence and a well-aligned instrument free of astigmatism and specimen drift. The ODP is the important experimental proof of these conditions. These cannot be predetermined with sufficient accuracy and so a series of pictures are taken at different defoci and ODP's from each negative are used to find suitable images.

Under these conditions white dots in the image correspond to minima in the projected potential i.e. tunnels in the structure provided that the specimen is properly oriented and less than $\xi_g/2$ thick (ξ_g is the extinction distance for the g reflection, e.g. for Si (111) at 200kV it is 290Å). This was demonstrated for Si and $CoSi_2$ in figure 1. $NiSi_2$ has slightly shorter extinction distances but is otherwise almost identical to $CoSi_2$ in this context. The specimen thickness and orientation constraints are equally critical and perhaps more difficult to achieve. If any of these conditions is not satisfied interpretation is difficult if not impossible.

In attaining these conditions, particularly correct orientation and thickness, it is easiest to study very thin or discontinuous films in which stress is minimal [5]. During preparation stress tends to distort specimens and can even cause films to separate from the substrate in thin areas. Sometimes it is necessary to study thicker films and then the number of suitable micrographs found is much reduced (e.g. $CoSi_2$ on (111)Si). The method used for thin cross-section specimen preparation is based on that of Sheng and Marcus[1] and entails 4kV Ar ion thinning at 20° incidence and about $30\mu A$ beam current. This causes no other problems for Ni and Co disilicides but can raise specimen temperatures sufficiently to cause problems with less stable silicide phases [4].

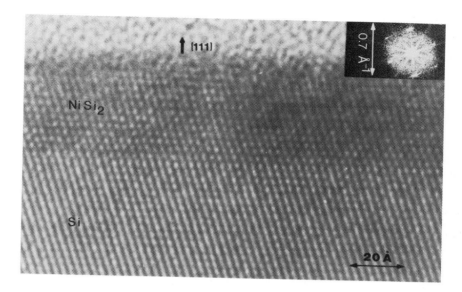

Fig. 5. An axial-bright field HRTEM image of an A-oriented (111) $NiSi_2$ film on Si. The image was taken sufficiently near the Scherzer focus, as demonstrated by the inset ODP which shows a broad ring extending at least to the (111) and (002) diffraction spots. The specimen thickness is also less than 90Å allowing direct interpretation of the white dots in this image as tunnel positions.

RESULTS

$NiSi_2$ on (111) Si

Figure 5 shows a [1$\bar{1}$0] cross-section HRTEM image of the result of rapid $450^\circ C$ in-situ annealing of an 18Å thick Ni layer deposited at UHV ($<10^{-9}\tau$) on atomically-clean (111) Si. We find a continuous 40Å thick layer which is pseudomorphic (i.e. free of misfit dislocations and commensurate with the Si lattice). This layer is found to be a single-crystal of $NiSi_2$ with the A orientation. Although the majority of $NiSi_2$ is contained within this layer, isolated islands with either A or B orientation are found atop this continuous layer. Figure 5 is from a thin area and demonstrates the uniformity and epitaxial orientation. This axial bright-field picture is taken near the Scherzer focus as evidenced by the inset optical diffraction pattern (ODP). The first bright band of this ODP includes both (111) and (200) diffraction spots from Si and $NiSi_2$ (the abrupt cut-off is due to an objective aperture used during electron exposure). The specimen thickness in this area was confirmed to be <90Å (by the position of reversals in image contrast) so that conditions are satisfied for direct interpretation as discussed above, and white dots ≡ tunnels in the atomic arrangement.

Fig. 6. Axial bright-field HRTEM image of a thicker area of the specimen depicted in figure 5. The defocus is no longer at the Scherzer (see inset ODP) and the thickness exceeds the extinction distance ξ_0, so that direct interpretation is not possible.

In contrast, figure 6 is from a much thicker specimen area (probably about 500Å thick) in which several reversals have occurred. The ODP also reveals that the image was not taken near the Scherzer focus. Note that in this thicker region the silicon/silicide boundary is far more distinct and obviously uniform. This is because of the difference in the mean potential between NiSi$_2$ and Si, which results in darker contrast within the silicide because of scattering outside of the objective aperture. This effect is only weakly seen in the thin area of figure 5. However, the interface "structure" seen in figure 6 is ambiguous because of fresnel fringes which result from electron refraction at this potential step. The white dots can no longer be related to atomic positions even far from the interface. Compare the shift in inclined $(11\bar{1})$ planes at the interface between figures 5 and 6. Figure 5 is, as stated, suitable for direct comparison with the image positions depicted in figure 2. The sense of shift in the inclined $(11\bar{1})$ planes is seen to be consistent with the 7-fold interfacial model. Furthermore the displacement in the (111) direction is also of the correct sense for this model (equivalent to an expansion of the (111) interfacial plane). It's magnitude of $.6 \pm .2$Å is in excellent agreement with the model's 0.7Å.

A single-crystal NiSi$_2$ layer of similar thickness and continuity but of the B orientation results if, for example, 20Å Ni followed by 30Å Si is deposited at room temperature on atomically-clean (111) Si and heat-treated in the same fashion as above. Figure 7 shows an axial-bright-field HRTEM image of a $[1\bar{1}0]$ cross-section of

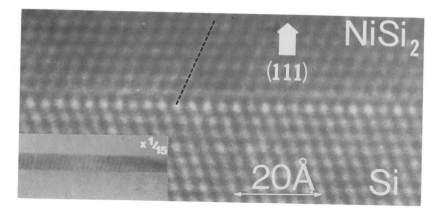

Fig. 7. Axial-bright-field HRTEM image from a thin $NiSi_2$ B layer on (111) Si; conditions are satisfied for direct interpretation. The dotted line enables comparison with figure 3. Inset is a lower magnification image of this layer.

this layer, whose ODP and thickness are seen to be suitable for direct interpretation. Because of the change in direction of $(11\bar{1})$ planes, this interface is more easily seen in thin areas and is clearly confined to a single (111) plane. The layer is again continuous and pseudomorphic. Inspection of figure 3 reveals that the behaviour of inclined $(11\bar{1})$ planes at the interface in figure 7 is characteristic again of the 7-fold model. The relative lattice displacement of (111) planes is again $.7 \pm .3\text{Å}$ and of the correct sense for this 7-fold model.

We will not comment here on the unusual and fascinating reversal of orientation that occurs in these very thin single-crystal layers, which is the subject of another recent publication [16] and is referred to elsewhere in this volume [15]. Suffice to say that one can heat these thin layers up to $\sim 700^\circ C$ and deposit Ni directly on them to yield thick, continuous single-crystal layers of $NiSi_2$ of whichever orientation desired [16]. For this reason we refer to these ultra-thin layers as "templates". In all (111) $NiSi_2$ cases which have been studied, both thin and thick, A and B layers, the interfacial co-ordination appears to be 7-fold as above. This is in agreement with the study due to Cherns et.al.[5] which was carried out on non-single crystalline discontinuous layers of $NiSi_2$ on (111) Si, grown under relatively "dirty" conditions. Thus we can conclude that the 7-fold interface structure is intrinsic to $NiSi_2$ on (111) Si and is not changed by wide variations in the conditions of nucleation and growth.

Before going on to discuss $CoSi_2$ on (111) Si we should highlight another application of HRTEM in determining, not the interface structure but the crystallography of $\sim$ monolayer thick films. When only 5Å of Ni is deposited on

(111) Si and reacted to form a "template" as above, LEED shows it to be NiSi$_2$, mostly of the B orientation[16]. A cross-section of this reveals a continuous layer only 1 or 2 monolayers in thickness. The remainder of the Ni appears to have migrated to A and B orientation NiSi$_2$ islands above this ~ monolayer thick single-crystal. Since we have no long-range over which to have order in such a thin layer we cannot expect to be able to do interferometry as discussed above. Indeed HRTEM images, such as figure 8, do not clearly reveal the structure of such thin layers which show up primarily through amplitude contrast. Figure 8(b) is an optical diffraction pattern which shows faint streaks parallel to the (111) surface normal which arise from the very thin "silicide" layer. These streaks display the periodicity expected of NiSi$_2$. Figure 8(c) shows an ODP from a B-NiSi$_2$ island as indicated - the doubled (11$\bar{1}$) and (002) spots can be seen to come (a) from the substrate and (b) from the B-NiSi$_2$ island. Figure 8(d), from a thicker specimen area, of the template and substrate confirms that the streaks are close to the B position thus giving the probable crystallographic orientation and composition of this very thin layer as B-NiSi$_2$. Here we are essentially carrying-out "microdiffraction" on an authentic replica of the object - the HRTEM image. This is another useful application of the HRTEM technique.

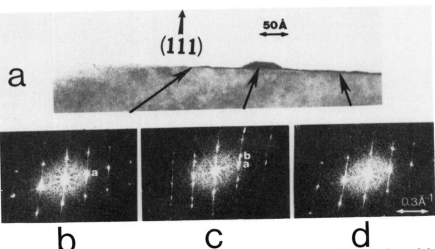

Fig. 8. Axial bright-field HRTEM image of a very thin layer of NiSi$_2$ formed by deposition of only 5Å Ni at room temperature on atomically clean Si and subsequent low-temperature anneal. The inset ODP's from the arrowed areas enable us to identify the probable layer structure and orientation as B-NiSi$_2$ (see text).

CoSi$_2$ on (111) Si

Thick (~ 1000Å) single-crystalline CoSi$_2$ layers on (111) Si can be grown by room temperature metal deposition and annealing at ~900°C for ~ 1/2 hour provided the initial surface is atomically clean and the whole process occurs in-situ under UHV conditions[11,14]. Figure 9 is a typical axial-bright-field HRTEM image from the interface of a 600Å (layer thickness) (111) CoSi$_2$ on Si sample fabricated in

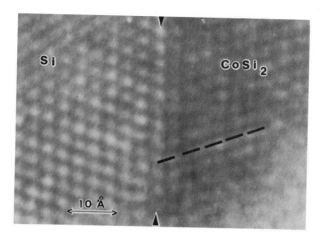

Fig. 9. Axial bright-field HRTEM image of the (111)CoSi$_2$/Si interface in [1$\bar{1}$0] cross-section. The dotted line is to aid comparison with figure 3.

this manner. The B orientation of these CoSi$_2$ single-crystals is obvious and since the imaging conditions and specimen thickness are suitable for direct interpretation we can compare this with figure 3. The dotted line demonstrates that the shift of inclined (11$\bar{1}$) planes is consistent in this case with the 5-fold model (figure 3 (b)). Furthermore the sense of displacement in the (111) direction is equivalent to a compression of the interfacial plane, as expected for this 5-fold model. Its magnitude is somewhat larger than the expected value of 0.7Å which may be evidence for some reconstruction at the interface. As this study of (111)CoSi$_2$ films on Si was restricted to thick, rather highly-stressed films the number of suitable thin areas found in specimens prepared for HRTEM observation was unfortunately small (three) compared with the NiSi$_2$ films described above and by Cherns et.al.[5]. It would be worthwhile to extend this study to thinner films, even if discontinuous and also to films prepared under "dirty" conditions which have both A and B regions [12]. Nonetheless the systematic observation so far is opposite to that for NiSi$_2$ and is that the metal atoms adjacent to the interface have 5-fold co-ordination.

Figure 10 shows another (111)CoSi$_2$/Si interface, this time prepared by metal-Si co-deposition at only 650°C[18]. The field-of-view includes a misfit dislocation of burger's vector $\sim 1/6(11\bar{2})$, the type invariably seen beneath these B-type single crystals. At the core of this dislocation (arrowed) one sees a step of height 2(111) planes. Models of this dislocation (far from the core) show that such a step is necessary to preserve interfacial co-ordination. Note that the core itself appears to lie within the Si lattice[6].

406

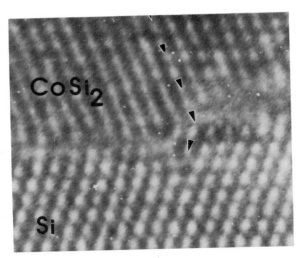

Fig. 10. Axial bright-field image of a misfit-dislocation core at the (111) $CoSi_2$/Si interface. The extra plane is arrowed.

$NiSi_2$ on (100) Si

We mentioned that thick layers of $NiSi_2$ grown on (100) Si tend to facet on (111) planes and have poor crystallography[2,3,13]. We have found that stable single-crystalline $NiSi_2$ layers can be fabricated when $10Å<t_{Ni}<30Å$ Ni is deposited on atomically-clean (100) Si and annealed at $\sim 400°C$. These very thin layers have flat interfaces confined to a single (100) plane, as seen in figure 11 (b). If Ni is subsequently deposited hot ($\sim600°C$) on one such template a thick single-crystal $NiSi_2$ layer is formed with great uniformity as seen in figure 11(c). The thickness of Ni needed to form a uniform template is critical. If, for example, less than $10Å$ is deposited the resulting silicide film is discontinuous and the interface is heavily faceted as in figure 11(a). Needless to say, hot deposition of Ni does not yield continuous uniform layers of $NiSi_2$ on a "template" such as this.

Figure 12 is an axial bright-field HRTEM image from a thin area of the (100) $NiSi_2$ layer in figure 11(b). This [110] cross-section image is taken under suitable conditions for direct interpretation. The image displays the local uniformity of the silicide layer and its interface. However it raises an important limitation in direct HRTEM interface imaging which is that the interface must be flat and lined up with the beam direction since we are at best given a projection in this direction. For "interferometry" this is not such a severe problem since one studies the lattice shifts "far" from the interface. Nonetheless one must assume the interfacial habit plane in order to model the interface, as in figure 4. In some areas the (100) interface, although uniform, seems to extend over several (200) planes, suggesting the possibility that it might be "micro-faceted" on (111) planes. Since the extent of the interface does not vary with specimen thickness in the beam-direction, we can rule

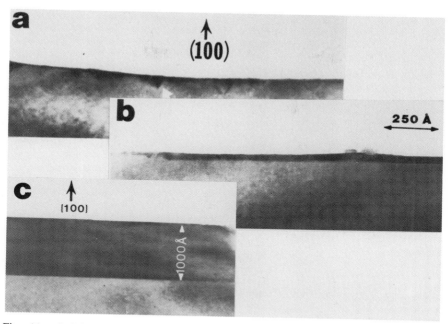

Fig. 11. Axial bright-field HRTEM images of NiSi$_2$ layers on (100) Si: (a) a "template" formed from 8Å Ni deposition and reaction; (b) a "template" formed from 11Å Ni deposition and reaction; (c) a thick single-crystalline NiSi$_2$ layer grown on (b).

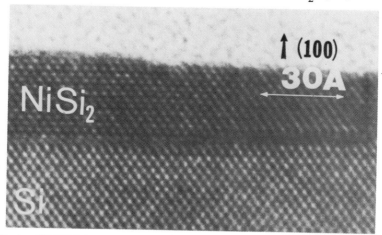

Fig. 12. An HRTEM image taken under conditions suitable for direct interpretation, of the (100) NiSi$_2$/Si interface from fig.11(b).

out gross facets. Only a fine "sawtooth" interface could be consistent. Interferometry might enable us to distinguish this if the (111) facets have the expected 7-fold co-ordination. In fact the displayed area does fit reasonably well with the 6-fold model and anti-phase boundaries are seen in the very thin films. However this study is still preliminary and detailed image simulations are probably required to distinguish these cases.

DISCUSSION

Our observation of 7-fold interfacial co-ordination for $NiSi_2$ continuous single-crystal layers is in accordance with the previous results of Cherns et.al.[5] for discontinuous non-single-crystalline layers. Thus despite the fact that the conditions of nucleation and growth are very different (e.g. in the "templates" it is unlikely that the nucleation proceeds through the Ni_2Si and NiSi phases as is known for thick reacted layers[8]), the interfacial state is the same. The 7-fold co-ordination model is the one in which the total number of interfacial bonds is maximised. Only one free electron remains per metal atom near the interface. The more surprising observation is that the (111) $CoSi_2$/Si interface appears to be only 5-fold co-ordinated. There would be three dangling bonds per metal atom at this interface. In fact the chemical bond strength of the Co-Si bond (at 66kcal/mole) is indeed smaller than that of the Ni-Si bond (76kcal/mole) [19] so the trend is right. We speculate that the key to understanding this difference may lie in the core of misfit dislocations. The Si-Si bond has energy of about 78 kcal/mole so that it would be more favourable if a single broken Co-Si bond lay at the core of a misfit dislocation, as opposed to one or more Si-Si bonds. (For Ni-Si less, if anything, would be gained by this.) Since we know that misfit dislocations have steps associated with them at the B interfaces, movement of such defects in the presence of excess metal could lead to the growth of the silicide[4,6]. From models it seems that if these dislocation cores contain broken Co-Si bonds, the 5-fold co-ordinated interface would result. Such a mechanism may also explain the dominance of the B orientation for high-temperature reaction of $CoSi_2$ on (111) Si, where the misfit stress is greater. For $NiSi_2$ formed through templates, dislocations probably do not play a role at the initial stages which determine final orientation[15].

In conclusion we have found a systematic difference in the co-ordination number of metal atoms adjacent to the interfaces of (111)$NiSi_2$ and (111)$CoSi_2$ on Si. The models we have determined for these (111) interfaces and perhaps the (100) interface should enable detailed calculations of the electrical properties e.g. Schottky barrier height[7]. They may also aid in understanding the fascinating phenomena of silicide growth and its dependence on preparation conditions. The success of our simple "ball-and-spoke" models demonstrates that these interfaces are indeed simple and regular. The techniques of HRTEM "interferometry" certainly have a much wider application in interface science.

ACKNOWLEDGMENTS

We wish to acknowledge the assistance of M. L. McDonald in preparing cross-section specimens.

REFERENCES

1. T.T. Sheng and R.B. Marcus, J. Electrochem. Soc. **127** ,737 (1980).

2. K.C.R. Chiu, J.M. Poate, J.E. Rowe, T.T. Sheng and A.G. Cullis, Appl. Phys. Lett. **38** , 988 (1981).

3. H. Foll, P. S. Ho and K. N. Tu, J. Appl. Phys. **52** , 250 (1981)

4. D. Cherns, D.A. Smith, W. Krakow and P. E. Batson, Phil. Mag. **A45** , 107 (1982).

5. D. Cherns, J. C. H. Spence, G. R. Anstis and J. L. Hutchison, Phil. Mag., in press (1982).

6. J. M. Gibson, J. C. Bean, J. M. Poate and R. T. Tung, Appl. Phys. Lett., Nov. (1982).

7. M. Schluter, Thin Sol. Films **93** ,3 (1982).

8. K. N. Tu, E. I. Alessandrini, W. K. Chu, H. Krautle and J. W. Mayer, Jpn. J. Appl. Phys., suppl. 2, part 1, 669 (1974).

9. H. Ishiwara, M. Nagatumo and S. Furukawa, Nucl. Inst. Meth. **149** , 417 (1978).

10. J. M. Gibson, J. C. Bean, J. M. Poate and R. T. Tung, Inst. Phys. Conf. Ser. **60** , 415 (1981).

11. R. T. Tung, J. M. Poate, J. C. Bean, J. M. Gibson and D. C. Jacobson, Thin Sol. Films **93** , 77 (1982).

12. J. M. Gibson, J. C. Bean, J. M. Poate and R. T. Tung, **ibid** , 99 (1982).

13. L.J. Chen, J.W. Mayer, K.N. Tu and T.T. Sheng, Thin Sol. Films **93** , 91 and **ibid** , 135 (1982).

14. R. T. Tung, J. C. Bean, J. M. Gibson, J. M. Poate and D. C. Jacobson, Appl. Phys. Lett., **40** , 684 (1982).

15. R. T. Tung, J. M. Gibson and J. M. Poate, this volume

16. R. T. Tung, J. M. Gibson and J. M. Poate, to be published

17. for a good review see : **Experimental High-Resolution Electron Microscopy** , J. C. H. Spence, Clarendon Press (Oxford 1981).

18. J. C. Bean and J. M. Poate, Appl. Phys. Lett. **37** , 643 (1982).

19. **Handbook of Chemistry and Physics** , Chemical Rubber Company, (1980).

ION IRRADIATION EFFECTS ON Pt CONTACTS TO Si WITH AND WITHOUT INTERFACIAL CHEMICAL OXIDE

THOMAS BANWELL, MANUELA FINETTI[a], ILKKA SUNI[b] AND MARC-A. NICOLET
California Institute of Technology, Pasadena, California 91125, USA

and

S. S. LAU AND DAVID M. SCOTT
University of California at San Diego, La Jolla, California 92093, USA

ABSTRACT

The electrical properties of ion irradiated metal-semiconductor contacts are investigated. Silicide contacts are fabricated by depositing Pt on chemically clean or slightly oxidized ($\sim$ 14 Å SiO_2) n^+-and n-type <111> Si, followed by a Si ion irradiation (10^{14} - 6×10^{15} Si/cm^2) through the metal-Si interface at various substrate temperatures, and a final thermal annealing in vacuum to form the silicide. Forward I-V measurements are employed for electrical characterization. Metal-Si interaction and substrate damage are measured by MeV ion backscattering and channeling, and interfacial oxygen monitored by nuclear $^{16}O(d,\alpha)^{14}N$ reaction.

Platinum contacts prepared on clean n-type substrates are Schottky diodes with a barrier height $\phi_{Bn} = 0.83$ eV. After Si irradiation, the forward I(V) is a power law whose form is largely independent of the dose. Subsequent thermal annealing induces silicide formation, but at a reduced rate compared to irradiated samples. The dc characteristics is roughly exponential again, but departures from the original Schottky characteristics remain and are largest for the highest Si doses. The effect is attributed to radiation damage in the Si that is not consumed by the silicide reaction.

Platinum contacts prepared on chemically oxidized samples behave differently for different substrate materials, although the total amount of interfacial oxygen is always the same. On n^+-type samples, the silicide formation at 400°C is laterally uniform for Si doses > 2 × 10^{14} cm^{-2}, but is nonuniform for all doses (< 2 × 10^{15} Si/cm^2) on n-type samples. For n^+-type samples at $\overline{2}50$°C, a dose of 2 × 10^{15} Si/cm^2 is required to induce (uniform) silicide formation; the kinetics displays a time delay compared with that of clean n^+ substrates. On oxidized n-type substrates, the I(V) characteristics of Pt contacts before irradiation is not Schottky-like, but power-law-type. After irradiation, the characteristic is the same as for the clean irradiated samples. Thermal annealing induces only incomplete recovery toward an exponential behavior.

[a] Permanent Address: LAMEL Laboratory, C.N.R., Via de Castagnoli 1, 40126 Bologna, Italy

[b] Permanent Address: Semiconductor Laboratory, Technical Research Centre of Finland, Otakaari 5A, SF-02150 Espoo 15, Finland

Mat. Res. Soc. Symp. Proc. Vol. 14 (1983) Published by Elsevier Science Publishing Co., Inc.

412

These results demonstrate that radiation damage in the
unreacted Si remains significant for the electrical behavior
of all.
These results demonstrate that radiation damage determines
the I(V) characteristics of as-irradiated Pt contacts to n-type
Si regardless of the presence of an interfacial oxide layer.
After annealing at 400°C for 30 min, radiation damage is still
significant, but the oxidized samples recover less than the
clean ones. The results are attributed to radiation damage
in the unreacted Si substrate.

INTRODUCTION

There is decided interest in ohmic and Schottky barrier contacts to Si using
metal silicides because of their stable and reliable characteristics. It is
desirable in VLSI technology to minimize the processing temperature employed
for silicide formation. Residual contamination at the metal-Si interface can
inhibit silicide formation [1] and results in laterally nonuniform silicide
formation at low annealing temperatures [2,3]. The room temperature formation
of several silicides induced by ion beam has been extensively investigated
[4,5], although the electrical properties were not examined. Recent studies
have demonstrated that a thin interfacial oxide diffusion barrier can be dis-
rupted by ion irradiation allowing subsequent thermal silicide formation [3].
The ion irradiation also affects the bulk transport process.
We have investigated the influence of Si ion irradiation on the kinetics of
Pt_2Si formation and on the forward I(V) behavior of Pt/Si Schottky diodes, both
on substrates with and without an interfacial oxide layer produced by wet chemical oxidation. We also briefly examine the role of Si substrate doping level
on the interfacial oxide layer's inhibition of Pt diffusion.

EXPERIMENTAL

Two types of Si wafers used in this study were n^+-type 0.005-0.020 $\Omega \cdot cm$
<111> Si, and a 10 μm 10-30 $\Omega \cdot cm$ n-type epilayer on a 0.003 $\Omega \cdot cm$ n^+ <111> sub-
strates. Four wafers each were consecutively cleaned ultrasonically in acetone
and methanol followed by etching in 12% HF and then slightly oxidized in a
boiling solution of $1:H_2O_2:NH_3(aq.):H_2O = 1:1:5$. One each of the oxidized
wafers were rinsed in distilled water, blow dried with N_2, and loaded into an
oil-free e-beam evaporation system. Clean substrates with only the unavoidable
native oxide were prepared by dipping the other two wafers in 3% HF prior to
loading. A metal mask was used to define square areas of 7.7×10^{-2} cm^2 on the
epitaxial wafers for contacts. A ∿ 800 Å Pt film was evaporated onto each substrate, during which the pressure was < 10^{-7} torr. The interfacial oxygen on
both of the oxidized substrates was measured using the $^{16}O(d,\alpha)^{14}N$ nuclear
reaction to be $6.5 \pm 0.8 \times 10^{15}$ atoms cm^{-2}, corresponding to ∿ 14 Å SiO_2. 0.1
μm Al contacts were evaporated onto the back surface of the epi wafers.
Silicon irradiations were made at sample temperatures of -196°C and 27°C with
180 - 200 keV Si^+ (R_p(Pt) ∿ 670 Å, ΔR_p(Pt) ∿ 420 Å) [6] to fluence of 0.2 - 2 ×
10^{15} cm^{-2}. The epi wafer samples were subsequently annealed at 400°C for 30
min in a quartz tube vacuum furnace. Standard forward dc I(V) measurements
were made on the contacts after the Si irradiation and after annealing.
Samples studied for thermal reactivity were isothermally annealed over the
temperature range 254 - 560°C for periods of 10 - 190 min. The samples were
analyzed by 2.0 MeV $^4He^+$ ion backscattering spectrometry. The Pt was etched
from selected unannealed samples with boiling aqua regia and the damaged Si
region analyzed by $^4He^+$ ion channeling measurements.

RESULTS AND DISCUSSION

Figure 1 shows a semi-logarithmic plot of typical forward I(V) curves for un
-annealed Pt contacts with and without Si irradiation on both clean and oxidized
n-type epi substrates. Schottky diode behavior is observed for as-deposited Pt
on clean Si. The Schottky barrier height calculated assuming thermionic emission is ϕ_{bn} = 830 mV, consistent with previously reported values [7]. The I(V)
relation for as-deposited Pt on oxidized Si is not exponential; rather, the
curve suggests a power law dependence $I_F \propto (V_F)^p$ for V_F > 50 mV with p = 2-3.
All of the contacts exhibit the same forward I(V) characteristics after 0.2 -
2×10^{15} Si cm^{-2} irradiation, independent of substrate temperature during irrad
-iation. Again, a power law relationship is suggested, with p = 2 ± 0.2, by
these characteristics. This uniform
behavior after irradiation indicates
that electrical transport is dominated
by radiation damage in the vicinity of
the contact. This is consistent with
the fact that a significant fraction of
the implanted Si penetrates the Pt-Si
interface at the energies employed [6].

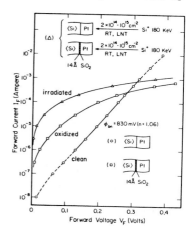

The left-hand side of Fig. 2 shows
semi-logarithmic plots of the forward
I(V) characteristics of typical Si
irradiated Pt contacts on the clean n-
type epi substrates after 400°C annealing for 30 min. It is evident that
diode-like behavior is restored by the
annealing. Some dependence on the Si
irradiation remains, however, in that
the contacts receiving the greatest Si
fluence differ most from the unirradiated contacts which exhibit Schottky
diode behavior (ϕ_{bn} = 842 mV, n = 1.05).
Backscattering spectra of all these
samples show laterally uniform silicide
formation with PtSi stoichiometry. We
believe that the Si damaged by irradiation
is not completely consumed during silicide
formation. This is consistent with the
observed influence of Si dose on the I(V)
characteristics after annealing. It is
not anticipated that the 400°C annealing
will entirely remove the radiation damage [8].

Fig. 1. Typical forward I(V)
characteristics of Pt contacts on
clean and slightly oxidized n-Si
before and after Si$^+$ irradiation.
All contacts exhibit the same I(V)
behavior after the Si irradiation.

The influence of the Si irradiation on silicide formation kinetics is shown
in Fig. 3. These measurements were made on the n$^+$ doped substrates, and per-
tain to the first silicide phase formed, namely Pt$_2$Si. Note the annealing
temperature is only 254°C. In contrast, Fig. 2 pertains to n-type substrates
and second phase formation at 400°C. Figure 3 shows that silicide formation
remains transport limited after Si irradiation with a reduction in the diffusion constant. This concurs with previous observations that Xe irradiation also
retards Pt silicide formation [3]. We note that the reaction rate decreases
monotonically with increasing Si fluence. Channeling measurements performed
on 2×10^{15} cm^{-2} Si irradiated samples, after Pt removal, show that the Si substrate is heavily damaged ("amorphized") to a depth of $\sim$ 2000 Å.

Only 530 Å of Si is consumed by the complete reaction of the Pt producing
Pt$_2$Si. 1100 Å of Si is consumed in the formation of PtSi. This supports our
contention that a residual underlying damaged Si layer is involved in the

414

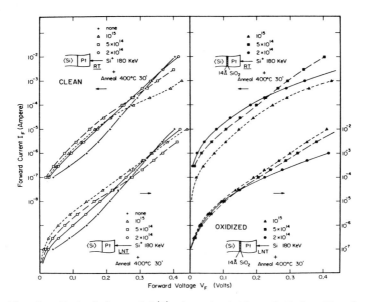

Fig. 2. Typical forward $I(V)$ characteristics of Si irradiated
Pt contacts on clean and slightly oxidized n-Si after 400°C
annealing for 30 min.

altered $I(V)$ characteristics of irradiated contacts after annealing (left-side
of Fig. 2).

We also show in Fig. 3 kinetic studies on oxidized n^+ substrates also made
at 254°C. No reaction is observed with as-deposited or 2×10^{14} cm^{-2} Si irradiated samples. Silicide formation is observed after a 27°C irradiation of $2 \times$
10^{15} Si cm^{-2} at 190 keV. In this case, the lateral uniformity of the moving
interface is sufficient to define a kinetic behavior. The first phase silicide
formation is again transport limited after a delayed initiation of ∿ 27 min.
The effective diffusion coefficient is only slightly lower than that observed
with clean substrates identically irradiated. We guess that the delay is due
to a slow permeation of the oxide barrier followed by its catastrophic failure
after which transport is dominated by bulk diffusion.

In contrast to the annealing results at 254°C, Pt silicide formation is
observed on the oxidized n^+ substrates after 400°C annealing, although the resulting Si-silicide interface is laterally nonuniform. A uniform interface is
obtained with this substrate after 190 keV Si irradiation at fluences of $2 \times$
10^{14} cm^{-2} or more. However, radically different behavior was observed with the
oxidized n-type epi wafers. In the later case, no reaction was evident by
backscattering spectrometry with unirradiated samples for 30 min annealing at
< 500°C. Irradiated samples on the oxidized n-type epi substrates annealed at
$\overline{4}$00°C for 30 min either did not react or reacted nonuniformly. Typical forward
$I(V)$ characteristics of these contacts are shown in the right-side of Fig. 2.
It is clear from these curves that Schottky diode behavior is not produced after
annealing as in the case with clean substrates. Since the interfacial oxygen
is known to be the same for both n and n^+ substrates, we conclude that the oxide
on highly doped material is structurally different from that formed on low doped

415

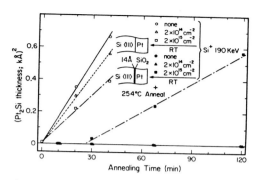

Fig. 3. Graph showing Pt$_2$Si formation at 254°C versus time for Pt films. Silicon irradiated and unirradiated on clean and slightly oxidized n$^+$ Si.

substrates. Silicon oxidation is known to be affected by high level doping [9].

CONCLUSIONS

It is evident from this study that ion irradiation has a detrimental effect on Schottky diode behavior. Structural modifications induced by ion irradiation readily dominate the electrical behavior of these contacts, and is not completely removed by annealing at temperatures employed for silicide formation. The singular shape of the I(V) curves after ion irradiation suggest that a single mechanism controls electrical current transport. Although ion irradiation has been shown to promote silicide formation on contaminated substrates in certain cases, the bulk transport of Pt is reduced by irradiation.

The mechanism of this later process deserves investigation. Oxide structure is a relevant factor in diffusion barrier behavior, as implied by the substantial substrate doping dependence observed.

ACKNOWLEDGMENTS

The authors thank R. Fernandez (Caltech) for assistance in the sample preparation and the General Products Division of IBM, Tucson, Arizona (T. M. Reith) for financial support.

REFERENCES

1. R. M. Andersom and T. M. Reith, J. Electrochem. Soc. 122, 1337 (1975).

2. H. Föll and P. S. Ho, J. Appl. Phys. 52, 5510 (1981).

3. L. S. Wieluński, C-D. Lien, B. X. Liu and M-A. Nicolet in: Metastable Materials Formation by Ion Implantation, S. T. Picraux and W. J. Choyke eds. (North-Holland, Amsterdam, 1982) p. 139.

4. J. W. Mayer, B. Y. Tsaur, S. S. Lau, and L. S. Hung, Nucl. Instr. & Meth. 182/183, 1 (1981).

5. L. S. Wieluński, B. M. Paine, B. X. Liu, C-D. Lien and M-A. Nicolet, phys. stat. sol. (a) 72, 399 (1982).

6. G. Dearnaley, J. H. Freeman, R. S. Nelson and J. Stephen, Ion Implantation (North-Holland, Amsterdam, 1973).

7. E. H. Rhoderick, Metal-Semiconductor Contacts (Oxford Univ. Press, Oxford, 1978) p. 53.

8. J. W. Mayer, L. Eriksson, and J. A. Davies, Ion Implantation in Semiconductors (Academic Press, New York, 1970) p. 113.

9. C. P. Ho and J. D. Plummer, J. Electrochem. Soc. 126, 1516 (1979).

FORMATION OF TITANIUM SILICIDE AT ATMOSPHERIC PRESSURE

PETER REVESZ, JENO GYIMESI AND JOZSEF GYULAI
Central Research Institute for Physics, H-1525 Budapest 114, P.O.B. 49, Hungary

ABSTRACT

Two problems connected with the growth of Ti-silicide have been investigated. It is shown if a silicon dioxide step on a single crystal of silicon covered with titanium is annealed then, following vertical growth on the silicon part, lateral growth of Ti-silicide takes place over the oxide layer. We also studied the problems of Ti-silicide growth on samples implanted with high doses of Sb, As, P, Ar and O prior to Ti evaporation.

EXPERIMENT

In our experiments n-type Si wafers of <100> orientation, 10 ohmcm resistivity were used as substrates. For the experiments connected with the lateral growth the wafers were oxidized to form a 100 nm thick SiO_2 layer. Using photolithography and SiO_2 etching, islands of SiO_2 were formed. Evaporation of the Ti was carried out in an oil-free UHV system. The thickness of the Ti was ~100 nm.

Annealing of the samples was carried out in a horizontal CVD reactor in forming gas. To minimize the interaction of the Ti-layer with forming gas the samples were placed on a dust-free Si wafer with Ti-layer facing the polished surface of the Si wafer. An additional Si wafer was placed on the sample to minimize the gas flow around the sample. Rutherford backscattering (RBS) measurements show that this "face-to-face" annealing procedure results in good quality of Ti-silicide layer. On the other hand, exposing the samples directly to the annealing gas TiO_2 is formed on the surface of the samples.

In the case, when the effect of ion-implantation on the growth of Ti-silicide was studied the samples were implanted with Sb(75 keV), As(80 keV), P(80 keV), Ar(80 keV) and O(40 keV) with dose of 1.25×10^{16} cm^{-2}. After implantation the wafers were cleaned by a plasma stripping procedure in a mixture of $CF_4 + O_2$ to remove the hydrocarbon contamination layer formed during implantation [1]. Composition and the thickness of the silicide layers were determined using the 1.5 MeV RBS technique [2]. Sheet resistance was measured by a standard 4-point probe method. In some cases X-ray diffraction was used to determine the phases formed after annealing.

LATERAL GROWTH OF TITANIUM SILICIDE

The kinetics of the lateral growth is shown in Figure 1 where the lateral growth as a function of the square root of the annealing time is displayed. Annealing of the samples was carried out at 750, 800, 850 and 950°C for various times up to 40 minutes. The lateral growth was measured using an optical microscope. It can be seen that the data points fall mostly on straight lines indicating that the lateral growth of Ti-silicide is a diffusion limited process. The activation energy of the lateral growth obtained on the basis of our measurements is $E_a = 1.89$ eV. A similar value for the case of the vertical

Mat. Res. Soc. Symp. Proc. Vol. 14 (1983) Published by Elsevier Science Publishing Co., Inc.

418

growth at temperatures below 700°C was obtained earlier [3].
In Figure 2, an SEM picture of a structure with lateral growth is displayed.

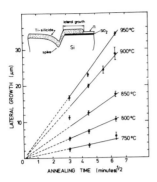

Fig. 1. The kinetics of the lateral growth of Ti-silicide.

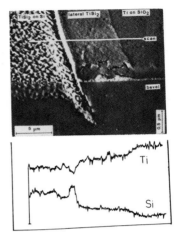

Fig. 2. SEM picture of a structure with lateral growth.

A low angle bevel at the edge of the SiO_2 layer was made on this sample to visualize the spiking phenomenon occurring during lateral growth.

Because of the silicon-consuming nature of Ti-silicide growth [3], in the case of lateral growth a large amount of silicon is consumed to form a lateral silicide layer tens of micron long. The silicon atoms incorporated in the lateral silicide layer originate from the closest source. As the transport of silicon atoms on to the oxide layer proceeds the edge of the Ti-silicide layer at the SiO_2 step sinks deeper and deeper into the silicon crystal forming a spike. The ratio of the length of lateral growth to the amount of silicon consumed during lateral growth (measured by Talystep surface profiler) indicates that the composition of the laterally grown silicide is $TiSi_2$. The electron microprobe measurements (see Fig. 2.) showed that during lateral growth the silicon atoms are distributed uniformly.

GROWTH OF TITANIUM SILICIDE ON ION-IMPLANTED SILICON

After high dose implantation the first hundreds of nanometers of the silicon are amorphous. This implanted amorphous layer can be recrystallized using heat treatment in the 500-600°C temperature range [4]. It is also known that the electrically active impurities increase the recrystallization velocity [5], while the annealing of high dose Ar or O implanted silicon results in stable defect complexes [6].

It would seem that two cases are distinguishable. In the case of Sb, As and P implanted samples the regrowth rate of the implanted amorphous layer under the Ti-film is so high that the regrowth process will be completed at a very early stage of silicide formation. When samples implanted with argon or oxygen are annealed with Ti-films on them the interaction of Ti with silicon occurs in a highly damaged region with a large number of defect complexes.

In Figure 3 the sheet resistance of the layers as a function of annealing

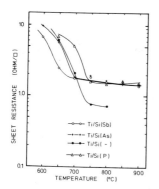

Fig. 3. Sheet resistance as a function of annealing temperature.

temperature is shown. For samples implanted with Sb, As and P the saturation value of the sheet resistance is about the same, but this value is approximately two times higher than for the nonimplanted case. This suggests that for the implanted samples the Ti-film is not fully converted even at temperatures as high as 900°C.

In Figure 4, RBS spectra of Sb implanted samples annealed at 650 and 850°C are shown. For comparison the spectrum of the nonimplanted sample annealed at 750°C is also shown. Although in all three cases the composition of the formed silicide is $TiSi_2$, for the Sb implanted samples annealed at 850°C there is still a relatively large amount of unreacted Ti left on the surface.

The RBS measurements show that for all implanted samples there is a characteristic saturation of the $TiSi_2$ growth. The thickness of the silicide layer on the implanted surfaces is less than half of the silicide thickness grown on the nonimplanted sample.

An interesting question is that relating to the redistribution of the implanted atoms during silicide growth.

In Figure 5 the depth profile of Sb atoms (in relative concentration units)

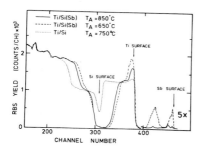

Fig. 4. RBS spectra of Sb implanted (and nonimplanted) samples after various annealing.

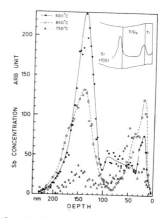

Fig. 5. Depth profiles of Sb atoms after annealing at different temperatures.

is shown for various annealing temperatures. Atoms of Sb start to diffuse through the $TiSi_2$/Ti layer already at 600°C. With increasing the annealing temperature the concentration of Sb atoms located in the silicon decreases steadily. The surface peak of Sb atoms first increases (up to 750°C) but when the source of Sb atoms in the silicon becomes exhausted, this surface peak starts to decrease too. Assuming a simple diffusion process from a limited source from the maximum RBS yield for the Sb peak in the silicon one can determine the diffusion coefficient of Sb atoms through the metal layer. We have found that the activation energy of diffusion

420

of Sb atoms in the silicide film is E_a=1.9 eV which is practically equal to the activation energy of the silicide growth [3]. This suggests that the diffusion process of Sb atoms through $TiSi_2$ has the same mechanism as the silicon atoms diffusing during the silicide growth.

The RBS measurements show that for As there is practically no pile-up at the surface and the As atoms are distributed homogeneously in the growing $TiSi_2$ layer. At 900°C the concentration level of As atoms decreases to $1.5x10^{20}$ atoms/cm^3.

Quite different behaviour is observed when we anneal argon or oxygen implanted samples covered with evaporated Ti layers. At about 700°C the metal film starts flaking off and after the 800°C annealing cycle practically no metal is left on the sample. At 700°C annealing for 30 minutes the sheet resistance of the argon and oxygen implanted sample is about 50% of their initial values, while for the nonimplanted sample it decreases to 9%.

In Figure 6, RBS spectra are shown for argon implanted samples annealed at 600°C and 700°C for 30 minutes. After 600°C annealing there is no change compared to the nonimplanted sample. At 700°C the Ti and the Si distribution changes greatly: the relative amount of Ti atoms increases up to 70% at a depth of 250 nm. The oxygen implanted samples exhibited the same behaviour.

A possible explanation might be that before annealing we have a structure containing an amorphous implanted layer with some percentage of Ar (or O) atoms in it; when annealed as described, this layer transforms into one containing a high concentration of stable defects [6]. We think that the interaction between Ti and Si in this enviroment differs from the Ti-silicide growth. In this case diffusion of Ti atoms can take place as well, mainly along the grain boundaries of the extended defects. The highly damaged region of the silicon may act as traps for the Ti atoms leading to the formation of Ti precipitates in the disordered region. This is analogous to the gettering effect of Au atoms by an Ar implanted amorphous layer [7].

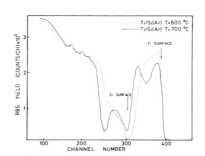

Fig. 6. RBS spectra of Ar implanted samples with Ti-film annealed at 600°C and 700°C.

It would seem that the presence of Ti precipitates weakens the adhesion between the metal layer and the silicon which results in the flaking off the metal film at higher annealing temperature.

To clarify the structure of the metallic layer on the argon and oxygen implanted samples annealed at 700°C X-ray diffraction measurements have been carried out. For these samples in agreement with the RBS measurements no $TiSi_2$ phase was found. The X-ray diffraction measurements indicated the presence of the unstable orthorombic TiSi phase for the case of argon and oxygen implanted samples annealed at 700°C. For the nonimplanted sample annealed at the same temperature we found a mixture of $TiSi_2$:TiSi=10:1 phases in good agreement with earlier experiments [3].

ACKNOWLEDGEMENT

The authors wish to thank prof. J.W. Mayer for useful discussions, G. Peto
for the Ti evaporation, E. Zsoldos for the X-ray diffraction measurements and
F. Bányai for assistance with experiments.

REFERENCES

1. T. Lohner, G. Vályi, G. Mezey, E. Kótai and J. Gyulai, Rad. Effects $\underline{54}$, 251 (1981)

2. Ion Beam Handbook for Materials Analysis, ed. J.W. Mayer and E. Rimini, Academic Press, Inc. New York-San Francisco-London 1977

3. S.P. Murarka and D.B. Fraser, J. Appl. Phys. $\underline{51}$(1), 342 (1980)

4. L. Csepregi, J.W. Mayer and T.W. Sigmon, Phys. Lett. $\underline{54A}$, 157 (1975)

5. L. Csepregi, E.F. Kennedy, T.J. Gallagher, J.W. Mayer and T.W. Sigmon, J. Appl. Phys. $\underline{48}$(10) 4234 (1977)

6. P. Révész, M. Wittmer and J.W. Mayer, J. Appl. Phys. $\underline{49}$, (10) 5159 (1978)

7. T.W. Sigmon, L. Csepregi and J.W. Mayer, J. El. Chem. Soc., $\underline{123}$, (7) 1117 (1976)

THE EFFECT OF OXYGEN IN COSPUTTERED (TITANIUM + SILICON) FILMS

R. Beyers and R. Sinclair, Department of Materials Science and Engineering,
Stanford University, Stanford, California 94305 and M. E. Thomas, Fairchild
Camera and Instrument Corporation, Palo Alto, California 94304

ABSTRACT

The effect of oxygen incorporation on the growth and
microstructure of $TiSi_2$ has been investigated. Cosputtered
films, with Si/Ti ratios between one and three, were
deposited on (100) Si substrates and reacted at
temperatures from 650° to 1050°C. Both Auger electron
spectroscopy, in conjunction with sputter profiling, and
Rutherford backscattering spectrometry indicate that oxygen
in the as-deposited films redistributes to the
silicide-silicon interface upon heating. Cross sectional
transmission electron microscope images show that the
oxygen is present as an amorphous oxide.

INTRODUCTION

Owing to their low resistivity and high thermal stability, refractory
metal silicides are currently being considered for use in integrated circuit
metallization schemes [1]. To use these materials reliably will require a
basic understanding of the effects of deposition technique, film composition,
heat treatment and impurity content on the resulting silicide microstructure.
This paper examines the effects of oxygen impurities on the growth and
microstructure of cosputtered $TiSi_2$ films. Resistance measurements, Auger
electron spectroscopy (AES) in conjunction with sputter profiling, Rutherford
backscattering spectrometry (RBS) and a variety of transmission electron
microscopy (TEM) techniques have been used to characterize the films.

EXPERIMENTAL

The substrates used in this study were three inch diameter (100) n-type
silicon wafers with 6 Ωcm resistivity. The wafers were cleaned with a
10:1 (H_2O:HF) solution until hydrophobic, spun dry and immediately placed into
a sputtering chamber. A 3120 Varian S-gun ® System with two independently
powered d.c. magnetron sources was used to deposit the films. The system was
pumped down to a base pressure of 7×10^{-5} Pa prior to backfilling with Argon at
0.2 Pa sputtering pressure. The substrates were not heated.
Thin films with atomic ratios from Si/Ti = 1 to Si/Ti = 3 were deposited
by simultaneously sputtering from silicon and titanium sources. The amount of
titanium deposited was held constant while the amount of silicon deposited was
varied in order to generate the equivalent of 0.1 μm of $TiSi_2$ in completely
reacted films. The composition of the as-deposited films was determined from
individual thickness calibrations from each gun using a Dektak profilometer
and, in some instances, was also checked by RBS measurements.
After deposition, the wafers were reacted in hydrogen at temperatures
between 650° and 1050°C for 30 minutes.

Mat. Res. Soc. Symp. Proc. Vol. 14 (1983) © Elsevier Science Publishing Co., Inc.

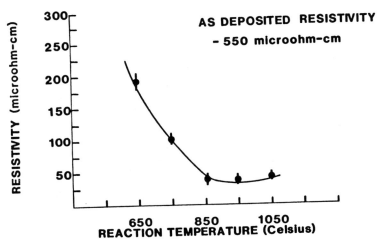

Figure 1.   Resistivity as a function of reaction temperature for films with
initial composition Si/Ti = 2.

RESULTS

While it was possible to produce films with oxygen contents on the order
of 1%, many of the as-deposited films contained oxygen concentrations between
10 and 16% as measured by RBS. The tendency for sputtered films to contain
substantial amounts of oxygen impurities has been noted previously [2];
however, the effects of oxygen have not been adequately described. For this
reason, the films with substantial oxygen content were studied.

The resistivity as a function of reaction temperature for films with
initial composition Si/Ti = 2 is shown in Fig. 1. Similar curves were
obtained for both silicon-rich and titanium-rich films. These results are
similar to those reported for cosputtered films on polysilicon [1],
coevaporated films on SiO_2 [4], and alloy sputtered films on polysilicon [2].

TEM micrographs of samples with Si/Ti = 2 reacted at 650°, 750° and 850°C
are shown in Fig. 2. The samples transformed at 650° and 750°C have an
average grain size of 0.2 μm. Although diffuse polycrystalline ring
diffraction patterns taken from these samples (Fig. 2a and 2b) could be
indexed consistent with the equilibrium phase $TiSi_2$, many of the
microdiffraction patterns taken from individual grains were not consistent
with reported structures for titanium silicides, titanium oxides or silicon
oxides. For the samples transformed at 850°C and above, both the ring
patterns and the microdiffraction patterns indicate that $TiSi_2$ was the only
silicide phase present. The samples reacted at 950 and 1050°C are similar in
appearance to those at 850°C, however, the average grain size increases from
0.3 μm at 850° to 0.7 μm at 1050°C.

AES sputter profiles from the films pictured in Fig. 2 are presented in
Fig. 3. The sample reacted at 650°C has a uniform oxygen concentration
through the film (as did the unreacted film). However, the oxygen in samples
reacted at 750° and 850°C has redistributed to the silicide-silicon
interface. This oxygen redistribution occurs in all films heated between 750°
and 950°C, regardless of the initial film composition. RBS spectra show
the same oxygen behavior.

425

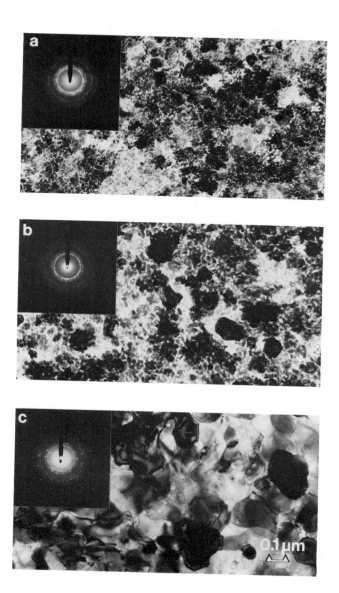

Figure 2.   TEM micrographs and diffraction patterns of samples with initial
composition Si/Ti = 2 reacted at (a) 650°, (b) 750° and (c) 850°C.

426

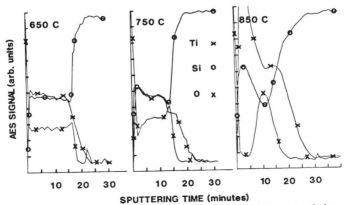

Figure 3.   Auger depth profiles of samples with initial composition Si/Ti = 2 reacted at (a) 650°, (b) 750° and (c) 850°C.

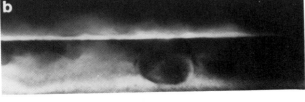

Figure 4.   Cross sectional TEM micrographs of samples with initial composition Si/Ti = 2 reacted at (a) 650°, (b) 750° and (c) 850°C.

427

Cross sectional TEM micrographs of samples with initial composition Si/Ti = 2 reacted at 650, 750 and 850°C are shown in Fig. 4. Two features of these micrographs are particularly noteworthy: first, protrusions of the silicide phase extend into the silicon substrate and, second, growth of an amorphous phase (presumed to be an oxide) has occurred at the original silicon-silicide interface. The protrusions, due to the diffusion of excess silicon into the films, indicate that titanium silicide has a range of stoichiometry. The same conclusion was reached by Murarka [3] using cosputtered films on polysilicon. The size and number of protrusions increase with annealing temperature.

High resolution images of the sample reacted at 650°C reveal patches of oxide approximately 2 nm thick at the original interface. These patches are probably native oxide on the wafers which formed between cleaning the wafers and inserting them into the vacuum system. The samples annealed above 650°C form a layer of oxide approximately 4 nm thick at the original interface. The oxide becomes rougher as the annealing temperature increases.

SUMMARY

The behavior of cosputtered (titanium + silicon) films containing oxygen may be described as follows. Upon heating, there is a rapid crystallization of the amorphous as-deposited film. Penetration of the silicide phase through the native oxide occurs due to diffusion of silicon from the substrate to the silicide. Concurrently, oxygen in the film diffuses to the native oxide. For films annealed at low temperatures (750°C) with relatively fewer protrusions, the oxide grows as a nearly continuous layer. For films annealed at higher temperatures, the number of protrusions through the native oxide is much greater, causing the subsequent oxide growth to be discontinuous.

ACKNOWLEDGEMENTS

The authors would like to thank Rhonda Cox for her help in depositing and reacting the thin films. Use of the facilities of the Stanford Materials Science and Engineering Department, of the Center for Materials Research at Stanford University, and of the Fairchild Camera and Instrument Corporation is gratefully acknowledged.

REFERENCES

1. S. P. Murarka, J. Vac. Sci. Technol. 17, 775 (1980).

2. R. F. Pinizzotto, K. L. Wang and S. Matteson, Proceedings of the 4th International Symposium on Semiconductor Silicon 1981, edited by H. R. Huff, R. Y. Kriegler and Y. Takeishi (Electrochemical Society, Princeton, 1981), p. 562.

3. S. P. Murarka and D. B. Fraser, J. Appl. Phys. 51, 350 (1980).

4. M. J. H. Kemper and P. J. Oosting, J. Appl. Phys. 53, 6214 (1982).

INVESTIGATIONS OF METAL-SILICON INTERFACES BY TIME-OF-FLIGHT ATOM PROBE.

CHRISTOPHER GROVENOR[*] AND GEORGE SMITH
Department of Metallurgy and Science of Materials, Parks Road,
Oxford, OX1 3PH, England.

ABSTRACT

The use of a Time-of-Flight Atom Probe in
the analysis of silicon surfaces, and the
interfaces between metals and silicon,
promises to provide very accurate chemical
analysis allied with structural information
from Field Ion Microscopy images. This paper
presents results on the analysis of silicon
surfaces by this technique, showing that good
spectra can be obtained without difficulty.
Some preliminary experiments on the structure
of such specimens after the deposition of thin
layers of Pd and Ni will be described, concen-
trating on the analysis of the stoichiometry
of the reacted layers.

INTRODUCTION.

In the last few years the growth of metal silicides on silicon
has become on area of considerable interest due to the potential
of this kind of system to provide contacts to silicon devices that
combine stability with satisfactory electrical character [1]. The
structure of the interface between Ni and Pd silicide layers and
the silicon substrate have been investigated by Transmission
Electron Microscopy [2,3], and the stoichiometry of these silicide
layers, and the indiffusion of the metal into the silicon, studied
by a variety of techniques including Rutherford Back Scattering
[4] and Ultraviolet Photoemission Spectroscopy [5]. Some questions remain concerning the precise composition at the silicide/
silicon interface itself [6], and changes in stoichiometry at this
interface may have substantial influence over the electrical
character of the contact as a whole. It is however difficult to
obtain information on the chemical composition of the few atomic
layers at the interface by use of the techniques mentioned above.
Recently it has been shown that it is possible to produce good
quality silicide layers by deposition of silicon onto metal Field
Ion Microscope specimens [7], and that the detailed structure and
stoichiometry of these silicides could be investigated by Time-of-
Flight Atom Probe analysis. This paper presents some preliminary
results on the preparation and analysis of metal layers deposited
onto silicon Field Ion Microscope specimens, showing the potential
of the technique of Atom Probe analysis [8] for the accurate
chemical analysis of the structure of the silicide/silicon
interface.

---

[*] Presently at the IBM TJ Watson Research Center, Yorktown Heights.

Mat. Res. Soc. Symp. Proc. Vol. 14 (1983) ©Elsevier Science Publishing Co., Inc.

EXPERIMENTAL DETAILS AND RESULTS

In order to prepare metal-on-silicon Field Ion specimens it is necessary to have a reproducible method of making sharp silicon tips. The silicon used in these experiments was single crystal phosphorus doped ($6.10^{15}m^{-3}$) material , and no attempt was made to control the orientation of the specimen axis. The details of the specimen preparation techniques have been described elsewhere [9]. These silicon tips have been analysed in the Oxford Time-of-Flight Atom Probe [10] in order to determine the specimen cleanliness , and the best conditions for extended analysis of this material which is of limited conductivity,(0.9Ωcm). Atom Probe spectra have been obtained from silicon specimens by a number of other workers [11,12] , but they have concentrated on preparing specimens from (111) oriented whiskers rather than bulk material, and only the experiments in which the desorption of the surface ions was achieved by laser pulsing [11] have been successful in the analysis of a large number of desorbed ions.

In this work Atom Probe spectra have been obtained from silicon tips under a wide variety of conditions [9], but here consideration will be limited to two examples; the operation of the Atom Probe in pulsed laser and voltage pulsed modes. Figure 1 shows a comparison of the mass range 12 to 16 AMU from spectra obtained from these two techniques. The laser desorption was stimulated by a JK Laser System 2000 Nd-YAG instrument with a 12ns pulse width operating at a wavelength of 355nm. The Si^{++} peak in the laser spectrum shows complete resolution of the silicon isotopes from a tip maintained at 16kV and at 78K. The voltage pulsed spectrum (electric) was obtained using a 15% pulse fraction applied to a tip held at 17kV and at room temperature. There are Si^{++} and $Si(H)_x^{++}$ peaks visible , but the resolution of the spectrum is relatively poor. These substantial hydride peaks are found on all voltage pulsed results. Possible explanations for the disparity between the ionic species observed in the two kinds of spectra have been discussed elsewhere [9]. However for both laser and voltage pulsed spectra only a low concentration of contamination species have been found , especially after field evaporation has cleaned the oxidised layer from the tip surface. These results indicate that it is possible to obtain clean spectra from randomly oriented silicon tips if reasonable care is taken over specimen preparation and the setting up of the analysis conditions.

Once the endform of the silicon specimens has been formed by field evaporation a thin metal layer can be deposited , and a heating cycle used to react the metal with the silicon surface. In the equipment used in this work it is currently impossible to deposit the metal layer in the UHV system of the Atom Probe. Thus the silicon tip must be exposed to , at best, low vacuum before the metallisation process can be carried out. The result of this is that some contamination is always included in the deposited films , the most common ionic species that have been analysed being C^+, H_2O^+ and complex ions containing these contaminating elements. Examples will be given of analysis from silicon tips coated with thin layers of Ni and Pd.

Nickel layers on silicon. A number of specimens have been investigated where Ni has been evaporated in relatively poor vacuum onto silicon tips. In general these tips have fractured as soon as they were returned to the Atom Probe. In order to try and strengthen the tips the specimens have been heated after evaporation to react the Ni with the silicon surface. An example of an Atom Probe spectrum from a specimen that has been heated to 350°C for 15 hours is shown in Figure 2. It is clear that the tip surface is contaminated , C^+, O^+ and water related peaks are obvious , but a large sharp peak can also be seen at about 35AMU. This peak can be tentatively analysed as being from the ionic specy $(Ni_2Si)^{4+}$, and this identification is strongly supported by the observation that the peak is split , the two subsiduary peaks separated by about half an AMU. Nickel has two primary isotopes at 58 and 60AMU , so that in the 4+ charge state the observed peak splitting can readily be explained. However it is surprising that no doubly or triply charged states of the disilicide ion have been detected. Whether it is possible from this kind of analysis to be sure that the surface of the specimen is covered with a reacted silicide layer (possibly Ni_2Si , the Ni silicide which should be stable at 350°C [2]) is uncertain. The Ne ion image of these surfaces show no evidence of the presence of a structured silicide layer.

Palladium layers on silicon. Pd has been deposited onto silicon tips under the same poor vacuum conditions as the Ni,described above , and contamination has proved a serious problem in the accurate analysis of the chemical composition of the specimens. In one experiment however useful data has been obtained from the analysis of a Pd layer nominally 5nm thick , and this specimen was not heated after deposition. Analysis was carried out in the Atom Probe using laser desorption , and a depth profile of the Pd in the silicon surface could be obtained. The Pd was found to desorb as single ionic species with charge states 3+,2+ and 1+ , and also as $PdSi^+$ and Pd_2Si^+ species. Figure 3 shows this result where the Pd/Si ion ratio is plotted on the vertical scale after subtraction of all ions identified as contamination. The horizontal scale is difficult to convert to a depth measurement without detailed knowledge of the blunting rate of the specimen. A gradual falling off of the Pd concentration can be seen as the atomic layers are removed , profiling through the top layers of the silicon. This concentration is very high given that the specimen has not been heated , and was maintained at 78K during the course of the experiment. Indiffusion of the Pd would not therefore have been expected to occur to produce Pd concentrations as high as 10%.

CONCLUSIONS

These two results are presented as early evidence that quantitative analysis of metal/silicon specimens can be carried out in an Atom Probe , giving information on the structure of a reacted layer , and diffusion profiles into the silicon. As yet the quality of the silicide layers that can be grown is poor , and contamination of the metal layers occurs during preparation , but new equipment in which all analysis and evaporation can be carried out in the same UHV system is under construction. It is hoped that with this equipment routine analysis of silicide/silicon interfaces will be possible , as has been shown for the case of metal silicides grown on metal Field Ion specimens [7] , and in the case of the silicon specimens themselves [9]. One of the first questions that will be investigated is that of the high

432

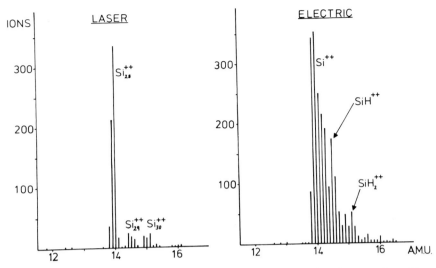

Fig.1. The Si^{++} peaks from laser and voltage pulsed experiments.

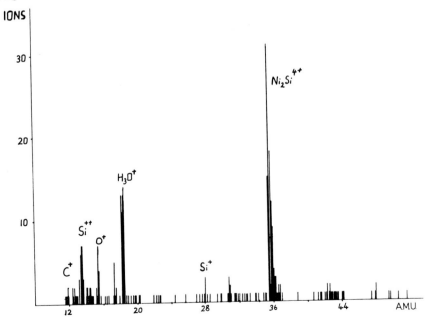

Fig.2. The spectrum from a Ni coated Si specimen, 15 hours at 350°C.

Pd concentration in the silicon under the evaporated layer analysed in the results presented above. Such a rapid indiffusion of Pd could have serious effects on the performance of any silicon device containing a Pd_2Si contact layer.

ACKNOWLEDGEMENTS

The authors are grateful to Mr.H.Spanner of JK Lasers for his assistance in the setting up of the laser facility , and Professor Sir Peter Hirsch FRS for provision of laboratory facilities. The development of the laser probe was supported by a grant from the Paul Instrument Fund of the Royal Society.

REFERENCES

1. K. N. Tu and J. W. Mayer in:Thin Films-Interdiffusion and Reactions, J. M. Poate,K. N. Tu and J. W. Mayer eds.(Wiley,New York 1978) pp. 359.

2. H. Föll,P. S. Ho and K. N. Tu, Philos.Mag.A 45, 31 (1982).

3. H. Föll, Phys.Stat.Sol.(a) 69, 779 (1982).

4. C. Canali,G. Majni,G. Ottaviani and G. Celotti, J.Appl.Phys, 50, 255 (1979).

5. J. L. Freeouf,G. W. Rubloff,P. S. Ho and T. S. Kuan, J.Vac.Sci Tech. 17, 916 (1980).

6. J. L. Freeouf, J.Vac.Sci.Tech. 18, 910 (1981).

7. O. Nishikawa et al. Paper presented to International Field Emission Symposium, Gothenberg 1982. in press.

8. E. W. Müller,J.A. Panitz and S. B. McLane , J.Rev.Sci.Inst. 39 83 (1968).

9. C. R. M. Grovenor and G. D. W. Smith, Letter to Surface Science. in press.

10. M. K. Miller,P. A. Beaven and G. D. W. Smith, Surface and Interface Analysis 1, 149 (1979).

11. G. L. Kellogg, Appl.Surface Science, 11/12,186 (1982).

12. A. J. Melmed et al. Surf.Sci. 103, L139 (1981).

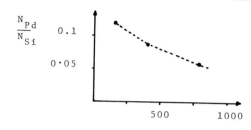

Fig.3. Plot of Pd concentration against the total number of ions collected. 5nm of Pd were evaporated onto the silicon surface.

The Growth of Epitaxial NiSi $_2$ Single Crystals on Silicon by the Use of   Template Layers

R. T. Tung
J. M. Gibson
J. M. Poate

Bell Laboratories
Murray Hill, New Jersey 07974

ABSTRACT

A novel crystal growth technique for NiSi$_2$ epitaxy is presented which utilizes thin silicide ($<$60Å) template layers to pin the subsequent growth under ultrahigh vacuum conditions. Single crystalline NiSi$_2$ films can be grown with either type A or type B orientations on Si(111). Continuous single crystalline NiSi$_2$ is grown on Si(100) with flat interface and uniform thickness.

The formation of silicides from the reaction between deposited thin metal films and silicon substrates has wide application in the semiconductor industry.[1] Virtually all silicides in practical use are polycrystalline in nature. The growth of CoSi$_2$ and NiSi$_2$ is particularly interesting because they offer the prospect of epitaxial silicide formation.[2-4] These two silicides have similar lattice structures (CaF$_2$) and close lattice matches with Si. Epitaxial silicide films have lower resistivities and higher thermal stability than polycrystalline films. Moreover, high quality Si films can be grown on top to form semiconductor-metal-semiconductor heteroepitaxial structures.[5,6]

Thin NiSi$_2$ films grown on Si(111) by molecular beam epitaxy (MBE) co-deposition are single crystalline and epitaxial. [7] These films share the surface normal $<$111$>$ axis with the substrate but are rotated 180° about that axis with respect to the substrate (type B orientation). Thin (500–2000Å) NiSi$_2$ films grown by Ni deposition at room temperature and subsequent heating at ~800°C (standard reaction technique) on Si(111) have good crystallinity, but contain mixed grains[8,9] of both type B orientation and the regular, unrotated, type A orientation, even under ultrahigh vacuum (UHV) conditions.[4] On Si(100), NiSi$_2$ films grown by standard reaction techniques have poor crystallinity.[4,8] The silicide-Si(100) interface is unstable and breaks up into large scale facets along inclined {111} planes.[8] Because the inclined facets reach the surface, these films are not continuous even for average film thickness as large as 1000Å. In this paper, we report a novel crystal growth method which allows the growth of continuous, uniform, single crystal, epitaxial NiSi$_2$ films on Si(111) of *either* type A or type B orientations. This method can be used, for the first time, to grow continuous, thin, single crystalline NiSi$_2$ films on Si(100) with flat interfaces and uniform thicknesses. Thin NiSi$_2$ template layers[10] are formed *in situ* to pin the subsequent silicide growth.

Polished Si(111) and (100) substrates of both n- and p-type were degreased and dipped in buffered HF before mounting on a sample holder. Samples are cleaned by resistive heating and, when necessary, sputtering in a UHV chamber with base pressure of 1×10^{-10} torr. At the onset of each experiment, samples were heated to 1100°C for ~2 minutes and allowed to cool slowly. Surfaces so prepared always displayed sharp (7×7) low energy electron diffraction (LEED) patterns for (111) substrates and sharp 2×1 patterns for (100) substrates, characteristic of clean Si. No impurities were detected by Auger electron spectroscopy (AES) except negligible amount of carbon.

Thin ($<$60Å) continuous layers of single crystalline NiSi$_2$ of either type A or type B orientation can be grown in-situ on Si(111) by deposition of Ni at room temperature and annealing to 450–550°C.[10] The variation of the orientation of thin NiSi$_2$ layers on Si(111) as a

Mat. Res. Soc. Symp. Proc. Vol. 14 (1983) © Elsevier Science Publishing Co., Inc.

436

function of deposited Ni thickness is shown in Fig. 1. The NiSi$_2$ orientation was determined by LEED symmetry studies[10] and subsequently verified by MeV ion channeling[2,4] and high resolution TEM imaging.[11] RBS channeling spectra along the substrate [110] and [114] directions of a type B NiSi$_2$ layer formed with 4Å of Ni are shown in Fig. 2(a). A comparison of the backscattering yields from Ni atoms in NiSi$_2$ crystals (peak near 1.5 MeV) along these two directions verifies that the vast majority of NiSi$_2$ is type B oriented.[2,4] Figure 2(b) shows spectra from a type A NiSi$_2$ layer formed with 16Å of deposited Ni and the orientation can be identified as type A.

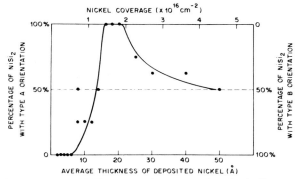

Fig. 1   The percentage of NiSi$_2$ volume with type A orientation in films grown by annealing room-temperature-deposited Ni on Si(111), as determined by LEED, RBS and channeling, and TEM. A 1Å thick Ni film has areal density of 9.1×10^{14} atoms/cm^2 and reacts with Si to form a NiSi$_2$ layer of 3.6Å average thickness.

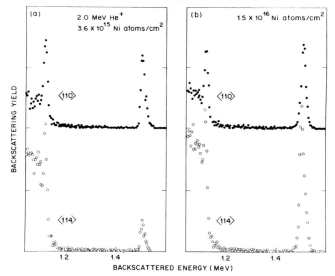

Fig. 2   MeV He ion channeling spectra of thin NiSi$_2$ layers on Si(111) formed by annealing (a) 4Å and (b) 16Å of Ni deposited at room temperature. The incident beam was aligned with <110> (closed circles) and <114> (open circles) of the substrate.

After a thin NiSi$_2$ template layer of specific orientation was formed, the temperature was raised to ~650—775°C and ~50—300Å Ni was deposited immediately onto the template surface at rates of .5—3Å/s. After the deposition, the LEED pattern of the new surface appeared to be identical as before. The formation of the NiSi$_2$ phase was checked by in-situ RBS measurement. The RBS channeling and random spectra of a type A and a type B oriented NiSi$_2$ layers are shown in Fig. 3. These two ~900Å thick layers were formed by depositing ~235Å of Ni at 775°C onto a ~40Å thick type B template, and a ~66Å thick type A template, respectively. The thickness of these layers is very uniform as is reflected in the sharp trailing edges of the two random spectra. Channeling minimum yield measurements (χ_{min}'s) along surface normal <111> axes for both layers are less than 3%, typical of perfect single crystals. The orientations are clearly shown to be type A and type B, respectively, for the two layers, from channeling spectra along the <110> and <114> directions.[12] TEM diffraction studies verify that both layers are single crystals of 100% type A and type B orientation, respectively. Misfit dislocations, of Burger's vector 1/6 (11$\bar{2}$) for type B interfaces and 1/2 (1$\bar{1}$0) for type A interfaces, are observed. Ultrahigh resolution TEM observations, with lattice imaging, of the cross-section of these interfaces show that both types of interface are extremely flat and locally well defined to ~3Å.

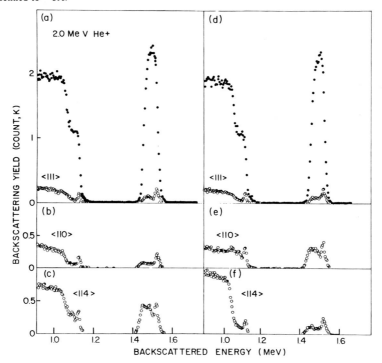

Fig. 3   Channeling (open circles) and random (closed circles) RBS spectra of ~900Å thick (a)-(c), a type A NiSi$_2$ layer and (d)-(f), a type B NiSi$_2$ layer on Si(111). All channel directions refer to the silicon lattice. See text for specific conditions under which these two layers are grown.

438

Thin (<100Å) continuous NiSi₂ template layers can also be formed on Si(100) under UHV by deposition of 10–30Å of Ni at room temperature and annealing to 450-550°C for a few minutes.[10] Figure 4(a) is a TEM image of the cross-section of such a template layer on Si(100). Layers of NiSi₂ formed by the same technique but with either <10Å or >40Å of deposited Ni are not continuous and have facetted interface with Si, such as illustrated in Fig. 4(b). The continuous and uniform NiSi₂ thin layers (Fig. 4(a)) have the same orientation as the substrate. Thick (>500Å) films of NiSi₂ can be grown from such a template layer by Ni deposition at elevated temperatures (~600–700°C). RBS and channeling spectra of a ~860Å thick NiSi₂ layers on Si(100) are shown in Fig. 5. This layer was formed by depositing ~220Å Ni at 650°C on a 55Å thick NiSi₂ (100) template. Channeling χ_{min} along the surface normal <100> axis is 4.5%, which is considerably better than any of the NiSi₂ layers previously grown on Si(100) and is close to that expected for a perfect single crystal (χ_{min} along the <110> axis of this sample is 3.4%). The sharpness of the trailing edge in the random spectrum shows that the thickness of the films is uniform.

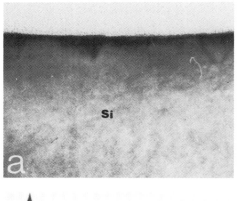

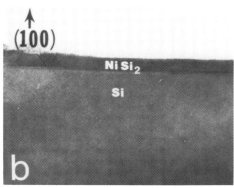

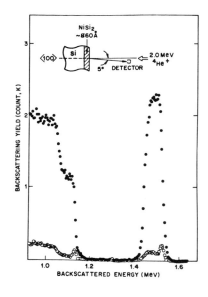

Fig. 4   Ultrahigh resolution TEM images of NiSi₂ on Si(100). These layers were formed with UHV annealing of (a) 6Å of Ni, and (b) 11Å of Ni. Dark "spikes" near the surface in (a) are NiSi₂ crystals.

Fig. 5   RBS channeling (open circles) and random (closed circles) spectra of a ~860Å thick NiSi₂ layer grown on Si(100).

The formation of thin ($\lesssim 60$Å) single crystal NiSi$_2$ template layers on the (100) and (111) surfaces of Si is discussed elsewhere.[10] The presence of the template layer prior to the growth of thick films is the key to controlled epitaxy during hot deposition. As an example, Ni deposition at $\sim$500–700°C on clean Si(111) results in NiSi$_2$ islands of both type A or type B orientations. The template layers pin the orientation and the continuity of the subsequent NiSi$_2$ epitaxy. The reaction of thick deposited Ni films on Si results in the formation of firstly Ni$_2$Si, followed by NiSi, and finally the growth of NiSi$_2$.[1] Moreover the growth of NiSi$_2$ does not occur by simple movement of a planar interface but rather by the lateral growth of NiSi$_2$ columns.[12,13] When Ni is deposited onto a hot NiSi$_2$ surface, the thick film reaction sequence is no longer followed. Rather, Ni reacts to form NiSi$_2$ directly either at the surface or at the NiSi$_2$–Si interface, depending on the Ni and Si diffusivities and the arrival rate of Ni. Growth is therefore planar. There are other variations of this technique such as the incremental deposition at room temperature of small amounts of Ni followed by annealing.[10] Undoubtedly, MBE co-deposition on (111) or (100) templates will also produce uniform single crystal films.

We have demonstrated that NiSi$_2$ epitaxy can be controlled by the use of novel template layers. Epitaxial NiSi$_2$ films grown on Si(111) can be either type A or type B single crystals. This is the first time the orientation of a double positioning system is controlled uniquely. Continuous single crystalline NiSi$_2$ is also grown on Si(100) with flat interfaces and uniform thickness, for the first time.

REFERENCES

[1] K. N. Tu and J. W. Mayer, in *Thin Films-Interdiffusion and Reactions*, edited by J. M. Poate, K. N. Tu, and J. W. Mayer, Wiley, New York, 1978.

[2] R. T. Tung, J. C. Bean, J. M. Gibson, J. M. Poate, and D. C. Jacobson, Appl. Phys. Lett. *40*, 684 (1982).

[3] S. Saitoh, H. Ishiwara, T. Asano, and S. Furukawa, Japan. J. Appl. Phys., *20*, 1649 (1981).

[4] R. T. Tung, J. M. Poate, J. C. Bean, J. M. Gibson, and D. C. Jacobson, Thin Solid Films, *93*, 77 (1982).

[5] J. C. Bean and J. M. Poate, Appl. Phys. Lett., *37*, 643 (1980).

[6] S. Saitoh, H. Ishiwara, and S. Furukawa, Appl. Phys. Lett., *37*, 203 (1980).

[7] T. R. Harrison, private communication, and J. C. Bean private communication.

[8] K. C. R. Chiu, J. M. Poate, J. E. Rowe, T. T. Sheng, and A. G. Cullis, Appl. Phys. Lett. *38*, 988 (1981).

[9] H. Föll, P. S. Ho, and K. N. Tu, J. Appl. Phys., *52*, 250 (1981).

[10] R. T. Tung, J. M. Gibson, and J. M. Poate, submitted to Phys. Rev. Lett..

[11] J. M. Gibson, R. T. Tung, and J. M. Poate, this volume.

[12] J. Baglin, F. d'Heurle, and S. Petersson, in *Thin Film Interfaces and Interactions*, edited by J. E. E. Baglin and J. M. Poate, The Electrochemical Society, Princeton, 1980.

[13] S. S. Lau and N. W. Cheung, Thin Solid Films, *71*, 117 (1980).

V
DEFECTS AND DEVICE PROPERTIES

Correlation of Structure and Properties of Silicon Devices

R. B. Marcus

Bell Laboratories
Murray Hill, New Jersey 07974

ABSTRACT

Three types of structure-related problems in VLSI materials and the impact of these problems on device properties are discussed: substrate defects and device leakage/breakdown, metallization linewidth and device failure, and SiO_2/polysilicon interface texture and device leakage/breakdown. In many cases transmission electron microscopy is the major source of information on structure.

1. INTRODUCTION

Operating characteristics of a silicon device may differ from the designed characteristics because of faulty design, errors during processing, or because of device materials interactions that occur during processing which affect the electrical behavior of the device. The third class of problems is most interesting from a materials scientists' point of view, since it generates opportunity for challenging materials research as well as practical application of new and existing concepts of materials science to an important technology.

Three examples of materials problems have been selected for discussion. These are taken from the area of n-channel MOS very large scale integration (VLSI) technology, and are (1) substrate defects and device leakage/breakdown, (2) metallization linewidth and device failure, and (3) SiO_2/polysilicon interfacial texture and device leakage/breakdown. These structure-related materials problems require some form of microscopy for their analysis.

Transmission electron microscopy (TEM) and scanning electron microscopy (SEM) are useful for these studies. The most efficient use of TEM in solving structure-related device problems occurs when the microscopist and device engineer are highly interactive, and fast feedback of TEM information is available. Thus, silicon device samples suitable for TEM study need to be prepared and studied quickly. A TEM test pattern chip has been devised for this purpose[1] and a small number of such chips is positioned on every processed silicon wafer. A TEM test pattern is designed to incorporate every morphological feature relevant to the particular device technology, and presents these features in an arrangement suitable for vertical cross-section TEM sample preparation.

2. SUBSTRATE DEFECTS AND DEVICE LEAKAGE/BREAKDOWN

The role of substrate defects in leakage/breakdown behavior can be determined by two methods. One is a generic approach, suitable for device development studies, and the other is an analysis of a specific device failure site. Although a large amount of

Mat. Res. Soc. Symp. Proc. Vol. 14 (1983) Published by Elsevier Science Publishing Co., Inc.

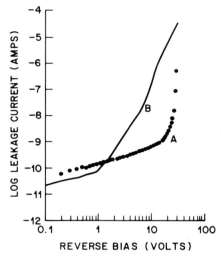

Fig. 1. Leakage currents for pn-junctions on two wafers which differ in thickness.

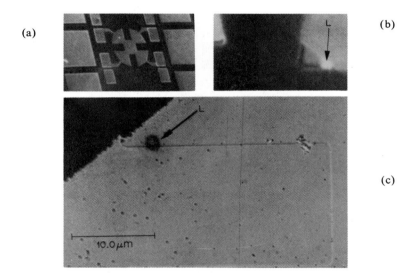

(a)

(b)

(c)

Fig. 2. SEM image of p⁺n test feature (a), EBIC image from part of test feature showing leakage site at L (b), and optical micrograph of sample after preparation for TEM study showing leakage site marked by polymerized hydrocarbon deposit (c).

443

generic work has been performed in this area (see for example Chapter 5 in Ref. 2), only a small number of published cases exist which use the latter approach.

In order to analyze the origin of device failure thought to be caused by a specific defect, the failure site must be located, marked, and then examined. The most generally useful methods for mapping regions of device leakage/breakdown are those employing the scanning electron microscope in the EBIC mode. A defect in the substrate of a Schottky barrier or pn-junction influences the local EBIC current through enhanced carrier recombination and generation.[2,3] Local EBIC current in a capacitor is modulated by a substrate defect either through a local perturbation of the electric field across the oxide[4] or a local perturbation of the space charge region, depending on applied bias.

A survey is needed to establish that the sites of discrete EBIC signals are also sites of device leakage/breakdown. Once this is done, the sites need to be marked in such a manner that they can be re-located after TEM sample preparation. Either irregularities on the device surface can be used as fiducial marks, or lines or spots of polymerized hydrocarbon can be written on the device surface with the SEM beam,[5,6] using residual contamination as the hydrocarbon source.

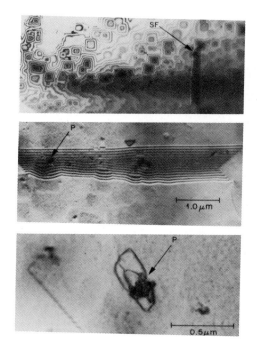

Fig. 3. TEM photos of leakage site showing presence of decorated stacking fault (SF) with a precipitate (P).

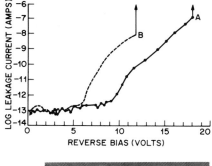

Fig. 4. Leakage current for two capacitors with 250Å SiO$_2$ and 3500 Å polysilicon field plates. The field plate is negative.

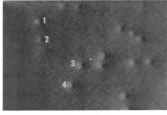

(a)

Fig. 5. Low field EBIC image of part of capacitor B (a), and high field EBIC image of Schottky barrier made at same site (b) showing a coincidence of EBIC spots.

(b)

One application of this procedure to a pn-junction leakage problem has been described in Ref. 6 and is summarized here. Consistent differences in pn-junction leakage had been found on n$^+$p-junction test features on two wafers of different thickness (Fig. 1). The leakage current was higher on the thicker 508 μm-thick wafer (B) than on the 355 μm-thick wafer (A). EBIC measurements showed that leaky diodes were correlated with the presence of sites of high EBIC current (Fig. 2b). One such site was marked by contamination and a sample was prepared for TEM study (Fig. 2c). TEM study showed a stacking fault intersecting the junction at that site (Fig. 3). The fault contained a dislocation tangle and a copper-containing precipitate. Subsequent work showed that a more extensive gettering treatment succeeded in removing precipitates from the active region and lowering the leakage current to acceptable values; the stacking faults remained. Thus, decoration of stacking faults, rather than the presence of faults, was found to dominate leakage current. This observation is consistent with the observation that the magnitude of the modulation of EBIC current caused by the presence of decorated crystallographic defects at junctions is related to the magnitude of decoration.[7,8]

A similar procedure has been applied to capacitor leakage/breakdown problems. In one example,[5] the distribution of leakage current from thin dielectric (250Å SiO_2) capacitors with polysilicon field plates was found to be bimodal; Fig. 4 shows typical leakage curves. EBIC spots were found in the inferior capacitors (Fig. 5a), and the number of EBIC spots correlated with both the leakage current and with the amount of lowering of the breakdown voltage. Schottky diodes made at the same site from which capacitors had been removed by chemical stripping showed EBIC sites at the same locations (Fig. 5b), thus showing that the cause of the appearance of EBIC spots was defects within the substrate. TEM study of selected EBIC spot sites showed the presence of decorated stacking faults (Fig. 6); microanalysis showed that the precipitates contained copper, nickel and tin.

Both these examples show that decorated stacking faults are responsible for junction and capacitor leakage. It is considered likely that in both cases device leakage is dominated by the strain field and chemical activity of the precipitate, not by the fault alone.

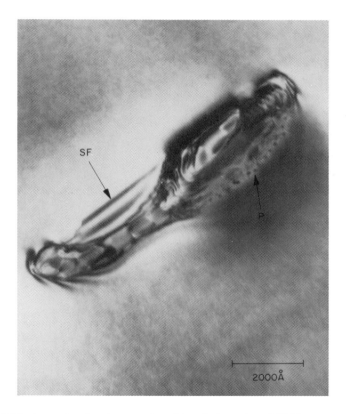

Fig. 6. TEM photo showing decorated stacking fault at site of EBIC spot 2.

3. METALLIZATION LINEWIDTH AND DEVICE FAILURE

Interconnection between device regions on a VLSI chip are made with narrow films of metallization, usually aluminum or polycrystalline silicon ("polysilicon"). A major reliability problem is the mass flow of metallization material in response to current flow which causes opens and shorts. This electromigration process is thermally activated, and high ambient temperatures or local joule heating due to locally high current densities can accelerate the process. Attention is focussed here on phenomena which relate device feature morphology to two types of device failure: opens in aluminum lines caused by aluminum electromigration, and pn-junction shorts. The parameter "mean time to failure" (t_{50}) is used in this discussion as a convenient measure of the relevant device property. That is, for electromigration, t_{50} is the mean time to failure of a population of metallization lines which were under continuity test at a specified current load, voltage, and ambient temperature; failure is a line resistance above a specified value. For pn-junction shorts, t_{50} is the mean time to failure of a population of pn-junction under similar test; failure is irreversible junction leakage above a specified value.

Aluminum Electromigration

Electromigration in lines of aluminum metallization has the effect of transporting metal in a direction parallel to the flow of electron current. Two cases are relevant to this discussion: where the linewidth w is larger than the average crystallite size s, and where it is smaller.

Until recently, most studies of electromigration in aluminum were concerned with the case where the linewidth (usually larger than 2-3 μm) is larger than the grain size. For this case[9]

$$t_{50} \propto I^{-2} \, e^{Ea/kT} \tag{1}$$

where I is the current per unit area. The activation energy E_a has the value[10] 0.5 eV which is also that of grain boundary self diffusion of aluminum.[11] Electromigration transports aluminum along grain boundaries. As w becomes smaller and approaches s, t_{50} also decreases[10] due to the increasing importance of morphological inhomogeneities within the aluminum line to electromigration and to device failure. A model describing the inhomogeneities and their role in electrotransport has not yet been developed, but t_{50} has been found empirically to relate to grain size, grain size variation σ (standard deviation), and degree of preferred orientation (at constant current density and temperature) as[10]

$$t_{50} \propto \frac{s}{\sigma^2} \, P_{(111)}^3 \, . \tag{2}$$

The quantity $P_{(111)}$ is a measure of the degree of (111) preferred orientation of the aluminum grains. This relation has been found for the case w = 2.0μm, T = 80°C and I = 1×10^5 A/cm^2. Both equations (1) and (2) apply when linewidth exceeds grain size.

A marked increase in t_{50} results when the metal linewidth becomes smaller than the mean grain size. For this case[10] (at constant current density and temperature)

$$\log t_{50} \propto -Aw \, . \tag{3}$$

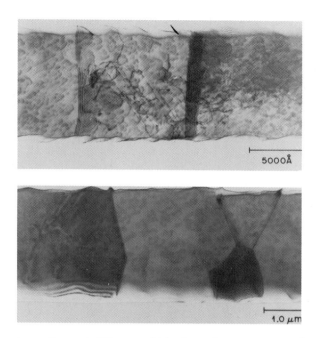

Fig. 7. Aluminum lines of different width show the appearance of only transverse grain boundaries in the narrower film (upper), causing a corresponding improvement in electromigration.

The parameter A has been found to have the value 1.6 for the case $s = 4.5\mu$m, $\sigma = 0.8\mu$m, T = 80°C and I = 1×10^5 A/cm². The improvement in t_{50} with decreasing w occurs because of the decreasing number of grain boundary triple points and grain boundaries running along the direction of the current as w becomes smaller than s, permitting less material to be transported by grain boundary diffusion. One and 5 μm wide lines etched out of the same aluminum film show a factor of five improvement in t_{50} for the narrower lines. Micrographs showing this essential difference in grain boundary configuration for lines of different widths are shown in Fig. 7.

Pn-junction Shorts

A major reliability problem with aluminum contacts to shallow junctions is caused by interdiffusion at the metal-semiconductor interface. Interdiffusion occurs during processing, and is often present before electromigration occurs. This problem can be avoided by using a two-layer metallization scheme of aluminum over doped polysilicon.[12] This metallization scheme has been applied to small area, shallow pn-junction device structures. A number of experiments have been made to determine t_{50}. In these experiments, the direction of electron current flow was from a metallization line into a window to an n^+ diffused region in a p-type silicon wafer, laterally along the diffused region, and out through another window into another metallization line. The mean time to failure of these structures was found to correlate with the junction depth

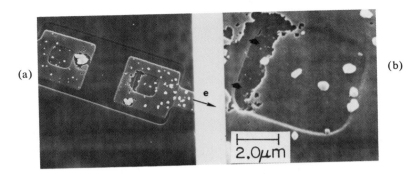

Fig. 8. (a) Pn-junction test structure after aging and after aluminum etch showing direction of electron flow, and (b) enlarged view of leading edge of positive window showing loss of polysilicon of leading edge and formation of etch pits (arrows). (After Vaidya and Sinha, Ref. 13).

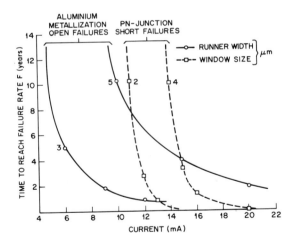

Fig. 9. Time required to obtain a failure rate F vs. current for aluminum metallization opens caused by decreasing linewidth, and pn-junction shorts caused by decreasing window diameter. F is one failure in 10^7 device hours. (After Vaidya and Sinha, Ref. 13.)

x_j and the current density I as[13]

$$t_{50} \propto (x_j)^2 \, I^{-10} e^{Ea/kT} \ . \tag{4}$$

The dependence of t_{50} on $(x_j)^2$ is due to the increasing current density within the n^+ region with decreasing junction depth. Increasing the current density increases the electrotransport of material and also increases the joule heating in the junction, which further accelerates electrotransport and other thermally activated processes.

An analysis of the failure mechanism shows[13] that silicon is transported into the aluminum at the positive electrode and precipitates downstream. Polysilicon under the aluminum at the contact is a major source of silicon; a secondary source is the silicon substrate. Out-diffusion of silicon forms pits in the substrate which eventually become filled with aluminum. An aluminum-silicon eutectic forms at a pit bottom, and a eutectic filament grows toward the negative electrode, eventually forming a short. Fig. 8 shows a test structure after aging and after etching off the aluminum with a partially denuded polysilicon at the leading edge of the positive electrode; silicon pits are also visible.

E_a (in Eqn. 4) has the value 0.9 eV which is the activation energy for electrotransport of silicon along aluminum grain boundaries. The effective current density I is given by $\dfrac{J}{w^{0.6}}$ where J is the current and w is the diameter of the window. Smaller window diameters and shallower junction depths increase the probability of device failure.

Although both types of device failure—metallization opens and pn-junction shorts—can occur, failure is dominated by aluminum electromigration. The dependence of both types of failure on morphological features of the device and on operating parameters is shown in Fig. 9 for the special case where T = 80°C, $x_j = 0.4 \ \mu$m, and the aluminum thickness is 1.0μm. Fig. 9 shows that pn-junction shorts only begin to be a major problem at higher current densities as window size becomes small.

4. SiO$_2$/POLYSILICON INTERFACE TEXTURE AND LEAKAGE/BREAKDOWN

The dielectric properties of thermal oxides grown on polysilicon are usually inferior to those of oxides grown on single crystal silicon. Breakdown field strengths of ~5 MV/cm (5×10^6 V/cm) for the former and 10 MV/cm for the latter are typical. Degradation of polysilicon oxides has been attributed to the appearance of texture at the oxide/polysilicon interface.[14,15] Locally high fields at interfacial protuberances and other features produce locally enhanced conduction,[15] and hence early breakdown. Four classes of relevant interfacial textural features have been found, and each is produced by a different growth mechanism.[16] The four classes are bumps, interfacial roughness, protuberances, and inclusions.

Anomalous grain growth at selective sites in the substrate during polysilicon deposition produces locally thick polysilicon at these sites. These local regions are usually less than 1 μm across and project as bumps above the mean surface by 10-20% of the nominal polysilicon film thickness. The area density of these features varies widely depending on the cleanliness of the procedure, and is usually less than 10^7/cm^2. Oxidation of such polycrystalline surfaces produces an oxide film of uniform thickness;

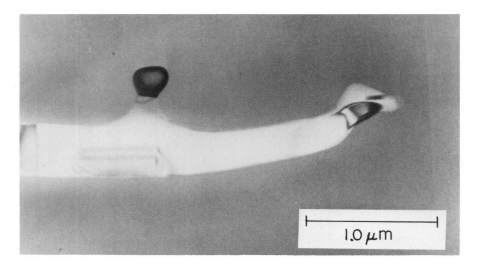

Fig. 10. Oxidized polysilicon film showing a protuberance on the upper polysilicon surface. The polysilicon is surrounded by oxide. The oxide below polysilicon was grown on the silicon substrate.

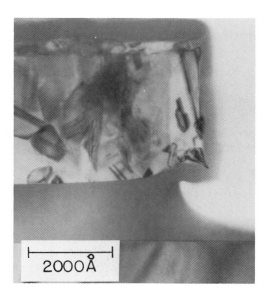

Fig. 11. "Horns" of silicon grown at the corners of polysilicon during oxidation.

the upper oxide surface mirrors the lower interface with some loss of detail.

The texture of the polysilicon surface is due to the grain size of the deposit and is typically in the range of 50-1000Å, depending on the deposition condition. PBr_3 doping and oxidation both result in grain growth, and although the period of interfacial texture increases, elevation changes from grain to grain also appear, due mostly to different oxidation rates of differently oriented crystallites. The final interfacial texture (after oxidation) is significantly less rough for films initially deposited in the amorphous state ($T_{dep} \lesssim 600°C$) and oxides grown from originally amorphous polysilicon show improved breakdown and leakage behavior.

A number of polysilicon crystallites undergo a marked inhibition of oxidation and appear as protuberances which project into the oxide. This inhibition is thought to be due to locally high stresses occurring during oxidation.[17] Oxide is significantly thinner over these protuberances. An example of a protuberance is shown in Fig. 10. As oxidation continues and a grain boundary at the base of the protuberance emerges into the oxide, oxidation proceeds rapidly at the grain boundary to pinch off the protuberance and leaves a silicon inclusion in the oxide.

Both protuberances and interfacial roughness are thought to be the major contributors to enhanced conduction and premature breakdown in capacitors made from polysilicon oxide, and device processing steps which reduce the appearance of these features are needed. Another manifestation of the protuberance problem is the appearance of "horns" of silicon at the corners of patterned polysilicon during oxidation, as shown in Fig. 11. The thermal oxide dielectric over this surface is thin at these corners, and enhanced leakage or breakdown to an overlaying metallization layer is likely to occur.

5. SUMMARY

Three classes of problems have been described which involved a correlation of structure with VLSI device properties:

1. Substrate defects and device leakage/breakdown. Methods for determining the origin of pn-junction and capacitor leakage/breakdown were described for two cases where decorated stacking faults were found to be the relevant defects.

2. Metallization linewidth and device failure. Opens in aluminum lines and pn-junction shorts caused by aluminum and silicon electromigration were both discussed. In both cases device failures were related to linewidth and other structural parameters.

3. SiO_2/polysilicon interface texture and leakage/breakdown. Four types of interfacial textures and their roles in the degradation of MOS dielectric properties were discussed.

ACKNOWLEDGMENTS
Discussions with S. Vaidya on this paper are appreciated.

REFERENCES

[1] T. T. Sheng and R. B. Marcus, J. Electrochem. Soc. *127* 737 (1980).

[2] K. V. Ravi, *Imperfections and Impurities in Semiconductor Silicon*, Wiley, 1981.

[3] C. Donolato, in O. Johari, Ed., *Scanning Electron Microscopy/1979*, IITRI, Chicago, 1979, vol. 2, p. 257.

[4] P. S. D. Lin and H. Leamy, to be published.

[5] P. S. D. Lin, R. B. Marcus and T. T. Sheng, to be published in J. Electrochem Soc.

[6] J. M. Dishman, S. E. Haszko, R. B. Marcus, S. P. Murarka and T. T. Sheng, J. Appl. Phys., *50*, 2689 (1979).

[7] H. Blumtritt, R. Gleichman, J. Heydenreich and H. Johansen, *Phys. Status Solidi.*, (a) *55*, 611, (1979).

[8] R. B. Marcus, M. Robinson, T. T. Sheng, S. E. Haszko and S. Murarka, *J. Electrochem Soc., 124,* 425, (1977).

[9] J. R. Black, Proc. 16th Annual Reliability Physics Symposium, New York, 1978, p. 233.

[10] S. Vaidya, D. B. Fraser and A. K. Sinha, Proc. 18th Annual Reliability Physics Symposium, Las Vegas, 1980, p. 165.

[11] F. M. d'Heurle and P. S. Ho, in J. M. Poate, K. N. Tu and J. W. Mayer, Eds., *Thin Films - Outerdiffusion and Reactions*, Wiley, New York, 1978, p. 243.

[12] J. Clemens and K. Locke, Bell Laboratories, private communication.

[13] S. Vaidya and A. K. Sinha, Proc. 20th Annual Reliability Physics Symposium, San Diego, 1982, p. 50.

[14] E. A. Irene, E. Tierney and D. W. Doug, *Journ. Electrochem. Soc.*, **127**, 705 (1980).

[15] D. J. DiMaria and P. R. Kerr, Appl. Physics. Lett., *27*, 505 (1975).

[16] R. B. Marcus, T. T. Sheng, and P. Lin, *J. Electrochem. Soc.* **129**, 1282 (1982).

[17] R. B. Marcus and T. T. Sheng, Journal Electrochem. Soc., **129** 1278 (1982).

DEFECTS, DISLOCATIONS AND DEGRADATION OF COMPOUND SEMICONDUCTORS

W. DEXTER JOHNSTON, Jr.
Bell Laboratories, Murray Hill, New Jersey 07974

ABSTRACT

Diode lasers and/or LEDs fabricated from the $Ga_{1-x}Al_xAs/GaAs$ or $In_{1-x}Ga_xAs_yP_{1-y}$ alloy system provide the basis for rapidly developing optical communications systems. These devices are operated at very high optical and electrical power densities, and the inevitable less-than-perfect efficiency results in intense local thermal and athermal lattice excitation. This in turn can lead to generation, motion and growth of extended defect structures including dislocation networks, precipitation of impurities, or phase separation.

For the $Al_xGa_{1-x}As$ material, a predominant degradation effect is the development of so-called <100> "dark line defects" (DLDs). These have been shown to arise from recombination enhanced climb of threading dislocations giving rise to dipole loops of a primarily interstitial character. A less common form, the <110> DLD, is associated with slip and typically arises from the strain associated with mechanical damage or careless handling or device processing.

The $In_{1-x}Ga_xAs_yP_{1-y}$ device material affords less energy per recombination event by virtue of its smaller band-gaps, but is likely to have more lattice mismatch strain than $Ga_{1-x}Al_xAs$ for typical compositions of device interest. Thus slip, <110> DLDs, and precipitates are the commonly observed features of $In_{1-x}Ga_xAs_yP_{1-y}$ degradation.

In this paper we will review the present understanding of recombination enhanced (and other) degradation processes, principally in InGaAsP, referring to the relatively well understood effects in $Ga_{1-x}Al_xAs$ for purposes of contrast and comparison.

INTRODUCTION

Transmission of optical pulses via fused silica fiber is rapidly being introduced into telecommunications systems around the world, and promises to replace or supplant existing wire and coaxial cable in new terrestrial and suboceanic installations. The present, first generation systems employ $Al_xGa_{1-x}As/GaAs$ laser diodes typically emitting at .82 µm, where a local loss minimum in the glass fiber guide coincides with a reliability maximum for the $Al_xGa_{1-x}As$ lasers as a function of active layer composition (and hence emission wavelength).[1] Dispersion and residual fiber loss for the .8-.9 nm wavelengths limit LED applications to low bit rates and/or relatively short data links on the order of one to several km, whereas the laser systems can operate over roughly 10 times the distance or at 10 times the bit rate as for LEDs.

Mat. Res. Soc. Symp. Proc. Vol. 14 (1983) © Elsevier Science Publishing Co., Inc.

454

At 1.3 or 1.55 μm, however, the fiber losses are reduced to a few tenths of a db per km, and dispersion can be adjusted to zero. Laser-based systems operating over many tens of km, or LED systems operating over 10-20 km (a typical maximum distance for many terrestrial systems) have been demonstrated at these longer wavelengths. LED and laser sources at 1.3 or 1.55 μm are based on $In_{1-x}Ga_xAs_yP_{1-y}$ epitaxial layers with x and y chosen to yield a lattice constant match to InP substrates as well as to define the desired band-gap energy appropriate for the system wavelength specification. For 1.3 μm, for example, an $In_{.78}Ga_{.22}As_{.6}P_{.4}$ composition is suitable. Double heterostructure devices consisting of a submicron thick quaternary active layer sandwiched between p and n-type InP cladding layers were prepared initially from material grown by liquid phase epitaxy (LPE).[2] Vapor-phase, metallorganic, and molecular beam epitaxial growth techniques (VPE, MOCVD, and MBE) have all now been used successfully as well.[3,4,5]

Obviously the lifetime of the laser diodes and LEDs is of critical impor- tance to the success of the lightwave communications systems. Understanding and elimination or control of the degradation mechanisms has been essential to bringing this technology out of the laboratory and into the field. While more sophisticated degradation modes are also important to systems engineers (i.e., loss of on-off contrast or change of modulation characteristic), the simplest mode of degradation involves loss of light-generating capability. In this "light-bulb failure" mode, lasers simply stop lasing or LEDs fall off in luminous efficiency. This behavior typically stems from the growth of localized regions having poor radiative efficiency, rather than an overall, uniform loss of efficiency. In $Al_xGa_{1-x}As$ lasers these regions often show up dramatically as dark lines oriented along <100> crystallographic directions across the laser or LED active region.[6] Identification and control of these dark-line defect (DLD) features was essential to the realization of usefully longlived $Al_xGa_{1-x}As$ lasers. The <100> DLD has been much studied and there is now general agreement that it results from recombination enhanced motion of threading dislocations.[7] These move by a combination of climb and glide leaving behind a dislocation dipole network of primarily intersti- tial character. Although the threading dislocation initially may be quite inactive as a recombination center, the dislocation dipole network is highly effective at promoting the nonradiative recombination of electrons and holes and correlates exactly in spatial extent with the DLD as observed in electro- or photoluminescence.[8] Strain, due to lattice mismatch between epilayers or from inappropriate device processing or mounting, vastly accelerates the DLD generation process. Hence, control of strain by careful device processing and use of compliant mounting techniques, as well as proper choice of layer compositions, is essential. Selection of low dislocation GaAs substrates is also important, since most dislocations which emerge from the substrate surface will continue to thread through the epitaxy and will serve as a source for <100> DLD generation.

A less commonly observed degradation of GaAs material takes the form of <110> oriented DLDs which result from dislocation glide along {111} planes.[9] These are typically associated with fresh, post-growth dislocations arising from plastic deformation, rather than from grown-in dislocations threading from the substrate. Since the strain required to produce plastic deformation at room temperature in GaAs is relatively large (substantially larger than in InP, for example), the <110> form of DLD is rarely seen in devices unless deliberately induced by scratching or bending, etc.

In $In_{1-x}Ga_xAs_yP_{1-y}$ and InP DLDs are also a commonly observed degradation phenomenon. However, the character of the degradation is the opposite from that observed in the GaAs-based material. The <100> oriented climb/glide DLD associated with motion of threading dislocations has been only rarely reported, and it seems likely that in those few instances what was actually being observed was inclined <011>-type DLDs which projected onto the (001) wafer plane to appear as <010> DLDs. Indeed, grown-in dislocations seem of secondary importance and may be largely irrelevant to InP-based device degradation. Strain, inclusions, and <110> oriented DLDs associated therewith are the predominant degradation modality.[10]   Often dark spot defects (DSDs) with no apparent macroscopic orientation appear following current injection or optical excitation of InP or quaternary device material. These are revealed on microscopic examination to consist of segments oriented along the several possible <110> directions, and are the result of recombination enhanced slip along the {111} planes.

Historically, the correlation between understanding of the nature of DLDs and improvement in reported laser and LED lifetimes is also quite different for the cases of the GaAs and InP-based material systems. The first $Al_xGa_{1-x}As$ lasers to operate continuously at room temperature had life-times of at most tens of seconds, whereas virtually from the first reports of successful operation, room temperature lifetimes for the quaternary lasers were in the thousand-hour range.[2]   As DLD processes in the $Al_xGa_{1-x}As$ material came to be understood, steady progress in laser reli-ability was made and systematically increasing lifetimes (now on the order of 10^5 hours at room temperature) were reported. In contrast, it seems that the more we have come to understand about the quaternary material, the more surprising it is that the long-wave light emitters last as long and work as well as they do. Study of the degradation mechanisms in the quaternary material is far from a moot point, in spite of the demonstrated long lives of both laser[11] and LED[12] devices.   Reliability assurance requires that accurate projection of component life be possible, and the scale of time involved requires that this be based on accelerated life-test data. This can only be provided after the degradation mechanisms are understood. It now seems clear that the degradation of quaternary material can include thermal effects[13] as well as recombination enhanced defect motion (REDM),[14] and that even the REDM effects are qualitatively different from those that dominate the degradation of the GaAs-based light emitters.

OCCURRENCE OF DARK-LINE AND DARK-SPOT DEFECTS IN $In_{1-x}Ga_xAs_yP_{1-y}$

A variety of dark-spot defects, or regions of enhanced nonradiative recom-bination for minority carriers, have been observed in $In_{1-x}Ga_xAs_yP_{1-y}$ epitaxial layers.[15]   These are associated with minority carrier injection[16] in forward biased diode devices, with optical excitation[17] of 'bulk' layers (i.e. relatively thick, single epitaxial layers) grown by either LPE or VPE techniques, or simply with aging at moderately elevated temperatures.[13]   In the case of complete device structures it is often the case that a variety of possibly contributing factors must be considered. These include strain due to contact metallization and alloying,[18] strain due to multiple layers slightly mismatched as to lattice constant relative to each other and to the substrate, migration of contact materials, heavy doping effects including microprecipitation or aggregation of dopant impurities as a second phase or on defect cores,[19] etc. These factors all tend to obscure 'intrinsic' defect properties of the pure quaternary layer, but of course are also highly relevant to practical studies aimed to improve real device performance.

As mentioned above, effects outwardly similar have been observed in $Al_xGa_{1-x}As$ devices and have been extensively studied for over a decade. We now feel that there is relatively little in common between the two compounds as to degradation mode, however. This can be rationalized in terms of several potentially important differences between the ternary and the quaternary materials.

The band-gap energy provides a measure of the energy available from a recombination event to enhance defect motion. This is about 1.4 eV in GaAs, and only two-thirds that amount in the 1.3 μm emitting quaternary. The $Al_xGa_{1-x}As$ ternary material is generally thought to be a well-behaved solid solution over the entire composition range. On the other hand it has recently been demonstrated that a long-suspected miscibility gap in the InGaAsP system is real and covers the region corresponding to 1.3 and 1.55 μm quaternary compositions.[20] Growth of compositions in this region is allowed only because of the stabilizing effect of strain relative to the lattice match condition, which for small mismatch produces again a concave, locally stable free energy surface. There is no reason to expect clustering in the $Al_xGa_{1-x}As$ alloy, so that we expect the Al and Ga atoms to be randomly arrayed on the A sublattice. In the quaternary, however, it is expected that there will be clustering effects and that there will be a nonzero correlation of the atoms on the A and B sublattices.

The effect of residual lattice mismatch is pronounced in both materials. There appears to be a critical level or threshold above which degradation is markedly enhanced. Lasers with binary GaAs active layers degrade rapidly as compared to those with an $Al_{.08}Ga_{.92}As$ active layer, which is of course more closely matched to the $Al_{.4}Ga_{.6}As$ compositions typically used as cladding layers. $In_{1-x}Ga_xAs_yP_{1-y}$ layers with lattice mismatch of a few tenths of one percent can be seen to degrade in seconds under moderate optical excitation, while layers matched to within .05% are highly resistant to optical degradation and are likely to 'burn' rather than degrade. Once thermal or mechanical damage is introduced, of course, slip and <110> DLD degradation proceed rapidly via recombination enhanced defect motion (REDM). The slip is not confined to the quaternary layers, but propagates readily through the InP as well and can be observed a a rectilinear grid by Nomarski interference contrast microscopy at the top surface of double heterostructure devices.

Degradation of macroscopically well-lattice-matched layers also occurs, although typically the very rapid glide and linear <110> orientation is not apparent. In this case, the initiating factor is a dislocation cloud emanating from precipitates or microinclusions.[21] This mode of degradation may set in suddenly or develop more or less steadily during minority carrier injection, depending on whether inclusions were initially present from imperfect growth or developed as a result of thermal damage or from some other cause during device aging. The problem of inclusions is all the greater in the quaternary material and in InP as well because of the tendency for these compounds to decompose by loss of phosphorus.[22] An InGa rich phase can typically be expected to have a lower melting point - a metallic inclusion being an obvious extreme - and under thermal or strain gradients could move through the crystal lattice as a microdroplet. Drastic changes in junction current-voltage properties are associated with such inclusion migration. Nonradiative recombination plays only a secondary role, serving as a source of local heating, and indeed inclusion related effects have been observed in 'shelf-life' degradation of LEDs stored at temperatures of 250C.

TEM examination of the thermally aged and degraded LEDs showed defects with inclusion-like, second phase contrast. X-ray fluorescence confirms that no nonassay impurities are involved, excluding the participation of contact metals.[23] Other studies have suggested that this need not always be the case, however, and that Au, in particular, may be involved in certain contact-mediated degradation modes.[24] These inclusions, from whatever source, give rise to a complex, fibrous dislocation structure. Far from the source of lattice disturbance the debris tends toward simple edge dislocations.

The possible relation of degradation mode and susceptibility to growth technique is an interesting subject which has not yet been adequately studied. Although quaternary lasers have been prepared from LPE, VPE, MOCVD and MBE material, nearly all of the life test data has been derived from LPE material. The other growth techniques have in principal the capability to access different regimes of 3-5 ratio in the growth medium and stoichiometry of the grown material, which might well affect degradation. In practice there is not as much room to alter stoichiometry as one might hope with these other growth techniques, since the quality of the epilayer tends to be highly sensitive to the growth variables, and nucleation problems become severe when any substantial departure from optimum growth is introduced. Again, this arises largely from the tendency of the InP based compounds to undergo thermal decomposition at temperatures above a few hundred C.

PHASE SEPARATION AND GROWTH-RELATED EFFECTS

Phase segregation is generally held to be undesirable if not fatal for lightemissive devices. Certainly this is justified in gross cases such as the In or Ga inclusions arising form poor wipe-off in LPE or substrate surface decomposition in VPE. Regions of an active layer with different composition than that intended generally broaden the optical emission and lead to reduced laser gain. An exception may be cited - the multiple quantum well laser structure in which thin layers, of order 100 A, of alternately wide and narrow gap material are interleaved deliberately to produce a confined density of states and enhanced inversion.[25] While such well-ordered and controlled phase separation can be beneficial, phase segregation which is not controlled or even not identified and recognized may be either good or bad. If it is of the quasiperiodic type resulting from spinodal decomposition, lattice hardening can be expected which should serve to constrain slip and pin dislocation motion. Local second phase inclusions will serve as a source of dislocations to feed REDM type degradation.

Microscopic In inclusions can melt and move as droplets, dissolving material at one face and resolidifying material at the other. The overall composition must stay about the same since no material may be lost in this process, but spinodal decomposition of metastable or conditionally stable phases may well occur if the effective temperature is significantly different from the growth temperature. What that temperature may be is not easy to specify, but given the high energy density and power dissipation density in lasers and high radiance LEDs it is reasonable to expect that metallic inclusions of micron size could be heated to temperatures in the 200-600C range.

In inclusions are not solely the result of LPE growth problems. CVD and MBE growth require that the substrate be heated to 650-700C in gas or vacuum ambient. This is well above the 300C limiting temperature for congruent evaporation of InP and an In rich surface is unavoidable. At best the development of macroscopic In droplets can be avoided or suppressed by overpressure with an appropriate phosphorus species. In LPE growth such In droplets are accommodated into the growth solution, but in CVD they contribute to an In rich interface layer and can result in lattice mismatch and the generation of misfit dislocations.

Metallorganic CVD has the additional problem that foreign second phase material of polymeric adducts of the group five hydrides and the group three metal alkyls may be incorporated as well. In fact, if measures are not taken to suppress this tendency, device quality InP layers cannot be grown by MOCVD. There is no data bearing on the question of whether such hydrocarbon complex phases participate directly in device degradation, but it is reasonable to expect that they would at least act like oxide or other refractory particles as a source of dislocations and local strain.

While the above problems are nontrivial in practice they are of the sort that one hopes will disappear once the right growth recipes are found. The effects which may arise from spinodal decomposition are of a more subtle and fundamental nature. Typically one would expect composition waves in the quaternary layer. Such have indeed been observed in both VPE[26] and LPE material,[26,27] showing that a true phase instability is involved and not for instance some melt instability of a hydrodynamic nature. The spatial extent of the periodic composition modulation seems to depend on the growth kinetics, surface mobilities, substrate crystallographic orientation, etc., and may lie in the 100-500A range. At the larger end of this range one would expect to see broadening of the spectrum of emitted light and increase of laser threshold, with possible changes in the temperature dependence of laser threshold as well.[28] The latter is a crucial parameter for the 1.3 and 1.55 µm lasers since thresholds rise rapidly just above room temperature. The rise is driven by Auger recombination and other hot carrier effects which increase superlinearly with increasing carrier density. This should be locally enhanced by the band-gap modulation to be expected from spinodal decomposition. The experimental results show that the periodic composition modulation is such as to produce regions of large band-gap and smaller lattice constant than that intended. Thus a periodic variation is strain as well as band-gap results. It is not now clear how the spinodal decomposition is related to degradation, but it is apparent that even if quaternary layers are grown without it (say by suitably elevating the growth temperature) these layers are indeed instable at room temperature and can be expected to separate at rates limited by diffusion and/or REDM processes. Further studies focussing on correlations between growth temperature and method, TEM examination for composition modulation, and device degradation susceptibility or resistance seem essential to improved understanding and the realization of consistently improved quaternary device reliability.

DISCUSSION

In the case of GaAs lasers the central role of radiation enhanced defect motion in dark-line degradation is well established. It is not clear in the case of the InP based devices that any one mode of degradation dominates, but it is apparent that REDM plays at least a contributory role. Continuous operation of quaternary lasers for periods exceeding 10^4 hours has been reported for both VPE and LPE material. LED aging experiments have shown that devices can be made which show no degradation below 200C for similar periods. At the same time, some quaternary devices fail at short times, and others deteriorate gradually under forward bias. Some degrade in hot storage without bias. In this confused situation it has been easy to associate the degrading material with 'poor' growth and the longlived devices with 'improved' growth techniques. This now appears to be an oversimplification. Certainly some of the factors associated with 'poor' growth are obvious and others have been identified, but there remains a good deal of mystery.

Strain, precipitates, substrate decomposition and inclusions are all to be avoided. The role of spinodal decomposition must be examined in more detail. It is probable that second generation lightwave devices will utilize VPE or MBE growth to an increasing extent to take advantage of scaling and throughput capabilities, and it will be important to understand the relation of growth conditions in these technologies to phase stability and dislocation generation and pinning.

The principal degradation mode (<100> DLD) of GaAs lasers is apparently unimportant for the long wavelength sources. This is particularly fortunate as the InP substrate material available is not yet of as high quality as GaAs substrates and threading dislocations are very difficult to avoid. They

appear to have no important effect on device lifetime, unlike the case in GaAs
where they dominate other factors. On the other hand, inclusions and
precipitates are rare in good-quality GaAs, but appear to be the norm in InP
and the quaternary material. Degradation resulting from recombination
enhanced motion of inclusion-generated dislocations is thus a principal cause
of quaternary device failure. It is not clear at this point whether all
inclusions are grown in or whether some form during device operation or hot
storage as a result of decomposition induced either thermally or triggered by
nonradiative recombination, nor do we know for certain that inclusion-free
material will not degrade. The present trends in optical communications
emphasizing quaternary sources make it clear that these are questions of very
substantial practical consequence.

REFERENCES

1. A. A. Bergh, J. A. Copeland and R. W. Dixon, Proc. IEEE 68, 1240 (1980).

2. C. C. Shen, J. J. Hsieh and T. A. Lind, Appl. Phys. Lett. 30, 353 (1977).

3. G. H. Olsen, C. J. Nuese and M. Ettenberg, IEEE J. Quant. Electronics
   QE-15, 688 (1979).

4. M. Razeghi, M. A. Poisson, P. Hirtz, B. deCremoux and J. P. Duchemin,
   Paper I-2, 1982 Electronics Materials Conference, Fort Collins, CO, USA.

5. W. T. Tsang, J. Appl. Phys. 52, 3861 (1981).

6. B. C. DeLoach, Jr., B. W. Hakki, R. L. Hartman and L. A. D'Asaro, Proc.
   IEEE 61, 1042 (1973).

7. P. M. Petroff and L. C. Kimerling, Appl. Phys. Lett. 29, 461 (1976).

8. P. M. Petroff, W. D. Johnston, Jr. and R. L. Hartman, Appl. Phys. Lett.
   25, 226 (1974).

9. K. Maeda, M. Sato, A. Kubo and S. Takeuchi, Tech. Rep. ISSP A 1243, U.
   Tokyo, Tokyo, Japan (July 1982) (submitted to J. Appl. Phys.).

10. S. Mahajan, W. D. Johnston, Jr., M. A. Pollack and R. E. Nahory, Appl.
    Phys. Lett. 34, 717 (1979).

11. H. Temkin, C. L. Zipfel, M. A. DiGiuseppe, A. K. Chin, V. G. Keramidas
    and R. H. Saul. Bell Syst. Tech. Journal (to be published, 1982).

12. T. Yamamoto, K. Sakai and S. Akiba, IEEE J. Quant. Elect. QE-15, 684
    (1979).

13. H. Temkin, C. L. Zipfel and V. G. Keramidas, J. Appl. Phys. 52, 5377
    (1981).

14. D. V. Lang, Ann. Rev. Mat. Sci. 12, 377 (1982).

15. W. D. Johnston, Jr. in "GaInAsP Alloy Semiconductors," T. P. Pearsall,
    Ed., J. Wiley & Sons, Sussex, England (1982).

16. K. Ishida, T. Kamejima, Y. Matsumoto and K. Endo, Appl. Phys. Lett. 40,
    16 (1982).

460

17. W. D. Johnston, Jr. G. Y. Epps, R. E. Nahory and M. A. Pollack, Appl. Phys. Lett. 33, 992 (1978).

18. A. K. Chin, M. A. DiGiuseppe and W. A. Bonner, Mat. Lett. 1, 19 (1982).

19. A. K. Chin, A. Temkin and S. Mahajan, Bell Syst. Tech. Journal 60, 2187 (1981).

20. M. Quillec, C. Duguet, J. L. Benchimol and H. Launois, Appl Phys. Lett. 40, 325 (1982). Also G. B. Stringfellow, J. El. Mat. 11, 903 (1982).

21. H. Temkin, S. Mahajan, M. A. DiGiuseppe and A. G. Dentai, Appl. Phys. Lett. 40, 562 1982, 81-52342-53.

22. P. K. Gallagher and S. N. G. Chu, J. Phys. Chem. 86, 3246 (1982).

23. S. Yamakoshi, M. Abe, O. Wada, S. Komiya and T. Sakurai, IEEE J. Quant. Electronics QE-17, 167 (1981).

24. A. K. Chin, C. L. Zipfel, F. Ermanis, L. Marchut, I. Camlibel, M. A. DiGiuseppe, and B. H. Chin, (to be publ. in Trans. IEEE, April, 1983).

25. N. Holonyak, Jr., R. M. Kolbas, R. D. Dupuis and P. D. Dapkus, IEEE J. Quant. Electronics QE-16, 170 (1980).

26. P. Petroff, Priv. Comm. 1982.

27. F. Glas, M. J. M. Treacy, M. Quillec and H. Launois. Paper K-2, 1982 Int. Symp. on GaAs, Albuquerque, NM, USA (Proc. to be publ., Inst. Phys. Conf. Series, 1983).

28. N. K. Dutta, submitted to J. Appl. Phys. (1982).

PHOTOPLASTIC EFFECTS IN II-VI CRYSTALS

SHIN TAKEUCHI, KOJI MAEDA AND KIYOKAZU NAKAGAWA*
Institute for Solid State Physics, University of Tokyo, Roppongi, Minato-ku,
Tokyo 106, Japan

ABSTRACT

A reversible change in the flow stress with the
illumination of a band gap light (photoplastic effect:
PPE) is observed commonly in II-VI semiconducting
compounds both with the zincblende and the wurtzite
structures. After reviewing the experimental results
accumulated so far concerning the usually observed PPE,
detailed microscopic experiments on CdTe (partly on CdS)
single crystals are described. In-situ TEM straining
experiments with a laser illumination system clarified
that the dislocation mobility, evidently controlled by the
Peierls mechanism, is not affected by light illumination.
The investigation of the dislocation glide behavior over a
longer range using the etch pit method and the cathodoluminescence microscopy with a SEM revealed that the dislocations are apt to become immobilized after traveling a
certain distance from the sources with a high velocity.
The dislocation loops and cusps observed by TEM suggest
that the immobilization is caused by jog formation along
screw segments. Those results suggest that the positive
PPE is caused by the decrease in the mean free path of
multiplied dislocations due to the enhanced jog formation
by illumination. This viewpoint is supported by the high
density of dislocation debris left in specimens deformed
under illumination, the enhanced work hardening rate under
illumination and the reduction of the PPE when the dislocation mean free path becomes determined by other photo-
insensitive factors such as forest dislocation cutting
and specimen size. The microscopic mechanism of the
enhanced jog formation is discussed in terms of electrostatic interaction between charged point defects and
dislocation charge which is increased with the illumination.

INTRODUCTION

A remarkable feature of plasticity of II-VI semiconducting compounds, both
with the zincblende structure and the wurtzite structure, is the photoplastic
effect (PPE) which is commonly observed in almost all II-VI combinations such
as CdS, CdTe and ZnO [1]. The PPE is the phenomenon in which the flow stress
in dynamic straining tests is affected by the illumination of a light near
the fundamental absorption edge of the material. The effect is almost
reversible and in usual cases causes hardening of the specimen (the positive
PPE).
It is well established that the plastic deformation of III-V compounds is
controlled by surmounting of the intrinsic Peierls barrier, while that of

* Now at Hitachi Central Research Laboratory, Kokubunji, Tokyo 185, Japan

Mat. Res. Soc. Symp. Proc. Vol. 14 (1983) © Elsevier Science Publishing Co., Inc.

I-VII, i.e. alkali halides, is governed, except at very low temperatures, by extrinsic barriers such as impurities and forest dislocations. The Peierls stress or the Peierls potential height is closely related to the bonding nature of the material, ionic or covalent. From the ionicity of II-VI compounds which intervenes between III-V compounds and alkali halides, it is easily guessed that the plastic deformation in II-VI compounds is controlled by a composite mechanism in which the Peierls mechanism and the defect controlling mechanism are both participating. Thus, the key problem to be solved before considering the mechanism of the PPE generally observed in II-VI compounds is to clarify the underlying basic mechanism govering the flow stress of these crystals.

Since the PPE was discovered by Osip'yan and coworkers [2], various experimental facts have been accumulated by several research groups, which will be summarized in the following section. The similar effect in which a light illumination affects the flow stress of materials has earlier been observed in alkali halides [3]. The difference from the case of II-VI compounds, however, is in the spectrum of light responsible for the effect; i.e., a light exciting F-centers in alkali halides [3] while the fundamental absorption light in II-VI compounds [2, 4-6]. Thus the cause of the PPE in alkali halides is obviously related to the change in the charge state of point defects, which results in a change in the retarding force against the motion of dislocations. The mechanism of the PPE in II-VI compounds was first conjectured to be analogous to the case of alkali halides.

However, later experiments suggested that the thermal activation process controlling the rate of plastic flow is the Peierls mechanism [5,7,8]. This fact led some people to conceive an illumination-induced change of the Peierls barrier as a cause of the PPE. In the measurements of electric current carried by dislocations, Osip'yan et al. showed that the dislocation charge estimated from the dislocation current is increased by illumination [5]. From this fact, they considered that the positive PPE is caused by a hightening of the Peierls barrier arising from the electrostatic interaction between the dislocation line charge and the ionic row of the lattice [5]. Gutmanas et al. on the other hand considered the double kink formation energy as a decreasing function of dislocation charge and explained the positive PPE as a result of decrease (not increase) of dislocation charge by illumination [9]. An earlier study by Carlsson et al. claimed that the positive PPE is caused by a decrease of mobile dislocation density rather than dislocation mobility [10].

The controversy is continuing which factor, dislocation velocity, mobile dislocation density or some other quantity, is really responsible for the PPE. In this respect, direct measurements of dislocation velocities with and without illumination may be crucial. In the later section following a survey of previous experimental facts in the next section, experimental results including microscopic experiments on CdTe (partly on CdS) single crystals performed by the present authors are described, and discussion is made on the mechanism of the PPE conceived on the basis of the experimental facts.

SUMMARY OF PREVIOUS EXPERIMENTS

In this section, the established results of the experiments concerning the positive PPE are briefly surveyed. The negative PPE that was recently found in some crystals will be mentioned in the later discussion.

The spectrums of light responsible for the positive PPE are summarized in Fig. 1 for various compounds [4,5,11]. The common feature of the spectrums is that they have a peak near the fundamental absorption edge of the crystal but their shapes are quite asymmetric tailing to the longer wavelength in comparison with the spectrum of photoconductivity. When the photon energy exceeds the band gap energy and the light does not penetrate into the interior of the crystal, the PPE is rapidly reduced. This fact indicates that the

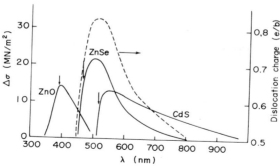

Fig. 1. Spectral dependence of the positive PPE for CdS [11], ZnSe [5] and ZnO [4]. The vertical arrows indicate the band gap energies of the respective crystal. The broken line for ZnSe shows the spectral dependence of dislocation charge [5]. (Reproduced with permissions from Publishers)

phenomenon is not a surface effect but related to an electronic excitation in the bulk.

The typical dependence upon illumination intensity is shown in Fig. 2. In CdS [2] and ZnO [4], the magnitude of the PPE increases with the illumination intensity and becomes saturated above several thousands lux (the order of 10^{14} photons/s/cm^2). In some CdTe, CdS and CdSe, however, the intensity dependence exhibits a peak due to a commencement of the negative PPE which becomes stronger with the intensity [12,13].

It is quite common that the absolute magnitude of the PPE is a decreasing function of the testing temperature as shown in Fig. 3 [5,9,15,16]. The temperature at which the PPE diminishes is paralleled systematically with the gap energy of the crystal.

As seen in Fig. 3, pronounced feature of the positive PPE is the orientation dependence; in crystals with the wurtzite structure such as CdS and ZnO, the positive PPE is not observed when the specimen is deformed in such an orienta-

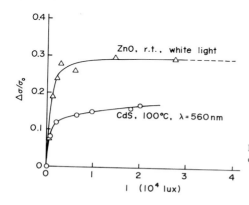

Fig. 2. Illumination intensity dependence of the positive PPE for CdS [14] and ZnO [4]. (Reproduced with permissions from Publishers)

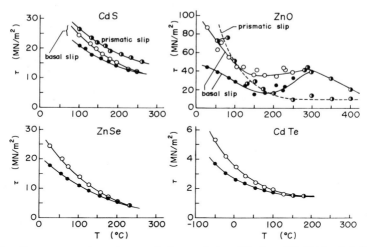

Fig. 3. Temperature dependence of the positive PPE for CdS [16], ZnO [15],
ZnSe [5] and CdTe [9]. Closed circles are for data in the dark and open ones
those under illumination. The half-filled circles indicate that the PPE is
not observed. (Reproduced with permissions from Publishers)

tion as to inhibit the basal slip. This fact indicates that the mechanism of
the positive PPE is related to the nature of the dislocation core.

The magnitude of the PPE, in most cases, depends on the plastic strain ε_p.
Figure 4 exemplifies typical ε_p dependences for CdS [14] and CdTe [10], in
which the PPE is reduced considerably with the plastic deformation. This fact
means that deformation-produced defects, dislocations and/or point defects, are
involved in determining the magnitude of the PPE.

An infrared (IR) irradiation in addition to the illumination of the band gap
light causes a decrease in the flow stress [13,17,18]. This IR-quenching effect
necessitates the superposition of the band gap light and the IR. Only minor

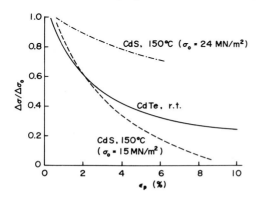

Fig. 4. Reduction of the PPE with
plastic strain for CdS [14] and
CdTe [10]. (Reproduced with permissions from Publishers)

effect is obtained by the IR alone. This fact together with the above mentioned spectral difference between photoconductivity and the PPE suggests that the cause of the positive PPE is a change in the charge state of some defects by the band gap excitation.

The dislocation charge which had been hypothetically supposed in semiconductors was directly measured by Osip'yan et al. [19]. It was found that the flow stresses of various II-VI compounds are closely correlated with the amount of dislocation charge [20]. Moreover, the dislocation charge was found to increase upon illumination in consonant with the increase in the flow stress, i.e. the positive PPE. The close relationship between the dislocation charge and the flow stress is demonstrated by Fig. 1, in which the spectral dependence of dislocation charge increase for ZnSe shown by a broken line is perfectly identical with the corresponding flow stress increase.

EXPERIMENTS

Experiments were performed in the following three levels: (1) the macroscopic level using compression tests to observe the ordinary PPE, (2) the semi-microscopic level using a scanning electron microscope in the cathodoluminescence mode (SEM-CL) [21] and the etch-pit method to observe the relatively long range motion of dislocations, and (3) the microscopic level using a transmission electron microscope (TEM) to observe in situ the short range motion of dislocations and the dislocation morphology in deformed specimens.

Undoped n-type CdTe single crystals ($n = 4 \times 10^{14} cm^{-3}$, $\mu = 100 cm^2/V/s$) were grown from the melt by the Bridgman method and cut to rectangular blocks for compression tests, to rectangular plates for bending tests in SEM-CL observation and to thin foils for TEM experiments. In every case, the stress axis was made parallel to [$\bar{1}23$] direction. For the TEM studies, thin tiny CdS ribbons grown from the vapor phase were also used. Those CdS crystals with the surface orientation of ($1\bar{1}00$) were thin enough for the TEM observation without thinning process. SEM observations were made with an accelerating voltage of 30 kV. Details of the in-situ SEM-CL experiments have been given in a previous paper [21]. TEM observations were performed with a 350 kV electron beam.

For light illumination, a monochromatized light from a 150 W halogen lamp was used for the macroscopic compression tests. For in-situ TEM straining experiments, special light illuminating systems were devised. A white light from the halogen lamp in case of CdTe and 514.5 nm line from a 4 W argon ion laser in case of CdS were used. The illumination intensity at the specimen position was confirmed to be sufficiently high to obtain the macroscopic PPE. The details of the experimental procedures have been given in previous papers [22,23].

In the in-situ straining tests in TEM, dislocations in CdTe and CdS were both observed to glide in a continuous manner which is characteristic of the dislocation motion controlled by the Peierls mechanism. Illumination of light, however, did not affect the velocity of individual dislocations. The result for basal dislocations in a CdS specimen is shown in Fig. 5, where the velocities in darkness are plotted with closed marks and those under illumination with open ones. The specimen being strained continuously was alternately kept in darkness and under illumination with the light for a prescribed period. In order to avoid possible influences of electron beam irradiation (heating effect, cathodoplastic effect [24], recombination enhanced dislocation glide [25], etc.), the TEM observation was made intermittently only to locate the dislocations that change their position during the intermission. In the successive measurements, no systematic difference in the velocity-loading time relation is recognized. Thus, no evidence for a velocity change due to illumination was obtained in the short range motion of dislocations probably controlled by the Peierls mechanism.

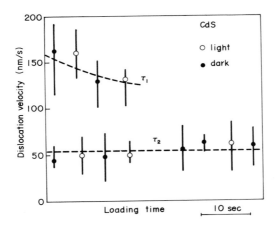

Fig. 5. Influence of light illumination on the velocity of basal dislocations in a CdS specimen measured by in-situ TEM straining experiment [23]. The closed marks and open marks are velocities in the dark and those under illumination, respectively. The upper series is for a test at a higher stress.

From analyses of stress relaxation curves in macroscopic compression tests, activation volumes V* were evaluated as a function of stress τ. Figure 6 shows the results for the tests in darkness and those under illumination. The magnitude of V* is strongly dependent on τ, ranging from 200 b^3 down to 10 b^3, where b is the magnitude of the Burgers vector. The small value of V* at high

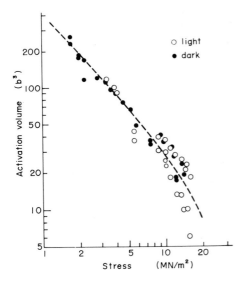

Fig. 6. Activation volume vs effective stress relation in CdTe [23] in the dark (closed marks) and under illumination (open marks).

stress suggests the involvement of the Peierls mechanism in the plastic flow. In contradiction to the result of Gutmanas et al. [9] but consistently with the result of the microscopic measurements described above, no significant difference in the stress dependence is recognized between the two conditions in the present study. Therefore, the previous hypotheses based on the notion that the cause of the PPE is a change of the Peierls potential are quite doubtful.

Although the short range glide motion of dislocations is evidently controlled by the Peierls mechanism, the trials to measure the dislocation mobility in the long range by the conventional etch-pit method and the novel method using the SEM-CL were unsuccessful unlike the case of more covalent materials such as Si and GaAs. This is mainly because the dislocations in II-VI compounds are apt to become immobile after traveling a certain distance from sources while other new dislocations are multiplied in other places. Furthermore, moving state of dislocations before they became immobilized was not able to be observed in the in-situ SEM-CL observations, indicating rather fast movement of dislocations above the stress level at which they were multiplied from sources. In fact, a direct in-situ observation of dislocation multiplication showed that a multiplying dislocation spent most of the time during its bow-out process at its Frank-Read source. An example is presented in Fig. 7 by a series of micrographs taken from a video-tape. The time required to emit a pair of dislocations from the source is more than 10 seconds while the time of travel of multiplied dislocations for a distance of 100 μm (a typical mean free path) estimated from the VTR is a few seconds. These observations suggest that the factor mainly controlling the macroscopic flow stress is not only the mobility of dislocations but also their traveling length before they are stopped.

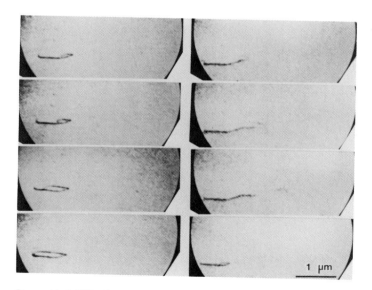

Fig. 7. Sequential TEM micrographs of dislocation loop emission at a multiplication source in CdTe. The process from the left top, through left bottom and right top, to right bottom repeats with a period of about 10 seconds.

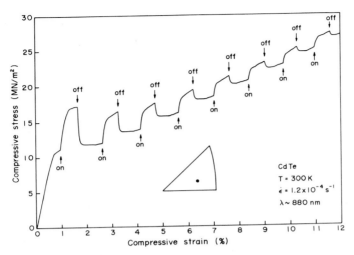

Fig. 8. A typical stress strain curve in CdTe under cyclic illumination of a light.

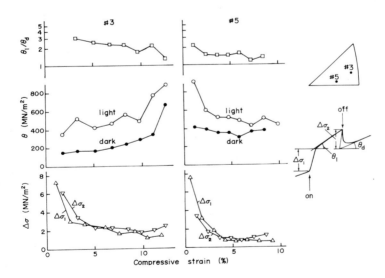

Fig. 9. Strain dependences of the PPE (bottom), of the work hardening rate with and without illumination (middle) and of their ratio (top). The specimens are CdTe crystals compressed at room temperature in two different orientations.

It is worthy to note that the effect of light illumination affects the rate of work hardening. The specimens with different compression axes were compressed with a strain rate of $1.7 \times 10^{-4} s^{-1}$ and cyclically subject to illumination. Figure 8 shows an example of the stress-strain curve. Figure 9 plots the work hardening rates in the two conditions together with the ratio between them and the magnitudes of the PPE, as a function of accumulated plastic strain. For all specimens, light illumination increases the work hardening rate by a factor of $2 \sim 3$. As a result, the flow stress after extinguishing the light never returns to the extrapolation of the stress-strain curve before the illumination but becomes higher just by an amount of increased work hardening during the illumination. This enhanced work hardening is due to the excess dislocations stored under illumination. In fact, much higher density of dislocations are left in the specimens deformed under illumination than in those deformed in darkness as shown by the etch-pit densities in Fig. 10. The dislocation

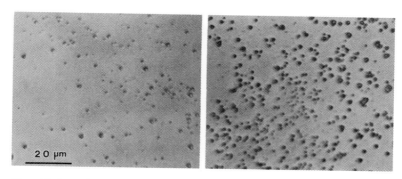

Fig. 10. Dislocation etch-pit pattern on CdTe deformed by 0.9 % in the dark (left) and that deformed by 0.6 % under illumination (right).

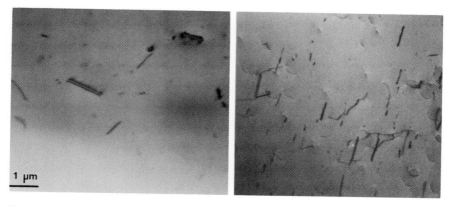

Fig. 11. Electron micrographs showing dislocation morphology in CdTe deformed by 1.4 % in the dark (left) and under illumination (right).

morphology observed in the TEM study (Fig. 11) for specimens deformed in bulk tests shows higher density of dislocation loops and more cusps on dislocation segments in the specimens deformed under illumination.

MECHANISM OF PPE

The above results strongly suggest that the important factors that control the macroscopic flow in II-VI compounds are the rate of dislocation multiplication $\dot{n}$,* and the mean free path $\bar{\ell}$. Therefore, rather than the conventional expression for the plastic strain rate in terms of the mean dislocation velocity $\bar{v}$ and the mobile dislocation density ρ_m

$$\dot{\epsilon}_p = b\rho_m\bar{v}, \tag{1}$$

an equation

$$\dot{\epsilon}_p = b\bar{\ell}\dot{n} \tag{2}$$

will be more appropriate in discussing the PPE and more generally the macroscopic flow behavior in II-VI compounds.

The dislocation multiplication rate is determined by the velocity of the source dislocation near its critical bow-out configuration. Since the short-range motion is governed by the Peierls mechanism, $\dot{n}$ will be expressed by

$$\dot{n} = (Nv_0/L)\exp\{-U(\tau_{eff})/kT\}. \tag{3}$$

Here, N is the source density and $U(\tau_{eff})$ the activation energy in the Peierls mechanism at the effective stress τ_{eff}, v_0 the prefactor and L the characteristic length relating to the source length. τ_{eff} may be smaller than the applied stress τ_a by the long-range internal stress τ_i due to other dislocations and the back stress τ_c due to the line tension of the dislocation at the critical bowing-out configuration.

The plastic strain rate is then given by

$$\dot{\epsilon}_p = (bNv_0\bar{\ell}/L)\exp\{-U(\tau_a-\tau_i-\tau_c)/kT\}. \tag{4}$$

If one considers a simple case that any of the quantities other than $\bar{\ell}$ in eq. (4) is not affected by illumination, the magnitude of the PPE $\Delta\tau$ (increment of τ_a upon illumination) will be expressed by

$$\Delta\tau = \ell n(\bar{\ell}_d/\bar{\ell}_1)(kT/V^*), \tag{5}$$

where $\bar{\ell}_d$ is the mean free path in darkness, $\bar{\ell}_1$ is that under illumination and V^* is the activation volume defined by $V^* = -\partial U/\partial\tau_{eff}$. In the case where the mean free path is less than the specimen size, immobilized dislocations are stored in the specimen. The stored density ρ is inversely proportional to $\bar{\ell}$ and hence the $\bar{\ell}$ can be evaluated from the value of ρ at the plastic strain ϵ_p by

$$\bar{\ell} = \epsilon_p/(b\rho). \tag{6}$$

From the etch-pit counting for the deformed specimens (Fig. 10),

---

* Direct observation by TEM indicates that the deformation rate is governed by the motion of screw dislocations. The deformation process, therefore, will be treated two-dimensionally taking only screw dislocations into account. Hence, $\dot{n}$ here has the dimension of $[L]^{-2}$.

$$\bar{\ell}_d/\bar{\ell}_1 \sim 6. \tag{7}$$

The stored immobilized dislocations give rise to internal stress. According to the Bailey-Hirsch relation

$$\tau_i = \alpha\mu b\sqrt{\rho}, \tag{8}$$

where μ is the shear modulus and α is a numerical constant of ~ 0.3. The work hardening rate due to the accumulation of primary dislocations is therefore expressed by

$$\theta = d\tau_i/d\epsilon_p = (\alpha\mu b/2)(1/\sqrt{\rho})(d\rho/d\epsilon_p). \tag{9}$$

In large specimens in which most multiplied dislocations are eventually immobilized, from eqs. (6) and (9)

$$\theta = \alpha\mu/(2\bar{\ell}\sqrt{\rho}). \tag{10}$$

At the same strain, hence for the same ρ, the ratio of work hardening rates in the two conditions is approximately given by*

$$\theta_1/\theta_d \sim \bar{\ell}_d/\bar{\ell}_1. \tag{11}$$

From Fig. 9 and eq. (11),

$$\bar{\ell}_d/\bar{\ell}_1 \sim 3. \tag{12}$$

If one uses experimental values of $V* = 30\sim50\ b^3$ at $\tau_{eff} = 5\sim10\ MN/m^2$ at $T = 300\ K$ and the value of $\bar{\ell}_d/\bar{\ell}_1 = 3\sim6$ as evaluated above, $\Delta\tau = 1\sim2\ MN/m^2$ is obtained from eq. (5), which is in fair agreement with experimentally obtained magnitude of the PPE.

In very small specimens in which $\bar{\ell}$ is limited by the specimen size, a light illumination is not expected to cause any change in $\bar{\ell}$. The size effect was actually observed when the specimen thickness becomes less than $\sim100\ \mu m$ [26]. As shown in Fig. 9, the magnitude of the positive PPE decreases with the plastic strain as the ratio θ_1/θ_d decreases. Such a correlation is readily explained by the above argument. The decrement of θ_1/θ_d is probably due to the situation that in the later stage of deformation the work hardening rate or the dislocation mean free path becomes more and more controlled by a different process such as the forest dislocation cutting that is not affected by light illumination.

DISCUSSION

The microscopic mechanism of dislocation immobilization is not known in detail. However, the existence of quite a few cusps and loops in the deformed specimens strongly suggests that the immobilization is caused by jog formations along screw dislocations. Actually, jog formations are observed quite frequently in the in-situ TEM observations. A quantitative measurement of slip band growth rate by the SEM-CL method in addition to the TEM observation showed that the dislocation component with lowest velocity is of screw type [21]. This may be related to the frequent jog formations along the screw component.

---

* In the actual situation, the operation of secondary slip system must contributes equally to both θ_1 and θ_d and hence the value $\bar{\ell}_d/\bar{\ell}_1$ must be larger than θ_1/θ_d.

Therefore, the macroscopic plastic deformation is governed mainly by screw dislocations and the PPE may be brought about by their retardation due to jog formation which is enhanced by illumination.

The increment of dislocation charge upon illumination [5] may be responsible for the enhancement. As mentioned above, the flow stress in II-VI compounds was found to have a strong correlation with the dislocation charge [20]. If the flow stress is determined mainly by the magnitude of dislocation charge, it will be most natural to attribute the increment of dislocation charge to the cause of the PPE.*

Although the dislocation charge on the screw component has not been investigated systematically, it was reported that the screw dislocations on the basal plane in CdS are not charged [28]. Jog formations are facilitated when a kink on the screw segment on the primary slip plane is stopped somehow before it glides out of the whole segment. Figure 12 schematically illustrates the sequence of the conceivable jog formation process. Let us consider the kink "K" being stopped by a point obstacle "P". If a cross slip occurs in another place on the same segment and the formed jog travels toward the kink, the jog will be blocked when it encounters the standing kink. As a result, the obstacle "P", even if it itself is not a strong barrier against the screw dislocation, produces a jog which acts as an obstacle that is strong enough to immobilize the dislocation.

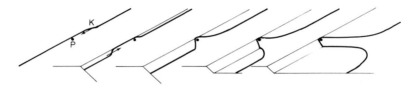

Fig. 12. Process of jog formation which retards the motion of screw segments on the primary slip system. The kink "K" stopped by a point defect "P" results in blocking of a jog formed on the other part of the screw segment.

The charge of the kink, which is of edge character, will be increased by illumination, although that of screw will not. If the obstacle P is charged, the electrostatic interaction between the kink and the P will be increased and consequently a jog formation will be enhanced, leading to the positive PPE. It was shown that the dislocations on the prismatic slip plane in the crystals of the wurtzite structure possess much less charge than those on the basal plane [28]. If the above argument is applied to the deformation of crystals oriented to inhibit the basal slip, the positive PPE will not arise. This explains the actually observed absence of the positive PPE in such orientation (Fig. 3).

So far, the discussion has been restricted to the positive PPE. Recently, however, it was found that the negative PPE is also observed in some crystals. The crystals that exhibit the negative PPE contain either specific centers [11, 29] or having been subject to a heat treatment that probably introduces some

---

* It is, at the same time, possible to interpret this dislocation charge increase as a result of the PPE rather than its primary cause, because the dislocation charge increases with the dislocation velocity [27], which should become higher when the stress increases as a consequence of the PPE. In the present paper, we tentatively adopt the viewpoint that the primary cause of the positive PPE is the increase of the dislocation charge.

point defects [12]. The negative PPE is observed in the conditions in which
the positive component of PPE is suppressed by some reasons; for example,
deformation in non-basal slip orientations in the wurtzite crystals such as CdS
and CdSe [13], in the later stage of plastic flow in CdTe [12] and in polycrystalline CdTe in which $\bar{\ell}$ is determined by the grain size [29]. The spectrum of
light responsible for the negative PPE has a peak near the fundamental absorption edge but its shape is clearly different from that of the positive PPE.
Furthermore, as shown in Fig. 13, the temperature dependence and strain dependence of the PPE indicate that the negative PPE operates also in the basal slip
orientation in an additive manner to the positive component. Thus, the negative
PPE is considered to be competitive with the positive one and seems to be independent in the mechanism of the latter.

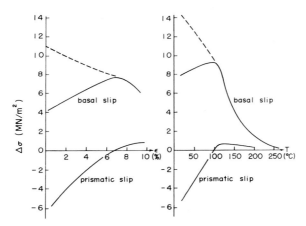

Fig. 13. Strain (left) and temperature (right) dependences of the PPE in CdS
[13] deformed in two different orientations. The broken lines were tentatively
drawn by the present authors as the positive component. Note that the positive
PPE is quite dependent on the deformation orientation but negative component is
not.

A plausible explanation of the negative PPE is a weakening of the interaction
between dislocations and point defects due to a change in the charge state of
the latter. This is in contrast to the case of the positive PPE where it is
dislocation that change their charge state. Since the negative PPE, if it
operates, has a similar magnitude regardless of the operating slip system (Fig.
13), the origin of the dislocation-point defect interaction should not be
electrostatic as in the case of the positive PPE but presumably elastic in
nature. The effect of the weakening of the elastic interaction can be incorporated either into the increase of $\bar{\ell}$ in a similar manner as the positive PPE
or into the decrease of τ in eq. (4). In the latter case, the critical stress
τ_c for a source dislocation to be multiplied (so-called the Orowan stress) must
be modified to a larger value taking account of the point obstacles dispersed
in the matrix. At present, no experimental evidence is available to infer
which is the real case. Further investigations, preferably using well-charac-
terized specimens in respect of contained point defects, are necessary to
clarify the detailed microscopic mechanism of the negative PPE.

474

ACKNOWLEGMENTS

The authors would like to thank K. Suzuki and M. Ichihara of ISSP, University
of Tokyo for their technical assistance in the electron microscope observation.
They also thank Prof. H. Iwanaga of Nagasaki University for supplying CdS thin
single crystals for the TEM in-situ observation.

REFERENCES

1. C. A. Ahlquist, M. J. Carroll and P. Stroempl, J. Phys. & Chem. Solids 33,
    337 (1972).

2. Yu. A. Osip'yan and I. B. Savchenko, JETP Lett. 7, 100 (1968).

3. J. S. Nadeau, J. Appl. Phys. 35, 669 (1964).

4. L. Carlsson and C. Svensson, Solid State Comm. 7, 177 (1969).

5. Yu. A. Osip'yan and V. F. Petrenko, Sov. Phys. JETP 48, 147 (1978).

6. E. Y. Gutmanas and P. Haasen, J. de Phys. 40, Suppl. C6-169 (1979).

7. K. Maeda, K. Nakagawa and S. Takeuchi, Phys. Stat. Sol. (a) 48, 587 (1978).

8. E. Y. Gutmanas, N. Travitzky, U. Plitt and P. Haasen, Scripta Met. 13,
    239 (1979).

9. E. Y. Gutmanas and P. Haasen, Phys. Stat. Sol. (a) 63, 193 (1981).

10. L. Carlsson and C. N. Ahlquist, J. Appl. Phys. 43, 2529 (1972).

11. Yu. A. Osip'yan and M. Sh. Shikhsaidov, Sov. Phys. Solid State 15, 2475
    (1974).

12. E. Y. Gutmanas, N. Travitzky and P. Haasen, Phys. Stat. Sol. (a) 51, 435
    (1979).

13. M. Sh. Shikhsaidov, Sov. Phys. Solid State 23, 968 (1981).

14. Yu. A. Osip'yan and V. F. Petrenko, Sov. Phys. JETP 36, 916 (1973).

15. L. Carlsson, J. Appl. Phys. 42, 676 (1971).

16. Yu. A. Osip'yan and I. B. Savchenko, Sov. Phys. Solid State 14, 1723 (1973).

17. Yu. A. Osip'yan, V. F. Petrenko and I. B. Savchenko, JETP Lett. 13, 442
    (1971).

18. N. V. Klassen, Yu. A. Osip'yan and M. Sh. Shikhsaidov, Sov. Phys. Solid
    State 18, 922 (1976).

19. Yu. A. Osip'yan and V. F. Petrenko, Sov. Phys. JETP 42, 695 (1976).

20. V. F. Petrenko and R. W. Whitworth, Phil. Mag. A 41, 681 (1980).

21. K. Maeda, K. Nakagawa, S. Takeuchi and K. Sakamoto, J. Mater. Sci. 16, 927
    (1981).

22. K. Nakagawa, K. Maeda and S. Takeuchi, J. Phys. Soc. Japan 49, 1909 (1980).

23. K. Nakagawa, K. Maeda and S. Takeuchi, J. Phys. Soc. Japan 50, 3040 (1981).

24. K. Maeda and K. Sakamoto, J. Phys. Soc. Japan 42, 1914 (1977).

25. K. Maeda and S. Takeuchi, Japan. J. Appl. Phys. 20, L165 (1981).

26. K. Nakagawa, K. Maeda and S. Takeuchi, J. Phys. Soc. Japan 48, 2173 (1980).

27. L. G. Kirichenko, V. F. Petrenko and G. V. Uimin, Sov. Phys. JETP 42, 389 (1978).

28. A. V. Zaretskii, Yu. A. Osip'yan, V. F. Petrenko and G. K. Strukova, Sov. Phys. Solid State 19, 240 (1977).

29. F. Buch and C. N. Ahlquist, Mater. Sci. & Eng. 13, 194 (1974).

DEGRADATION BEHAVIOR OF OPTOELECTRONIC DEVICES

JUNJI MATSUI
Fundamental Research Laboratories, Nippon Electric Co., Ltd.,
Miyazaki Yonchome, Miyamae-ku, Kawasaki, 213 Japan

ABSTRACT

Various degradation modes and features of crystalline defects associated with the degradation observed both in GaAlAs/GaAs and InGaAsP/InP double heterostructure light emitting sources (LED's and Lasers) are reviewed, noticing similarities and differencies between those two material systems. Non-existence of rapid degradation in the quaternary caused by DLD formation (dislocation motion) will be discussed in terms of atomic rearrangements around the dislocation core.

INTRODUCTION

A number of reports [1] have appeared on work involving the GaAlAs/GaAs and InGaAsP/InP double heterostructure (DH) light emitting diodes (LED's) or laser diodes (LD's) for applications in about 0.8 μm and 1.3-1.5 μm range optical communications, respectively. In an early stage of the CW GaAlAs/GaAs DH LD development, a great variation in life time was observed, i.e., some LD's degraded within an hour and others lived for more than 10^{3-4} hours without apparent deterioration of optical output power. The performance was also found to be much influenced by generation of various kinds of grown-in and fabrication-induced crystal defects. In order to realize highly reliable devices, much effort on elimination of the crystal defects and also on improvement of device fabrication technologies has been paid and in these days estimated life time of the GaAlAs/GaAs DH LD's in excess of 10^5 hours has been obtained.

As for the InGaAsP/InP DH LED's and LD's, they have been much developed under the situation that wavelength of light absorption minimum of fibers moves to the region longer than 1.0 μm. In addition, they usually do not degrade so rapidly as compared with GaAlAs/GaAs material systems as far as they are not operated with exclusively high CW current density. This is much favorable in practice for optical communication.

Utilizing each of the above two material systems, carrier-injection LD's and LED's degrade more or less only during the current passage. Various observations have been made to investigate mechanisms governing the degradation in these materials.

As is well known in the case of LD, the degradation which is accompanied by an increase in threshold current density for CW operation can be separated into three categories; (1) rapid degradation which occurs after relatively short operating time, being usually related to the generation of so-called dark-line-defects (DLD's), (2) slow degradation which is gradually revealed during long term operation, and (3) other degradation modes which include degradation due to mirror facet damage or erosion and catastrophic failure occurring under high optical power density operation. In the case of LED's, the degradation also occurs with a rapid (category (1)) or gradual (category (2)) decrease of light output power.

In this paper, experimental behavior which characterizes above categories is described taking account of similarities and differences between those two

478

materials. Although many methods for evaluation of the materials are availabe, electron-beam-induced current (EBIC) scan using an SEM and transmission electron microscopy (TEM) are mainly employed for the observation of degraded regions, the former technique being conventional for visualizing a full view of degraded regions and the latter technique for revealing crystalline defect structures associated with the degradation.

The difference in degradation rate of LD's and LED's between those two material systems will be discussed in terms of atomic radii and their arrangements around a dislocation core.

DLD DEGRADATION

It appeared in an early time that the GaAlAs/GaAs DH LD's degraded very rapidly if the material having dislocations was used for device fabrication. DLD's associated with the degradation in an LD were observed in a photoluminescence pattern and also in an EBIC image using an SEM. By means of TEM, the DLD's were also found to propagate in the form of a dislocation network. The start of dislocation movement usually delayed after the start of

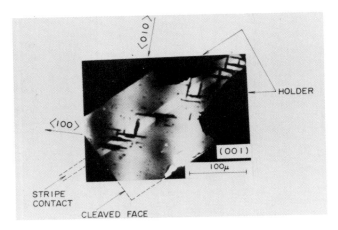

Fig. 1(a)   EBIC image of ⟨100⟩ DLD's observed in a rapidly degraded LD. [T]

Fig. 1(b)   TEM micrograph showing a ⟨100⟩ elongating dislocation dipole developing from a threading dislocation connected to that in the substrate. [T]

TABLE I
Feature of DLD degradation of light emitting devices.

Material	Half life	Appearance	Defect structure	Origin	Mechanism
GaAlAs/ GaAs	~10^{0-1}H	DSD	helix	threading dislocation	climb
		<100> DLD	dipole ($b=\frac{a}{2}$<101>)	threading dislocation (from substrate)	climb
			dipole ($b=\frac{a}{2}$<101>)	threading dislocation (from inside epilayer)	climb
			dipole ($b=\frac{a}{2}$<101>)	misfit dislocation	climb
		<110> DLD (along stripe)	dipole ($b=\frac{a}{2}$<110>)	threading dislocation (from inside epilayer)	climb
		<110> DLD	misfit dislocation or half loop	lattice mismatch or fab-induced stress	glide
	~10^3H	<110> DLD (clustered)	dipole ($b=a$<001>) (branch-like)	clustered dislocation	climb
InP/ InGaAsP	~10^{3-4}H	DSD	precipitate	unknown	
		<100> DLD	dipole ($b=\frac{a}{2}$<101>)	misfit dislocation	climb
		<100> DLD	dipole	precipitate(?)	climb
	~10^{2-3}H	<110> DLD (cross-hatched)	misfit dislocation or half loop	lattice mismatch or fab-induced stress	glide
		<110> DLD (along stripe)	half loop	fab-induced stress	glide
		<110> DLD (normal to stripe)	stacking fault	misfit dislocation	glide

operation. Once started, however, the dislocation moved fast mostly in a ⟨100⟩ direction projected onto (001).

It is well known that the degradation occurrence is not dependent upon whether carriers are induced by current passage or optical excitation, and also that, only when electron-hole pairs non-radiatively recombine resulting in energy release into the lattice, the dislocations are enhanced to move or develop in the active layer (recombination-enhancement effects). Some features of the degradation due to formation of dark spot defects (referring as DSD's) or DLD's usually observed both in the GaAlAs/GaAs and InGaAsP/InP devices are summarized in Table I.

In the case of GaAlAs, most ⟨100⟩ DLD's initiate from DSD's which have been identified to be dislocations threading through the active layer [2] and sometimes to be stacking faults [3]. In Figs.* 1(a) and (b) are shown ⟨100⟩ DLD patterns appearing in an EBIC image [4] and a TEM image, respectively, taken with a degraded LD. The threading dislocations are grown-in (1) being connected to those in substrate or (2) being generated inside the first epitaxial layer. DLD features associated with the latter type of dislocations are demonstrated in Figs. 2(a) and (b) [5]. Petroff et al. [2] and Hutchinson et al. [6] made intensive TEM observations of the dislocation network which expanded from the threading dislocations. They reported that the network consisting of many dislocation dipoles grew by a dislocation climb process on an inclined {110} planes.

Another feature that is frequently observed with rapid degradation is a development of the ⟨110⟩ DLD's caused by generation of the dislocations gliding from the surface down to the vicinity of the interface between the active layer and the upper cladding layer. This type of dislocation is easily induced under the existence of stress due to lattice mismatch, poor metal alloying or bonding, scratch and so on. Such ⟨110⟩ DLD's in GaAlAs/GaAs LD's are shown in

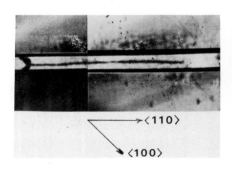

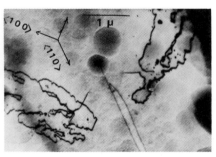

Fig. 2(a) EBIC image of ⟨100⟩ and ⟨110⟩ DLD's developing from a dislocation cluster generated inside the first epitaxial layer. [T]

Fig. 2(b) TEM micrograph showing a dislocation cluster and ⟨100⟩ and 110 dipoles developing from the dislocation cluster. [T]

---

*A symbol [T] which is placed after each figure caption denotes that the photograph is taken for a light emitting source made of GaAlAs/GaAs material system, and a symbol [Q] for that made of InGaAsP/InP material system.

Fig. 3(a). It is clearly seen that well-known ⟨100⟩ DLD's also develop from the ⟨110⟩ DLD's. Figure 3(b) shows a microscopic view obtained by TEM, revealing ⟨100⟩ dipoles expanding from a ⟨110⟩ dislocation.

Even after living more than 10^3 hours, some of GaAlAs/GaAs LD's often degraded quickly with generation of many ⟨110⟩ DLD's being distributed over the whole carrier-injected area, as shown in Fig. 4(a). TEM observation reveals dislocation dipoles expanding in the two ⟨110⟩ directions parallel to the (001) specimen surface, like tree branches, and that they are located at the heterointerface. Figure 4(b) reproduces such dipoles of which Burgers vector is, in appearance, a[001] normal to the surface [7].

Fig. 3(a) EBIC image of ⟨110⟩ DLD's expanding in ⟨110⟩ normal to the stripe direction and ⟨100⟩ DLD's developing from these ⟨110⟩ DLD's. [T]

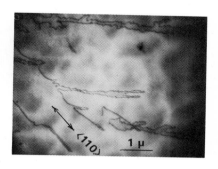

Fig. 3(b) TEM micrograph showing ⟨100⟩ dipoles developing from ⟨110⟩ dislocations lying parallel to the interface. [T]

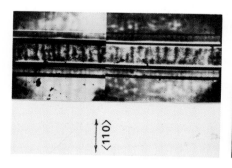

Fig. 4(a) EBIC image of ⟨110⟩ DLD's developing over the whole area of the stripe region. [T]

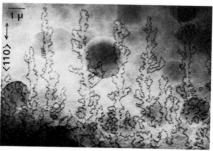

Fig. 4(b) TEM micrograph showing branch-like dislocation dipoles developing in two ⟨110⟩ directions parallel to the surface. [T]

With InGaAsP/InP devices, in contrast, the ⟨100⟩ DLD generation occurs much more slowly as compared with the GaAlAs/GaAs devices. At the start of degradation in InGaAsP, in addition, DSD's or DLD's are not clearly recognized, the situation being quite different from that in GaAs. The DLD's sometimes observed in InGaAsP/InP material system after a considerably long (10^3 hours) term of aging are mostly ⟨110⟩ glide dislocations expanding parallel to the stripe direction, not in the active layer but in the vicinity of the interface between the active layer and the upper cladding layer. These dislocaions result from the existence of lattice misfit or stress induced during fabrication process, e.g., selective diffusion or metal contact process. An example of the ⟨110⟩ DLD's due to fabrication-induced stresses are shown in Figs. 5(a) and (b) which are observed in an LD [8]. This type of dislocations sometimes can be a source of ⟨100⟩ DLD's , as clearly seen in Fig. 5(b), the dislocation configuration being quite similar to that in the GaAlAs/GaAs devices (Fig. 3(b)). However, since the ⟨100⟩ DLD's grow very slowly in the InGaAsP, they are not considered to belong to the degradation category (1). Figure 6 shows another type of dislocation half loops observed in an LED, which glide deeper into the thickness on an inclined {111} plane. They probably are the same type as was observed in an optically excited InGaAsP LD by Mahajan et al. [9]. Although the DLD degradation occurs both in the GaAlAs/GaAs and the InGaAsP/InP systems, it should be emphasized that, in the GaAlAs/GaAs system,

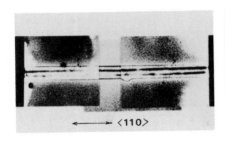

Fig. 5(a) EBIC image of ⟨110⟩ DLD parallel to the stripe direction. [Q]

Fig. 5(b) TEM micrograph showing ⟨110⟩ dislocation generated by slip and ⟨100⟩ dipoles developing from the ⟨110⟩ dislocation. [Q]

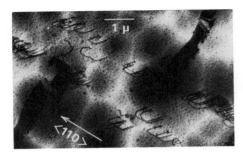

Fig. 6 TEM micrograph showing dislocation half loops gliding deeper into thickness. [Q]

climb process or probably combination of climb and glide processes gives rise to the occurrence of the rapid degradation, whereas in the InGaAsP/InP system only glide process due to excessive lattice mismatch or fabrication-induced stresses are responsible for the "fairly" rapid (10^{2-3} hours) degradation. The fact that the quaternary InGaAsP withstands to undergo $\langle 100 \rangle$ DLD degradation will be discussed later.

GRADUAL DEGRADATION

By employing high-quality substrates with low dislocation density (especially for the GaAlAs/GaAs) or by suppressing the generation of lattice misfit or fabrication-induced defects (especially for the InGaAsP/InP), high performance light emitting devices have a long-term operating life with relatively slow degradation. There has been a common feature that a uniform increase in darkness in the carrier-injected area is recognized as well as a gradual decrease in the quantum efficiency. Table II summarizes features of the gradual degradation which appeared in GaAlAs/GaAs and InGaAsP/InP material systems.

A dark area (referring as DA) containing small dislocation loops of interstitial character has often been seen in any devices degraded by the accelerated aging at an elevated temperature, as shown in Fig. 7. The fact suggests that the degradation is strongly dependent upon migration and aggregation of point defects [10]. Although it has been found by DLTS [11] measurement that hole traps with an activation energy of 0.54 eV for LD's [12] or 0.43 eV for LED's [13] increase to some extent with the degradation, they seem to be short of explaining the amount of degradation and, therefore, some other effects on the gradual degradation are expected. Wakefield [14] showed an interesting result that stresses arising at oxide film edge affected migration of point defects which may be responsible for the appearance of DA in GaAlAs/GaAs LD's.

As previously described, the InGaAsP/InP devices for 1.0 μm wavelength range do not show so fast degradation as in the GaAlAs/GaAs (~1 hour), but they do

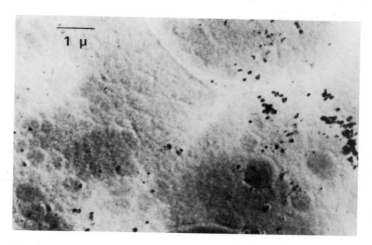

1 μ

Fig. 7 TEM micrograph of an LD degraded by accelerated aging, showing a lot of small loops. [T]

TABLE II
Features of gradual degradation of light emitting devices.

Material	Half life	Appearance	Defect structure	Origin	Mechanism
GaAlAs/ GaAs	$\sim 10^{5-6}$ H	DSD (DA)[a)	dislocation loop stacking fault	point defect	climb
	$\sim 10^{4}$ H	DSD (DA)	dislocation loop	point defect migration by strain enhancement	climb
	$\sim 10^{3-4}$ H	DA (in the vicinity of mirror)	mirror erosion and dislocation loop	oxygen point defect	oxidation climb
InP/ InGaAsP	$> 10^{6}$ H	DSD (DA)	precipitate	impurity atom	
		DSD (DA)	multiple dislocation	precipitate	climb

a) DA = Dark area

show a slow degradation. From a fabrication point of view, however, a slow degradation due to reaction between semiconductor materials and electrode metal or liquid metal used for heat-sink adhesion is one of the severe problems. Ueda et al. [15] reported the generation of precipitate-like defects in the InGaAsP/InP LED's appearing as DSD's (or DA), which are considered to be generated during metal alloying. Multiple dislocation loops developing from the precipitate, or dislocation networks containing precipitates inside the network are sometimes exhibited. Some other complicated structures appearing as a rosette pattern are also observed in an InGaAsP LED, as shown in Fig. 8, the pattern being similar to those in a GaAs LD reported by Dobson et al. [16].

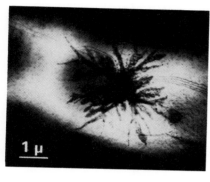

Fig. 8 TEM micrograph showing complicated structure defect appearing as a rosette pattern. [Q]

OTHER DEGRADATION MODES

Many workers have reported significant mirror erosion to occur in unprotected mirror GaAs DH LD's and they can be a source of gradual degradation for LD's run CW. Yuasa et al. [17] found, using Auger electron spectroscopy, that an oxide layer was built up on the mirror surface during CW lasing. In Fig. 9 are shown Auger electron spectroscopy (AES) profiles for Ga, As and O after the mirror is eroded. Even when operated in N_2 atmosphere or in an sealed package, the oxidation of Ga and As proceeded, giving rise to an

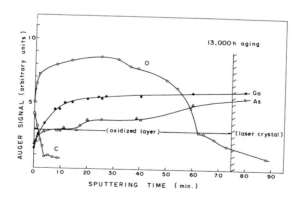

Fig. 9 AES profiles of Ga, As and O along the mirror normal. [T]

increase of surface recombination velocity. Dark area is also recognized inside the mirror as shown in Fig. 10. A number of very small dislocation loops are observed in the dark area, the feature being similar to that in Fig. 7. In contrast, InGaAsP/InP DH LD's do not suffer from the mirror oxidation.

Another serious problem with which we meet in the GaAlAs/GaAs DH LD's is a catastrophic optical damage (COD) that occurs under extremely high optical power density operation (several MW/cm^2). The COD is caused by the mirror surface damages due to local heating up and is accompanied by instantaneous $\langle 110 \rangle$ DLD propagations inwards, normal to the mirror facet [18], as shown in Fig. 11(a) and (b). They are generated via the process of a local melting, moving and quenching due to thermal runaway. In contrast to COD's in GaAlAs, those in InGaAsP have been shown to be due to localized melting at material defects and not at the mirror facet [19].

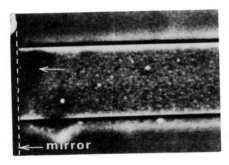

Fig. 10   EBIC image of dark area in the vicinity of the mirror. [T]

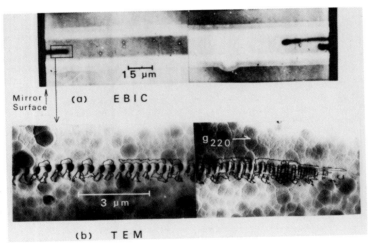

Fig. 11(a)   EBIC image of $\langle 110 \rangle$ DLD growing very fast from the mirrors due to COD generation. [T]

Fig. 11(b)   TEM micrograph showing dislocation networks generated under high power repetetive pulsed operation. [T]

DISCUSSION

From a view point of degradation problems due to dislocation network propagation, InGaAsP materials have an advantage for the practical use as compared with GaAs. It is well recognized that the degree of degradation in every mode (DLD formation, gradual degradation and mirror damage occurrence) is quite different between those two material systems. An empirical point to be emphasized is that, although dislocation climb motion is observed in the both material systems, their moving speed in the InGaAsP is orders of magnitude lower than that in the GaAs.

There exists a general understanding that a site of non-radiative recombination is a dislocation core (a dangling bond) or a point defect such as vacancy or interstitial or an antisite defect existing in the active region of the device. However, the density of dislocations or of grown-in point defects does not seem to be so largely different as the degradation rate between the two materials. Therefore, some other difference of the easiness in dislocation motion must be taken into account.

a) Point defect migration which is necessary for the dislocation climb motion seems to occur much less frequently in the InGaAsP than in the GaAs. This may be partly because that, for example, vacancy migration which is necessarily accompanied by site changes of host atoms will be more restrained in such a crystal that host atom radii are considerably different from each other as in the InGaAsP.

b) It will be reasonable to say that the rate of non-radiative recombination at the dislocation core or at the point defects state in the InGaAsP is much smaller than that in GaAs [10]. Besides, as pointed out by Ueda et al. [15], trap levels associated with the dislocation or with the surface state in InGaAsP cannot be expected to be deep. The fact that mirror erosion does not easily occur in the InGaAsP seems to indicate that surface state energy levels are not deep [20]. Energy release into lattice, i.e., lattice vibration coming from non-radiative recombination of excited carriers is localized at the lattice points with a deep trap level. Such a local lattice vibration will result in heating up to a temperature high enough for the atoms to move out. Such an energy dissipation process will not be prominent in the quaternary.

c) In addition to the above-described situations, it is speculated that the difficulty of dislocation motion in InGaAsP is attributed to the relatively small strain energy around the dislocation core, as will be explained more in detail as follows.

(i) Ga and As atoms have the same value of rationalized atomic radii [21] r_{Ga} $r_{As} \approx 1.225$ A, while those of In and P are $r_{In} \approx 1.405$ A and $r_P \approx 1.128$ A, respectively. Each of four kinds of crystals, GaAs, GaP, InAs and InP has a lattice parameter nearly equal to be $a_O \sim 4(r_A+r_B)/\sqrt{3}$, where r_A is an rationalized atomic radius of the III-group atom A (Ga or In) and r_B that of V-group atom B (P or As), for example, $4(r_{Ga}+r_{As}) \sim 9.80$ A $\sim \sqrt{3}$ a_O ($a_O=5.653$ A) for GaAs. This means that most stable atomic construction without strain for those zincblende type crystals is established in such a fashion that, as shown in Fig. 12, the A and B atoms stack sequentially as if they are hard spheres with the radii r_A amd r_B, respectively.

(ii) The quaternary crystal $In_{1-x}Ga_xAs_yP_{1-y}$ is considered to consist of four types of A-B pair bonds, i.e., Ga-As, Ga-P, In-As and In-P with their compositions of xy, $x(1-y)$, $(1-x)y$ and $(1-x)(1-y)$, respectively. Although it has not been reported how the real quaternary crystal is constructed atomically by the above four types of bonds, it is unlikely that every A-B pair bond in the quaternary lattice-matched to InP holds stable having the fixed value, $l_{AB}=r_{In} + r_P \sim 2.54$ A, for the distance between the centers of A and B atoms.

(iii) Generally speaking, when the crystal is dislocated, the lattice parameter around an 60° type dislocation core is smaller than a_O (=lattice parameter for an undistorted crystal) at one half side above the glide plane

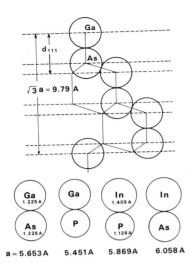

Fig. 12 ⟨110⟩ projection of atomic arrangements in GaAs crystal and rationalized atomic radii of Ga, In, P and As [21].

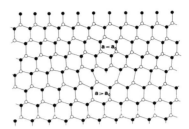

Fig. 13 ⟨110⟩ projection of atomic arrangements around a dislocation core showing smaller lattice parameter above the glide plane and larger one under the glide plane. Note that the glide plane is slightly curved in the vicinity of the dislocation core.

where an extra plane is inserted (i.e., the lattice is contracted) and is larger than a_o at the other half side under the glide plane (i.e., the lattice is dilated), as schematically shown in Fig. 13, thus, bringing about strain energy around the dislocation core. This is true not only for the lattice parameter in the direction parallel to the Burgers vector b, but also for that in the direction normal to b, resulting in a slight curvature of the glide plane in the close vicinity of the dislocation core. As is well known, the curvature of glide plane is a cause of faint dislocation contrast appearing in an X-ray or electron diffraction image, due to g(b x u) = 0 even if g·b = 0 (g = a diffraction vector in the reciprocal lattice space and u = unit vector parallel to the dislocation line).

(iv) In the quaternary, the above-described lattice strain or instability around the dislocation core would be minimized if the A-B pair bonds with a small lattice parameter, such as Ga-P, increases locally above the glide plane and the A-B pair bonds with a larger lattice parameter, such as In-As, increases under the glide plane. Such an atomic rearrangements may be possible during an epitaxial growth or cooling immediately after the growth, being similar to Kirkby's model for the impurity atoms around a dislocation core [22]. The situation will not be the same for GaAs and GaAlAs, since $r_{Ga} \simeq r_{As}$ and $r_{Al} = 1.230$ A which is nearly the same as r_{Ga} or r_{As}.

(v) Thus, the dislocations in the quaternary will resist to be moved from such stable (less strained) region. Furthermore, in order for jogs to move along the dislocation line necessarily for the dislocation to climb, rearrangements of a large amount of the A-B pair bonds will be needed, which might be difficult in the quaternary where atomic radii are somewhat different from each other. Although the above-mentioned atomic rearrangements around the dislocation core in a mixed III-V semiconductor material seem to be speculative at the present stage, it will be safely said that among more than two different

A-B pair bonds, any one can be selected to stack so that the strain energy may be locally minimized around the dislocation core especially during the crystal growth.

Above description does not refer to binding energies and acceptability of distortion from tetrahedron for each A-B pair bond. It is likely, however, that addition of In atoms into GaAs or GaP increases ionicity and that bond angles of tetrahedron become more changeable [21]. A local concentration modulation of host atoms at the dislocations has been observed in the quaternary by Seki et al. [23], resembling a spinodal phase separation. However, further microscopic investigation on the relationship between the content variation and the dislocation structure is required in the mixed III-V semiconductors such as the quaternary InGaAsP.

CONCLUSION

In the present paper, comparisons of degradation behavior between GaAlAs/GaAs and InGaAsP/InP DH light emitting sources have been made in terms of structural defects. Following items associated with the degradation are described.
1) Although DLD formation occurs in the both material systems, its growth speed in the InGaAsP is orders of magnitude slower than in the GaAs especially for the DLD's formed via climb process.
2) <100> DLD generation from <110> glide dislocation is common for the both materials. However, <100> DLD's in the InGaAsP are not responsible for rapid degradation.
3) Gradual degradation proceeds in the InGaAsP as well as in the GaAs, though the features of structural defects associated with the gradual degradation are different from each other, i.e., dislocation micro-loops are major defects in the GaAs, while precipitates with multiple dislocation loops are often observed in the InGaAsP.
4) Other degradation modes such as mirror facet erosion due to surface oxidation and catastrophic optical damage occurring under extremely high optical power operation are not serious in the InGaAsP.
5) Concerning the slow DLD formation in the InGaAsP, a model in terms of atomic rearrangements around a dislocation core in that material is presented; preferable incorporation of atoms with a small atomic radius (such as P) into a compressive region and those with a large atomic radius (such as In) into a dilated region can lower strain energy around the dislocation core.

ACKNOWLEDGEMENTS

The author would like to express his thanks to Drs. T. Kamejima, K. Ishida, H. Yonezu, M. Ogawa, Y. Matsumoto, K. Endo, T. Yuasa, and T. Uji for their contribution to the studies presented in this review paper, and also to all of workers having been engaged in the device fabrication in Fundamental Res. Labs. and Optoelectronic Res. Labs., NEC. He also wish to acknowledge Dr. I. Hayashi[*] for valuable discussions. Drs. F. Saito and T. Kawamura are appreciated for their encouragements.

REFERENCES

1. For referring a lot of papers on lasers and LED's published until 1977, see, for example, H. Kressel and J.K. Butler, Semiconductor Lasers and

---

[*] Present address; Optoelectronic Joint Res. Lab., Nakahara-ku, Kawasaki, 211 Japan.

490

   Heterojunction LEDs (Academic Press, New York, 1979) or H.C. Casey, Jr. and M.B. Panish, _Heterostructure Lasers_ (Academic Press, New York, 1978).

2. P. Petroff and R.L. Hartman, Appl. Phys. Lett. _23_, 469 (1973).

3. R. Ito et al., IEEE J. Quantum Electronics QE-11, 551 (1975).

4. H. Yonezu et al., Appl. Phys. Lett. _24_, 18 (1974).

5. K. Ishida and T. Kamejima, J. Electronic Materials _8_, 57 (1979).

6. P.W. Hutchinson et. al., Appl. Phys. Lett. _26_, 250 (1975).

7. P.W. Hutchinson and P.S. Dobson, Phil. Mag. _32_, 745 (1975).

8. K. Ishida et al., Appl. Phys. Lett. _40_, 16 (1982).

9. S. Mahajan et al., Appl. Phys. Lett. _34_, 717 (1979).

10. I. Hayashi, _Proc. 15th Int'l. Conf. Physics of semiconductors_, Kyoto, 1980.

11. D.V. Lang, J. Appl. Phys. _45_, 3023 (1974).

12. T. Uji et al., Appl. Phys. Lett. _36_, 655 (1980).

13. K. Kondo et al., Jpn. J. Appl. Phys. _19 Suppl._ 437 (1979).

14. B. Wakefield, J. Appl. Phys. _50_, 7914 (1979).

15. O. Ueda et al., J. Appl. Phys. _53_, 2991 (1982).

16. P.S. Dobson et al., _Proc. Int'l. Conf. of GaAs and Related Compounds_, pp. 419, 1977.

17. T. Yuasa et al., Appl. Phys. Lett. _32_, 119 (1978).

18. T. Kamejima and H. Yonezu, Jpn. J. Appl. Phys. _19 Suppl._ 425 (1979).

19. H. Temkin et al., Appl. Phys. Lett. _40_, 562 (1982).

20. W.E. Spicer et al., J. Vac. Sci. Technol. _16_, 1422 (1979).

21. J.C. Philips, _Bonds and Bands in Semiconductors_ (Academic Press, New York, 1973).

22. P.A. Kirkby, IEEE J. Quantum Electronics QE-11, 562 (1975).

23. M. Seki et al., Appl. Phys. Lett. _40_, 115 (1982).

VI
SPECIAL PROCESSES AND MATERIALS

LASER ANNEALING OF ION IMPLANTED SEMICONDUCTORS*

J. NARAYAN

Solid State Division, Oak Ridge National Laboratory, Oak Ridge, TN 37830

ABSTRACT

Photon energy from laser beams can be used to rapidly heat and melt localized regions of semiconductors with a high degree of spatial and temporal selectivity. Pulsed lasers have been successfully used to anneal displacement damage and to remove other defects. However, the number density of trapped defects increases with velocity of solidification and finally thin layers turn directly amorphous after laser-melt quenching. Annealing characteristics are found to be a strong function of ion implantation variables, which determine optical properties of materials. By both solid- and liquid-phase crystallization, supersaturated solid solutions can be formed. Residual defects in SPE grown layers primarily consist of dislocation loops. Device applications utilizing these transient thermal processing techniques are reviewed briefly.

INTRODUCTION

Laser-solid interactions involve a rapid transfer (in $< 10^{-13}$ s) of photon energy from the laser beams into the electronic system of the material. The energy contained in the electronic system is transferred to the lattice in less than one nanosecond. This energy is then subsequently utilized to heat and melt thin (< 1 μm) layers of semiconductors and metals [1,2]. Some authors, notably Khaibullin et al. [3] and Van Vechten et al. [4], have attempted to explain laser annealing of displacement damage on the basis of a plasma model. According to the plasma model, the energy transfer from the electronic system to the lattice is sufficiently delayed (> 200 ns) through the screening of phonon emission, and the high concentration of electrons and holes (plasma) produced during pulse laser irradiation may cause annealing via enhanced diffusion of point defects, and glide and climb of dislocations. It is speculated that at a still higher plasma density ($\sim 8 \times 10^{21}$ cm^{-3}) the electrons excited into antibonding states lead to a second-order phase transition resulting in softening of the lattice. The crystal is then in a fluid-like state with energy retained primarily in the electronic system. This energy would then be distributed over a much greater depth as carrier diffusion becomes important. As the plasma density declines, the material passes back through the phase transition leading to crystallization. Recently Lo and Compaan [5] claimed to have measured the lattice temperature of silicon during pulsed laser irradiation by Raman scattering within 10 to 15 ns of a dye laser pulse ($\lambda = 0.485$ μm, $\tau = 9$ ns) in the energy range 0.7 to 1.1 Jcm^{-2}. They reported a temperature rise of $\sim 300°$C for a 1.0 Jcm^{-2} laser pulse, thus lending support to the plasma model. However, other "insitu" measurements such as time-resolved reflectivity [6], transmission [6], and

*Research sponsored by the Division of Materials Sciences, U. S. Department of Energy under contract W-7405-eng-26 with Union Carbide Corporation.

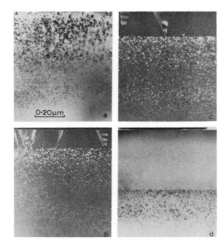

Fig. 1. Cross-section TEM
micrographs showing annealing of
dislocation loops in boron implanted
(^{11}B$^+$, 25 keV, 1.7 x 10^{15} cm^{-2}; 50
keV, 2.5 x 10^{15} cm^{-2}; 100 keV, 3.5 x
10^{15} cm^{-2}; 200 keV, 4.65 x 10^{15} cm^{-2},
and laser (λ=0.308 µm) annealed
(100) silicon: (a) 1.0 Jcm^{-2} (70 ns)
(b) 1.0 Jcm^{-2} (25 ns), (c) 1.5 Jcm^{-2}
(70 ns), (d) 2.0 Jcm^{-2} (25 ns).

strain [7] are consistent with the melting model. The laser-irradiated samples
used for Raman measurements contain microstructural modifications, which are
entirely consistent with first-order phase transition or melting induced during
pulsed laser irradiation [8]. Lower temperatures in Raman scattering are
believed to be due to the pulse to pulse variation in energy density and spatial
inhomogeniety, including some fundamental discrepancy related to equilibration
of phonons. In this review, we cover annealing of displacement damage
(amorphous as well as layers containing only dislocation loops) and associated
dopant redistribution after pulsed laser irradiations. These results provide a
convincing evidence for the melting above a certain pulse energy density. This
threshold for melting depends upon ion implantation conditions which affect
absorption of laser energy [9]. After laser-melt quenching, it is shown that
vacancies are retained in metals and interstitials in semiconductors.
Metastable supersaturated solid solutions can be formed by both solid- and
liquid-phase crystallization [10,11]. In this context, some recent work on germanium alloys is reviewed. The details of solid-phase-epitaxial growth involved
in cw laser annealing are discussed. The residual damage in SPE grown samples
consists of primarily defect clusters and dislocation loops. Some of the device
applications of pulsed and cw laser processing are very briefly reviewed [12,13].

ANNEALING OF DISLOCATION LOOPS

The specimens used in this investigation were implanted with ^{11}B$^+$ ions of
graded energy up to 200 keV and fluences, which produced a uniform distribution
of boron up to a depth of 0.65 µm followed by an exponential fall. These specimens contained dislocation loops to a depth of 0.70 µm. Figure 1 shows cross-
section micrographs revealing the annealing of loop-type damage as a function of
energy density of a XeCl-Excimer laser (λ = 0.308 µm) with 25 and 70 ns pulse
durations. These gas lasers have been found to be superior to solid state
lasers such as ruby and Nd:YAG in terms of spatial homogeniety and controls over
the shape and the duration of the pulse. The thresholds for annealing were
found to be 1.0 and 0.5 Jcm^{-2} for 70 and 25 ns laser pulses respectively.
Figure 1 shows the increase in depth of annealing with increasing pulse energy
density and decreasing pulse duration. The annealed regions were found to be

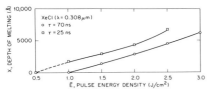

Fig. 2. Depth of melting vs. pulse
energy density.

completely free from "visible damage". A complete removal of dislocation loops
in the annealed regions along with a sharp transition between the annealed and
unannealed regions are entirely consistent with first-order phase transition
involving melting. Thermal annealing of dislocation loops requires activation
of 5.6 ± 0.5 eV [14]. Since the maximum energy available from ionization processes is approximately equal to the band gap ($\sim$1 eV), it is highly unlikely
that loop annealing will be significantly affected by the presence of the plasma
($< 4 \times 10^{20}$ cm^{-3}). Under solid state conditions the times required for
annealing even close to the melting point, are at least of the order of several
seconds which are much higher than the pulse duration. Figure 2 shows the plots
of melting depth versus energy density for two different pulse durations. These
results including the thresholds for melting are in good agreement with the
melting model (HEATING-5) calculations. These calculations were carried out
using the macroscopic diffusion equation, cast into finite difference form for
numerical solution on a closely spaced mesh of points in space and time to
describe the heat flow. The program incorporates the temperature-dependent
optical properties of material including the phase change occurring upon melting
[15].

In the following, we cover annealing of amorphous layers and compare these
results with those obtained from layers containing dislocation loops and
tangles, but having identical dopant distribution profiles. The amorphous
layers were produced by 100 keV As$^+$ implantation at a dose rate ($\leq$ 10 μamp
cm^{-2}), while the implanted layers containing only loops and tangles resulted
from ion implantation at a higher dose rate ($\geq$ 200 μamp cm^{-2}). The as-implanted
specimens [Fig. 3(a)] contained $\sim$1500 Å thick amorphous layer followed by a band
of dislocation loops. After irradiation with 0.5 Jcm^{-2}, the top $\sim$900 Å thick

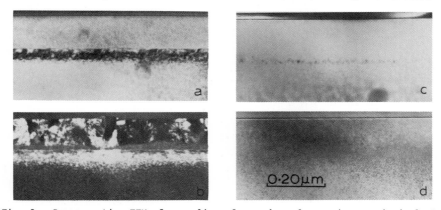

Fig. 3. Cross-section TEM of annealing of amorphous layers in arsenic implanted
silicon (100 keV, 1.0×10^{16} cm^{-2}): (a) as-implanted amorphous layer, (b) 0.5
Jcm^{-2} (λ=0.308 μm, τ=70 ns), (c) 1.0 Jcm^{-2}, (d) 1.5 Jcm^{-2}.

494

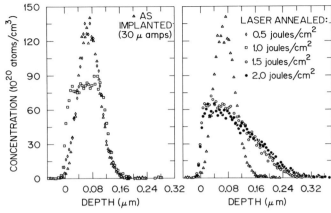

Fig. 4. Arsenic
concentration
(determined by ion
channeling) vs.
depth corresponding
to Fig. 3. (Ref. 16)

layer turned polycrystalline as shown in Fig. 3(b). This was followed by the
original band of dislocation loops. The large grains in the recrystallized
region were observed to protrude above the surface approximately 5% of its
length. This may indicate that these grains have completely recrystallized compared to the neighboring regions. After irradiation with 1.0 Jcm^{-2}, only a part
of the original band of dislocation loops remains. At 1.0 Jcm^{-2}, the melt-front
has clearly penetrated the amorphous layer and reached near the end of the
underlying dislocation band. After irradiation with 1.5 Jcm^{-2}, a "complete"
annealing of damage is observed because the melt-front penetrates the entire
damage layer so that underlying "defect-free" substrate can provide a seed for
crystal growth. Figure 4 shows results on dopant redistribution after annealing
with different pulse energy densities corresponding to Fig. 3. The as-implanted
arsenic profile is Gaussian with peak around 580 Å. After 0.5 Jcm^{-2} pulse, some
broadening is observed in the top 200-300 Å region. From this dopant profile
broadening, we obtain the solidification velocity of 10-12 ms^{-1}. At 1.0 Jcm^{-2},
the specimens show redistribution of dopants in the top 1450 Å, which is consistent with the depth of melting in Fig. 3(c). It should be noted that at this
energy density the melting of crystalline silicon substrate has occurred. As
the energy density of the pulse is further increased, more redistribution occurs
because of slower velocity of solidification and longer times for diffusion are
available. By fitting this distribution profile, the diffusion coefficient of
arsenic was determined to be 3.0 x 10^{-4} cm^2 s^{-1}, which is in good agreement with
the literature value of diffusion coefficient of arsenic in liquid silicon.

The as-implanted specimens in Fig. 5 contain arsenic distribution profiles
similar to that of Fig. 4 but they contain dislocation loops and tangles instead
of amorphous layers. In these specimens ∿700 Å thick top layer is free from
visibile damage and it is followed by 750 A thick layer of loops and tangles.
It was found that 0.5 Jcm^{-2} pulse is below the threshold for melting. This was
found to be consistent with dopant profile distribution because as-implanted
profile and the profile after 0.5 Jcm^{-2} irradiation were virtually indistin-
guishable (Fig. 6). In a specimen treated with a pulse of 1.0 Jcm^{-2}, the
depth of melting was determined to be 1100 Å (Fig. 5). Figure 6 shows
corresponding redistribution of dopants in the top 1100 Å. A lower depth of
melting in these specimens compared to those containing amorphous layers [Fig.
3(c)] is probably due to lower heat of crystallization or melting point of
amorphous silicon. The irradiation with 1.5 and 2.0 Jcm^{-2} produce a complete
removal of dislocation tangles because melt front has penetrated the defective

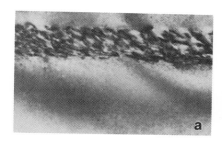

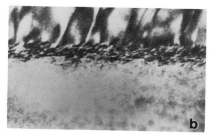

Fig. 5. Cross-section TEM of annealing dislocation loops and tangles in arsenic implanted specimen (100 keV, 1.0×10^{16} cm^{-2}) (a) as-implanted specimen, (b) 1.0 Jcm^{-2} ($\lambda = 0.308$ μm, $\tau = 70$ ns) (c) 1.5 Jcm^{-2}.

layer ($\sim$1550 Å thick). The dopant redistribution of Fig. 6 suggest depths of 2500 and 3500 Å after irradiation with 1.5 and 2.0 Jcm^{-2} pulses respectively. Figure 7 shows melting thresholds after irradiation with a pulsed ruby laser ($\lambda = 0.693$ μm, $\tau = 15 \times 10^{-9}$ s) as determined by time-resolved reflectivity measurements. The plot shows the melt-duration (high reflectivity phase corresponding to molten silicon) as a function of pulse energy density. These results show that samples similar to those shown in Fig. 3, melt at 0.2 Jcm^{-2}, while arsenic implanted samples similar to those shown in Fig. 5 have melt threshold of 0.5 Jcm^{-2}. The TEM results using ruby laser irradiation are entirely consistent with these reflectivity measurements of threshold for melting [9].

DEFECT TRAPPING

Trapping of defects after liquid-melt quenching is of particular interest. Since these defects can act as trap or recombination centers, their concentration and the mechanisms of formation should be understood for the control and elimination of defects. It should be remarked that metals expand whereas semiconductor contract by $\sim$10 % upon melting. We have observed vacancies after melt quenching of aluminum at room temperature. It is envisaged that after quenching from melt, trapped vacancies cluster to form dislocation loops, as shown in Fig. 8(a). From a detailed contrast analysis, these loops were identified to be vacancy type. The concentration of vacancies in these loops was estimated to be $\sim$200 ppm, however, this concentration represents a lower limit [17] because a large number of defects either escape to the surface or annihilate at the dislocations.

The nature and concentration of defects trapped during laser melt quenching of semiconductors have been a subject of extensive investigations recently [18]. Defect states A and B, both acting as donors, have been identified by DLTS deep in laser processed materials. The defects state A, E (0.19 eV) has been observed following both liquid- and solid-phase crystallization. The defect

496

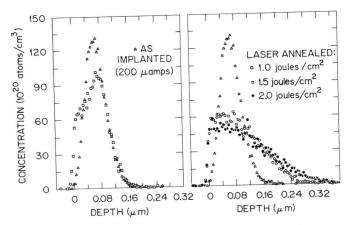

Fig. 6. Arsenic concentration profile as a function of depth corresponding [corresponding to Fig. 5(a)], to Fig. 5. crystalline silicon. (Ref. 16)

state B, E(0.33 eV) is observed only after liquid-phase processing and its depth distribution correlates with the depth of melting. However, the nature of these defects has been intriguing many researchers in this field. We have determined the nature and concentration of quenched-in defects as a function of velocity of solidification. The laser annealed layers were subsequently heated to induce clustering of isolated defects. The defect clusters in the form of dislocation loops were identified to be interstitial type. The concentration of interstitials in arsenic-doped specimens was investigated as a function of velocity of solidification. The number density of interstitials is 10^{15}-10^{16} cm^{-3} in the velocity range of 3-4 ms^{-1}, however; as the velocity increases to 10 ms^{-1} the concentration of defects reaches $\sim 10^{19}$ cm^{-3}, as shown in Fig. 8(b). At ~ 18 ms^{-1}, (100) silicon directly turns amorphous after melt quenching. In the case of (111) silicon twins start forming above ~ 8 ms^{-1}, as shown in Fig. 9. This is because there is not enough time available for 3 atom nuclei to assume normal position. The twinning mistakes lead to the formation of microtwins, which propagate during growth. It is interesting to note that at these high velocities dislocations in Fig. 9 are not able to grow along the energy minimum directions [19].

METASTABLE ALLOYING

Under rapid laser-melt quenching conditions, it is possible to greatly exceed retrograde solubility limits [11]. As the solidification velocity increases, the amount of solute which can be incorporated into substitutional sites increases. Under these conditions the distribution coefficient of solutes (ratio of solubility in the solid to that in the liquid) approaches near unity. Table I summarizes the equilibrium values of distribution coefficient k_0, k' values at 4.5 ms^{-1}, retrograde solubility limits, and observed solute concentrations at 4.5 and 6.0 ms^{-1}. In the case of bismuth the observed concentration at 6.0 ms^{-1}, exceeds retrograde solubility limits by a factor of 1200. The absolute thermodynamic limits, corresponding to the intersection of solid and liquid free energy curves in the free-energy vs composition diagram, were approximated by C_s^o/k_0 where C_s^o is the retrograde solubility limit. In the

TABLE I
Retrograde solubility limits (c_s^o) and distribution coefficients (k_o)*, observed and predicted maximum solubility limits in Si

Dopant	k_o at M.P.	c_s^o (cm^{-3})	Predicted Maximum c_s^o/k_o	LPE Observed Solubility Limits[11]		k' V=4.5 ms^{-1}
Group III						
B	0.80	6.0×10^{20}	7.5×10^{20}	V=4.5 ms^{-1}	V=6.0 ms^{-1}	
Al	0.0020	2.0×10^{19}	1.0×10^{22}			
Ga	0.0080	4.0×10^{19}	5.0×10^{21}	4.5×10^{20}	8.8×10^{20}	0.2
In	4×10^{-4}	8×10^{17}	2×10^{21}	1.5×10^{20}	2.8×10^{20}	0.15
Tl				(100)	(100)	
IV						
Si	1	--				
Ge	0.33	--				
Sn	0.016	4.5×10^{19}	2.8×10^{21}			
Pb	--	--				
V						
P	0.35	1.3×10^{21}	3.4×10^{21}			
As	0.3	1.8×10^{21}	6.0×10^{21}	6.0×10^{21}	6.0×10^{21}	1.0
Sb	0.023	6.5×10^{19}	2.8×10^{21}	1.3×10^{21}	2.5×10^{21}	0.7
Bi	7×10^{-4}	8.0×10^{17}	1.1×10^{21}	4.0×10^{20}	1.1×10^{21}	0.4

*F. Trumbore, BSTJ, p. 205, Jan.1960.

case of arsenic and bismuth observed concentrations approach the absolute thermodynamic limtis while in other cases the observed concentration is still less than the predicted maxima. The value of k' are much higher than k_o, in some cases k' values approach 1 while in others they seem to approach a constant value, not necessarily unity. The values of k' were found to be higher in <111> than <100> orientation, this leads to enhanced trapping in <111> compared to <100> orientation.

Substitutional concentrations of solutes in germanium were investigated after pulsed laser annealing and were found to far exceed retrograde solubility limits [20] similar to the case of silicon. Unlike the case in silicon, however, ion implantation in germanium exhibits a unique damaged state under irradiation, which has not been observed so far in silicon. Germanium turns amorphous, similar to silicon, above a certain dose during ion irradiation; however, as the ion dose is increased further the amorphous germanium becomes unstable. Small craters near the surface form first, which develop into large voids or craters as irradiation is continued [21]. The mechanism of laser annealing in germanium is the same as in silicon. When the melt-front exceeds the thickness of the damage layer, liquid-phase-epitaxial growth occurs with underlying substrate acting as a seed for crystal growth. The solidification velocity of the order of several meters per second is attained under these laser irradiation conditions. The substitutional concentration of solutes in rapidly solidified layers are observed to greatly exceed the retrograde solubility limits as described in Table II. In the case of Ge-Pb system, the observed concentration exceeds retrograde solubility limit as much as a factor of 770. It was interesting to note that the substitutional concentration in (111) layers is considerably higher than in (100) layers. It is envisaged that increased concentration is

TABLE II
Retrograde solubility limits (c_s^o) and distribution coefficients (k_o)*, observed and predicted solubility limits in Ge

Dopant	k_o at M.P.	c_s^o (cm^{-3})	Predicted Maximum c_s^o/k_o	LPE Observed Solubility Limits (cm^{-3})
Group III				
B	17	--	--	
Al	0.073	4.0×10^{20}	5.5×10^{21}	
Ga	0.087	5.0×10^{20}	5.7×10^{21}	
In	0.001	3.0×10^{20}	3.0×10^{21}	3.0×10^{20} (100) $>4.0\times10^{20}$ (111)
Tl	4×10^{-5}	--	--	
IV				
Si	5.5	--	--	
Ge	1	--	--	
Sn	0.020	5.0×10^{20}	2.5×10^{22}	6.0×10^{20} (100) $>7.0\times10^{20}$ (111)
Pb	1.7×10^{-4}	5.5×10^{17}	3.1×10^{21}	4.0×10^{20} (100) 4.0×10^{20} (111)
V				
P	0.080	--	--	
As	0.02	1.8×10^{20}	9.0×10^{21}	
Sb	0.0030	1.2×10^{19}	4.0×10^{21}	3.5×10^{20} (100)
Bi	4.5×10^{-5}	6.0×10^{16}**	1.3×10^{21}	5.0×10^{20} (111)

*F. Trumbore, BSTJ, p. 205, Jan. 1960.
**F. Trumbore et al., J. Electro. Chem. Soc. 109, 734 (1962).

due to enhanced trapping (larger k') in [111] compared to [100] orientation. Thermodynamic limits corresponding to maximum solute concentrations were calculated and the results are summarized in Table II. In germanium, the observed concentrations have not yet achieved the predicted thermodynamic limits. It is expected that higher velocity of solidification would lead to higher substitutional concentration.

CW LASER AND FURNACE ANNEALING

Continuous wave (CW) Ar$^+$ ion lasers have been primarily used for annealing under solid-phase-epitaxial growth conditions. However, recently higher powers have been used to induce melting and to obtain lateral and seeded crystallization of thin semiconductor layers on insulating substrates. In this paper, we concentrate on the characteristics of annealing under solid-phase crystallization. In solid-phase regime, the annealing is similar to that achieved in the furnace. Figure 10 shows the movement of crystalline-amorphous interface as a function of time at 525°C in a furnace for antimony implanted specimen. The as-implanted specimens (dose = 4.4×10^{15} cm^{-2}) contain a 1580 Å thick amorphous layer followed by a band of dislocation loops. After 5, 10, 15, and 20 minutes, the amorphous layer moves successively toward the surface. The thickness of regrown layer as a function of temperature was found to follow a growth kinetics behavior described by $V = V_o \exp (2.5 \pm 0.2 \text{ eV}/kT)$. The value of the constant V_o was found to be a function of crystal orientation, a factor of 25 greater in [100] compared to [111] orientation. During SPE growth, the structure of interface is of particular interest. The detailed atomic structure at the interface plays an important role in the crystal-growth kinetics. The dislocation loop-size distributions in the underlying dislocation band remains quite unchanged up

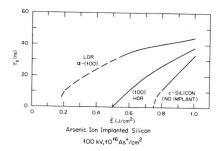

Arsenic Ion Implanted Silicon

$100 \text{ kV}, 10^{16} \text{As}^+/\text{cm}^2$

Fig. 7. Duration of high-reflectivity phase (melting) as a function of pulse energy density in arsenic implanted (100) silicon: LDR [specimens corresponding Fig. 3(a)], HDR [corresponding to Fig. 5(a)], crystalline silicon.

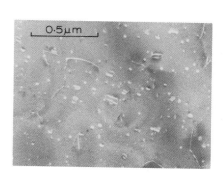

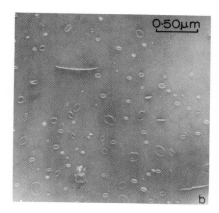

Fig. 8. Trapping of defects after laser-melt quenching: (a) vacancy loops in aluminum of irradiation with ruby laser ($\lambda=0.693$ µm, $\tau=15$ns, $\overline{E}=3.3$ Jcm^{-2}), (b) interstitial loops in arsenic-doped silicon after laser irradiation ($E=1.5$ Jcm^{-2}) and thermal annealing at 950°C for 20 min.

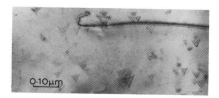

Fig. 9. Formation of twins and growth of dislocations in undefined directions after laser irradiation of (111) Si.

500

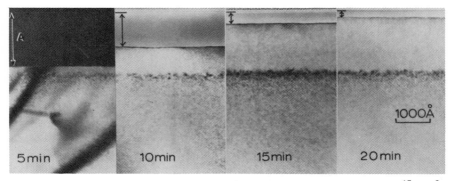

Fig. 10. Solid-phase-epitaxial growth Sb$^+$ implanted (200 keV, 4.4 x 10^{15} cm^{-2}), (100) silicon after different isothermal heat treatment at 525°C in a furnace.

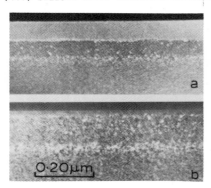

Fig. 11. SPE growth in ion implanted (125 keV, 1.0x10^{15} cm^{-2}) and CW laser annealed (100) silicon: a) bright-field, 8 watts, CW Ar$^+$ laser, 1 scan; b) weak-beam, dark field, 8 watts, CW laser, 5 scans.

to annealing temperature of around 600°C. Above this temperature, coarsening of dislocation loops occurs via loop coalescence. Figure 11 shows annealing by the process of solid-phase-epitaxial growth during CW laser annealing. A single scan of an 8 watt CW laser leads to ∿500 Å SPE growth, further scans lead to a complete growth. The dislocation loops in the underlying band and a small number density of loops in the SPE grown regions are present. In the case of indium implanted specimens, indium precipitates are also observed in the SPE grown region. The defects in CW laser annealed specimens primarily consist of defect clusters or dislocation loops. Figure 12 shows cross-section micrographs across a single laser scan. Annealing occurs by SPE growth up to the surface in the central regions. Near the edges SPE grown region tapers off till it meets the as-implanted amorphous layer. The presence of an underlying dislocation band indicates that growth has occurred by solid-phase-epitaxial growth. Figure 11 shows also the presence of defect clusters in the SPE grown layer. The dislocations are generated in the near surface region, as shown in Fig. 12(b). The plastic deformation leading to the formation of slip dislocations is caused by large thermal gradients. The surface steps act as sources of dislocations. The plastic deformation can be avoided by heating the substrate.

Fig. 12. SPE growth near the edge (top) and the central (bottom) regions of a single trace Ar^+ laser: $V = 10$ cms^{-1}, power = 11 Watts, spot size = 70 μm. Specimens: (100) Si, ^{75}As$^+$, 5.0×10^{14} cm^{-2}.

0.20μm

FORMATION OF METASTABLE ALLOYS BY SOLID-PHASE-EPITAXIAL GROWTH

Formation of supersaturated alloys is also possible by solid-phase-epitaxial growth. The observed concentrations by SPE growth below 600°C are found to greatly exceed retrograde solubility limits as much as 560 in the Si-Bi system. SPE growth at higher temperatures leads to dopant precipitation and therefore substitutional concentrations decrease at these temperatures. Table III summarizes the observed concentrations for various dopants and compares them with predicted solubility limits. The predicted limits are obtained by equating crystallization free-energy to size misfit energy. The maximum limits of solid solubility, corresponding to the intersections of free-energy versus composition curves for amorphous and crystalline silicon, were calculated for Si-Sb, Si-In, Si-Bi, Si-Ga, and Si-As systems. This approach is similar to that of Cahn et al [22] who calculated maximum concentrations during rapid solidification corresponding to liquid and solid free-energy intersections. The values of 0.1 eV/atom for crystallization enthalpy at 1014 K, and 0.2 k for the entropy at 0 K were used in the calculations. The maximum limit of solute concentration is reached when gain in free energy due to amorphous-crystalline transformation is equal to the increase in strain energy associated with crystalline silicon resulting from differences in covalent radii of dopants and the host. A comparison of the calculated maximum values and the observed concentrations is given in Table III. The observed concentrations approached the calculated maximum concentrations in the Si-Sb and Si-As systems. The higher substitutional concentrations in liquid-nitrogen temperature implants is consistent with their lower free-energy state compared to high current, room temperature implants. Different free-energy states for amorphous silicon have been reported by Fan and Anderson [23]. For other silicon-impurity systems, the formation of defects and solute segregation at the interface during SPE growth prevented the maximum achievable concentrations. For the details of impurity segregation and redistribution including the role of extended defects in these phenomena during SPE growth, the reader is referred to the original papers [10,21].

TABLE III
Comparison of observed and calculated solubility limits in Si-In, Si-Sb, Si-Bi, Si-As, and Si-Ga systems

Dopants	C_O Retrograde Maximum Solubility	C_S Limiting Conc. under SPE Growth Conditions (<600°C)	Predicted Maximum Solubility Limits
Sb	7×10^{19} cm^{-3}	1.3×10^{21} cm^{-3} 1.6×10^{21} cm$^{-3}(1)$ 2.5×10^{21} cm$^{-3}(2)$	3.0×10^{21} cm^{-3}
In	8×10^{17}	5.5×10^{19}	1.5×10^{21}
Bi	8×10^{17}	4.5×10^{20}	1.0×10^{21}
Ga	4.5×10^{19}	2.5×10^{20}	6.0×10^{21}
As	1.5×10^{21}	6.0×10^{21}	$1.5 \pm 0.5 \times 10^{22}$

(1) and (2) represent lower free energy states.
(1) Low-current implant at room temperature.
(2) Liquid nitrogen temperature implant.

DEVICE APPLICATIONS

Pulsed laser processing has been extensively applied in the fabrication of solar cells [24]. High quality p-n junctions have been produced either by ion implantation and laser annealing, or by laser-induced diffusion of dopants [25]. In the case of laser-induced diffusion, a thin layer of dopant is deposited on the surface and then the laser melting is used to alloy these layers and form p-n junctions. Recently, low energy (1 KV) dc glow discharge BF$_3$ implantation has been used to form shallow p-n junctions in the front part of the solar cells. The back surfaces of these cells were made degenerate by laser-induced diffusion of Sb. The best excimer laser processed solar cell [12,26] had the following parameters: open circuit voltage, V_{OC} = 610 mV; short circuit current, J_{sc} = 34.7 mV/cm^2; fill factor, FF = 0.79. This resulted in a record solar cell efficiency of 16.7% (AM$_1$ at 28°C). The excimer lasers have spatial homogeniety within 5% and high repetition rate to cover a large area for the manufacture of solar cells. This process is rather inexpensive and can be easily adapted to assembly-line fabrication of solar cells. The above solar cells contained double layers (Ta$_2$O$_5$ and MgF$_2$) of antireflection coatings. Pulsed laser processing is found to be particularly suitable for forming p-n junctions in polycrystalline materials because rapid quenching does not allow dopant segregation at the grain boundaries and dislocations.

CW scanned lasers are finding wide-range applications for crystallization of thin semiconductor layers on foreign substrates. It has been also realized that annealing of ion implantation damage ∿1200°C requires times of several seconds instead of minutes and hours. These basic ideas are beginning to be utilized in the fabrication of transistors [13].

503

CONCLUSIONS

In conclusion, laser annealing or melting is a strong function of ion implantation conditions, which influence optical properties of materials. Point defects can be quenched-in during liquid to solid transformation: vacancies in metals (Al); interstitials in semiconductors (Si). The number density of point defects increases with velocity of solidification. Properties of dislocations and formation of twins during laser annealing provide information on the interfacial rearrangement of atoms during crystal growth. Metastable alloys can be formed by both solid- and liquid-phase crystallization. These results shed light on the physics of nonequilibrium rapid solidification. The nature of residual damage during SPE growth is dominated by defect clusters and dislocations. Large-area devices such as solar cells have been fabricated by pulsed laser annealing using rather "inexpensive processing" with record efficiency 16.7%. CW lasers are expected to find more uses in multilayer devices.

ACKNOWLEDGMENT

The author is thankful to C. W. White, O. W. Holland, and S. J. Pennycook for useful comments on the manuscript.

REFERENCES

1. See e.g., Laser and Electron Beam Interactions with Solids, ed. by B. R. Appleton and G. K. Celler, North Holland, New York, 1982.
2. See e.g., Laser-Solid Interactions and Transient Thermal Processing of Materials, ed. by J. Narayan, W. L. Brown and R. A. Lemons, North Holland, New York, 1983, to be published.
3. I. B. Khaibullin, B. I. Shtyrkov, M. M. Zaripov, R. M. Bayazitore and M. F. Galjautdinore, Radiat. Eff. 36, 225 (1978).
4. J. A. Van Vechten, R. Tsu, F. W. Saris and D. Hoonhout, Phys. Lett. A 74, 417 (1979); J. A. Van Vechten, R. Tsu and F. W. Saris, Phys. Lett. A 74, 422 (1979).
5. H. W. Lo and A. Compaan, Phys. Rev. Lett. 44, 1604 (1980).
6. D. H. Lowndes, G. E. Jellison and R. F. Wood, Phys. Rev. B (Dec. 15, 1982).
7. B. C. Larson, C. W. White, T. S. Noggle and D. Mills, Phys. Rev. Lett. 48, 337 (1982).
8. J. Narayan, J. Fletcher, C. W. White and W. H. Christie, J. Appl. Phys. 52, 7121 (1981).
9. J. Narayan and D. H. Lowndes, to be published.
10. J. Narayan and O. W. Holland, Appl. Phys. Lett. 41, 239 (1982).
11. C. W. White, S. R. Wilson, B. R. Appleton and F. W. Young, Jr., J. Appl. Phys. 51, 739 (1980).
12. R. T. Young, J. Narayan, D. E. Rothe, G. van der Leeden and J. I. Levatter, Ref. 2.
13. T. J. Stultz, J. Sturm and J. F. Gibbons, Ref. 2.
14. W. K. Wu and J. Washburn, J. Appl. Phys. 48, 3747 (1977).
15. R. F. Wood and G. E. Giles, Phys. Rev. B 23, 2923 (1981).
16. J. Narayan and O. W. Holland, unpublished.
17. J. Narayan, p. 389, Ref. 1.
18. L. C. Kimerling and J. L. Benton, p. 385 in Laser and Electron Beam Processing of Materials, ed. by C. W. White and P. S. Peercy, Academic Press, New York, 1980.
19. J. Narayan and F. W. Young, Jr., Appl. Phys. Lett. 35, 330 (1979).
20. O. W. Holland, J. Narayan, C. W. White and B. R. Appleton, Ref. 2, to be published.
21. O. W. Holland, B. R. Appleton and J. Narayan, J. Appl. Phys. (in press).

504

22. J. W. Cahn, S. R. Coriell and W. J. Boettinger, p. 89 in Ref. 18.
23. J.C.C. Fan and C. H. Anderson, J. Appl. Phys. **52**, 4003 (1981).
24. R. T. Young, R. F. Wood, J. Narayan, C. W. White and W. H. Christie, IEEE Trans. ED-27, 807 (1980).
25. J. Narayan, R. T. Young, R. F. Wood and W. H. Christie, Appl. Phys. Lett. **33**, 338 (1978).
26. R. T. Young, G. van der Leeden, J. Narayan, W. H. Christie, R. F. Wood, D. E. Rothe and J. I. Levatter, IEEE Elect. Dev. Lett. EDL-3, 280 (1982).

MEASUREMENTS OF ION-IMPLANTATION DAMAGE IN GaP*

D. R. MYERS, P. S. PEERCY, P. L. GOURLEY
Sandia National Laboratories, Albuquerque, New Mexico 87185

ABSTRACT

We have applied stress measurements using the cantilever beam technique and Raman spectroscopy to characterize the dose dependence of damage production for He^+, C^+, or Ar^+ implants into GaP. Stress increases monotonically with dose until a species-dependent critical dose is reached. Above that dose, the material yields at an integrated lateral stress of $\sim 2 \times 10^5$ dynes/cm^2, corresponding to an expansion of $\sim 1\%$ in the implanted volume. The dose dependence of stress scales well with the volume density of ion energy deposited into atomic collisions. Raman measurements indicate that the material is still crystalline when the yield stress is reached.

INTRODUCTION

Studies of ion-implantation damage in III-V compound semiconductors are of fundamental importance not only for the basic understanding they provide, but also for their use in developing new technologies. This need is especially important in GaP, which is finding increasing application for semiconductor devices for high-temperature (>400 C) operation [1].

It is well known [2] that ion implantation of crystalline semiconductors produces an increase in volume of the implanted region. In a bulk crystal with a surface implanted layer, this swelling is constrained by the substrate, thereby producing a biaxial compressive stress in the implanted layer. We have investigated the dose dependence of this lateral stress and have examined surface crystallinity by Raman spectroscopy, as recent work in GaAs [3] has suggested that the annealing behavior of ion-implanted compound semiconductors is strongly affected by damage-induced stresses exceeding the elastic limit before the formation of a surface amorphous layer.

EXPERIMENTAL PROCEDURE

Two different n-type, (100)-oriented GaP surface layers were used in these studies. One was cut from an ingot doped with tin to a density of $\sim 3 \times 10^{17}$/cm^3, while the other consisted of a $\sim 5\mu m$ layer of nominally undoped GaP ($\sim 5 \times 10^{16}$ net donors/ cm^3) grown by liquid phase epitaxy on the above substrates. Surfaces of bulk material were polished in a bromine-methanol solution; the epitaxial layers were examined with their as-grown surfaces. We expect these measures to minimize residual stress from poor surface preparation from interfering with stresses produced by implantation. Measurements of lateral stress integrated over the implant depth were obtained by a cantilevered beam technique [4] for 25 keV He^+, 150 keV C^+, or 200 keV Ar^+. Implantations were performed

*This work performed at Sandia National Laboratories supported by the U.S. Department of Energy under contract number DE-AC04-76DP00789.

506

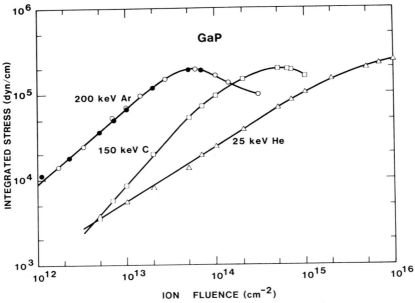

Figure 1   Dose dependence of integrated stress production for implantation into
            GaP for the three species shown.

at room temperature in a minimum channeling alignment [5] with typical beam
current densities of less than 100 nA/cm². To help interpret the stress data,
computer modelling of the implants was performed using the TRIM code [6].   In
our calculations, the displacement energies for both Ga and P atoms were taken
to be 25 eV [6], ion trajectories were stopped when ion energy dropped below 5
eV, and 10,000 histories were followed for each implant to obtain good statistics.   Raman spectra were obtained from the implanted material at room temperature using 457.9 nm-wavelength probe light in a near-backscattering geometry.
Details of the optical system are provided elsewhere [7].

STRESS MEASUREMENTS

    In Figure 1 we present the dose dependence of lateral stress integrated over
the implanted depth for three implant conditions.   Lateral stress increases
monotonically with dose until a species-dependent critical dose is reached.
For all three cases, the critical integrated stress is approximately 2×10^5
dynes/cm, although the lengthy irradiation time prevented an exact determination
for the He implant.   Above the critical dose, the implanted region yields, reducing the net stress.
    In Table I, we summarize the results of the TRIM simulations of the implant
conditions.   As the ion mass increases, more of the incident ion's energy is
deposited into atomic (collisional) processes.   Since it is this energy that
is primarily responsible for displacing substrate atoms--and thereby expanding
the lattice to produce stress in the implanted volume--we have chosen to nor-
malize the dose dependence of lateral stress to the energy density per unit

Table I: Summary of Implant Modelling and Derived Parameters

Ion	Energy keV	Deposition Depth (nm)	Critical Dose (cm^{-2})	Ave. E_a [a] at Critical Dose (eV/cm^3)	Volume Increase at Critical Dose
He	25	285	$>1 \times 10^{16}$	$>1.4 \times 10^{24}$	.014
C	150	470	5×10^{14}	3.9×10^{23}	.0072
Ar	200	235	7×10^{13}	3.7×10^{23}	.014

[a] Energy deposited into atomic collisions/unit volume

volume deposited into atomic processes. For simplicity, we have assumed that the energy is deposited uniformly over a deposition depth at which 95% of the energy lost to atomic collisions has been transferred to the target. These calculations reveal (Figure 2) that a good correlation is obtained between measured stress and the calculated volume density of energy deposited into atomic collisions. While the He$^+$ implant essentially follows the same relationship as the C$^+$ and Ar$^+$ implants for stress production at low deposited energies, it departs at higher deposited energy (higher doses). We speculate that this departure may be due either to the high He concentration (roughly two orders of magnitude greater peak atomic density at the critical dose than for C or Ar), or from a self-annealing effect resulting from a greater fraction of incident ion energy going into ionization processes for He. From the known sample dimensions, stresses, implant conditions, and elastic constants [8], an implantation-induced volume increase of ~1% at the critical dose is obtained.

RAMAN SPECTROSCOPY

Raman spectra from crystalline semiconductors are characterized by narrow discrete lines, while spectra from amorphous semiconductors are characterized by broad continuous bands [9]. For the orientation studied here, LO-phonon-related scattering dominates the first-order Raman spectra from crystalline GaP (see, e.g., [7]). With increasing dose, the LO-phonon line in the Raman spectra of undoped GaP remains narrow and stationary up to the critical dose. Thus, for this orientation, the spectral position of the LO-phonon line is insensitive to stress [10]. However, at doses near or above the critical dose, the LO-phonon line shifts to lower Raman energies and broadens on the low-energy side; at the critical dose the full width at half-maximum (FWHM) is on the order of 10 cm^{-1}. This FWHM at the critical dose is far below the FWHM of ~100 cm^{-1} observed for amorphous GaP [9], and indicates that GaP is still crystalline at doses that produce stresses which exceed the elastic limit. This result is similar to x-ray rocking curve measurements of the stress behavior of ion implanted GaAs [3], which also yields before an amorphous surface layer is formed.

508

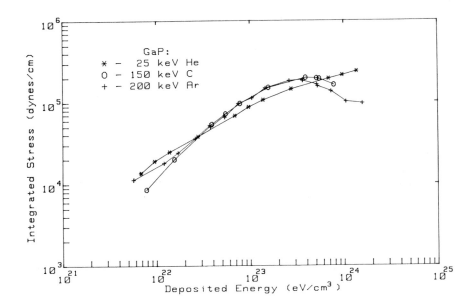

Figure 2 Dose dependence of integrated stress production for implantation into
GaP normalized to energy deposited into atomic collisions for the
three ion species shown.

CONCLUSIONS

We have found that stress in implanted GaP increases monotonically with
dose until GaP exceeds its elastic limit at an integrated stress of 2×10^5 dynes/
cm. The stress production rates scale well with the volume density of energy
deposited into atomic collisions for the different species. Raman spectroscopy
indicates that GaP is still crystalline at the implant dose that produces stress
that exceeds the elastic limit. While for (100)-GaP, Raman lines are insensitive to stress, above the critical dose, rapid broadening of Raman lines indicates significant changes in material parameters for GaP that has exceeded the
yield stress. The yield behavior of implanted GaP is thus similar to that of
implanted GaAs [3].

ACKNOWLEDGEMENTS

The authors would like to thank L. R. Dawson and R. Chavez for the epitaxial GaP, D. P. Wrobel and J. B. Snelling for technical assistance, and one
of the authors (DRM) wishes to thank D. K. Brice for his help with the TRIM
code, and G. C. Osbourn for useful discussions.

REFERENCES

1. D. R. Myers, R. M. Biefeld, T. E. Zipperian, and L. R. Dawson, Electron. Letters 18, 323 (1982).

2. R. E. Whan and G. W. Arnold, Appl. Phys. Lett. 17, 378 (1970).

3. V. S. Speriosu, B. M. Paine, M-A. Nicolet, and H. L. Glass, Appl. Phys. Lett. 40, 604 (1982).

4. E. P. EerNisse, Appl. Phys. Lett. 18, 581 (1971).

5. D. R. Myers, R. G. Wilson, and J. Comas, J. Vac. Sci. Technol. 16, 1893 (1979).

6. J. P. Biersack and L. G. Haggmark, Nucl. Instr. Methods 174, 257 (1980).

7. D. R. Myers and P. L. Gourley, J. Electrochem. Soc., to be published.

8. G. Simmons and H. Wang, Single Crystal Elastic Constants and Calculated Aggregate Properties, 2nd ed. (MIT Press, Cambridge MA 1971) p. 28, p. 186.

9. B. L. Crowder, J. E. Smith Jr., M. H. Brodsky, and M. I. Nathan in: Ion Implantation in Semiconductors, I. Ruge and J. Graul, eds. (Springer Verlag, NY 1971) pp. 255-261.

10. I. Balslev, Phys. Stat. Solidi b 61, 207 (1974).

DAMAGE AND IN SITU ANNEALING DURING ION IMPLANTATION

D. K. SADANA AND J. WASHBURN
Lawrence Berkeley Laboratory and Department of Materials Science and Mineral
Engineering, University of California, Berkeley, CA 94720

P. F. BYRNE AND N. W. CHEUNG
Department of Electrical Engineering and Computer Sciences, University of
California, Berkeley, CA 94720

ABSTRACT

Formation of amorphous (α) layers in Si during ion
implantation in the energy range 100 KeV-11 MeV and temperature range liquid nitrogen (LN)-100°C has been investigated.
Cross-sectional transmission electron microscopy (XTEM) shows
that buried amorphous layers can be created for both room
temperature (RT) and LN temperature implants, with a wider
100 percent amorphous region for the LN cooled case. The
relative narrowing of the α layer during RT implantation is
attributed to in situ annealing. Implantation to the same
fluence at temperatures above 100°C does not produce α
layers. To further investigate in situ annealing effects,
specimens already containing buried α layers were further
irradiated with ion beams in the temperature range RT-400°C.
It was found that isolated small α zones ($\leq$ 50 Å diameter)
embedded in the crystalline matrix near the two α/c interfaces dissolved into the crystal but the thickness of the
100 percent α layer was not appreciably affected by further
implantation at 200°C. A model for in situ annealing during
implantation is presented.

INTRODUCTION

Although doping of semiconductors by ion implantation has been utilized
extensively over the last two decades, the damage structures created during the
implantation and annealing are not yet fully characterized. Neither are the
mechanisms by which crystalline (c) materials transform into the amorphous (α)
state by radiation damage understood. For most ion implants, the dose
routinely used by device manufacturing industry is sufficient to form a
continuous α layer in silicon if the implant temperature is at room temperature
(RT) or below. However, non-uniform heating due to poor thermal contact of the
wafer to the substrate holder has been shown to cause unanticipated defect
distributions in the implanted region [1]. In this paper the effect of wafer
heating on the formation and regrowth of amorphous layers and in situ annealing
of amorphous regions by the ion beams has been investigated.

The experiments for this work were designed such that controlled heating
of wafers occurred during implantation. The mechanisms of in situ annealing
suggested by Washburn et al. [2] have been further classified. Two step

Mat. Res. Soc. Symp. Proc. Vol. 14 (1983) ©Elsevier Science Publishing Co., Inc.

implantations were also performed to investigate in-situ annealing effects. For example, some wafers were first implanted at LN temperature followed by another implantation at high temperature. Ion beam induced annealing studies have been reported earlier in the literature, but the emphasis was on the annealing of already existing amorphous layers [3]. No serious attempt was made to understand the in situ annealing effects during implantation.

High energy (MeVs) ion implantations have also been carried out here to investigate the electronic contribution on the formation of amorphous layers and accompanying in situ annealing.

EXPERIMENTAL

(100) and (111) oriented Si wafers were implanted with P, Si or As at 120, 700 or 11000 keV in the temperature range LN-100°C. The doses used in the three cases were 3×10^{14}, 5×10^{14} and 1.9×10^{15}cm^{-2}, respectively. Details of 11000 keV As implantations have been described elsewhere [4]. Some of the wafers implanted at LN with P and Si were subsequently implanted in the temperature range RT-400°C. Cross-sectional transmission electron microscopy (XTEM) was utilized to obtain the widths of damage regions and study the in situ annealing effects.

RESULTS

Figure 1 illustrates the effect of implantation temperature on the formation of an amorphous layer in (111) Si. Phosphorus ions of 120 keV energy were implanted at LN, RT and 100°C, respectively, to a dose of 3×10^{14} cm^{-2} in each case. The dark bands (Figs. 1a and 1b) that represent α or heavily damaged regions appeared in LN and RT implanted cases while no such band was present for 100°C implant. It is well known that solid phase regrowth of implantation induced α Si does not occur until temperatures $\geq 500°C$ are reached. However, in Fig. 1c, a substrate temperature of only 100°C was enough for annealing out of any small α zones that were created during the implantation. This radiation assisted annealing has been referred to as in situ annealing earlier in the text.

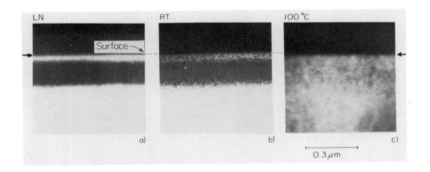

Fig. 1. The effect of implantation temperature on the formation of an amorphous layer in P implanted (111) Si. Dose: 3×10^{14} cm^{-2}; Energy: 120 keV; Implantation Temperature: a) LN, b) RT and c) 100°C.

Figure 2 shows a series of micrographs corresponding to two step implantations. The reference sample (Fig. 2a, first step implantation) is the same as shown in Fig. 1a. The LN implanted reference samples were subsequently further implanted with equal doses ($3x10^{14}cm^{-2}$) of P at RT, 100, 200 and 400°C, respectively. Figure 2b shows that additional RT implantation produced more damage as is evident from the width of the α layer from LN+RT sample is 2100 Å as compared with 1425 Å of the reference sample. However, LN + 100°C and LN + 200°C implantations produced α layers of widths 1650 and 1380 Å (Fig. 2c and 2d), respectively. Also the upper and lower α/c interfaces became much more sharply defined in the LN+200°C case (Fig. 2d). The α layer did not completely recrystallize even when the second implantation was carried out at 400°C (Fig. 2e).

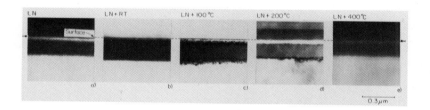

Fig. 2. Further implantation of the sample of Fig. 1a at higher temperature: a) LN only, b) LN+RT, c) LN+100°C, d) LN+200°C and e) LN+400°C. Additional dose: $3x10^{14}cm^{-2}$; Energy: 120 keV.

In situ annealing was also sutided in (100) Si using the two step implantation procedure. (100) Si samples were self implanted at 700 keV to a dose of $5x10^{14}cm^{-2}$ at LN followed by further implantations of equal doses ($5x10^{14}cm^{-2}$) at elevated temperatures (RT-200°C). The results from these samples were qualitatively similar to those discussed above in that the initial damage structure did contain a buried α layer, the small α zones in crystalline matrix near that α/c interface dissolved and the width of the α layer remained almost unchanged after subsequent implantation at 200°C.

Figure 3 shows the XTEM micrographs from (100) Si implanted with As at LN and RT, respectively, to a dose of $1.9x10^{15}cm^{-2}$. Buried α layers were created in both cases with mean widths of 3.5 μm and 1.5 μm, respectively (Fig. 3a and 3b). A large reduction in the width of the RT implanted sample occurred which was strikingly different from the low energy implantation results (Figs. 1a and 1b). Rapid recrystallization of α zones in crystalline matrix near the α/c interfaces and slow regorwth of 100 percent α layer was also observed when a 1.2 MeV electron beam was placed near the two interfaces.

DISCUSSION

From the results of Figs. 1, 2 and 3 it is clear that the formation of α layers is inhibited drastically when the implantation temperature is increased. However, once an α layer is formed, its regrowth during further ion damage even at high temperature is not significant. The following correlates the experimental observations made so far.

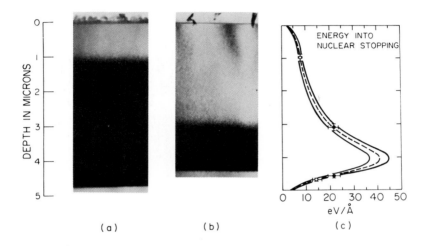

Fig. 3. Comparison of XTEM measured α–c interface depths with calculated displacement energy absorption vs. depth at LN and RT.

Each ion during implantation is expected to create a damage cascade containing interstitials and vacancies along its track in the crystalline (c) matrix. Small α zones are eventually formed either directly near the end of a heavy ion track or by nucleation within heavily damaged regions [5]. The boundaries between the α zones and surrounding crystal have been found to be sharp within two to three atomic distances [3]. The position of the boundary of the α zone therefore probably corresponds to the surface inside of which the critical concentration of point defects necessary for damaged crystal to transform into an amorphous state has been exceeded. If point defects are mobile, a small α zone surrounded by crystal is inherently unstable. At LN temperature or below where the mobility of the point defects created by energetic ions is limited, a buried α layer will be created when the dose is such that the α zones overlap either due to continuous formation of new α zones or due to their growth as perhaps the point defects concentrations in remaining crystalline volumes exceed the critical level of 10-15 percent necessary for c ---> α transformation [6]. This mechanism is believed to account for the α layer of Fig. 1a. In Fig. 1b intense interpenetratin of α and c regions occurs near the two α/c interfaces indicating that elimination of the interpenetrating crystalline volumes becomes more difficult.

In the case of high energy As implantation broad regions of interpenetration occurred at the upper α/c interface even for LN implants. This can be understood from the energy deposition vs depth curve which shows only a slow rise in the amount of energy deposited from the surface to 2.5 μm (Fig. 3). The α/c interface was located in the center of this slow rising region. Furthermore, at 11 MeV, electronic stopping may also contribute to in situ annealing by excitation of atoms at the α/c interface near to the surface.

Small α zones in crystalline matrix should be metastable so long as the point defect concentration in the surrounding crystal stays below the critical level (10-15 percent) of c ---> α transformation. Furthermore, α/c interfacial energy favors a reduction in the size of the α zone. Therefore, α zones should tend to shrink by a net transfer of atoms across the interface from the α side to the crystalline side. Concomitant with the formation of amorphous zones, however, interstitials and vacancies will be continuously created in the crystalline material. All the previously formed α-c interfaces should act as sinks for these defects because they are mobile at room temperature. Their recombination at the interface and other radiation induced transfers of energy to atoms at the interface could provide the necessary activation energy for continuous shrinkage of all existing amorphous zones. It has been shown that MeV electron irradiation at slightly above RT causes regrowth of α zones.[2]

In the case of the 100°C implantation, it is believed that the point defect concentration never exceeded the critical 10-15 percent and the α zones that may have been formed directly at the end of the ion tracks shrank and disappeared faster than new ones were being formed due to high mobility of the surrounding point defects. As a result, an α layer was never formed.

The results of two step implantation can also be explained by the above model. The regions above and below the α/c interfaces where α and c zones co-existed initially were strongly affected by the second higher temperature implant. For the 200°C second implant small α zones surrounded by crystal shrank and disappeared leaving sharper α/c interfaces. There was therefore some shrinkage of the previously formed amorphous volume. However, the width of the fully amorphous layer remained almost unchanged. A minimum rate of migration of the α/c interface sufficient to eliminate small α zones of the order of 50 Å in diameter within the time of the second irradiation would cause only an insignificant reduction in thickness of the 100 percent amorphous layer.

CONCLUSIONS

In conclusion, the details of the formation of an amorphous layer are sensitively dependent on implantation temperature. It is suggested that this happens because small amorphous zones surrounded by crystalline matrix can undergo radiation induced shrinkage at temperatures $\geq$ 100°C where elementary point defects are mobile. At very high implant energies, electronic stopping may also contribute to in situ annealing by excitation of atoms at α/c interfaces.

516

ACKNOWLEDGEMENTS

The authors would like to thank the Heavy Ion Linear Accelerator staff of Lawrence Berkeley Laboratory for 11 MeV As implantations and Ilka Suni of Caltech, Pasadena, CA, for 700 keV Si implantation. One of us (DKS) would also like to acknowledge useful discussions with Simon Prussin of TRW, Lawndale, CA. This work was supported by the Director, Office of Energy Research, Office of Basic Energy Sciences, Material Science Division of the U.S. Department of Energy under Contract No. DE-AC03-76SF00098.

REFERENCES

1. D. K. Sadana, M. Strathman, J. Washburn and G. R. Booker, J. Appl. Phys. 51 5718 (1980).

2. J. Washburn, C. S. Murty, D. Sadana, P. Byrne, R. Gronsky, N. Cheung and R. Kilaas, Nucl. Inst. Meths. (in press).

3. I. Golecki, G. E. Chapman, S. S. Lau, B. Y. Tsaur and J. W. Mayer, Phys. Lett. 71A 267 (1979).

4. P. F. Byrne, N. W. Cheung and D. K. Sadana, App. Phys. Lett. 41 537 (1982).

5. F. F. Morehead Jr. and B. L. Crowder, Rad. Eff. 6 27 (1970).

6. L. A. Christel, J. F. Gibbons and T. W. Sigmon, J. App. Phys. 52 7143 (1981).

NON DESTRUCTIVE OPTICAL ANALYSIS OF IMPLANTED LAYERS IN GaAs BY RAMAN SCATTERING AND SPECTROSCOPIC ELLIPSOMETRY

JOSEPH BIELLMANN, BERNARD PREVOT AND CLAUDE SCHWAB
Laboratoire de Spectroscopie et d'Optique du Corps Solide
(L.A. N° 232 - CNRS) 5, rue de l'Université, 67000 Strasbourg, FRANCE

JEAN-BERNARD THEETEN AND MARKO ERMAN
L.E.P., B.P. 15, 94450 Limeil-Brévannes, FRANCE

ABSTRACT

Non destructive analysis by Raman Scattering (RS) and Spectroscopic Ellipsometry (SE) is demonstrated on B^+ and Se^+ shallow implanted GaAs. Qualitative informations are obtained from 1st and 2nd order RS spectra. The former are analysed using the intensity ratio of the TO and LO modes, which defines a lattice potential perfection scale. The SE analysis of the E_1, $E_1 + \Delta_1$ structure in the imaginary part of the dielectric function confirms the RS results and its multilayer analysis yields the depth profile of the implanted ions.

INTRODUCTION

Optical methods are useful for investigating semiconductors due to their non-destructive nature. This paper reports the use of Spectroscopic Ellipsometry (SE) and Raman Scattering (RS) as complementary tools for assessing thin implanted layers ($\sim300\text{Å}$) on semiinsulating GaAs. Both implantation damages and lattice recovery steps, including the electrical activation of the implanted ions, are easily evidenced. Further analysis of the experimental data allows for the determination of the depth profile of implants by SE and for a quantitative lattice disorder scale by RS.

EXPERIMENTAL

Semiinsulating GaAs (100) wafers have been implanted either with 10 keV-B^+ ions with doses ranging from 8×10^{11} to 1×10^{14} ions.cm^{-2} or with 50 keV-Se^+ ions at a dose of 10^{14}/cm^2. According to LSS theory [1], the depth profile should be Gaussian like with a maximum at about 200 Å, whereas the projected range straggling is 200 and 80 Å for the 10 keV-B^+ and 50 keV-Se^+ ions respectively. Boron implantation is interesting because it keeps the semiinsulating character of GaAs inducing only crystalline disorder defects. Conversely, Se implantation leads to a n-type layer after a suitable thermal treatment. A fraction of each wafer was masked during implantation to be used as a reference for the optical investigations. The thermal treatments were made in an atmospheric H_2 furnace with the surface of the implanted wafer covered by a second one in order to avoid As evaporation. The temperature was raised from 200 to 900°C by steps of 12 min duration.

SE measures the complex reflectance ratio ρ defined as :

$$\rho = r_p \cdot r_s^{-1} = \tan(\Psi)\exp(i\Delta) \qquad (1)$$

Mat. Res. Soc. Symp. Proc. Vol. 14 (1983) © Elsevier Science Publishing Co., Inc.

where r_p and r_s are the reflection coefficients of an interface for incident light with polarization either in the plane of incidence (r_p) or perpendicular to it (r_s), yielding the dielectric function ε through the relation :

$$\varepsilon = (1-\rho)^2 . (1+\rho)^{-2} \sin^2\theta . \tan^2\theta + \sin^2\theta \qquad (2)$$

where θ is the angle of incidence. Thus, a spectroscopic determination of ρ leads to a structural (and possibly chemical) analysis of the material under study. The details of the experimental set-up have been described elsewhere [2]. A spectral range extending from 1.6 to 5.4 eV was covered.

RS yields informations about the vibrational properties of the crystal through the coupling of its phonons to the incoming photons. The k-momentum conservation rule allows only $k \sim 0$ resulting wavevector phonon participations : thus phonon modes either at the center of the Brillouin zone (1st order RS) or overtones and combination modes (2nd order RS) can be evidenced. Experiments were performed in the Brewster backscattering configuration using the 4880 or 5145 Å lines of an Ar^+ ion laser.

Both SE and RS measurements were performed at room temperature on the same samples, all wafers being cut and polished from the same GaAs ingot.

RESULTS

Implanted substrates. Figure 1 shows a comparison of the RS spectra of a 10 keV-B implanted area and of the reference area of the same substrate. Upon implantation, the dominating first order TO and LO modes display an intensity reduction of the LO mode accompanied with a linewidth increase and a minor peak position shift. Figure 2 gives the variation of the ratio R (TO/LO) of the intensities of the TO and LO modes as a function of the B^+ fluence. It also reports the corresponding variation of the width (FWHM) of the LO line, including a spectrometer bandwidth of 3.2 cm^{-1}. The highly structured 2nd order RS spectrum resulting from numerous combination and overtone processes covering the 160 to

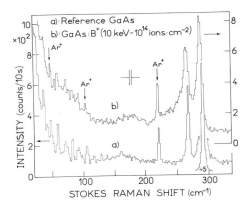

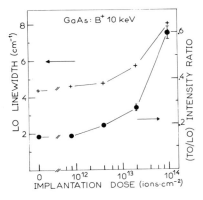

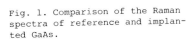

Fig. 1. Comparison of the Raman spectra of reference and implanted GaAs.

Fig. 2. LO mode linewidth (+) and 1st order mode intensity ratio (•) versus the B^+ implant dose.

520 cm^{-1} range converts into a spectrum consisting mainly in three broad bands
at about 75, 180 and 250 cm^{-1}. This RS spectrum of implanted GaAs is well described using the one phonon density of states ; the analysis of the phonon spectrum of GaAs by the Shell Model formalism [3] leads to a determination of the
vibrational density of states which compares well with the present observations.
This suggests that the main effect of implantation is the destruction of the
long-range crystalline order as already observed in GaAs [4]. Incidentally, one
can notice that the Raman signature of a polycrystalline sample looks quite different since it results from an angular average over all possible polarization
configurations.

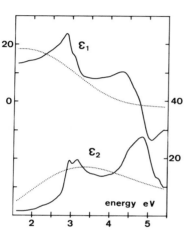

Fig. 3. Dielectric functions of
reference (full line) and amorphous
GaAs (dotted line).

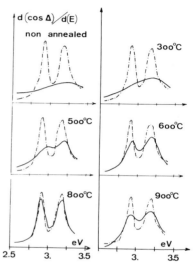

Fig. 4. d(cosΔ)/dE plot versus energy
for several thermal treatments for
10 keV-Se$^+$ implanted GaAs at 10^{15}
ions/cm^2.

Figure 3 compares the dielectric functions determined by SE on a reference
crystalline GaAs sample and on amorphous GaAs (dose > 10^{15} ions/cm^2). Marked differences appear at 3 and 4.5 eV, energies characteristic of the E_1, $E_1+\Delta_1$ and E_0'
optical transitions. Already at the lowest implantation dose, these features are
affected by an amplitude reduction whereas the other parts of the spectrum are
still unchanged. Although the two techniques involve different physical mechanisms, they have about the same sensitivity to implantation defects (as low as
8.10^{11} atoms/cm^2).

Lattice perfection recovery. With the lattice perfection recovery, one has
also to consider the electrical activation of the Se implants. Figure 4 shows
the evolution of the SE spectra of 50 keV-Se$^+$ implanted GaAs as a function of
the annealing steps. In order to increase the sensitivity, the derivative of
cos Δ versus energy is reported. The crystalline order is restored at about 800°C
as shown by the almost full recovery of the E_1, $E_1+\Delta_1$ structure. Beyond this temperature, this structure degrades again. This is a consequence of the electrical

520

activation of the Se⁺ implanted species which are inserted into substitutional sites in the lattice. This electrical activation is confirmed by transport measurements.

Figures 5-a and 5-b give the full RS spectra of GaAs:Se⁺ after the annealing steps till 400 and 900°C. If an important recovery is already noticed on the first spectrum, with even some of the 2nd order structures present, a degradation is definitively observed at 900°C. Conversely this latter effect does not appear on a GaAs:B⁺ sample submitted to a similar thermal treatment as shown in Figure 6. This confirms that the thermal treatment at 900°C induces the electrical activation of the implanted Se⁺ ions while Boron remains interstitial and leaves the GaAs semiinsulating.

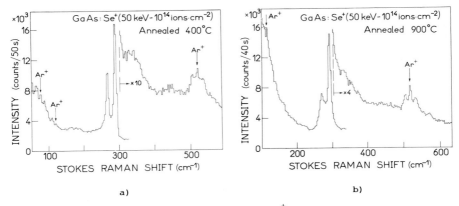

a)   b)

Fig. 5. Raman spectra of annealed GaAs:Se⁺ samples prior a) and after b) the electrical activation of the implants.

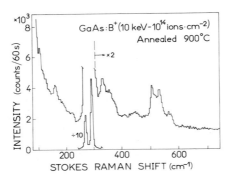

Fig. 6. Raman spectrum of GaAs:B⁺ annealed till 900°C. No electrical activation occurs (Cf. Fig. 5-b).

DISCUSSION

The former results show that SE and RS data lead to similar conclusions as far as the damages introduced by ion implantation and the lattice perfection recovery are concerned.

Recently [5], we performed an analysis of the relative intensities of the 1st order Raman TO and LO modes in zinc blende compounds. It could be demonstrated that two factors govern this ratio at a given energy of the excitation light :

$$R(TO/LO) \simeq G(TO/LO) . H(T,C) \tag{3}$$

G(TO/LO) is a geometrical factor taking into account the directions of the electric field unit vectors of the incoming and scattered light beams relative to the crystal orientation ; H(T,C) describes the microscopic properties of the crystal at a temperature T as a function of C, the Faust-Henry coefficient which expresses the relative strength of the electron-phonon interactions [6]. Hence a true bulk value of R(T/L) may be defined and used as a lattice perfection scale. By further analyzing the basic mechanisms involved in light scattering, it has been predicted that R(TO/LO) → 1 with increasing disorder, as is indeed confirmed by Figure 2. With the 5145 Å Ar$^+$ laser line, 90% of the scattered light is issuing from the top 300 Å of the sample, which includes the full implantation perturbed region ; for a deeper implantation depth, a suitable wavelength adjustement is certainly required.

SE measurements generate enough independent data in order to obtain a depth profile of the defects in the implanted region. We assume a multilayer structure each layer having a thickness d_i and a chemical composition volume fraction x_i and $(1-x_i)$ of crystalline and amorphous GaAs respectively. The dielectric function of such a mixture is evaluated in the effective medium approximation from the bulk values of amorphous and crystalline GaAs [7], (figure 3). Data from implanted samples have been analysed using a multilayer model with eight layers of 50 Å in thickness with only one adjustable parameter, namely the volume fraction x_i of each layer. Figure 7 shows the percentage of amorphous GaAs as a function of depth ; for comparison, the corresponding LSS distribution function is also reported. The uncertainties of the calculated ion damage profile increase as a function of depth since the medium is strongly absorbing (figure 3). The maximum in the amorphous percentage is found at around 200 Å in good agreement with the LSS theory. However, the two profiles differ markedly near the surface ; the

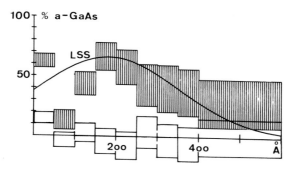

Fig. 7. Amorphous phase profile calculated from the data of Fig. 3 for reference (open rectangles) and implanted (crossed rectangles) GaAs compared with the corresponding LSS profile.

522

amorphous top layer and following dip are reproducible features, which cannot be ascribed neither to polishing damages nor to a mathematical artefact. The corresponding absorption increase is likely to be due to surface damage during the ion implantation while the dip could be due to partial annealing by the ion beam itself [8].

CONCLUSION

In summary, both SE and RS are sensitive methods for assessing non-destructively thin layers. The macroscopic parameters determined by SE can be modeled so as to give the depth profile of the implantation damages. Conversely, the microscopic informations derived from RS support the hypothesis of increasing amorphization with the implantation dose illustrating the complementary aspects of the two techniques.

ACKNOWLEDGEMENTS

We would like to thank M. Berth and C. Venger for providing us with the implanted GaAs wafers and for discussions of the SE results.

REFERENCES

1. J. F. Gibbons, W. S. Johnson and S. W. Mylroie, in: Projected Range Statistics, 2nd Ed., (Halstead Press 1975).

2. J. B. Theeten and M. Erman, J. Vac. Sci. Technol., 20, 471 (1982).

3. G. Dolling and J. T. L. Waugh, in: Lattice Dynamics, R. F. Wallis Ed., (Pergamon Press, Oxford 1965) p. 19.

4. J. E. Smith, Jr., M. H. Brodsky, B. L. Crowder and M. I. Nathan, in: Proc. 2nd Intern. Conf. Light Scattering in Solids, Paris, France 1971 (Flammarion Paris 1971) p. 330.

5. J. Biellmann, B. Prévot and C. Schwab, J. Phys. C: Sol. State Phys., in press (1982).

6. M. V. Klein, in: Light Scattering in Solids, M. Cardona Ed. (Springer, Berlin 1975) p. 158.

7. M. Erman and J. B. Theeten, Surface and Interface Analysis, 4, 98 (1982).

8. M. Erman, Thesis, Univ. Paris Orsay (March 1982).

A COMPARISON OF ELLIPSOMETER AND RBS ANALYSIS OF IMPLANT DAMAGE IN SILICON

W. M. PAULSON AND S. R. WILSON
Semiconductor Research and Development Laboratory, Motorola, Inc.
5005 E. McDowell Road, Phoenix, Arizona, USA

C. W. WHITE AND B. R. APPLETON
Solid State Division, Oak Ridge National Laboratory, Oak Ridge, Tennessee, USA

ABSTRACT

The purpose of this study is to analyze ion implant damage profile using RBS and ellipsometry. Silicon wafers were implanted with ^{75}As at 100, 200 or 300 keV; doses were chosen to generate constant peak impurity concentrations at each energy. The samples were then analyzed using RBS to obtain damage-depth profiles and ellipsometry to obtain Δ, Ψ parameters. At light doses decreasing Δ values correspond to increased scattering yield; at higher doses Ψ increases rapidly as the scattering yield approaches the random value. The higher energy implants shift the Δ-Ψ curves to larger Ψ values. Multilayer structures, that include lightly damaged silicon on either side of the project range as well as more damaged near the projected range, are required to model the ellipsometer parameters.

INTRODUCTION

Ion implantation generates damage in the substrate material; this damage changes the optical properties of these materials. The index of refraction and the absorption coefficients in the implanted layers can change depending upon the dose, energy and implanted species [1]. The implantation induced damage has been well characterized using RBS analysis [2]. For crystalline substrates and a given implant specie, increasing the dose or energy results in more crystal damage until the layer becomes amorphous.

Implant damage has been characterized using either optical reflection [3] or optical absorption [4]. Optical reflectance measurements have been used to monitor the laser-assisted regrowth of implanted layers [5,6]. Ellipsometry in conjunction with anodized stripping techniques was used by Adams and Bashara [7] to determine depth profiles of the index of refraction. They showed that both the index of refraction and the extinction coefficient varied with depth. Chemical etching and ellipsometry were utilized by Ohira and Itakura [8] to show that the damage peak position is on the surface side of the dopant peak position. Nakamura, Gotoh, and Kamoshida [9] characterized 50 keV P-implanted Si using ellipsometry and calculated the effective index of refraction using a two phase model.

The crystal damage for implanted P was modeled by Motooke and Watanaba [10] using a thin surface SiO_2 layer, an amorphous Si-layer, followed by a transition region to the single crystal substrate. They showed that increasing the layer thickness causes the Δ-Ψ curves to spiral toward values for thick, amorphous layers. The ellipsometer model by Delfino and Razouk [11,12] included air, native oxide, an implant damaged layer and the substrate. This model yielded the effective implant layer thicknesses and optical constants for As or B implants. Ellipsometry has also been used to analyze implant damage in GaAs [13] and in silica glasses [14]. All of these studies show the sensitivity of ellipsometer analysis to implant damage. The purpose of this study is to analyze the ion implant damage profiles using Rutherford backscattering measurements and optical ellipsometry. In addition, we have modeled the optical constants using multilayer structures.

EXPERIMENTAL PROCEDURES

The <100> single crystal silicon substrate wafers were implanted with [75]As at energies of 100, 200 or 300 keV. Four different doses were implanted at each energy to generate constant peak impurity values at all energies. The samples were then analyzed using Rutherford backscattering (RBS) and ellipsometry. A 2 MeV He^+ beam was aligned parallel to the [110] axis to obtain channeling spectra, which yields the damage-depth profiles of the Si. The samples were briefly etched in 1:10, HF:H_2O and rinsed in water prior to the ellipsometer analysis. The ellipsometer parameters Δ and ψ were obtained using a Rudolph Research "Auto El" ellipsometer. A He-Ne laser was used as a light source at a wavelength of 0.6328 μm and an angle of incidence of 70°. An index of refraction [1] of 3.858-0.018j was used for single crystal silicon at this wavelength. The index of refraction for amorphous sputtered silicon films that were 1.0 μm thick was measured to be 4.56-0.44j. This value was utilized in calculating the ellipsometer parameters using the computer program by McCrackin [15]. The implant damage was modeled using mutlilayer structures and assuming values for the complex index of refraction and thickness of each layer.

EXPERIMENTAL RESULTS

The backscattering spectra for the 100 keV [75]As implants to doses of 1, 2, 7 or 10 x 10^{13}/cm^2 are shown in Fig. 1. The peak of the damage distribution is located closer to the surface than the peak for the implanted As (58 nm), in agreement with the calculations of Brice [16]. The damage increases with increasing dose and for a dose of 7 x 10^{13}/cm^2 (broken line) the peak height almost reaches the random value, indicating that most crystal order is gone. With a higher dose of 1 x 10^{14}/cm^2 (dotted line) the amorphous layer is approximately 35 nm wide, but does not extend all the way to the surface. Fig. 2 presents the RBS data for 100, 200 or 300 keV [75]As implants to doses of 3, 5.4, or 7.7 x 10^{13}/cm^2, respectively. The peaks in the damage distributions occur at 45, 85 and 150 nm for the 100, 200 and 300 keV implants, respectively. In addition, the scattering yield increases for higher implant energies. None of the scattering peaks in Fig. 2 reach the amorphous level. The extent of the crystal damage depends upon dose, energy substrate temperature and dose rate, and, in addition, varies with depth into the crystal. Less crystal damage is observed near the surfaces, as indicated by χ_{min} values of 21, 16 or 13% obtained just behind the surface peak, for the 100, 200 or 300 keV implants, respectively.

The ellipsometer parameters Δ and ψ are plotted in Fig. 3 for 100, 200, or 300 keV implants. The circles show the Δ, ψ values for doses of 1, 3, 7 or 10 x 10^{13}/cm^2 at 100 keV. For the first two doses Δ decreases from an initial 176° to 163° while ψ is nearly constant. However, for the higher two doses ψ increases rapidly but Δ is approximately constant. Similar trends are observed with the 200 keV implants (triangles) for doses of 2.8, 5.4, 13.0 or 18.0 x 10^{13}/cm^2, but the Δ-ψ curve is displaced to larger ψ values. Again, with the 300 keV implants to doses of 2.6, 7.7, 18.0 or 26.0 x 10^{13}/cm^2, the Δ-ψ curve (squares) is shifted to larger ψ values, but follows the same trends.

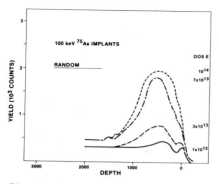

Fig. 1. RBS spectra for 100 keV arsenic implants to four different doses.

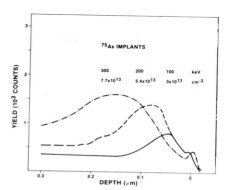

Fig. 2. RBS spectra for 100, 200 or 300 keV arsenic implants.

The highest dose implant at each energy causes the crystal to become amorphous over some region. For example, at 300 keV and $26.0 \times 10^{13}/cm^2$ the amorphous region is about 250 nm wide but at 100 keV and $10 \times 10^{13}/cm^2$ the width is only 35 nm wide. For the lower three doses at each energy, however, there is a significant near-surface region that is not amorphous.

CALCULATIONS

Models of the implant damage need to include the gradual changes in crystal damage, in order to account for the optical properties. The ellipsometer parameters Δ and ψ can be calculated by assuming values for the complex index of refraction, $N = n-kj$. The computer program by McCrackin [15] was used, which could include the effects of multiple layer structures. For all of these calculations a 2.0 nm thick SiO_2 layer was included.

The first calculation assumed a single layer of damaged silicon as sketched in Fig. 4. The index of refraction n was maintained constant while the extinction coefficient k was increased from the single crystal value of 0.0193 to a maximum of 0.540. The calculated Δ-ψ curves are shown in Fig. 4 for three different layer thicknesses. With a 100 nm thick layer (circles) Δ decreases from 173 to 159 while the corresponding ψ increases from 10.4 to 12.8. When the layer thickness was increased to 200 nm, the Δ-ψ curves shifts to the left as shown by the X's in Fig. 4. The Δ-ψ curve for a 300 nm thick layer (triangles) is displaced even further to lower ψ values. This calculation shows two important characteristics. First, increasing the extinction coefficient k results in lower Δ values, which agrees with the experimental results for the lower two doses (Fig. 3). Second, this calculation shows the Δ-ψ curve shifting to the left when the damage layer increases in width. However, this trend is the opposite of that observed experimentally (Fig. 3.) Therefore, this single layer model of implant damage does not adequately describe the experimental results.

A multilayer representation of the implant damage, as sketched in Fig. 5, was used to calculate the Δ-ψ parameters. Three layers that were 30, 60 and 30 nm thick were used to model the 100 keV ^{75}As implants. First, n was held constant and k for each layer was increased using steps of 0.08. In addition, K_2 was always larger and the difference between k_2 and k_1 or k_3 was 0.08, which simulates the RBS damage profiles. These calculated (Δ,ψ) values are shown by

526

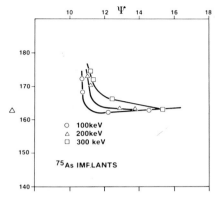

Fig. 3. Ellipsometer results for four different arsenic doses at implant energies of 100, 200 or 300 keV.

the circles in Fig. 5 at (173.4, 10.34), (168.5, 10.61) and (164.5, 10.9) for k_2 values of 0.15, 0.23 and 0.31. Next, for each layer n was increased from 3.858 to 5.00 along with slowly increasing the corresponding k from 0.31 to 0.40, while maintaining n_2 larger than n_1 or n_3. The calculated (Δ,Ψ) values change from (161.0, 11.19) to (163.7, 16.05), as shown by the circles in Fig. 5. These results can be compared with the experimental values (Fig. 3) and show the same trends.

Similar multilayer calculations were done to model the 300 keV implants. Four layers, that were 60, 50, 100 and 50 nm thick, were assumed along with $n_3 > n_2 = n_4 > n_1$. Increasing the k values with k_3 at least 20% larger than k_1, k_2 or k_4 yields the four (Δ,Ψ) values (shown by the squares in Fig. 5) from (174.3, 11.07) to (167.8, 11.73). Then, by increasing the n values from 3.858 to 4.60 in 0.1 steps gives the sequence of five points with increasing Ψ values at a nearly constant Δ of 167. The calculated values (squares, Fig. 5) compare well with the experimental results (squares, Fig. 3). Therefore calculations that include lightly damaged layers on both sides of a more heavily damaged layers on both sides of a more heavily damaged central layer generate Δ-Ψ

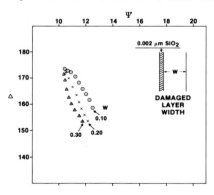

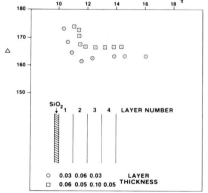

Fig. 4. Single layer model of implant damage used to calculate Δ and Ψ.

Fig. 5. Multilayer model of implant damage used to calculate ellipsometer parameters.

curves that fit the experimental results. In addition, the calculated Δ-Ψ
curves shift to larger Ψ values as the damage distribution moves deeper into
the crystal.   Furthermore, increasing the extinction coefficients k primarily
decreases Δ, while increasing the index of refraction n increases Ψ.

SUMMARY
    The RBS analysis shows a damage distribution with a peak located below the
surface. The scattering yield increases with higher doses or energies until
amorphous layers are generated. The corresponding ellipsometer measurements
show that for light doses the primary change is in Δ while heavier doses
primarily affect Ψ. Multilayer representations of the implant damage including
lightly damaged layers on both sides of a more heavily damaged layer are required to model the optical parameters of implanted layers.

ACKNOWLEDGMENTS
    We appreciate the assistance of B. Lorigan, G. Tam and R. Gregory for pre-
paring samples and obtaining measurements. Thanks are due to M. Scott for the
work in preparing this manuscript.

REFERENCES

1. P. D. Towsend and S. Valette, "Optical Effects of Ion Implantation," in Ion
    Implantation, ed. J. K. Hirvonen, Chap. 11, Academic Press, NY, 1980.

2. J. W. Mayer, L. Eriksson and J. A. Davies, Ion Implantation in Semiconduc-
    tors, Academic Press, N.Y., 1970.

3. T. C. McGill, S. L. Kurtin and G. A. Shifrin, J. Appl. Phys. 41, 246 (1970).

4. B. L. Crowder et al., Appl. Phys. Lett. 16, 205 (1970).

5. G. L. Olson et al., in Laser and Electron-Beam Solid Interactions and Mate-
    rials Processing, ed. J. F. Gibbons, L. D. Hess and T. W. Sigmon, Elsevier
    North Holland, N.Y., 1981, p. 125.

6. J. F. Ready, et al., ibid, p. 133.

7. J. R. Adams and N. M. Bashara, Surface Science 49, 441 (1975).

8. F. Ohira and M. Itakura, Jap. J. Appl. Phys. 21, 42 (1982).

9. K. Nakamura, T. Gotoh and M. Kamoshida, J. Appl. Phys. 50, 3985 (1979).

10. T. Motooka and K. Watanabe, J. Appl. Phys. 51, 4125 (1980).

11. M. Delfino and R. R. Razouk, J. Appl. Phys. 52, 386 (1981).

12. M. Delfino and R. R. Razouk, J. Electrochem. Soc. 129, 606 (1982).

13. Q. Kim and Y. S. Park, J. Appl. Phys. 51, 2024 (1980).

14. A. R. Bayley and P. D. Townsend, J. Phys. D:Appl. Phys. 6, 1115 (1973).

15. F. L. McCrackin, NBS Technical Note 479, 1969.

16. D. K. Brice, Ion Implantation Range and Energy Deposition Distribution,
    IFI/Plenum, New York, 1971.

THE STUDY OF DAMAGE PROFILE OF ION IMPLANTED LAYER
ON SI BY SPECTROSCOPIC ELLIPSOMETRY

JINSHEN LUO,* P.J. MC MARR AND K. VEDAM
Materials Research Laboratory, The Pennsylvania State University,
University Park, PA 16802

ABSTRACT

We have determined the dielectric function of silicon
samples which were implanted with 100-150 KeV P, As, Si ions
to doses of $2 \cdot 10^{14}$-$1 \cdot 10^{16} cm^{-2}$, by a rotating analyser Auto-
mated ellipsometer in the spectral range 1.77 - 4.59 eV. These
data have been analyzed using a simplified three layer model.
The results are compared with an earlier ellipsometric investigation [2].

INTRODUCTION

Damage profiles induced by ion implantation are usually determined by the
backscattering technique which is powerful but cumbersome. Ellipsometry, however, is a simple optical characterization technique. There are several papers
[1], [2], [3], [4], which reported the ellipsometric measurement of damage profiles induced by ion implantation. These measurements were made with fixed
wavelength ellipsometers operating at 5461 Å or 6328 Å. Recently, J. Cortot
et al. reported their work on the characterization of damage using a spectroscopic ellipsometer in the spectral range 3250 Å - 4000 Å. In this paper, we
studied the determination of damage profiles induced by P^+, AS^+ and Si^+ implantation in the dose range of $2 \cdot 10^{14} cm^{-2}$ to $1 \cdot 10^{16} cm^{-2}$ and in spectral ranges
of 1.77 - 4.59 eV using a spectroscopic ellipsometer.

THE EXPERIMENTS AND COMPUTATION RESULTS

The wafers for P^+ implantation were <111> P-type Si with resistivity of 8 to
10 Ωcm, and those for As^+ implantation were <100> P-type Si with resistivity of
20 - 30 Ωcm. Before ion implantation the surfaces of wafers were mechanochemi-
cally polished. Samples were implanted with P^+ or As^+ at an ion energy of 100
keV, and the doses varied from $2 \cdot 10^{14} cm^{-2}$ to $1 \cdot 10^{16} cm^{-2}$. Both P^+ and As^+ implantation were performed at room temperature. The implanting condition for
Si^+ implanted sample was $1 \cdot 10^{15} cm^{-2}$ (80KeV) + $2 \cdot 10^{15} Si^+ cm^{-2}$ (150KeV), which
induced an amorphous layer of 3400 Å in the top surface layer. During the ion
implantation, samples were misoriented by 7° - 8° with respect to the incident
ion beam to minimize the channeling of implanted ions.
Ellipsometric measurements were carried out by using an automatic spectroscopic ellipsometer of the rotating analyser type at a 70° angle of incidence.
The ellipsometric parameters were measured from 1.77 eV to 4.59 eV on each of
the ion implanted samples. From the ellipsometric measurements, the spectra of
the dielectric functions ε_1 and ε_2 were determined. To do the theoretical calculation, a simple three layer model as shown in Fig. 1 was used. In this
model we defined a factor

Department of Electronic Engineering, Xian Jiaotong University, Xian, Shaanxi,
People's Republic of China.

Mat. Res. Soc. Symp. Proc. Vol. 14 (1983) ⓒ Elsevier Science Publishing Co., Inc.

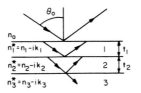

Fig. 1. Three layer model for theoretical calculation.

$$K_0 = \frac{n^* - n_3^*}{n_a^* - n_3^*},$$ (1)

where n^*, n_3^* and n_a^* are the complex refractive indices in the layer, c-Si and a-Si, respectively. The complex reflection coefficient of this three layer system for S polarized wave is given by

$$r_s = \frac{(m_{11} + m_{22}P_3)P_0 - (m_{21} + m_{22}P_3)}{(m_{11} + m_{22}P_3)P_0 + (m_{21} + m_{22}P_3)}.$$ (2)

Here $P_0 = \cos\theta_0$, $P_3 = n_3^* \cos\theta_3$, and the matrix

$$\begin{bmatrix} m_{11} & m_{12} \\ m_{21} & m_{22} \end{bmatrix} = \prod_{j=1}^{2} \begin{vmatrix} \cos(D_j) & \frac{i}{P_j}\sin(D_j) \\ iP_j\sin(D_j) & \cos(D_j) \end{vmatrix},$$ (3)

where $P_j = n_j^* \cos\theta_j$, $D_j = 2\pi t_j\, n_j^* \cos\theta_j/\lambda_0$, λ_0 is the light wave length in vacuum, and t_j is the thickness of the j-th layer. The equations for computing the complex reflection coefficient for P polarized wave can be obtained by replacing p_j by $q_j = \cos\theta_j/n_j^*$ in (2) and (3). The ellipsometric parameters satisfy the following equation:

$$tg\psi e^{i\Delta} = \frac{r_p}{r_s}.$$ (4)

Fig. 2 shows the results of backscattering measurements for four P^+ implanted samples of doses $2 \cdot 10^{14}$ cm^{-2}, $5 \cdot 10^{14}$ cm^{-2} and $1 \cdot 10^{16}$ cm^{-2}. The ε_2 spectra determined by ellipsometry are shown in Figs. 3 - 6.

In the case of lower dose, i.e. $2 \cdot 10^{14}$ P^+ cm^{-2}, the damage profile determined by backscattering measurement was a Gaussian distribution with the peak located at a depth of 1000 Å from the surface. In the calculation, we assumed a simplified model consisting of an oxide film of 10 Å, a uniform damaged layer of 1500 Å with $K_0 = 0.2$ and a single crystal Si substrate. The computed ε_2 spectrum is shown by the solid line in Fig. 3 and is in good agreement with the experimental data. Comparing this ε_2 spectrum with that of single crystal Si, which is also shown in Fig. 3, we find that after ion implantation of the lower dose the peak values of ε_2 are lowered and the spectrum becomes wider. Fig. 4 shows the measured and computation results for the cases of $1 \cdot 10^{15}$ P^+ cm^{-2}, $1 \cdot 10^{15}$ As$^+$ cm^{-2} and Si$^+$ implanted samples. The model for calculation consisted of an oxide film of 10 Å, an amorphous Si layer and single crystal Si

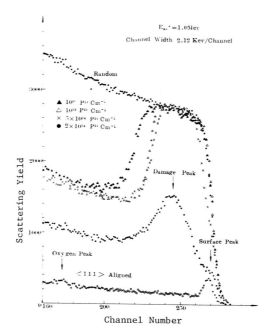

Fig. 2. The backscattering measurement.

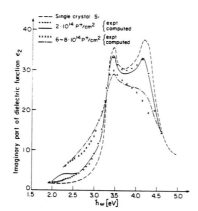

Fig. 3. ε_2 spectra in the
case of lower dose.

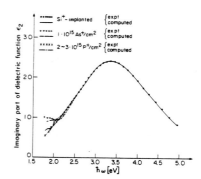

Fig. 4. ε_2 spectra for different thickness of amorphous layer.

532

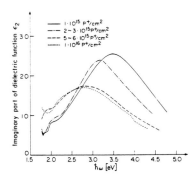

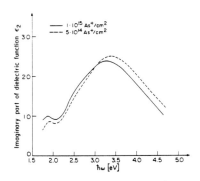

Fig. 5. ε_2 spectra in the high
dose case.

Fig. 6. ε_2 spectra in the As$^+$
implantation samples

substrate, and the computation results approximately agreed with the measured
results. The thickness of the amorphous layer in each sample is listed in
Table 1.

TABLE I

Sample No.	Implanted ions	Dose (cm^{-2}) and ion energy (KeV)	Thickness of amorphous layer (Å)
1	Si^+	$1 \cdot 10^{15}$(80KeV) + $2 \cdot 10^{15}$(150KeV)	3400
2	P^+	$1 \cdot 10^{15}$ (100KeV)	1650
3	As^+	$1 \cdot 10^{15}$ (100KeV)	1150

It can be seen that the differences among these spectra near the long wavelength region are due to the differences in the thicknesses of amorphous layers.
    Fig. 5 shows the ε_2 spectra in the case of high doses. It can be seen that
the peak value decreases with increasing implantation doses, and at the same
time its position shifts to the long wavelength region. This shows that the
structure of amorphous layer induced by ion implantation changes with the implantation dose in the case of very high dose. The change could not be observed by the backscattering method. For example, for $1 \cdot 10^{15}$ P^+ cm^{-2} and
$1 \cdot 10^{16}$ P^+ cm^{-2} implanted samples backscattering measurement only gave the difference in thickness of amorphous layer.
    In the case of $5 \cdot 10^{14}$ P^+ cm^{-2} implantation, Fig. 2 shows that the amorphous
layer is located from the depth of about 400 Å to about 1400 Å. The layer from
the surface to the depth of 400 Å was damaged but did not become amorphous.
The measured data for this sample are remarkably different from those of the
high dose case. The model we suggested for computation included these parts:
(1) a damaged layer of thickness with optical constants

$$n_1 = n_3 + k_0(n_2 - n_3),$$
$$k_1 = k_3 + k_0(k_2 - k_3), \tag{6}$$

where n_2, k_2 and n_3, k_3 respectively represent the optical constants of amorphous and single crystal Si, (2) an amorphous layer of thickness t_2, (3) single crystal Si substrate. As shown in Fig. 3, the computed line in the case $K_0 = 0.5$, $t_1 = 400$ Å, and $t_2 = 1000$ Å, approximately agrees with the measured data. Fig. 7 shows the computed results for different values of K_0, t_1, and t_2.

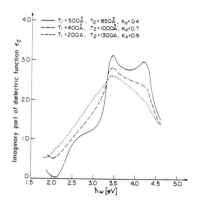

Fig. 7. The computed ε_2 spectra for different values of K_0, t_1, and t_2.

It can be seen that when K_0 and t_2 increase and t_1 decreases, corresponding to the increase of damage, the peak value near the long wavelength region moves down, another peak disappears, and eventually the ε_2 spectrum approaches that of amorphous Si.

CONCLUSION

1. In the case of lower dose, not causing the formation of amorphous layer in the damaged region, the peak values move down and the ε_2 spectrum becomes wider with the increasing implantation dose.
2. For such an implantation dose that the amorphous layer is only induced in the middle part of the damaged region, an increase in dose will cause the peak in the long wavelength region to move down, and the other one to disappear.
3. In the case of high doses, when the amorphous layer is induced in the whole surface layer, the peaks in the ε_2 spectra move down and shift to the long wavelength region.

As a nondestructive method, a spectroscopic ellipsometer can be used to determine the damage profile induced by ion implantation.

REFERENCES

1. J.R. Adams, Surface Science 56, 307 (1976).
2. J.S. Luo and M.Q. Chen in: Insulating Films on Semiconductors (M. Schulz and G. Pensl, Springer-Verlag, Berlin, Heidelberg, New York) 1981, pp. 174-178.
3. K. Nakamura, T. Gotch, and M. Kamoshida, J. Appl. Phys. 50, 3985 (1979).
4. T. Motooka and K. Watanabe. J. Appl. Phys. 51, 4128 (1980).
5. J.P. Corbot, Ph. Ged., Appl. Phys. Lett. 41, 93 (1982).

COMPARISON BETWEEN THERMAL AND LASER ANNEALING IN
ION-IMPLANTED SILICON.

A.BLOSSE
Laboratoire de Physique des Solides, I.S.E.N., 3 rue F. Baes,
59046 Lille, France

and

J.C. BOURGOIN*
Centro de Investigación y de Estudios Avanzados del IPN. Apartado
Postal 14-740, 07000 México, D. F.

ABSTRACT

N-type, 10^{15}-10^{16} cm^{-3} doped, Fz, Silicon has
been implanted with 1 to 4 x 10^{12} cm-2, 100 or 300
keV, As ions. The nature and concentration of the
defects has been monitored using Deep Level transient
Spectroscopy as a function of the thermal treatment
(in the range 500-900°C) and of the energy of the
pulse (15 ns) of a ruby (0.69 μm) laser (in the range
0.3 to 0.6 J cm^{-2}). The defects resulting from annealing by the two treatments are found to be the
same. Only, for energies higher than 0.5 J cm^{-2}, the
laser treatment introduced new defects (at E_c-0.32 eV),
presumably resulting from a quenching process. Thus
a laser energy below the theshold for melting and
epitaxial recrystallization is able to anneal the
defects produced by implantation, demonstrating
that the annealing process induced by the laser
pulse is not a purely thermal process but probably
involves an ionization enhanced mechanism.

INTRODUCTION

Although it is important to know the concentration, distribution
and electrical characteristics of the defects present in implanted
and laser annealed layers, there have been only few and rather
short studies of the annealing behaviour under pulse irradiation
of the point defects produced by ion implantation in silicon (1).
In this paper, we study the annealing of the defects produced by
As implantation in n-type Si due to a laser pulse. The aim of the
study is to monitor defect annealing as a function of the laser
energy and to compare the annealing behaviours induced by laser and
thermal treatments. In this way, it is possible to see if the
laser process induces new defects. Moreover, since the temperature of the laser irradiated region can be evaluated, this comparison
allows to know if the annealing induced by the laser pulse is
caused by a purely thermal process. In order to do such comparison,
the laser energy is chosen so that annealing occurs in the solid

* Permanent address: Groupe de Physique de Solides de l'E.N.S.
Universite Paris 7, Tour 23, 2 place Jussieu, 75221 Paris.

Mat. Res. Soc. Symp. Proc. Vol. 14 (1983) ©Elsevier Science Publishing Co., Inc.

phase regime; indeed, when the energy is such that the implanted
layer melts, annealing occurs through epitaxial recrystallization
independently of the nature and concentration of the original defects.

MATERIAL AND METHODS

The material studied was Fz, n-type, Si containing 10^{15} free ca
rriers at room temperature. This doping was chosen to allow the
investigation by capacitance a voltage (C-V) methods of a depth
($\sim 1 \mu$m) larger than the implanted region. With the energies of the
As implantation used, 100 or 300 keV, the depth distribution of the
original damage extend to 0.05μm and 0.15μm for, respectively,
the 100 keV and 300 keV implantations. However, defects can be
found for depths up to $\sim 1 \mu$m.
Low doses of implantation have been used (10^{12} to 4×10^{12} cm^{-2}) in
such a way that the damage produced do not compensate totally the
free carrier concentration, so that C-V and deep level transient
spectroscopy (DLTS) can be used to determine the profiles of the
defects which are detected. These techniques were applied on
Schottky barriers made by Au or Al evaporation after the annealing
treatment has been performed. The ohmic contacts on the back of
the wafers were realized, prior to implantation, by 10^{12} cm^{-2},
20 keV, P implantation followed by a 800°C, 1 h. treatment.

The capacitance measurements were performed with a model 410 PAR
capacitance-meter. The transient were analyzed with the help of a
double lock-in (6). Except in case where the annealing is sufficient, the defect concentration being large, the C-V characteristics and the DLTS spectra have to be corrected for the following
effects: 1) in case the defects compensate totally the free carriers',
the insulating layer corresponds to a constant capacitance in series with the capacitance of the space charge layer situated behind it; the thickness of this insulating layer can be deduced (7)
and only the defects which lie behind it are observed; 2) partial
compensation of the carriers leads to a resistance R in series
with the capacitance C of the space charge layer; when this resistance is large enough so that RCw (w is the pulsation of the drive
signal of the capacitance-meter) is not small compared to 1, the
amplitude of the capacitance transient must be corrected. (8) (the
sign of the transient can even be reversed when RCw>1); the procedure used to apply the correction is described in ref (8); 3) the
transition region between the depleted and the neutral zones, of
the order of five times the Debye length ($\sim 0.04 \mu$), is not small
compared to the projected standard deviation (0.02 and 0.05 μ for
100 and 300 keV implantations, respectively) and the defect profiles can not be obtained accurately; moreover, the defect concentration present in this transition region being non uniform, leads
to non uniform capture and emision rates and the capacitance
transient are not exponential , i.e. the DLTS spectra are distorted.

Thermal annealings were performed from 500 to 900°C under vacuum. Laser annealing was done using a pulse (15 ns) of a ruby
(0.69μm) laser with energies ranging from 0.3 to 0.6 J cm^{-2}. According to refs. (2-5) melting followed by epitaxial recrystallization occurs only for energies above ~ 1 J. cm^2.

RESULTS

Before any annealing five traps were observed behind the implanted region which is insulating (i.e. for depths larger than 0.5 μm in case of 300 keV implantation). They originate from the diffusion of vacancies out of the implanted layer where they are produced and from their interaction with impurities or between themselves. They are identified as the A center (vacancy-oxygen pair, labelled E_1) at $E_C-0.19$ eV, the E-center (vacancy-As pair, labelled E_4) at $E_C-0.42$ eV, the divacancy at $E_C-0.23$ eV (E_2) and $E_C-0.42$ eV (E_4'). The other levels we observed at $E_C-0.45$ eV (E_5) and at $E_C-0.34$ eV (E_3) have not been identified.

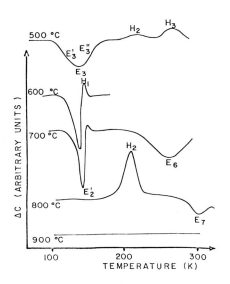

Thermal annealings from 500 to 900°C by steps of 100°C, 1 h. resulted in drastic changes of the DLTS spectra (Fig. 1). New levels were observed, whose detailed characteristics (variation of the emission rate with temperature, capture cross-section) will be given elsewere.

The original traps E_1, E_2, E_4, E_5 disappear after the first annealing step at 500°C and new traps are observed: E_2' at $E_C-0.26$ eV and a group of two traps (E'_3 and E''_3) whose overlapping with E_3 prevent their detailed characterization. A trap E_6 ($E_C-0.58$ eV), present after 600 and 700°C, disappears at 800°C to be replaced by the trap E_7 (which is actually a minority carrier trap).Other minority traps are observed (H_1 to H_3, see Fig. 1), observable because of the large leakage currents of the Schottky diodes.

Fig. 1   Uncorrected DLTS spectra (emission rate 62s^{-1}) obtained after various annealing temperatures. The reverse bias and pulse amplitudes are adjusted so that the same region of the layer is probed at each temperature. All the defects are not observed together under the same experimental conditions.

The energy levels of the electron traps observed following laser irradiation are given in Table I where they are compared with the traps produced by the thermal treatments. The annealing behaviour of the main ones is presented in Fig. 2. This figure has been drawn in the following way: the amplitudes of the differents peaks which compose a DLTS spectrum vary from one diode to another for a given energy of the laser pulse, indicating that this energy is not uniformly distributed over the irradiated region. In order to get an effective value of the laser energy, we assume that it is proportional to the annaling rate of trap E_f which anneals completely only for 0.6 J cm^{-2}.

TABLE I

Energy levels and estimated capture cross-sections of the various traps observed following laser (letter subscripts) and thermal (number subscripts) annealing.

Laser Annealing	Thermal Annealing	Energy Level (E_c-(?) eV)	Cross-Section (cm^{-2})
E_a	E'_3	0.32	2×10^{-16}
E'_a	-	0.32	2×10^{-15}
E_b	E_1	0.19	3×10^{-14}
E'_b	-	0.19	1×10^{-14}
E_c	E_2	0.22	5×10^{-16}
E_d	-	0.28	7×10^{-15}
E_e	E_3	0.34	3×10^{-16}
E_f	E_4	0.42	4×10^{-15}
E_g	E'_4	0.42	6×10^{-16}
E_h	E_5	0.45	2×10^{-16}
E_i	E_6	0.58	4×10^{-14}

Fig. 2 Annealing rates of the defects observed following laser annealing (letter subscripts) versus the energy of the laser pulse. The temperature scale indicates roughly the temperature at which the same defects observed following thermal treatments (number subscripts),anneal. The fact that the same trap level (E_e) is found to decrease and then appears again with further annealing suggests it should be associated with two different defects.

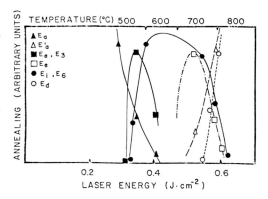

DISCUSSION

As shown in Table I, with the exception of two levels E'_b and E_d, the defects produced by laser and thermal treatments are the same. The fact that two defects E_f and E_i are observed to anneal completely only at 0.6 J cm^{-2} indicates that, indeed, the layer is not melted for energies $\sim$ 0.6 J cm^{-2}. However, for energies higher than $\sim$ -0.5 J cm^{-2}, new defects are created by the laser pulse (E_d, E'_a) which should thus be attributed to a quenching process.

The fact that we observe defect annealing occuring for laser energies below the threshold is apparently in contradiction with the results presented in ref. 1 where it is shown that Q-switched Nd:Yag (1.06 μm, 40 ns) laser in a single spot mode do not change the defect concentration for energies below the melt threshold. Since defects anneal thermally in Si with an activation energy of the order of 1 eV, significant annealing should not be expected to occur following a laser pulse which induce a temperature increase below the melting point for short times Δt_ℓ (<1 μs): the temperature $T\ell$ to be reached to induce the same annealing rate as a thermal process at temperature T_t during a time Δt_t, given by:

$$\frac{T_\ell}{T_t} = \ell n \ (\frac{\Delta t_t}{\Delta t_\ell})$$

is larger than the melting temperature ($T_\ell \sim 2$ x 10^4 K with T_t = 1000 K, Δt_t = 10^3s and Δt_ℓ= 1 μs). Thus, the fact that we observe a significant annealing following subthrehold energy irradiation indicates that the annealing do not occur through a simple thermal process. Presumably, the ionization which accompanies laser irradiation play a role in enhancing this annealing (9).

REFERENCES

1. L.C. Kimerling and J. Benton in Laser and Electron Bean Processing of Materials eds. C.W. White and P.S. Peercy (Academic Press, New York 1980) p. 385.

2. C.W. White, W. H. Christie, B.R. Appleton, S.R. Wilson and P.P. Pronko, Appl. Phys. Letter 33, 662 (1978).

3. P. Revesz, G. Farkas, G. Mezey and J. Gyulai. Appl. Phys. Letters 33, 431 (1978).

4. J. Narayan Appl. Phys. Letters 34, 312 (1979).

5. L. Jastrebski, A. E. Bell and C.P. Wu Appl. Phys. Letters 35,608 (1979).

6. D. Pons, P. Mooney and J.C. Bourgoin. J. Appl. Phys. 51, 2038 (1980).

7. P. Mooney, J.C. Bourgoin and J. Icole in Laser and Electron Beam Solid Interactions and Materials Processing ed J.F. Gibbons, L.D. Hen and T.W. Sigmon (North Holland, New York, 1981).

8. A. Broniatowski, A. Blosse, P.C. Srivastava and J.C. Bourgoin J. Appl. Phys. To be published.

9. J.C. Bourgoin and J.W. Corbett, Radiat. Eff. 36, 157 (1978).

Si-ON-SAPPHIRE AND Si IMPLANTED WITH Zr IONS: LATTICE LOCATION, SOLID PHASE
EPITAXIAL REGROWTH, AND ELECTRICAL PROPERTIES

I. GOLECKI[a,b] AND I. SUNI[b]
a) Rockwell International Corporation, Microelectronics Research and Development Center, 3370 Miraloma Avenue, Anaheim, CA 92803.
b) California Institute of Technology, Mail Code 116-81, Pasadena, CA 91125.

ABSTRACT

   Zr ions have been implanted at 300 keV (R_p = 1400Å) and doses of 3×10^{12}-
$3\times10^{15} Zr/cm^2$ into Si-implanted, amorphous Si layers on (100) bulk Si and Si-
on-sapphire. Rutherford backscattering and channeling spectrometry was used to
study the Zr distribution and lattice location during solid-phase regrowth of
the Si layers. The regrowth at 500-550°C stops at $3.4\times10^{20} Zr/cm^3$, and Zr
exhibits interface trapping and surface segregation effects. In this temperature range, Zr is essentially non-substitutional, and inactive electrically.

INTRODUCTION

   The present study is concerned with the structural and electrical properties
of Si crystalline layers containing Zr atoms. Our interest in Zr, Y, and other
metallic impurities in Si is related to the electrical properties of heteroepitaxial Si films grown by chemical vapor deposition on yttria-stabilized,
cubic zirconia single-crystal substrates at 950-1075°C [1]. In those studies
we found an unintentional n-type doping of the Si films which increased with
the Zr/Y ratio in the substrates for yttria mole fractions in the range 0.12 -
0.33. Although Zr belongs to the fourth column in the periodic table, Si doped
with ZrO_2 from the melt was reported to exhibit a shallow donor level at 70 meV
below the conduction band [2], and had a somewhat lower sensitivity to ionizing
radiation [3]. An interaction between group IV impurities and P leading to
electron capture was postulated [4]. Zr introduced into Si by ion implantation
to a shallow depth (≈200Å) was found to be non-substitutional even if implanted
at 350°C and after furnace annealing; however, damage by the 1.8 MeV ^{12}C analyzing beam could not be ruled out in that case [5]. Structurally, the presence of impurity atoms in Si is known to affect the solid-phase epitaxial regrowth rate of a self-implanted, amorphous layer. Electrically active impurities, such as B, P, or As, when present alone, increase the regrowth rate,
while non-substitutional species, such as C, N, O and the noble gases slow it
down [6]. The magnitude of the effect depends on the concentration of the
particular impurity. Exceeding the "meta-stable" solid-solubility limit generally reduces the regrowth rate and results in precipitation of fast diffusing
species or interface segregation of slow diffusers [7]. Similar levels of donors and acceptors (e.g., P and B), when present together, cancel each other's
effect [8].
   In this study, we have implanted Zr over a wide concentration range into
bulk Si crystals and Si-on-sapphire (SOS) heteroepitaxial films. The SOS layers
were chosen because of their intrinsically high resistivity (150-3000 Ωcm), and
because the presence of planar and linear lattice defects, and the variation of
the defect level with depth more closely resemble the situation found in Si
films on cubic zirconia.

EXPERIMENTAL PROCEDURE

Bulk, (100)-oriented Si crystals (100-300Ωcm, p-type, B-doped) and 0.45μm thick, undoped (100) SOS epitaxial films were implanted with ^{28}Si ions, so as to form 0.55μm and 0.35μm thick amorphous surface layers, respectively. The implantation conditions for Si were: 80 keV, 1x10^{15}Si/cm^2 + 140 keV, 8x10^{14} Si/cm^2 + 260 keV, 4x10^{15}Si/cm^2 at liquid-nitrogen temperature, and for SOS: 80 keV, 2x10^{15}Si/cm^2 + 200 keV, 2.5x10^{15}Si/cm^2 at room temperature (RT). ^{90}Zr ions were then implanted at RT into these samples at 300 keV to the following doses: 3x10^{12}, 3x10^{13}, 3x10^{14}, 3x10^{15}Zr/cm^2, using a ZrCl$_4$ solid source. All implantations were performed with the sample surface normals misaligned by 7° with respect to the ion beam. Ion current densities were kept below 0.5 μA/cm^2 to avoid sample heating. The samples were furnace-annealed in vacuum or flowing N$_2$ at temperatures of 500-840°C, and analyzed by MeV ^{4}He$^+$ Rutherford backscattering (RBS) and channeling (for crystalline structure, Zr concentration profiles and lattice location), by reflection electron diffraction, RED (for surface crystallinity), and by surface electrical resistivity measurements with a sensitive four-point probe instrument (Model 101, Four Dimensions, Inc., San Mateo, CA).

RESULTS AND DISCUSSION

The structural results for bulk Si and SOS were found to be generally the same, while measurement of the high resistivities in the implanted layers was simpler for SOS. We, therefore, present here only the findings in the SOS films. It was also observed from RBS measurements that implanted Fe was present at a similar depth as the Zr in most Zr implanted samples; the Fe was presumably introduced as ^{56}Fe^{35}Cl$^+$. Only samples for which the Fe (and presumably Cl) dose was less than 5% of the Zr dose are included in this paper.

The behavior of the SOS films implanted with the highest Zr dose, $\approx$3x10^{15} Zr/cm^2, was most affected. Figure 1 shows RBS and channeling spectra for a set of such samples. The measurements were performed with a variable-angle-detection system, to allow enhanced depth resolution when required. The as-implanted Si film (spectrum B) had a 3500Å thick amorphous surface layer. The peak of the Zr atomic profile was located at a depth of 1440Å, with a FWHM = 940Å. The implanted dose, found by integrating the net area under the Zr peak, was 2.6x10^{15}Zr/cm^2. After a 1h furnace annealing at 550°C, the reference sample (without Zr) regrew almost completely, and had a surface channeling yield χ_0 = 0.085. The regrowth started at the buried amorphous/crystal interface and proceeded towards the Si surface. After an additional 0.5h, the surface yield dropped further to 0.067 (spectrum E). The rapid increase in the Si channeling yield with depth is characteristic of SOS films, and the Si-implant amorphization and regrowth procedure used in this study reduces the level of planar defects only slightly compared to the as-deposited state [9]. By contrast, after 1h the Zr-implanted sample still had a 1200Å thick amorphous surface layer. A slight shift was noted in the position of the Zr peak to 1370Å, and the FWHM was reduced to 710Å. The position of the interface between the unregrown amorphous surface layer and the underlying single-crystal was thus on the surface side of the Zr peak. Upon an additional 0.5h annealing at 550°C (spectrum C), the width of the amorphous Si surface layer was reduced somewhat further to 1100Å and the Zr peak position (1190Å) correspondingly shifted towards the surface, without further change in width. The solid-phase epitaxial regrowth in the Zr-implanted sample essentially stopped at a Zr concentration of 3.4x10^{20}Zr/cm^3. This is in the range where such growth retardation effects had been found to occur [6,7] with other impurities exceeding their maximum solid solubilities.

These samples (C and E) were subsequently annealed at 840°C for 1h. The

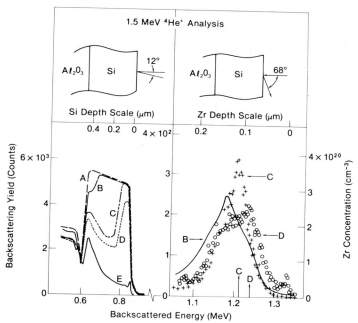

Fig. 1. Rutherford backscattering and channeling spectra of 0.45μm thick,(100) oriented Si-on-sapphire films, pre-amorphized to a depth of 0.35μm by Si ion implantation and then implanted with 300 keV, $3 \times 10^{15} Zr/cm^2$ ions. The Si spectra on the left were measured at a scattering angle $\theta = 168°$, while the Zr concentration profiles on the right were measured at $\theta = 112°$. A: random, B - E: channeled, [100] axis. B: as-implanted; C: annealed 1.5 h, 550°C; D: annealed 1.5 h, 550°C + 1 h, 840°C; E: Si implant only, annealed 1.5 h, 550°C or 1.5 h, 550°C + 1 h, 840°C. Zr profiles are for random orientation. The arrows marked C & D indicate the position of the interface between the unregrown Si surface layer and the underlying single-crystal

channeling spectrum of the reference sample (E) having no Zr, remained unchanged. The Zr-implanted sample, on the other hand, underwent several major changes. The width of the unregrown Si surface layer continued to decrease to ≈820Å. In addition, the channeling yield from this layer was reduced to χ = 0.8. This indicated that the layer had become polycrystalline, with some preferred orientation in the [100] direction [10]. RED measurements indicated that the top ≈50Å consisted of a mixture of single-crystal, polycrystalline, and amorphous material. The Zr distribution became wider, with FWHM = 1060Å. About 3% of the Zr was found to be at the surface. The center of gravity the remaining 97% was at the same depth (1190Å) as after 1.5h at 550°C. From the broadening of the Zr profile, the Zr diffusivity in polysilicon at 840°C is estimated to be on the order of $1 \times 10^{-15} cm^2/s$. The position of the poly/single-crystal interface was again on the surface side of the Zr distribution (see arrow D in Figure 1), with the Zr concentration there being $2.2 \times 10^{20} Zr/cm^3$, lower than after the low-temperature annealing. This indicates that either

544

the equilibrium solid solubility of Zr in Si decreases from 550 to 840°C, and/ or that the level measured at 550°C was metastable. The total measured Zr doses for all spectra were within 13% of each other (2.8x10^{15}Zr/cm^2 in spectrum C and 2.5x10^{15}Zr/cm^2 in D).

The Zr channeling yield, χ(Zr), was found to equal unity at all depths for the samples annealed at 550°C, i.e., even the deeper portion of the Zr distribution, located in the regrown, single-crystal Si, was not substitutional within our experimental accuracy. After the additional 840°C annealing, χ(Zr) dropped to ≈0.9 at all depths, indicating that possibly 20-30% of the Zr was now located in substitutional lattice sites. However, an accurate estimate of the substitutional Zr fraction in the *regrown* layer would require a remeasurement without the top, unregrown layer; also, the interpretation of impurity channeling yields in a polycrystalline layer (the *unregrown* layer) is not as developed as for single-crystals and needs corroboration by complementary characterization techniques, such as secondary ion mass spectrometry (SIMS) and transmission electron microscopy.

Measurements on bulk Si samples implanted with ≈3x10^{15}Zr/cm^2, and annealed at 500°C for 1-13 h, confirmed the above results and showed that the Si regrowth rate was unaffected for [Zr] ≤1.4x10^{20}/cm^3. SOS samples implanted with ≈3x10^{14}Zr/cm^2, and annealed at 525°C (3h) + 550°C (1.5h) regrew completely, but had a surface channeling yield of χ_o = 0.10, indicating some residual defects compared to Si-implanted, Zr-free samples, where χ_o = 0.067. Although the RBS level for Zr in these lower dose implanted samples was 6-9 times lower than in Figure 1, approximately half the total Zr dose appeared to reside at Si surface. If this effect is confirmed by more sensitive SIMS measurements, it would mean that Zr atoms were probably swept towards the surface by the moving amorphous/single-crystal interface. After an additional annealing of 1h at 840°C, χ_o dropped to 0.08, close to the value for Zr-free Si. Curiously, no Zr could be found at the surface of these latter samples; all the Zr was in depth. More extensive measurements are needed to clarify these interesting observations. Finally, at a lower implanted dose of 3x10^{13}Zr/cm^2, the Si regrowth was completely unaffected, with χ_o = 0.07 after an annealing of 3h at 525°C + 1.5h at 550°C. At this dose, the Zr RBS level was too low for meaningful measurements.

The *electrical* properties of the Zr-implanted SOS films are summarized in Table I. All of the samples were essentially of high resistivity. Even in the most conductive film, with $\bar{\rho}$ = 20 Ωcm (n-type), the average carrier concentration would only have been of the order of 1x10^{15}e/cm^3 (assuming an electron mobility in SOS of 300 cm^2/Vs). The small electrical activities seen at doses of 3x10^{12}- 3x10^{14}Zr/cm^2 after annealing at 840°C do not correlate with these Zr doses, and may well be related to secondary impurities and possibly defects. Thus, our results agree with the very low electrical activities reported in ref. 2. Additional measurements, e.g., the temperature-dependence of the carrier concentration and deep level transient spectroscopy techniques, are needed to obtain information on energy levels.

SUMMARY

In summary, we have found that Zr is not electrically active in Si over a very wide concentration range, although it may have limited substitutionality after annealing at 840°C. In fact, as a column IV element, Zr is not expected to be electrically active in Si, even if present in a substitutional lattice site, except in combination with other impurities or defects. The solid-phase epitaxial regrowth of Zr-implanted, (100) oriented Si films at 500-550°C is unaffected for [Zr] ≤1.4x10^{20}Zr/cm^3, but stops at [Zr] = 3.4x10^{20}Zr/cm^3. This essential lack of Zr electrical activity is correlated with the Si regrowth results, since electrically active, substitutional impurities (B,P,As)

Zr Dose (cm^{-2}) / Annealing Temp.	550°C $\bar{\rho}$ ($\Omega \cdot$cm)	550 + 840°C $\bar{\rho}$ ($\Omega \cdot$cm)	El. Activity
3×10^{12}	2×10^5	600	0.2%
3×10^{13}	2×10^3	20 , n	0.5%
3×10^{14}	600	60 , (n)	0.02%
3×10^{15}	1.5×10^{4}*	1.3×10^{4}**	—
Si Only	1.5×10^5	6×10^4	
Unimplanted	$150 - 3 \times 10^3$	$50 - 200$	

Table I. Results of surface electrical measurements on 0.45μm thick, initially un-doped Si-on-sapphire films. All samples were dipped in 10% HF prior to measurement. The samples marked * and ** had an amorphous or a poly-crystalline surface layer, respectively.

increase the regrowth rate of Si for concentrations $>5 \times 10^{19}$/cm^3[7,8]. Zr exhibits interface trapping and surface segregation effects due to its low diffusivity in Si. In conclusion, Zr can probably be ruled out as a major sole source of n-type doping in Si films grown on Zr-containing substrates.

ACKNOWLEDGMENT

We thank R. E. Johnson (Rockwell International) for RED measurements and M-A. Nicolet (California Institute of Technology) for valuable discussions.

REFERENCES

1. H.M. Manasevit, I. Golecki, L.A. Moudy, and J.E. Mee, Electrochem. Soc. Extended Abstracts Vol. 82-1, p. 332 (1982); I. Golecki, H.M. Manasevit, L.A. Moudy, J.J. Yang, and J.E. Mee, 24th Electronics Materials Conference, Fort Collins, CO, June 1982, abstract #D-8.

2. V.V. Voronkov, G.I. Voronkova, M.I. Iglitsyn, and A.G. Salmanov, Sov. Phys. Semicond. 8, 1277 (1975).

3. S. Mayer, U.S. Patent # 3,444,100 (1969).

4. N.T. Bagraev, L.S. Vlasenko, A.A. Lebedev, I.A. Merkulov, and P. Yasupov, Phys. Stat. Sol. (b) 103, K51 (1981).

5. B. Domeij, G. Fladda, and N.G.E. Johansson, Radiation Effects 6, 155 (1970).

6. L. Csepregi, E.F. Kennedy, T.J. Gallagher, J.W. Mayer, and T.W. Sigmon, J. Appl. Phys. 48, 4234 (1977); E.F. Kennedy, L. Csepregi, J.W. Mayer, and T.W. Sigmon, J. Appl. Phys. 48, 4241 (1977).

7. For a recent review, see J.S. Williams, Nucl. Instrum. and Methods (in press, 1983, Proc. International Conference on Ion Beam Modification of Materials, Grenoble, France, September 1982).

8. I. Suni, G. Göltz, M-A. Nicolet, and S.S. Lau, Thin Solid Films 93, 171 (1982).

9. I. Golecki, H.L. Glass, and G. Kinoshita, Appl. Phys. Lett. 40, 670 (1982).

10. D. Sigurd, R.W. Bower, W.F. van der Weg, and J.W. Mayer, Thin Solid Films 19, 319 (1973).

QUENCHED-IN DEFECTS IN CW LASER IRRADIATED VIRGIN SILICON.

A. CHANTRE, M. KECHOUANE AND D. BOIS.
CNET/CNS - BP : 42 - 38240 MEYLAN - FRANCE.

ABSTRACT

Quenched-in defects in cw laser irradiated silicon have been identified using deep level transient spectroscopy. Four among the five dominant defect states arise from transition metal impurities (iron, chromium) present in precipitates in the as-grown material and dispersed into the crystal upon heat treatment. Native defects are involved in the form of phosphorous-vacancy complexes, which account for the remaining level.

INTRODUCTION

Whereas a great deal is known on radiation damage in silicon, and many of the defects produced by irradiation have been identified on the microscopic scale, the situation regarding defects created by thermal processes such as fast cooling or quenching is far from clear, and the nature of these defects is poorly understood (l). A clearer picture becomes highly desirable, at the time where transient thermal processes are being explored as alternatives to standard furnace heat-treatments in device fabrication technology. The identification of defects in quenched silicon is also of great fundamental interest, as a step towards the understanding of the properties of native defects at high temperatures (2). However, quenching experiments are very difficult to carry out in practice, and extreme care has to be taken to avoid any contamination during the process. The use of directed energy beams to produce localized, short time heat treatments drastically reduces these difficulties, while the accessible high temperatures and fast cooling rates approach the ideal conditions of the "perfect quench" experiment one would expect to trap thermal intrinsic defects (3). We have combined these advantages with the sensitivity of capacitance transient techniques to study quenched-in defects in silicon. The role of transition metal impurities is pointed out, and the extent to which informations on native defects can be obtained is discussed.

EXPERIMENTAL DETAILS

The starting material consisted of phosphorous-doped (1.1-2.5 Ω cm) and boron-doped (1-5 Ωcm), (100)-oriented, Czochralski grown silicon wafers. The thermal quenching experiments were performed using a scanning cw argon laser (4). The main features of the process and the induced heat treatments have been described elsewhere (3). Briefly, the power level (P) is adjusted to be just below (typically 95 %) the value (P_M) leading to local melting (the so-called "solid phase regime"), and small areas (1 cm^2) are treated by overlapping repetitive single scan laser lines. With the laser parameters used in this work (scan speed 10 cm/s - P = 0.94 P_M), the characteristics of the thermal treatment can be summarized as follows : surface temperature ~ 1550 K, dwell-time ~ 1 ms, quench rate ~ 10^6 K/s. After laser irradiation, Schottky barrier structures were constructed by evaporation of gold (for n-type material) or aluminum (for p-type material) through a metal mask. Quenched-in defect states were investigated from the dopant freeze-out temperature (~ 30 K) to above room

Mat. Res. Soc. Symp. Proc. Vol. 14 (1983) © Elsevier Science Publishing Co., Inc.

temperature, using an experimental set-up which has already been described
(3,5). Electronically controlled defect reactions during in situ low
temperature thermal treatments (below 150°C) were used to probe the nature of
these defects.

RESULTS

Fig. 1 shows a defect state spectrum observed in the surface region (1.5 µm
thick) of an n-type material following cw laser irradiation. The two major
electron traps, E(0.22) and E(0.45), are reproducibly detected in all samples
processed in the solid phase regime (5). Several properties of these defects
are noteworthy. Fig. 2 displays the results of a test for donor character for
E(0.22). Reduction of the reverse bias voltage removes the higher electric
field regions from the measurement. Motion of the DLTS peak to higher
temperatures (Fig. 2a) depicts the corresponding reduction in the electron
emission rate at constant temperatures. The effect is analyzed quantitatively
on Fig. 2b, which shows the relationship between the apparent thermal activation
energy for electron emission (E_a) and the junction field (F). The data points
fit the straight line :

$$E_a = E_i - q \left(\frac{qF}{\pi\varepsilon}\right)^{1/2} \text{ with } E_i = 0.225 \text{ eV,}$$

expected for a Poole-Frenkel emission process (6), confirming the donor
character of the defect.

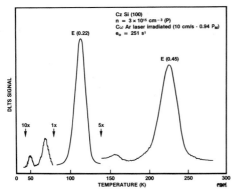
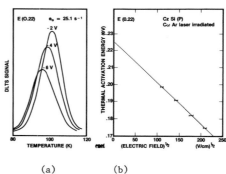

Fig. 1 : DLTS spectrum following cw
laser irradiation of P-doped silicon.

Fig. 2 : Electric field effect on
electron emission from E(0.22).

Fig. 3 demonstrates the charge state controlled stability of these
quenched-in defects. The change in charge state was accomplished by placing the
Fermi level either above (unbiased device) or below (reverse biased device) the
defect energy levels during annealing. E(0.45) is shown to anneal with first
order kinetics, with an annealing rate at 353 K almost two orders of magnitude
higher for the defect in the space charge region than in the bulk. The same but
weaker effect is observed for E(0.22), the enhancement being only by a factor of
six in this case. It should be pointed out however that E(0.22) is only
partially occupied at 353 K in the bulk of such a lightly doped sample ($N_D = 3 \times 10^{15} \text{ cm}^{-3}$), so that an admixture of the annealing rates in both charge states is
expected in the data for the unbiased device.

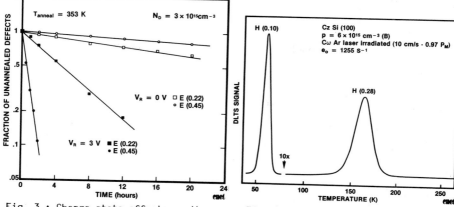

Fig. 3 : Charge state effects on the annealing kinetics of E(0.22) and E(0.45).

Fig. 4 : DLTS spectrum following cw laser irradiation of B-doped silicon.

Fig. 4 shows a capacitance transient spectrum of the defect states present in p-type silicon ~ 50 days after cw laser irradiation. Two hole traps, denoted as H(0.10) and H(0.28) are observed. Defect reactions during in situ low temperature thermal treatments are illustrated on Fig. 5. Upon annealing at 110°C for 3 hrs under zero bias conditions, a new hole trap H(0.44) is detected. However, the concentration of this defect level gradually decays if the sample is stored at room temperature, and after 12 hrs, it can no longer be detected. It should be pointed out that an accompanying reaction can be observed on Fig. 5 for H(0.28). But this defect level behaves oppositely, i.e. it decays upon annealing at 110°C, then recovers at room temperature. An additionnal piece of information concerning this defect is that its hole emission rate appears to be electric field-enhanced, as one would expect for a positive charge leaving a negative potential.

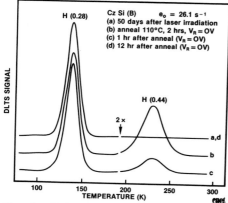

Fig. 5 : Defect reactions during low temperature thermal treatments.

TABLE I : Quenched-in defect states in cw laser irradiated virgin Si.

LEVEL	CAPTURE CROSS SECTION (cm^2)	PROPOSED IDENTITY	
E(0.22)	2×10^{-15}	Cr_i	$(0/+)$
E(0.45)	8×10^{-15}	$P-V$	$(-/0)$
H(0.1)	1×10^{-14}	Fe_iB_s	$(+/0)$
H(0.28)	1×10^{-14}	Cr_iB_s	$(0/-)$
H(0.44)	8×10^{-16}	Fe_i	$(+/0)$

550

We have summarized on Table I the thermal signatures of the five quenched-in defects studied in this work, together with a proposed identity as discussed below.

DISCUSSION

We identify E(0.45) to the phosphorous-vacancy complex P-V (E-center) on the basis of the following observations : (i) the level signature matches data from many workers for this radiation induced defect (7), (ii) the defect state increases as a result of MeV electron irradiation (3), and (iii) the charge state effects on the stability of the E-center (8) are operative for the annealing of E(0.45) (Fig. 3).

The H(0.44) and H(0.10) defect states arise from interstitial iron (Fe$_i$) and iron-boron pairs (Fe$_i$B$_s$) respectively. The occurrence of these two levels in quenched iron-doped silicon has been reported by several workers (9), and a detailed study of the electronically controlled reactions of iron and boron has been published recently (10). Briefly, Fe$_i$ observed immediately following quenching slowly reacts with B acceptors at room temperature to form Fe$_i$B$_s$ pairs ; these pairs dissociate and Fe$_i$ recovers during annealing at moderate temperatures. Although we did not directly monitor the transmutation between H(0.44) and H(0.10) during our low temperature thermal treatments, the behaviour of H(0.44) strongly supports the above assignments.

The E(0.22) defect state has been demonstrated to be different from the vacancy-oxygen complex V-O (3,5), contrarily to previous assignments (11). We alternatively propose that E(0.22) arises from a chromium atom at an interstitial site (Cr$_i$). The EPR spectrum of interstitial chromium Cr$_i^+$ has already been detected in quenched as-grown Czochralski silicon (12). The electronic level of Cr$_i$ was determined by correlating the EPR signal of Cr$_i^+$ to Hall measurements on identical samples ; the donor level was thus located at E$_c$ - 0.222 eV (13). This value is in good agreement with DLTS results from Graff and Pieper (14), which report a donor level at E$_c$ - 0.20 eV for Cr$_i$. These workers also pointed out a decreasing concentration of this trap with the time elapsed after quenching, which they assigned to chromium-boron pairs formation. Although we observe the E(0.22) level in n-type material, we believe that reaction of Cr$_i$ with a background acceptor concentration may be responsible for the slow decay of this level at room temperature. Such a pairing reaction would be expected to be driven by the Coulomb attraction between Cr$_i^+$ and B$_s^-$. The enhanced annealing rate of the ionized E(0.22) defect (Fig. 3) reinforces this interpretation.

Interstitial carbon (C$_i$) has been suggested (11) to account for the H(0.28) level which is commonly observed in beam annealed n$^+$/p junctions (11, 15). This interpretation is untenable since one remarkable property of C$_i$ is to anneal out at room temperature (7). Moreover, the above mentionned electric field effect on H(0.28) is inconsistent with the charge state assignment (+/0) of the C$_i$ related H(0.27) level (7). Other possible candidates would be the silicon di-interstitial Si$_i$-Si$_i$ (7) and the vacancy-oxygen-boron complex (16), since both involve simple intrinsic defects. Although we cannot completely rule them out, we believe that H(0.28) more likely arises from the chromium- boron pairs (Cr$_i$B$_s$) discussed above. Graff and Pieper have indeed correlated a E$_v$ + 0.29 eV level to Cr$_i$B$_s$ pairs from the transmutation between this level and the Cr$_i$ related E$_c$ - 0.20 eV state (14). The behaviour of H(0.28) in response to low temperature thermal treatments supports this assignment : as established for iron (9, 10), dissociation of pairs upon annealing at 110°C, and subsequent ion pairing at room temperature explain the defect reactions illustrated on Fig. 5.

The DLTS spectra of cw laser irradiated silicon appear to be dominated by transition metal impurities (Fe, Cr). This conclusion is in agreement with previous results from more conventional quenching experiments (12). However, the localized, short time nature of the laser induced heat treatment makes it very unlikely that the impurity levels reported in the present work arise from any contamination during the process. The experimental observation (17) that the E(0.22) defect concentration does not fall off from the irradiated surface, but rather is maximum some distance ($\sim$ 1 μm) below supports this assertion. We conclude alternatively that iron and chromium were present in the starting material in the form of precipitates and moved to interstitial sites upon heat treatment.

The production of P-V centers by the laser treatment has been analyzed in some details and shown to provide interesting informations on the properties of vacancies at high temperature (3). The extent to which data on self-interstitials can be obtained as well from the present results is less clear, in so far as the generation of interstitial metal impurities by an exchange of substitutional atoms with silicon self-interstitials is still highly controversed (18). Clearly, well known interstitial-related defects would be better candidates for such a purpose.

CONCLUSION

Cw laser irradiation and deep level transient spectroscopy have been combined to study the nature of quenched-in defects in silicon. Two promising features of such experiments lie in their ability to yield a very sensitive determination of transition metal impurity content in device grade material, and to provide experimental informations on intrinsic defects at high temperature.

REFERENCES

1. J. W. Chen and A. G. Milnes, Ann. Rev. Mater. Sci., 10 (1980) 157, and references herein.
2. J. A. Van Vechten and C. D. Thurmond, Phys. Rev. B, 14 (1976) 3551.
3. A. Chantre, M. Kechouane and D. Bois, 12th Int. Conf. Defects in Semi-conductors, Amsterdam (Sept. 1982), to be published in the Proceedings of the Conference.
4. G. Auvert, D. Bensahel, A. Georges, V. T. N'Guyen, P. Henoc, F. Morin and P. Coissard, Appl. Phys. Lett., 38 (1981) 613.
5. A. Chantre, M. Kechouane and D. Bois, Laser and Electron Beam Interactions with Solids 1981, edited by B. R. Appleton and G. K. Celler (North Holland, 1982) 325.
6. J. Frenkel, Phys. Rev., 54 (1938) 647.
7. L. C. Kimerling, Radiation Effects in Semiconductors, 1976 ; Inst. Phys. Conf. Ser. n° 31 (1977) 221.
8. L. C. Kimerling, H. M. De Angelis and J. W. Diebold, Solid State Commun., 16 (1975) 171.
9. K. Graff and H. Pieper, J. Electrochem. Soc., 128 (1981) 669, and references herein.
10. L. C. Kimerling and J. L. Benton, in Ref. 3.
11. N. M. Johnson, J. L. Regolini, D. J. Bartelink, J. F. Gibbons and K. N. Ratnakumar, Appl. Phys. Lett., 36 (1980) 425.
12. Y. H. Lee, R. L. Kleinhenz and J. W. Corbett, Defects and Radiation Effects in Semiconductors, 1978 ; Inst. Phys. Conf. Ser. n° 46 (1979) 521.
13. H. Feichtinger and R. Czaputa, Appl. Phys. Lett. 39 (1981) 706.
14. K. Graff and H. Pieper, Semiconductor Silicon 1981, edited by H. R. Huff and R. J. Kriegler (The Electrochemical Society, N. J., 1981) 331.
15. N. H. Sheng and J. L. Merz, in Ref. 5, 313.
16. P. M. Mooney, L. J. Cheng, M. Süli, J. D. Gerson and J. W. Corbett, Phys. Rev. B, 15 (1977) 3836.
17. A. Chantre, M. Kechouane, G. Auvert and D. Bois, to be published.
18. E. Weber and H. G. Riotte, J. Appl. Phys., 51 (1980) 1484.

AUTHOR INDEX

SUBJECT INDEX

Plasma stripping 417
Plastic deform. 344
Poly-Si 52, 325, 343, 357
Poly-Si CVD 357
Poole-Frenkel effect 110, 548
PPE:IR quenching 464
PPE=Photoplastic effect
Ppt.=Precipitation
Ppt. hardening 307
Ppt. softening 307
Prismatic punching 103, 104, 182, 307
Protrusion=Spike
Protuberance=Spike
Pseudo-binary phase diagrams 209
Pseudomorphic layer 401
Pseudo-Richardson factor 346
Pulse ann. 491
Quartz 201
Radiation-enhanced diff=RED
Raman modes 517
Raman scattering in Si 42
Raman:GaAs 517
Recomb.-enhan. climb 453
RED=radiation-enhanced diff.
REDM=Recomb.enhan.defect motion
Refl.High Energy Elec. Diff.=RHEED
REM=Recomb.enhan. motion
Retarded diff.,see also Oxidation-
Retrograde solub. 19, 498
Ribbon to ribbon:Si 357
Rippled surf. morph 277
Rydberg spectra 36, 38, 42, 107, 115, 169
Scanning TEM=STEM
Scanning elec. micro.=SEM
Scherzer focus 396
Schottky diode 411
Screw disl. 332
SDLTS:scanning DLTS 332
Secco-etch 360
Self diffusion:Si 46
SF pyramids 327
SF shrinkage 73
SF=Stacking faults
Shear-force effects 389
Shockley partial 303
Shockley partial:CdTe 297, 298
Shockley-Read-Hall form. 348
Si:Admittance 336
Si:AES 42, 423

Si:Al diff. 57
Si:Amorphous=A-Si
Si:As 33, 42, 523, 529, 535
Si:As diff. 65, 68
Si:As+P diff 65, 73
Si:As+Ti 417
Si:Au 8, 12, 19
Si:Au+Au pair 29
Si:Au diff. 52, 141
Si:Au+Fe pair 28
Si:B 6, 7, 10, 11, 13, 47, 57
Si:B diff. 57
Si:Backside oxid. 63, 72
Si:Bi 36, 42
Si:Bicrystal 343, 363
Si:B+O 110, 169
Si:Bound exciton 165
Si:Burger's vector 103
Si:C 8, 9, 11, 12:
Si:C+O 32, 110, 111, 117, 125, 146, 147, 188, 195
Si:C+disl. 307
Si:Carrier lifetime 363
Si:CAST 357
Si:Chalcogenides 33, 42
Si:Channeling 42, 541
Si:Co 19
Si:Co-silicide 395
Si:Co diff. 20
Si:CoSi2 42
Si:Co solubility 24
Si:Cr 19, 547
Si:Cr diff. 20
Si:Cr elec.level 27
Si:Cr solubility 24
Si:Cr+B 547
Si:Cr+B pair 28
Si:Cu 1, 3, 10, 19
Si:Cu+Cu pair 28
Si:Cu-decoration 1, 93
Si:Cu diff. 20
Si:Cu ppt. 444, 445
Si:Cu solubility 24
Si:D+A pair 165
Si:Dendritic web 357
Si-Denuded zone 181
Si:Device degrad. 52
Si:Dielectric func. 529
Si:Disl. 323, 383, 437
Si:Disl. dipole 182

19 73 89

RETURN TO: PHYSICS-ASTRONOMY LIBRARY
351 LeConte Hall

LOAN PERIOD 1 **1-MONTH**	2	3
4	5	6

ALL BOOKS MAY BE RECALLED AFTER 7 DAYS
Books may be renewed by calling 510-642-3122

DUE AS STAMPED BELOW

JUL 3 1 2013		

FORM NO. DD 22
1M   7-11

UNIVERSITY OF CALIFORNIA, BERKELEY
Berkeley, California 94720–6000